Erratum for *Magnetic Resonance Microscopy: Instrumentation and Applications in Engineering, Life Science, and Energy Research,* ISBN 9783527347605 (2022)

On page *iii* of the front matter, the editors should have been listed as:

Edited by
Sabina Haber-Pohlmeier
Aachen
Bernhard Blümich
Aachen
Luisa Ciobanu
Paris

On page *iv* of the front matter

Luisa Ciobanu's correct postal address is

NeuroSpin, CEA-Saclay
Bat 145, Point Courrier 156
91191 Gif sur Yvette France
France

The Library of Congress Card No. should have been printed as:

Library of Congress Card No.:
Names: Haber-Pohlmeier, Sabina, editor. | Blümich,
Bernhard, editor. | Ciobanu, Luisa, editor.

The Dedication on page *v* in this book was erroneously included.

There is an error in the preface on page *xi*:

The last line of the second paragraph should have been

However, we would like to point out that only a small number of applications could be addressed in this book. Finally, the last chapter advocates that theory and applications should not be treated separately, because much can be gained from their complementarity.

Magnetic Resonance Microscopy

Magnetic Resonance Microscopy

Instrumentation and Applications in Engineering, Life Science, and Energy Research

Edited by

Sabina Haber-Pohlmeier
RWTH Aachen

Bernhard Blümich
RWTH Aachen

Luisa Ciobanu
Paris

WILEY-VCH

Editors

Sabina Haber-Pohlmeier
RWTH Aachen University
Worringerweg 2
52074 Aachen
Germany

Bernhard Blümich
RWTH Aachen University
Worringerweg 2
52074 Aachen
Germany

Luisa Ciobanu
Point Courrier 156
Bat 145
91191 Gif sur Yvette
France

Cover Image: Courtesy of Denis Wypysek, Jan Korvink, Henk Van As, and Luisa Ciobanu

Library of Congress Card No.:

Names: Haber-Pohlmeier, Sabina, 1962- editor. | Blümich, Bernhard, editor. | Ciobanu, Luisa, editor.
Title: Magnetic resonance microscopy : instrumentation and applications in engineering, life science and energy research / edited by Sabina Haber-Pohlmeier, Bernhard Blümich, Luisa Ciobanu.
Description: Hoboken, New Jersey : John Wiley & Sons, [2022] | Includes bibliographical references and index.
Identifiers: LCCN 2021061399 (print) | LCCN 2021061400 (ebook) | ISBN 9783527347605 (hardback) | ISBN 9783527827237 (pdf) | ISBN 9783527827251 (epub) | ISBN 9783527827244 (ebook)
Subjects: LCSH: Magnetic resonance microscopy. | Magnetic resonance microscopy—Industrial applications.
Classification: LCC QC762.6.M34 M344 2022 (print) | LCC QC762.6.M34 (ebook) | DDC 502.8/2--dc23/eng20220215
LC record available at https://lccn.loc.gov/2021061399
LC ebook record available at https://lccn.loc.gov/2021061400

British Library Cataloguing-in-Publication Data
A catalogue record for this book is available from the British Library.

Bibliographic information published by the Deutsche Nationalbibliothek
The Deutsche Nationalbibliothek lists this publication in the Deutsche Nationalbibliografie; detailed bibliographic data are available on the Internet at http://dnb.d-nb.de.

Print ISBN: 978-3-527-34760-5
ePDF ISBN: 978-3-527-82723-7
ePub ISBN: 978-3-527-82725-1
oBook ISBN: 978-3-527-82724-4

Cover Design: Wiley

Typesetting: Set in 9.5/12.5pt STIXTwoText by Integra Software Services Pvt. Ltd, Pondicherry, India
Printing and Binding: CPI Group (UK) Ltd, Croydon, CR0 4YY

Printed on acid-free paper

C104863_140422

To my parents

Contents

Foreword

This book is the fourth in the Wiley-VCH series on *Magnetic Resonance Microscopy*, a series linked in spirit to the *International Conference on Magnetic Resonance Microscopy* (ICMRM). Winfried Kuhn and Bernhard Blümich organized the first meeting of this biannual conference in 1991 in Heidelberg, which led to the first book in this series. That Heidelberg meeting is also when Paul Callaghan burst upon the scene (with his student Yang Xia, who remains active in these meetings) with his new, but now classic, book *Principles of Nuclear Magnetic Resonance Microscopy* (Oxford University Press) – what timing! The wider appearance of magnetic gradient fields in the portfolio of magnetic resonance methods for imaging and studies of molecular transport phenomena was an exciting prospect. It motivated the still ongoing ICMRM conference series and the associated books, which summarize the progress in this field with chapters written by leading experts, among them Nobel Prize awardees Paul Lauterbur and Sir Peter Mansfield as well as Sir Paul Callaghan, who shaped that community like a force of nature from then on until his untimely death in 2012. Also, our brilliant colleague Robert Blinc from Slovenia attended the first ICMRM but had to leave early following an announcement during one of the sessions effectively saying "Professor Blinc, you are needed back in your country," at which Robert Blinc stood up and left to facilitate the independence of Slovenia from Socialist Federal Republic of Yugoslavia. We were witnessing the birth of a country, a unique experience for most of us. At the third meeting in Würzburg, the brave suggestion to hold a meeting in North America was accepted. Thus, the fourth meeting was in Albuquerque and ICMRM has now a truly international presence, having ventured as far away as Utsunomiya and Beijing. These meetings, originally dubbed the Heidelberg Meetings, have been at the forefront of amazing developments and accompanying applications of magnetic resonance. Despite the inclusion of the word *microscopy* in their name, they represent the much broader area of magnetic resonance with spatial resolution, which is expressed by the title of the second book from the Albuquerque meeting, *Spatially Resolved Magnetic Resonance*, as well as the organizing *Division of Spatially Resolved Magnetic Resonance* of the *AMPERE Society*. Thanks to the advances over three decades, we have micrometer spatial resolution in magnetic resonance imaging today, while in the early days the word microscopy was understood as a tool to see things hard to visualize just by eye. We believe the broad range of the science and applications of magnetic resonance represented in these meetings is unique to all science and the field displays no hint of imminent stagnation – welcome news to all of us. I would like to close this

Foreword with the observation that Bernhard Blümich, who with Winfried Kuhn founded these meetings 30 years ago, is still actively involved here as one of the editors of this book. I salute him for his continued contributions to the field and support of this conference.

Eiichi Fukushima
Albuquerque, 2021

Preface

Magnetic resonance microscopy (MRM) has focused on magnetic resonance imaging (MRI) applied to objects of smaller scale and higher spatial resolution for more than three decades. After the pioneering work by Eccles, Callaghan, Aguayo, Blackband, Johnson et al. in 1986, MRM quickly spread to, among other fields, chemistry, histology, and materials research. Since 1992, the edited book series *Magnetic Resonance Microscopy* has provided an important voice describing the latest developments in spatially resolved magnetic resonance methods and their applications far beyond the scope of medical diagnostics. An excellent introduction to MRM, focusing on the practical aspects of high magnetic fields and on the study of biological systems, was authored in 2017 by Luisa Ciobanu: *Microscopic Magnetic Resonance Imaging: A Practical Perspective* (Pan Stanford, Singapore, 2017). Our book complements this monograph by showing the use of MRM and related techniques in a much broader area and on a wider scale, which extends from chemical engineering to plant research and battery applications, highlighting the interdisciplinary nature of MRM.

The book opens with a section on hardware and methodology, covering aspects of micro-engineering, magnet technology, coil performance, and hyperpolarization to improve signal-to-noise ratio, a major bottleneck of MRM. Specific pulse sequences and developments in the field of mobile nuclear magnetic resonance are further topics of this first chapter. The following parts, 2 and 3, review essential processes such as filtration, multi-phase flows and transport, and a wide range of systems from biomarkers via single cells to plants and biofilms. Part 4 focuses on energy research, which is becoming increasingly important due to the globally growing environmental problems. It reports on battery types and their developments and how battery states can be recorded and characterized with MRM. However, we would like to point out to the reader that only a small sample of applications could be addressed in Chapters 1 to 4. Finally, the last chapter advocates that theory and applications should not be treated separately, because much can be gained from their complementarity.

The main aim of this book is to convince aspiring and established scientists from all fields that MRM is a versatile nuclear magnetic resonance (NMR) method that is capable of answering many questions from both the laboratory and everyday life. The book seeks to inspire a new readership from industries and innovative research directions to create synergies by adding MRM to their expertise.

The editors thank all the authors for contributing their invaluable knowledge to this book during a time challenged by COVID-19. Our thanks also go to the kind staff of the Wiley books department, who helped us with advice and support throughout the whole editing process.

Sabina Haber-Pohlmeier
Luisa Ciobanu
Bernhard Blümich
Summer 2021

Part I

Developments in Hardware and Methods

1

Microengineering Improves MR Sensitivity

Neil MacKinnon, Jan G. Korvink, and Mazin Jouda

Institute of Microstructure Technology, Karlsruhe Institute of Technology, Eggenstein-Leopoldshafen, Germany

1.1 Introduction

Thirty years have passed since 1991 when Paul Callaghan published his book on magnetic resonance microscopy [1], and many works have subsequently appeared that have made numerous advances in this exciting field possible. Our goal for this chapter is to (informally) revisit some of Callaghan's analysis, to reflect on it, and then take account of some of the advances and insights that have been reported since then.

1.1.1 Comparative Electromagnetic Radiation Imaging

Paul Callaghan's book [1] is perhaps the first publication to consider magnetic resonance imaging (MRI) in the same light as optical microscopy. This will also be our starting point.

Until the advent of super-resolution microscopy, refractive optical microscopy was essentially a radiation scattering method, in which a beam of photons from an independent light source was sent on its way to scatter off objects, followed by traversal of the beam through a focusing objective on its way back to a detector, to thereby reveal the structure and composition of the scattering object. The limitations of this approach, in terms of resolution, is known as the Abbe limit $\delta = \lambda/(2\ n \sin \theta)$, where n is the refractive index, θ the half-angle of the spot subtended by the lens, and λ the radiation wavelength.

Using radio waves taken for convenience at 300 MHz, a thus interpreted refractive MRI system would have a resolution of ~500 mm, which is a dire prospect for applications of MRI. In a seminal paper, Mansfield et al. [2] reported on a form of nuclear magnetic resonance (NMR) diffraction, in which they considered a solid-state periodic lattice of spins in a macroscopically sized lattice, revealing diffraction patterns on the order of the lattice. As a follow-up to this idea, Blümler et al. [3] and Bernhard Blümich [4] reported (the latter in a paper dedicated to Paul Callaghan) on an interesting intertwining of concepts of the *k*-space vector of refractive MRI and the spatial periodicity of a lattice-like diffractive structure, further exploring diffractive imaging. Blümich's paper contains a few more gems worth discussing, but would distract us too far from the optical viewpoint we are considering here.

Near-field effects can be further exploited to increase the resolution of an imaging system. At optical wavelengths, one is hardly able to extend beyond 200 nm of resolution with currently available light sources. "The diffraction limit of light is 100 times the size of structures that cell

Magnetic Resonance Microscopy: Instrumentation and Applications in Engineering, Life Science and Energy Research, First Edition. Edited by Sabina Haber-Pohlmeier, Bernhard Blümich, and Luisa Ciobanu.

biologists study as they characterise events in organelles or membranes," Hari Schroff (NIH/NIBIB) is quoted to say in [5], yet below 200 nm "is where most cellular action is," the author notes. The alternative is to avoid scattering as an imaging paradigm, instead, to image photon sources (also known as quantum emitters).

Interestingly, deep space astronomy always worked this way around by observing photon emitters, so that astronomers only consider objects that were once themselves sources of radiation, such as stars and their predecessors and descendants. In astronomy, the limit of resolution is therefore not dominated by the wavelength of the radiation, which can be very small when compared to the size and distance of the astronomical objects, but rather by the measuring instrument's principle of operation, its detection sensitivity, and in particular, its effective aperture.

When imaging radiation sources, such as single photon emitters in molecules, we now know that we can greatly improve on the Abbe limit, by about a factor of 10, especially when combined with techniques of stimulated emission and depletion, and one of the numerous variations based on fluorophore emission dynamics. These techniques, which have revolutionized cellular biology and won its inventor Stefan Hell the Nobel Prize in 2014, are of course not accessible to astronomers, who would have to wait too long for excitation signals to pass from observer to object and back again. But for cell biology this is not problematic. Although at currently ~30 nm, the resolution is still far from the desired 1 nm limit, advances in image processing present a feasible route to achieve further improvements. But the technique also raises some questions. Sample preparation is very difficult, and imaging is indirect as fluorophores have to be invasively attached to interesting molecules, almost certainly modifying their behavior.

Magnetic resonance microimaging is a noninvasive technique that is clearly more closely related to stimulated emission depletion (STED) microscopy than to conventional scattering light microscopy. A localized atomic nucleus' spin is a quantum absorber/emitter. By localizing the excitation field spatially or by frequency, a sub-selection of the spins in a sample can be prepared to absorb radiation. Further localization can ensure that emission of radiation energy is again sub-selected, for example along the geometrical intersection of two orthogonal manifold slices, the one for excitation, the other for emission. A range of further techniques, such as available through relaxation contrast, or phase accumulation, can again further sub-select spins before readout, thereby improving resolution in direct analogy to the techniques of fluorophore emission dynamics. Noninvasive Faraday-detected MRI has been reported down to ~3 μm resolution [6], which is already five orders of magnitude below the radiation wavelength. Nevertheless, even though MRI records radio frequency emissions, this is done almost exclusively through near-field interactions, i.e. by Faraday induction, and not from a beam or ray that requires a lens for focusing.

One of the limitations in NMR is certainly that a single quantum emission event is not yet readily observable as it would be in photonics, even though Dan Rugar showed that a single spin can be observed [7]. Thus all Faraday induction-acquired MRI images have to resort to averaging of a vast number of emission events, and over extended time, to yield useful information. If detection sensitivity were to be increased, fewer emitters could be used, and could perhaps be averaged over shorter times.

1.1.2 Limit of Detection

The statistical polarization level, i.e. that proportion of the total spin population that is available for quantum emission, is an additional cause of lack of signal. Proton spins for example are indistinguishable fermions, with a level occupation that follows

$$\langle N_i \rangle = \frac{1}{e^{(\varepsilon_i - \mu)/kT} \pm 1} \tag{1.1}$$

and which collapses to Maxwell–Boltzmann statistics when $e^{(\varepsilon_i-\mu)/kT} \gg 1$, because the energy of a proton flip $\gamma\hbar B_0 = 3.3 \times 10^{-25}$ J is tiny compared with the thermal energy $kT = 4.11 \times 10^{-21}$ J. Thus at typical equilibrium polarization levels at 11.7 T, a factor of only 10^{-4} in excess in population difference with respect to the Fermi level μ can contribute to the signal. An imaging voxel size is therefore principally limited by polarization, because we find – for microcoils at their limit of detection – a sample containing around 10^{13} spins is needed to form an observable signal. Clearly, this sets a lower concentration limit once the voxel size is specified. For example, at the average size of a single eukaryotic cell of (10 μm)^3, containing the required nuclei, implies a concentration of at least 1.66 μM. By increasing polarization, the voxel size is thus principally reduced, or the lower concentration limit is reduced, which could be achieved by resorting to out-of-equilibrium polarization techniques such as parahydrogen-induced polarization (PHiP), signal amplification by reversible exchange (SABRE), or dynamic nuclear polarization (DNP), all of which are rather hard to perform noninvasively, and hard to selectively localize too. We will return to this point shortly. One of the key advantages of MR-based microscopy is the ability to noninvasively reveal molecular composition, correlated with morphology. From the perspective of biological systems, this can be leveraged to monitor, for example, spatially resolved metabolism. To estimate the best achievable spatial resolution (voxel size), signal-to-noise ratio (SNR) should be considered in the context of the metabolically active system. Key parameters are the molecule abundances (concentrations) and timescale that are targeted. Consider a spatially resolved fluxomic investigation: can one estimate a realistic MRI spatial resolution taking into consideration the expected biological dynamics? Alternatively, what is the smallest biological structure with which metabolic flux can be measured – thus, is it possible to monitor flux at the level of an organelle, single cell, cell cluster, or tissue?

Water is the most abundant molecule in biosystems and can be used as a useful reference from which scaling based on metabolite concentrations can be made. Using only the physical volume of a water molecule (0.03 nm^3), and an optimistic limit of detection (LOD) of 10^{13} spins, then an order of magnitude estimate of the smallest voxel is 0.1 pl (approximately 4.5 μm isotropic resolution). This is approximately the volume of a single red blood cell. Intracellular metabolite concentrations vary over several orders of magnitude, with the most abundant molecules typically in the tens of millimolar regime. The best-case scenario scaling factor is then 10^4 relative to water (assuming [water] = 55 M), and thus the smallest voxel volume increases to 1000 pl (100 μm isotropic resolution). For reference, this would correspond to 10 000 red blood cells or 2 fat cells (volume 600 pl per cell). Can the resolution be improved by signal averaging as a means to enhance SNR? Assuming metabolism is active during the measurement then one must consider the turnover rate of the target metabolite(s) relative to the time over which signal averaging is performed. Enzyme catalytic (second-order) rate constants span several orders of magnitude (k_{cat}/K_M ~10^1–10^9 s^{-1} M^{-1}), with a median of ~10^5 s^{-1} M^{-1} [8]. If the metabolite concentration is 100 mM, then the metabolite will encounter the "median enzyme" with a rate of 10^4 s^{-1}. At this concentration and a volume of 1000 pl, the metabolite concentration would drop below the LOD (~10 mM) in 6000 days giving more than sufficient time for signal averaging. At the diffusion limit $10^9\ s^{-1}\ M^{-1}$ then, 6 days are required before the signal is not observable. This rough estimate takes many liberties in the assumptions (catabolic and anabolic reactions, multiple pathways, enzyme performance, cell cycle, etc. are neglected) and simply suggests that signal averaging is reasonable, most likely limited by technical factors like long-term sample maintenance and spectral resolution accounting for magnetic susceptibility effects. Interesting, it is revealed that single cell metabolic monitoring is challenging yet possible as long as (i) large cells are selected; (ii) the metabolite is among the most abundant in the cell at millimolar concentration; and (iii) the cell can be maintained in an active state during the measurement. Spatial resolution and/or detected concentrations can be improved if hyperpolarization strategies are used where the effective LOD can be improved by orders of magnitude with percent (instead of ppm) levels of polarization. Using the same set of assumptions, now with percent levels of polarization, it is estimated that on the order of a few minutes is required ("median enzyme" kinetics) before the observation of hyperpolarized metabolic products, consistent with observations [9].

Currently, the only other means known to further increase resolution in cell biology, is based on the higher sensitivity and simultaneously higher localization that arises from proximity to the spin. When detector and spin are separated on the order of 1 nm, the dipolar coupling is strong. The techniques, such as magnetic resonance force microscopy [7], or nitrogen-vacancy (NV) centers in nanodiamonds [10], are either invasive (magnetic resonance force microscope [MRFM]) or require similar sample preparation as for STED (NV centers), namely to introduce a mobile quantum emitter that is tagged to a molecule to wander through the cell.

1.1.3 Limit of Imaging Resolution

Paul Callaghan [1] and others [11–14] showed that the imaging resolution of MR microscopy is fundamentally limited by three factors: the diffusion coefficient of molecules within the sample, the line broadening due to magnetic susceptibility effects, and the specified SNR per voxel. While the limits placed by the first two factors can be pushed by stronger field gradients and dedicated pulse sequences [15], the rather poor SNR of the MR signal remains as the ultimate fundamental limit of resolution. This can be clearly seen from the following equation [16,17], which summarizes the factors that determine the achievable resolution for a specified SNR of the image:

$$V_{\mathrm{voxel}} \propto \frac{d}{\sqrt{t_{\mathrm{acq}}} \cdot B_0^{7/4}}, \qquad (1.2)$$

where V_{voxel} is the voxel volume, d is the coil diameter, t_{acq} is the total acquisition time, and B_0 is the field strength. According to this equation, maintaining the image SNR while, for instance, halving the isotropic resolution (reducing the voxel volume by a factor of 8) necessitates either miniaturizing the coil by a factor of 8, increasing the acquisition time by a factor of 64, or increasing the B_0 field by a factor of 3.28. This explains why high-resolution MR images take excruciatingly long to acquire, and why most groups decrease coil diameter. However, at room temperature, coil diameter cannot be reduced indefinitely without disadvantageously increasing coil resistance, so that quality factor Q will ultimately limit this strategy.

1.2 High Resolution From Enhanced Sensitivity

1.2.1 Coil Miniaturization

In Equation 1.2, d represents the sensitivity of the NMR detector, which has been proven to increase with decreasing d [18]. Numerous reports have been published on how to improve the NMR detection sensitivity by miniaturizing the detection coil [19–22]. The vast majority of those papers targeted NMR spectroscopy applications, and far fewer talked about microscopy [23–25]. One of the papers that targeted both applications though is the work reported by [26]. In this paper a novel high-resolution NMR/MRI Helmholtz microcoil was introduced (Figure 1.1). The coil that was manufactured using a combination of standard and home-developed micro fabrication technologies featured an extremely user-friendly sample-handling approach that allows easy loading/unloading of the sample.

The coil was manufactured by stacking three layers, the top and the bottom of which are made of glass, each featuring a wire-bonded 1.5-winding coil using a 25-μm diameter copper wire. The coils were encapsulated in SU-8 epoxy-based photoresist. The glass layers also contained copper traces for the feed and return paths of the current. These layers were spaced by a poly(methyl methacrylate) (PMMA) layer, which was U-shaped to allow the sample-handling microfluidic chip to slide in the sensor. The Helmholtz microcoil was designed and optimized to achieve an extremely uniform B_1 field (92% ratio of signal intensity at flip angles of 810/90) while maintaining the high B_0 homogeneity (1.79 Hz achieved linewidth of a water sample).

The exceptional performance of the microcoil in terms of B_1 uniformity and local field homogeneity allowed high-resolution microimaging. Figure 1.2 shows the optical (left) and MR (right) microimages of a deionized water sample of 154 nl volume. The sample contains 50-μm diameter polymer beads to show some contrast. The MR image that is a sum of 80 acquisition was recorded over a total scan time of 11 h, 22 min, and 40 s. With a matrix size of 256 × 256 and covering a 2.5 × 2.5 mm field of view (FoV), the MR image exhibited an in-plane resolution of 10 × 10 μm for a slice with a 100-μm thickness.

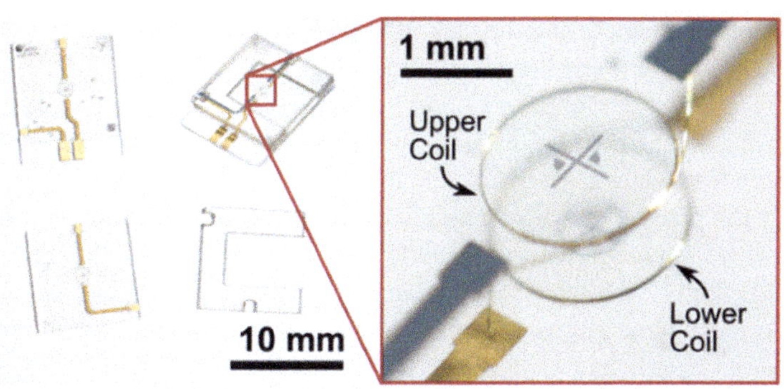

Figure 1.1 A micro Helmholtz coil manufactured using a mixture of standard and home-developed micro-manufacturing technologies. [26] N. Spengler et al. (2014), figure 04 [p.05]/with permission from IOP Publishing Ltd.

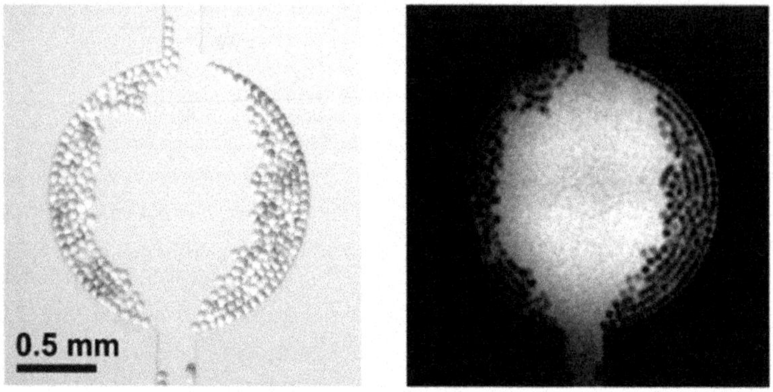

Figure 1.2 MR microimaging of a 154-nl deionized water sample with 50-μm diameter polymer beads. (left) Optical micrograph. (right) MR image. [26] N. Spengler et al. (2014), figure 10 [p.07]/with permission from IOP Publishing Ltd.

1.2.2 The Lenz Lens: A Tool to Boost Sensitivity

The quest for higher coil sensitivity, SNRcoil ∝ B1/(iR), made researchers invest huge efforts in the design and optimization of the NMR detector. These efforts resulted in numerous publications covering a wide diversity of coil geometries and applications. Nevertheless, none of these efforts went beyond two major pathways: (i) optimizing the coil geometry to better confine the sample and thus achieve a higher filling factor, which, in turn, results in a higher B_1 per unit current seen by the sample; and (ii) reducing coil resistance, R, and thus its noise contribution either by a careful selection of coil materials, or by cooling the coil windings. In certain cases, however, these strategies might not be applicable as, for example, if one wants to study the inner body parts. For this particular case, implantable inductively coupled inductor-capacitor (LC) resonators have been shown to enhance the imaging SNR remarkably [27–32]. Nevertheless, despite their advantages, LC resonators suffer from limited frequency accessibility due to their narrowband nature, which originates from the resonance phenomena, and the difficulty in tunability due to their sensitivity to both the sample and the pick-up coil [33].

An alternative to the LC resonator is the Lenz lens (LL) shown in Figure 1.3. The LL consists of two electrically connected loops: the inner and outer loop, where the outer loop collects the magnetic flux of the pick-up coil and converts it to a current. As this current passes through the inner loop of the LL, it generates a larger magnetic field in the sample than the field that would otherwise be produced by the pick-up coil only. The LL, though less powerful in terms of field amplification compared with the LC resonator, is superior to it in terms of tunability and wide frequency accessibility [33]. Figure 1.4 depicts a simulated comparison of the SNR as the coil geometry increases with respect to the sample volume of a single-loop planar wired coil (blue), a single-loop planar coil with a LL (red), and a single-loop planar coil with an LC resonator (green).

The concept of LL was applied by Spengler et al. [34] to his previously reported NMR micro Helmholtz detector. Figure 1.5 shows an LL with 1 mm outer diameter and 0.2 mm inner diameter inserted inside the micro Helmholtz LL2 coil. In the paper, four different lenses were tested, namely plate lenses LL1 and LL2 with inner diameters of 0.2 mm and 0.4 mm, respectively, and wire lenses LL3 and LL4 with inner diameters of 0.2 mm and 0.4 mm, respectively. The performance of the lenses was evaluated via a series of MRI spin echo imaging experiments on a

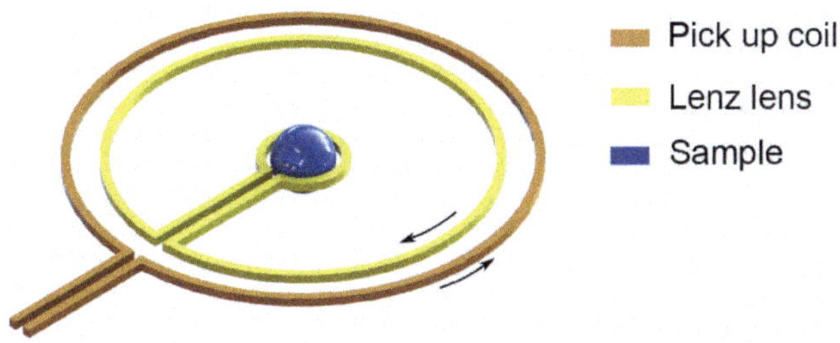

Figure 1.3 The Lenz lens (LL) collects the magnetic flux of the pick-up coil and focuses it into the sample region resulting in an enhanced sensitivity of the measurement.

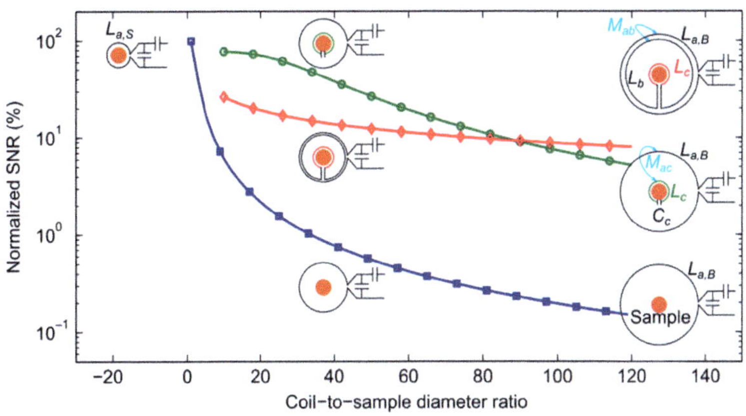

Figure 1.4 A comparison of the LL performance versus the LC resonator for a planar single-loop pick-up coil. The first point of the blue curve is the optimum situation where the coil geometry perfectly matches the sample volume (coil-to-sample diameter ratio = 1). As the coil geometry increases in comparison with the sample volume the SNR decreases. In this suboptimal situation, the SNR can be improved by using either an LC resonator (green curve) or a LL (red curve). In this figure, La,S is the inductance of the small pick-up coil, La,B is the inductance of a big pick-up coil, Lb is the inductance of the outer loop of the LL, Lc is the inductance of the inner loop of the LL, and M is the mutual inductance.

deionized water sample. All experiments were preceded by flip angle adjustment routines to ensure that all the measurements are conducted at the same flip angle, namely 90°. Figure 1.6 calculates the SNR enhancement due to the various lenses (LL1–LL4) and compares it with the reference SNR of the micro Helmholtz coil without lenses (the red curve). According to the results, exploiting LLs can significantly enhance the coil sensitivity and thus the achievable imaging resolution. The SNR enhancement varies largely with the inner diameter of the lens (increases as the inner diameter decreases) but changes slightly with the topology used (plate/wire). The maximum SNR enhancement reported in the paper was 2.8, which would allow reducing the voxel volume by 64% (reducing the voxel length by 29%) while maintaining the acquisition time and the SNR per voxel unchanged. A system that combines the concepts of both the LL and the LC resonator has been introduced by Kamberger et al. [35]. The so-called resonant LL benefits from the field-focusing feature of the LL and the signal amplification advantage due to resonance. The resonant LL was integrated with an MR-compatible incubation platform designed to cultivate organotypic hippocampal slice cultures (OHSCs), to perform in vitro MR microscopy of brain tissues. The wireless LL was purposely employed to avoid the direct

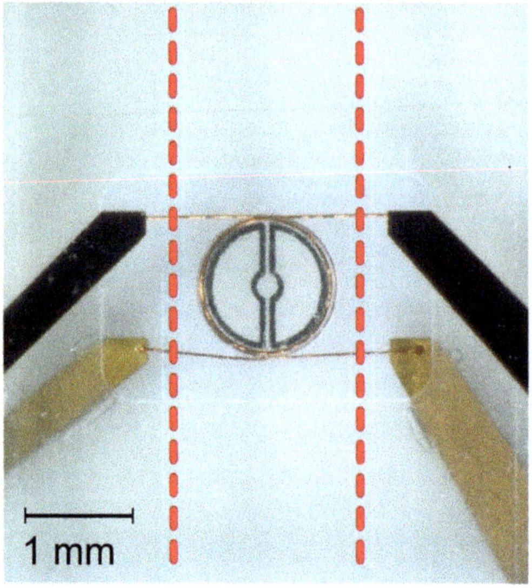

Figure 1.5 A Helmholtz micro coil with a wire Lenz lens. [33] Nils Spengler et al. (2017), figure 03[p.008]/Public Library of Science (PLoS)/CC BY 4.

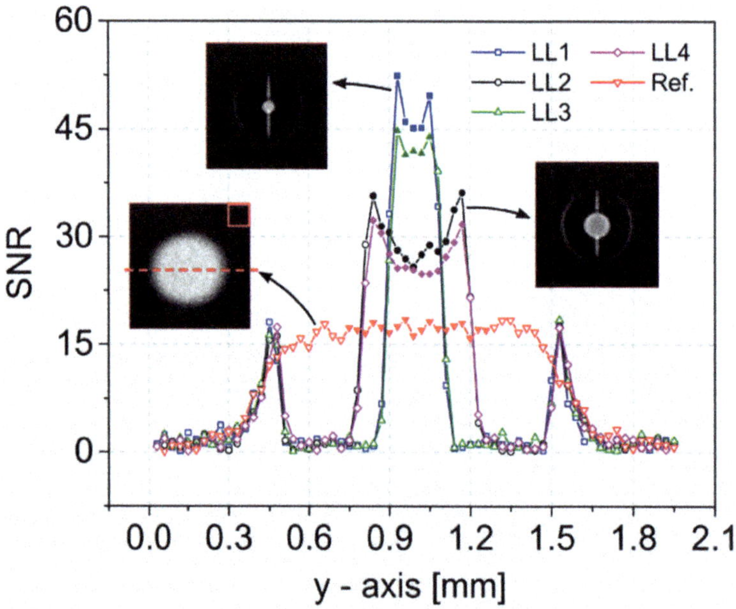

Figure 1.6 Sensitivity enhancement of the micro Helmholtz coil due to Lenz lenses. The red curve indicates the reference signal-to-noise ratio (SNR) of the Helmholtz coil along the center line of the image when no Lenz lenses are used. The other curves show how the use of Lenz lenses boosts SNR of the sample region in the inner loop of the lens. The SNR enhancement ranges from 1.6- to 2.8-fold. [34] Nils Spengler et al. (2017), figure S1/Public Library of Science (PLoS)/CC BY 4.0.

connections of wired RF coils, thereby giving the incubation platform more flexibility and freedom. Moreover and more importantly, avoiding wired coils reduces the number of different-materials interfaces, which consequently diminishes the susceptibility-mismatch-based imaging artifacts. In fact, the latter issue is an extremely important aspect when considering microscopy. For this reason, the LL introduced in [35] was carefully designed and manufactured by patterning thin copper tracks on a slender polymer foil so as to minimize the susceptibility-mismatch effects. The proposed LL was tested in a 9.4-T horizontal bore Bruker small animal MRI scanner. The scanner is equipped with a 72-mm-diameter volume coil. Three measurement scenarios were used to enable a comprehensive comparison of the performance. These scenarios are:

1) MR imaging using the bare volume coil without any add-ons. This measurement served as the baseline to which the other scenarios can be compared.
2) MR imaging with a broadband nonresonant LL integrated with the incubation platform and inserted in the volume coil.
3) MR imaging of a resonant LL tuned, by a discrete capacitor, at the Larmor frequency and integrated into the incubation platform.

In all experiments, a T_1-weighted Flash sequence was applied to obtain an MR image of a mouse brain slice with a 0.5-mm thickness and an in-plane resolution of 100 × 100 μm from 16 averages over 8 min. The results, as demonstrated in Figure 1.7, show a significant enhancement in the

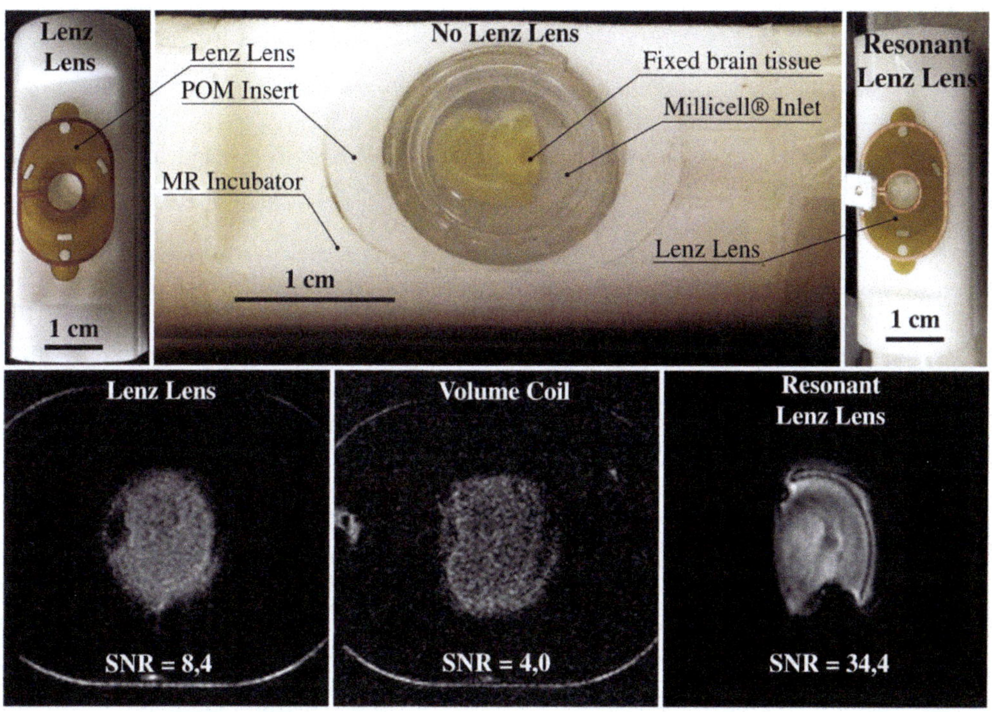

Figure 1.7 Top: Magnetic resonance (MR) compatible incubation platform for cultivating mouse brain slices. The platform was integrated with a broadband nonresonant Lenz lense (LL) (left), and a resonant LL (right). Bottom: MR images of a brain slice using the Bruker volume coil (middle), the broadband LL (left), and the resonant LL (right). [35] R. Kamberger et al. (2018), figure 06[p.13]/with permission from Elsevier.

imaging SNR due to LLs. More specifically, a broadband nonresonant LL achieved more than double the SNR of the volume coil, while an 8.5-fold SNR enhancement was obtained by the resonant LL.

1.3 MR Microscopy and Neurotechnologies

"One cubic millimeter of cerebral cortex contains roughly 50,000 neurons, each of which establishes approximately 6,000 synapses with neighbouring cells" [36]. Thus any attempt to shed light on brain function using MRI or NMR will necessarily have to apply MR microscopy. A number of aspects, especially in brain research, are amenable to MR, especially those associated with possible brain interventions, and those involved in functional magnetic resonance imaging (fMRI) of the brain. We briefly consider these aspects next.

1.3.1 Tissue Scaffolds and Implants

Neurotechnologies rely on long-term implantable technical systems, in which technical materials come into direct contact with brain tissue, which mainly consists of neurons and a permeating vasculature. Two questions are of central concern:

1) Do neurons permit intimate contact with the technical system?
2) Do the materials of the technical system disturb subsequent MRI?

The realization that carbon, despite its hardness, is readily accepted by many cell types, also by stem cells, led to the exploration of microstructurable carbon as a tissue scaffold for studies in cell migration, network formation, and cell response [37]. In this study, an aqueous cryogel was steadily cooled, causing ice nucleation and crystal growth with morphological control, which upon thawing remained separated so that after drying a 3D polymer network remained. The network was pyrolized to yield an interconnected 3D carbon lattice, which in turn could be populated by neuronal stem cells, which allowed medium- to long-term studies of cell viability, confirmed by gradient echo MRI, see Figure 1.8.

Moreover, the low susceptibility to warping of the magnetic field caused by the carbon scaffold at 11.74 T opened the door for the exploration of carbon as a neuronal implant [38]. In this study, the authors produced carbon brain implant microelectrodes on Kapton foil, by lithographically structuring a photopolymer followed by pyrolysis and embedding in durimine before release. The implants were then investigated for their MR properties, including force-induced vibrations using a specially constructed dynamic force sensitive probe head [39]. Compared with platinum electrodes, which sufficiently warped the MRI signal coming from neighboring voxels, carbon electrodes permitted the acquisition of unwrapped images right up to the microelectrodes, thus paving the way for studies in postoperation tissue recovery and wound healing.

1.3.2 The Case of Epileptogenesis: Ex Situ Brain Slices and in Situ Histology

Brain-implanted microelectrodes target devastating diseases such as Parkinsons or epilepsy, and one of the primary questions in the field is whether it is possible to use noninvasive MRI to diagnose the disease, monitor recovery and temporal evolution of disease progress, and confirm correct system function of the implant. The hope is to translate diagnostic findings from mouse models to humans, despite certain differences, but with the benefit of using higher fields and hence higher-resolution techniques on model organisms than are generally available for humans. The gold standard for epileptogenesis is a kainate-initiated temporal lobe epilepsy mouse model established by Bouilleret et al. [40], for which a dedicated MR-compatible tissue slice environment was developed [35]. In this device, a freshly extracted brain slice 5 mm diameter, 500 μm thickness is maintained at physiological conditions (36°C, perfused, oxygenated) during MRI (Figure 1.7). Additional studies, which tracked the morphological changes of the brain slice over longer timeframes, were performed using a cryogenic vendor-supplied small animal coil [41], for the first time yielding confirmation of histological correlates with various MRI modalities. Using the same cryo-imaging coil, translation of these techniques to the full animal model yielded a comprehensive picture of the disease progression [42] and established a number of new measurement and postprocessing techniques, including high-resolution diffusion tensor imaging, see Figures 1.9 and 1.10.

1.4 Augmented MR Microscopy

In situations in which microscopy is desired of objects that themselves have small dimensions, then there is an MR sensitivity advantage when using micro detectors. The challenge of sample handling at these dimensions, often achieved by microfluidic systems, can be turned into a feature when one considers the library of lab-on-a-chip (LOC) technologies already available, offering additional degrees of freedom in terms of sample management and interaction. Integration of NMR and microfluidics has continued to advance, starting from the earliest microcoil reports [43,44] to recent efforts taking advantage of LOC principles to enable

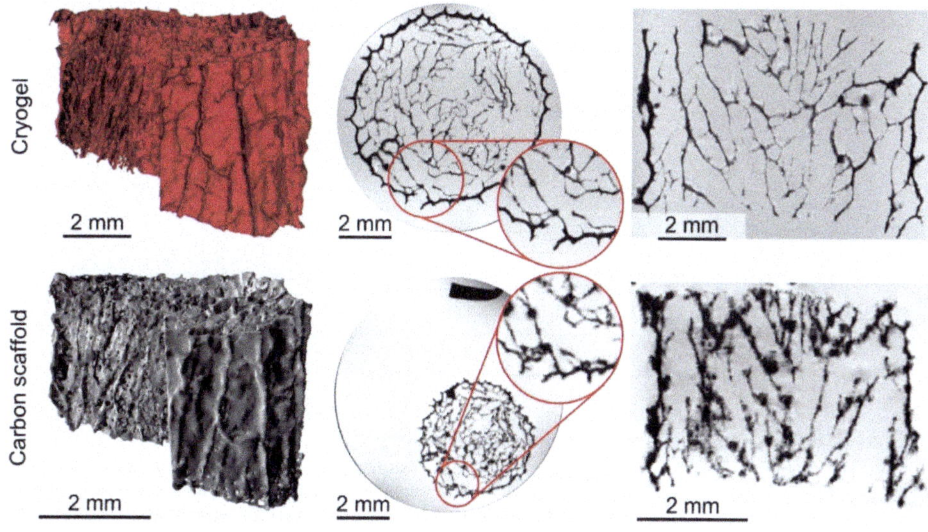

Figure 1.8 Porosity and connectivity analysis of a neuronal scaffold using magnetic resonance imaging microscopy, comparing a cryogel with its pyrolized and hence shrunk carbon counterpart. [37] Erwin Fuhrer et al. (2017), figure 02[p.03]/with permission from John Wiley & Sons, Inc.

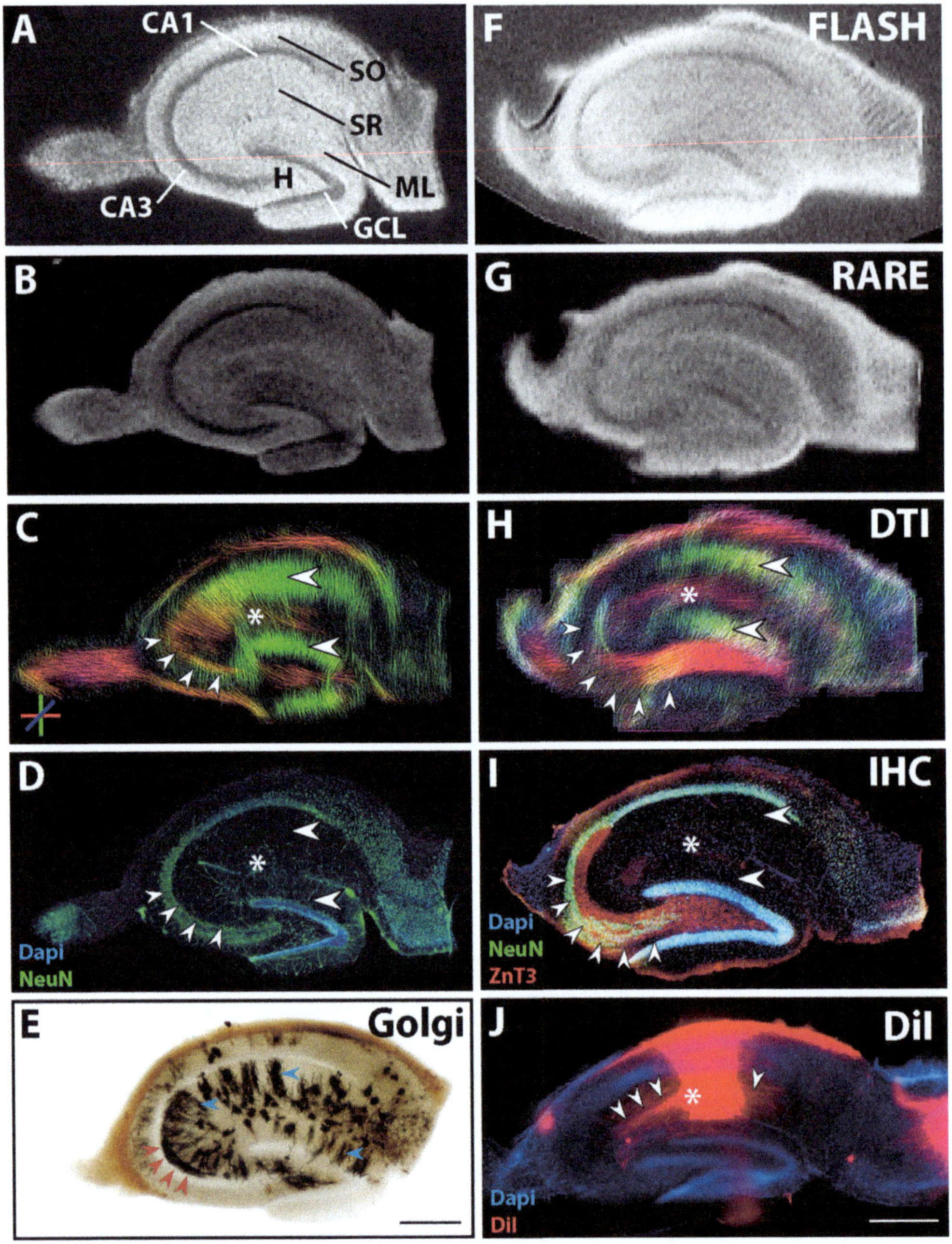

Figure 1.9 Magnetic resonance (MR) and histological images of fixed hippocampal sections of two control animals (Section 1: A–D, with an adjacent section used for Golgi staining: E; Section 2: F–I). (A, B, F, G) Structural magnetic resonance imaging (MRI) depicts the main neuronal cell layers and tissue architecture. Comparison of diffusion tractography images (C, H) to corresponding NeuN and ZnT3 immunostaining (D, I) shows that regions containing parallel extending dendrites of principal neurons evoke corresponding diffusion-weighted imaging (DWI) streamlines. (E) Golgi staining of an adjacent section depicts the localization and orientation of principal cell dendrites (blue arrowheads in SR and ML) and parts of mossy fibers in CA3 (red arrowheads). (J) DiI crystal placed into the CA1 stratum radiatum, showing the orientation of CA1 pyramidal cells as well as of innervating CA3 Schaffer collaterals (asterisk and arrowheads, respectively). Scale bar: 500 μm. [41] Katharina Göbel-Guéniot et al. (2020), figure 02[p.06]/Frontiers Media S.A./CC BY 4.0.

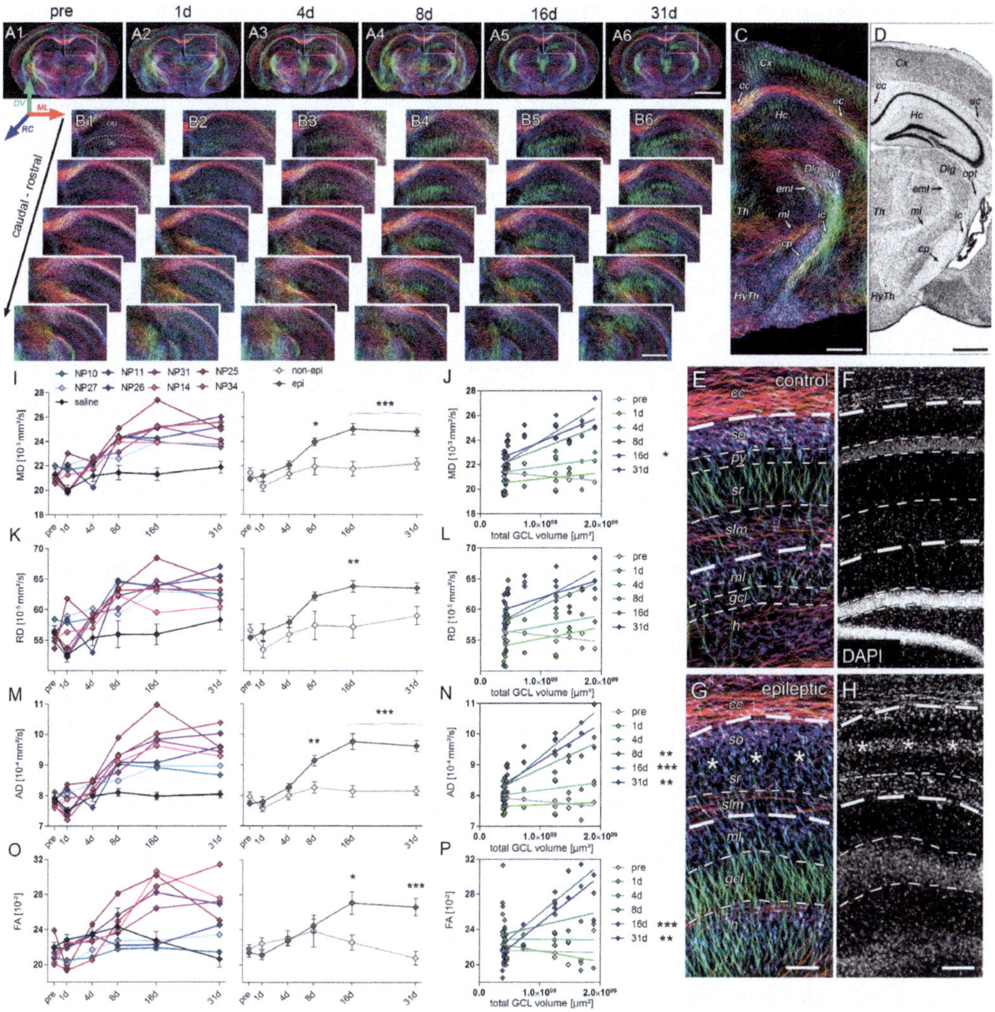

Figure 1.10 Microstructural reorganization quantified by diffusion-weighted imaging (DWI) during epileptogenesis predicts disease progression. (A1–6) Representative coronal sections from diffusion-weighted tractography at different time points during epileptogenesis (before injection = pre; 1 day, 4 days, 7 days, 14 days, and 31 days following SE). (B1–6) Enlarged images. (C–D) Representative tractography image and a Nissl-stained section of corresponding brain regions for anatomical comparison. Computed fibers relate to major axonal pathways and brain regions exhibiting highly oriented dendrites (cc, corpus callosum). (E, G) Enlarged tractography images demonstrating the distinct orientation of streamlines in different hippocampal layers. (F, H) Corresponding 4′,6-diamidino-2-phenylindole (DAPI)-stained sections. Scale bars in A, 2 mm; B–D, 500 mm; H (left), 100 mm (I, K, M, O). Quantitative analysis of DWI metrics in the dentate gyrus (DG), plotted for individual mice (left panel; controls, black, n = 5; kainate-injected animals color-coded) and for groups during epileptogenesis. [42] Philipp Janz et al. (2017), figure 05[p.10]/eLife Sciences Publications Ltd./CC BY 4.0.

microfluidic-based perfusion [35,45–49], electrochemistry [50], and hyperpolarization [51–53]. As discussed extensively in this chapter, micro-NMR coupled with MR microscopy is established and continues to evolve, with one branch of this evolution now extending toward fully exploring the augmented micro-MR system to further enhance the applicability of MR microscopy.

1.4.1 Perfusion

Perfusion can be considered a subclass of flow-based methods. In the context of this discussion, the definition of perfusion is relaxed slightly to include the passage of a fluid through microfluidic systems for the purpose of transporting nutrients and waste. Therefore, such systems can be used to maintain biological samples under conditions conducive to normal behavior, enabling long-term measurements of the system under normal and stimulated situations. Systems may include cell populations/layers/clusters and may increase in complexity up to tissue slices, organ-on-a-chip, and small organisms. Long-term measurement of such systems while using MR-compatible technical systems enables spatially resolved, longitudinal monitoring of morphology as a function of interesting stresses.

Starting from spectroscopy, perfusion-enabled microfluidic devices for long-term monitoring of biological systems have been used to monitor metabolic flux. Kalfe et al. [49] monitored a single tumor spheroid with diameter 500 μm over 24 h to characterize the transition from oxidative to glycolytic metabolism. Given the small volumes, it is important to ensure that the metabolic activity is reflective of the actual biological system and not perturbed, for example, by a lack of oxygen in the culture medium. This issue has been directly addressed by Yilmaz and Utz [47], who used a gas permeable membrane in combination with a micro-stripline [46] to ensure adequate oxygen supply for in situ cell culture NMR spectroscopy. Using 3D printing to produce MR-compatible microfluidic systems Montinaro et al. [45], demonstrated spectroscopy of small organisms, achieving detection volumes of 100 pl and therefore substructure spectroscopy of *Caenorhabditis elegans* in their work.

Incubator systems have been implemented for improved tissue slice MR microscopy. To enable long-term microscopy of a biologically viable tissue, incubator systems must not only manage perfusion but also gas concentration and temperature control. Flint et al. [48]. developed such an incubator system compatible with a 600-MHz NMR spectrometer. They demonstrated diffusion-weighted imaging of rat cortical slices with 31.25 μm isotropic resolution (1.5 h measurement time) over 21 h. The challenge for soft tissue incubation in vertical bore NMR systems is preventing tissue deformation caused by gravity. This can be addressed, for example, by physically clamping the tissue taking care not to unnecessarily perturb the tissue function. This challenge can be circumvented by ensuring gravity is perpendicular to the tissue surface, easily achieved in horizontal bore systems. Kamberger et al. [35] implemented an incubator for mouse brain slice imaging under this condition, with the added feature of a LL for magnetic field-focusing and improved SNR [33]. Using a 9.4-T MRI system, T_1-weighted images could be obtained in 8 min with 0.5 mm slice thickness and in-plane resolution of 0.1 × 0.1 mm, importantly, with a factor of 10 improved SNR yielded by the LL (Figure 1.7). In a clever use of capillary forces, tissue–air interfaces perpendicular to B_0 were avoided by allowing the perfusion medium to slightly overfill the tissue chamber thereby eliminating magnetic susceptibility-induced imaging artifacts.

1.4.2 Electrochemistry

The microfluidic LOC and micro total analysis system (μTAS) communities have recognized the added value of implementing electro-manipulation capabilities. From the perspective of biological samples, the electrical degree of freedom enables new methods for sample manipulation and sample analysis [54]. From a chemical perspective, electric fields can be used for selective analyte transport or to drive electrochemical reactions. Simultaneous integration with microscopy has the potential to spatially localize the electro-response of the system, noninvasively and label-free in the case of MR microscopy.

MR-coupled electrochemical methods are often used to study electrophoretic behavior of electrolytes [55–57]. In these experiments, the electrodes used to generate the electric field are located outside of the NMR-sensitive volume. In the case of monitoring electrochemical reactions, it would be advantageous to localize the reaction within or in the immediate vicinity of the NMR detection volume. This has been demonstrated outside of the microfluidic domain, with recent efforts aiming to retain high-resolution NMR spectra at high magnetic fields using standard sample tubes [58–61]. NMR-compatible microfluidic systems with integrated electrodes are rare; in one example, a digital microfluidic approach has been demonstrated as a means to manipulate droplets and deliver sample to the NMR detection volume [62,63]. An in situ electrochemical system has recently been reported, including an analysis of the appropriate electrode properties that maintain acceptable NMR spectral resolution [50] (Figure 1.11). In this report, MR microscopy is additionally demonstrated, although in this case for the purpose of B_0 and B_1 field characterization.

1.4.3 Hyperpolarization

Micro-NMR detectors feature a sensitivity advantage for mass-limited samples, yet still suffer in terms of absolute sensitivity as dictated by the total number of signal-contributing spins in the sample. Hyperpolarization methods successfully circumvent this limitation and have proven extremely fruitful under "standard" NMR measurement conditions. Extending these capabilities to the microfluidic NMR domain was inevitable, yet progress has not been as dramatic as in

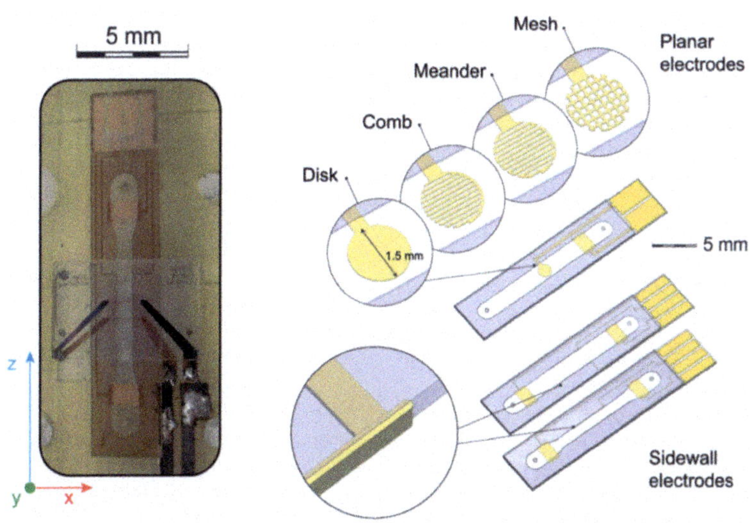

Figure 1.11 Photograph of a microfluidic insert featuring integrated electrodes (left) compatible with the micro Helmholtz detector. A variety of electrode geometries are possible from a fabrication standpoint (right), but care must be taken when considering MR compatibility. Davoodi et al. (2020). An NMR-compatible microfluidic platform enabling in situ electrochemistry. Lab on a Chip, 20(17), 3202–3212. Licensed under CC-BY-NC 3.0.

macro-scale NMR.[1] This observation can be explained by considering first, the microfluidic NMR community is certainly small compared with the NMR community as a whole; and second, several technical challenges must be addressed to bring hyperpolarization to the micro-scale (material interfaces, scaling effects, integration, etc.). Nevertheless, all versions of NMR hyperpolarization have been explored at small scales.

Hyperpolarized gases offer profiling of both physical structures and chemical space. Using a remote sensing approach, both Xe and H_2 gases with enhanced polarization levels were used together with microfluidic systems to characterize on-chip gas flow profiles [64] and chemical micro-reactor systems [65]. Production of Xe at polarization levels approaching 10% at the micro-scale using microfabricated hyperpolarization cells significantly enhances portability together with local production for microfluidic applications [66,67]. To access the chemical space, the hyperpolarized gas must typically be brought into solution. At the macro-scale this can be achieved by bubbling a sample and pressuring the tube with the gas prior to the NMR experiment. Under microfluidic conditions a bubbling procedure is somewhat disastrous – bubbles easily disrupt microfluidic flow and destroy NMR spectral resolution and thus must be either managed by fluidic design, or preferably avoided altogether. This has been addressed by using membrane contacting systems (high-field examples in Figure 1.12), enabling bubble-free parahydrogen

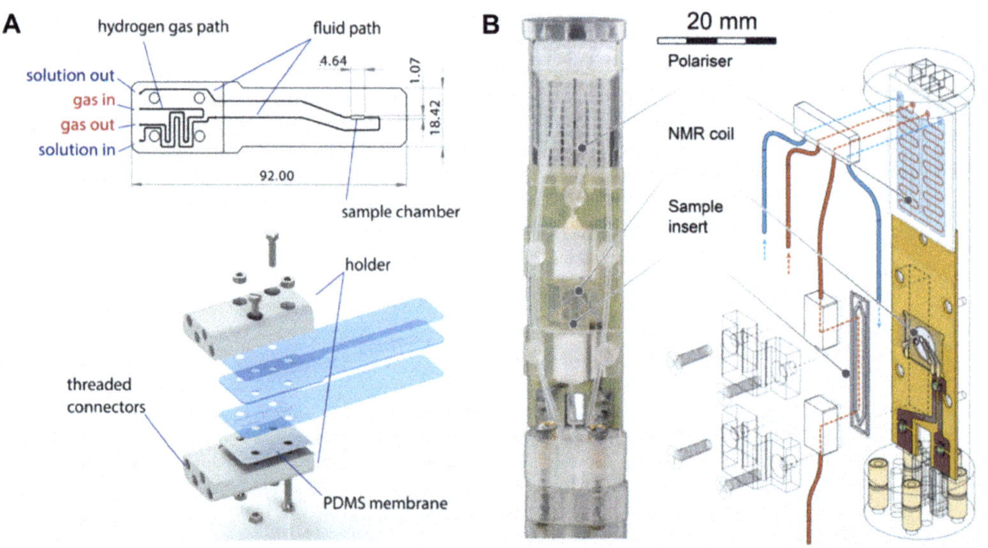

Figure 1.12 Membrane-contacting devices for high-field microfluidic parahydrogen-induced polarization (PHiP). (A) Drawing of the contact chip (top), and overview of the assembly (bottom). The region of the sample chamber slides into a micro-stripline detector. (B) Photograph of the gas contacting and NMR detector assembly (left) and schematic of the exploded assembly (right). The fluidic insert is compatible with a micro Helmholtz detector, with liquid lines in blue and gas lines in orange. [51] Lorenzo Bordonali et al. (2019), figure 01[p.03]/with permission from Royal Society of Chemistry.

1 A Web of Science literature search (2 October 2020) using the Topic keywords (microfluidic or microfluidics) and (NMR or nuclear magnetic resonance) and (hyperpolarisation or hyperpolarisation) revealed 11 results. The authors are aware of an additional contribution in 2020, perhaps not yet indexed at the day of searching, yielding 12 results. An identical search but removing the microfluidics keywords yielded 1014 results.

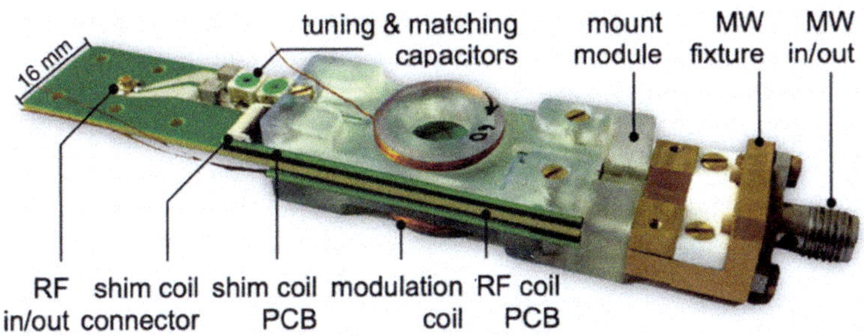

Figure 1.13 Photograph of a fully integrated Overhauser dynamic nuclear polarization (ODNP) probe head, designed to operate in a palm-held 0.5-T permanent magnet. [72] Sebastian Kiss (2019)/figure 05.26 [p.125]/with permission from University of Freiburg.

transport from the gas to solution phase. This has been leveraged for flow chemistry using both hydrogenative [52] and nonhydrogenative [51,53] PHiP experiments, with picomole sensitivities achieved under high-magnetic-field conditions.

DNP has made enormous advances in academic spectroscopy and imaging research, even to the point of commercialization and clinical studies [68]. Leveraging this potential for microfluidic-compatible MR microscopy has only begun, first taking root primarily in metabolic monitoring. Jeong et al. [69] used the relatively long-lived ^{13}C enhanced state of pyruvate and lactate to monitor cancer cell metabolism. The hyperpolarized pyruvate was added to a microcoil accommodating 2 μl of sample and subsequently moved to a 1.05-T NMR magnet for measurement. Metabolic flux was measured for a population of only 104 cells, three orders of magnitude better than what was typically required. Mompéan et al. [70] used the photo-inducible version of DNP (photo-chemically induced DNP, photo-CIDNP) on a micro-NMR flow device to enable sub-picomole sensitivity of a 1-μl sample at 9.4 T. Uberrück et al. [71] demonstrated liquid-state Overhauser DNP (ODNP) at 0.342 T, while Kiss [72] performed ODNP at 0.5 T using a fully integrated microsystem approach (Figure 1.13). A key challenge to be faced with microfluidic-compatible DNP-based hyperpolarization at high magnetic fields is handling high-frequency microwaves for electron excitation. This is alleviated in the case of photo-CIDNP where microwaves are not necessary, at the cost of limited applicability. This challenge is general, and is a focus for the entire high-field DNP community.

1.5 Conclusions

MR microscopy can benefit greatly from engineering approaches to bridge the gap between the sample and the spectrometer, especially in the application areas of cell and small organism biology, and (electro) chemistry. In this chapter we considered the contribution that microengineering can make, covering custom resonators and sample holders, which provide more ideal SNR and sample conditions, thus facilitating increased experimental flexibility. In essence, Faraday induction NMR detection has arrived at the LOD, so research work is currently mainly focusing on system issues, dealing with complex samples, achieving better field shimming, hyphenation of measurement modalities, and hyperpolarization of the spin population, and of course achieving spectrometer agility and proximity. Certainly, quantum sensors may provide more sensitivity in the future, but also require proximity of the detection system to the spin of interest, which for many applications is tantamount to becoming invasive. It is also not yet clear how hyperpolarization can be made less invasive, or less poisonous. The future will see small detectors becoming easier to use, not least because of better integration into the NMR workflow and equipment. Already some micro-NMR and hyperpolarization startups are pushing the field in this direction.

References

1 Callaghan, P.T. (1991). *Principles of Nuclear Magnetic Resonance Microscopy*. Oxford, UK: Clarendon Press.
2 Mansfield, P. and Grannell, P.K. (1973). NMR "diffraction" in solids. *Journal of Physics C* 6: L422–L426.
3 Blümler, P., Greferath, G., Blümich, B. et al. (1993). NMR imaging of objects containing similar substructures. *Journal of Magnetic Resonance A* 103: 142–150.
4 Blümich, B. (2016). k and q dedicated to Paul Callaghan. *Journal of Magnetic Resonance A* 267: 79–85.
5 Marx, V. (2013). Is super-resolution microscopy right for you? *Nature Methods* 10: 1157–1163.
6 Ciobanu, L., Seeber, D.A., and Pennington, C.H. (2002). 3D MR microscopy with resolution 3.7 μm by 3.3 μm by 3.3 μm. *Journal of Magnetic Resonance* 158 (1–2): 178–182.
7 Rugar, D., Budakian, R., Mamin, H. et al. (2004). Single spin detection by magnetic resonance force microscopy. *Nature* 430: 329–332.
8 Bar-Even, A., Noor, E., Savir, Y. et al. (2011). The moderately efficient enzyme: Evolutionary and physicochemical trends shaping enzyme parameters. *Biochemistry* 50 (21): 4402–4410.
9 Chen, H.-Y., Aggarwal, R., Bok, R.A. et al. (2020). Hyperpolarized 13C-pyruvate MRI detects real-time metabolic flux in prostate cancer metastases to bone and liver: A clinical feasibility study. *Prostate Cancer and Prostatic Diseases* 23 (2): 269–276. doi: 10.1038/s41391-019-0180-z.
10 Staudacher, T., Shi, F., Pezzagna, S. et al. (2013). Nuclear magnetic resonance spectroscopy on a (5-nanometer)^3 sample volume. *Science* 339 (6119): 561–563. doi: 10.1126/science.1231675.
11 Cho, Z.H., Ahn, C.B., Juh, S.C. et al. (1988). Nuclear magnetic resonance microscopy with 4-μm resolution: Theoretical study and experimental results. *Medical Physics* 15 (6): 815–824. doi: 10.1118/1.596287.
12 Ahn, C.B. and Cho, Z.H. (1989). A generalized formulation of diffusion effects in μm resolution nuclear magnetic resonance imaging. *Medical Physics* 16 (1): 22–28. doi: 10.1118/1.596393.
13 McFarland, E.W. and Mortara, A. (1992). Three-dimensional NMR microscopy: Improving SNR with temperature and microcoils. *Magnetic Resonance Imaging* 10 (2): 279–288. doi: 10.1016/0730-725X(92)90487-K.
14 Brandl, M. and Haase, A. (1994). Molecular diffusion in NMR microscopy. *Journal of Magnetic Resonance. Series B* 103 (2): 162–167.
15 Ahn, C.B. and Chu, W.C. (1969). Optimal imaging strategies for three-dimensional nuclear magnetic resonance microscopy. *Journal of Magnetic Resonance* 94 (3): 455–470. doi: 10.1016/0022-2364(91)90132-D.
16 Blackband, S.J., Buckley, D.L., Bui, J.D. et al. (1999). NMR microscopy – Beginnings and new directions. *Magnetic Resonance Materials in Physics, Biology and Medicine* 9 (3): 112–116.
17 Mansfield, P. (1982). *NMR Imaging in Biomedicine: Supplement 2 Advances in Magnetic Resonance, Volume 2*. Amsterdam: Elsevier.
18 Korvink, J.G., MacKinnon, N., Badilita, V. et al. (2019). "Small is beautiful" in NMR. *Journal of Magnetic Resonance* 306: 112–117. doi: 10.1016/j.jmr.2019.07.012.
19 Webb, A.G. (2013). Radiofrequency microcoils for magnetic resonance imaging and spectroscopy. *Journal of Magnetic Resonance* 229: 55–66. doi: 10.1016/j.jmr.2012.10.004.

20 Fratila, R.M. and Velders, A.H. (2011). Small-volume nuclear magnetic resonance spectroscopy. *Sensors and Actuators A: Physical* 4 (1): 227–249. doi: 10.1146/annurev-anchem-061010-114024.

21 Kentgens, A.P.M., Bart, J., Van Bentum, P.J.M. et al. (2008). High-resolution liquid-and solid-state nuclear magnetic resonance of nanoliter sample volumes using microcoil detectors. *Journal of Chemical Physics* 128 (5): 052202. doi: 10.1063/1.2833560.

22 Badilita, V., Meier, R.C., Spengler, N. et al. (2012). Microscale nuclear magnetic resonance: A tool for soft matter research. *Soft Matter* 8 (41): 10583–10597.

23 Ciobanu, L., Seeber, D.A., and Pennington, C.H. (2002). 3D MR microscopy with resolution 3.7 μm by 3.3 μm by 3.3 μm. *Journal of Magnetic Resonance* 158 (1–2): 178–182. doi: 10.1016/S1090-7807(02)00071-X.

24 Gruschke, O.G., Baxan, N., Clad, L. et al. (2012). Lab on a chip phased-array MR multi-platform analysis system. *Lab on a Chip* 12 (3): 495–502. doi: 10.1039/c2lc20585h.

25 Roffmann, W.U., Crozier, S., Luescher, K. et al. (1996). Small birdcage resonators for high-field NMR microscopy. *Journal of Magnetic Resonance. Series B* 111 (2): 174–177.

26 Spengler, N., Moazenzadeh, A., Meier, R. et al. (2014). Micro-fabricated Helmholtz coil featuring disposable microfluidic sample inserts for applications in nuclear magnetic resonance. *Journal of Micromechanics and Microengineering* 24 (3): 034004. doi: 10.1088/0960-1317/24/3/034004.

27 Schnall, M.D., Barlow, C., Subramanian, V.H. et al. (1969). Wireless implanted magnetic resonance probes for in vivo NMR. *Journal of Magnetic Resonance* 68 (1): 161–167. doi: 10.1016/0022-2364(86)90326-4.

28 Wirth, E.D., III, Mareci, T.H., Beck, B.L. et al. (1993). A comparison of an inductively coupled implanted coil with optimized surface coils for in vivo NMR imaging of the spinal cord. *Magnetic Resonance in Medicine* 30 (5): 626–633. doi: 10.1002/mrm.1910300514.

29 Silver, X., Ni, W.X., Mercer, E.V. et al. (2001). In vivo 1H magnetic resonance imaging and spectroscopy of the rat spinal cord using an inductively-coupled chronically implanted RF coil. *Magnetic Resonance in Medicine* 46 (6): 1216–1222. doi: 10.1002/mrm.1319.

30 Elshafiey, I., Bilgen, M., He, R. et al. (2002). In vivo diffusion tensor imaging of rat spinal cord at 7 T. *Magnetic Resonance Imaging* 20 (3): 243–247. doi: 10.1016/S0730-725X(02)00493-9.

31 Quick, H.H., Kuehl, H., Kaiser, G. et al. (2002). Inductively coupled stent antennas in MRI. *Magnetic Resonance in Medicine* 48 (5): 781–790. doi: 10.1002/mrm.10269.

32 Ginefri, J.-C., Rubin, A., Tatoulian, M. et al. (2012). Implanted, inductively-coupled, radiofrequency coils fabricated on flexible polymeric material: Application to in vivo rat brain MRI at 7T. *Journal of Magnetic Resonance* 224: 61–70. doi: 10.1016/j.jmr.2012.09.003.

33 Jouda, M., Kamberger, R., Leupold, J. et al. (2017). A comparison of Lenz lenses and LC resonators for NMR signal enhancement. *Concepts in Magnetic Resonance. Part B, Magnetic Resonance Engineering* 47B (3): e21357. doi: 10.1002/cmr.b.21357.

34 Nils Spengler, P.T., While, M.V., Meissner, U.W. et al. (2017). Magnetic Lenz lenses improve the limit-of-detection in nuclear magnetic resonance. *PLoS ONE* 12 (8): e0182779. doi: 10.1371/journal.pone.0182779.

35 Kamberger, R., Göbel-Guéniot, K., Gerlach, J. et al. (2018). Improved method for MR microscopy of brain tissue cultured with the interface method combined with Lenz lenses. *Magnetic Resonance Imaging* 52: 24–32. doi: 10.1016/j.mri.2018.05.010.

36 Lichtman, J., Pfister, H., and Reid, C. (2020). Connections in the brain https://www.rc.fas.harvard.edu/case-studies/connections-in-the-brain (accessed 25 October 2020).

37 Fuhrer, E., Bäcker, A., Kraft, S. et al. (2018). 3D carbon scaffolds for neural stem cell culture and magnetic resonance imaging. *Advanced Healthcare Materials* 7 (4): 1700915. doi: 10.1002/adhm.201700915.

38 Nimbalkar, S., Fuhrer, E., and Silva, P. (2019). Glassy carbon microelectrodes minimize induced voltages, mechanical vibrations, and artifacts in magnetic resonance imaging. *Microsystems & Nanoengineering* 5: 61. doi: 10.1038/s41378-019-0106-x.

39 Jouda, M., Klein, C.O., Korvink, J.G. et al. (2019). Gradient-induced mechanical vibration of neural interfaces during MRI. *IEEE Transactions on Bio-medical Engineering* 67: 915–923. doi: 10.1109/TBME.2019.2923693.

40 Bouilleret, V., Ridoux, V., Depaulis, A. et al. (1999). Recurrent seizures and hippocampal sclerosis following intrahippocampal kainate injection in adult mice: Electroencephalography, histopathology and synaptic reorganization similar to mesial temporal lobe epilepsy. *Neuroscience* 89 (3): 717–729. doi: https://doi.org/10.1016/S0306-4522(98)00401-1.

41 Göbel-Guéniot, K., Gerlach, J., Kamberger, R. et al. (2020). Histological correlates of diffusion-weighted magnetic resonance microscopy in a mouse model of mesial temporal lobe epilepsy. *Frontiers in Neuroscience* 14: 543.

42 Janz, P., Schwaderlapp, N., Heining, K. et al. (2017). Early tissue damage and microstructural reorganization predict disease severity in experimental epilepsy. *ELIFE* 6: e25742. doi: 10.7554/eLife.25742.

43 Olson, D.L., Peck, T.L., Webb, A.G. et al. (1995). High-resolution microcoil 1H-NMR for mass-limited, nanoliter-volume samples. *Science* 270 (5244): 1967–1970.

44 Massin, C., Boero, G., Vincent, F. et al. (2002). High-Q factor RF planar microcoils for micro-scale NMR spectroscopy. *Sensors and Actuators A: Physical* 97: 280–288. doi: 10.1016/s0924-4247(01)00847-0.

45 Montinaro, E., Grisi, M., Letizia, M.C. et al. (2018). 3D printed microchannels for subnL NMR spectroscopy. 13 (5): e0192780. doi: 10.1371/journal.pone.0192780.

46 Finch, G., Yilmaz, A., and Utz, M. (2016). An optimised detector for in-situ high-resolution NMR in microfluidic devices. *Journal of Magnetic Resonance* 262: 73–80. doi: 10.1016/j.jmr.2015.11011.

47 Yilmaz, A. and Utz, M. (2016). Characterisation of oxygen permeation into a microfluidic device for cell culture by in situ NMR spectroscopy. *Lab on a Chip* 16 (11): 2079. doi: 10.1039/c6lc00396f.

48 Flint, J.J., Menon, K., Hansen, B. et al. (2015). A microperfusion and in-bore oxygenator system designed for magnetic resonance microscopy studies on living tissue explants. *Scientific Reports* 5 (1): 18095. doi: 10.1038/srep18095.

49 Kalfe, A., Telfah, A., Lambert, J. et al. (2015). Looking into living cell systems: Planar waveguide microfluidic NMR detector for in vitro metabolomics of tumor spheroids. *Analytical Chemistry* 87 (14): 7402–7410. doi: 10.1021/acs.analchem.5b01603.

50 Davoodi, H., Nordin, N., Bordonali, L. et al. (2020). An NMR-compatible microfluidic platform enabling in situ electrochemistry. *Lab on a Chip* 20 (17): 3202–3212. doi: 10.1039/d0lc00364f.

51 Bordonali, L., Nordin, N., Fuhrer, E. et al. (2019). Parahydrogen based NMR hyperpolarisation goes micro: An alveolus for small molecule chemosensing. *Lab on a Chip* 19: 503–512. doi: 10.1039/C8LC01259H.

52 Eills, J., Hale, W., Sharma, M. et al. (2019). High-resolution nuclear magnetic resonance spectroscopy with picomole sensitivity by hyperpolarization on a chip. *Journal of the American Chemical Society* 141 (25): 9955–9963. doi: 10.1021/jacs.9b03507.

53 Lehmkuhl, S., Wiese, M., Schubert, L. et al. (2018). Continuous hyper-polarization with parahydrogen in a membrane reactor. *Journal of Magnetic Resonance* 291: 8–13. doi: 10.1016/j.jmr.2018.03.012.

54 Hiramoto, K., Ino, K., Nashimoto, Y. et al. (2019). Electric and electrochemical microfluidic devices for cell analysis. *Frontiers in Chemistry* 7 (396): 396. doi: 10.3389/fchem.2019.00396.

55 Jayawickrama, D.A. and Sweedler, J.V. (2004). Dual microcoil NMR probe coupled to cyclic CE for continuous separation and analyte isolation. *Analytical Chemistry* 76 (16): 4894–4900. doi: 10.1021/ac049390o.

56 Grass, K., Böhme, U., Scheler, U. et al. (2008). Importance of hydrodynamic shielding for the dynamic behavior of short polyelectrolyte chains. *Physical Review Letters* 100 (9): 096104. doi: 10.1103/physrevlett.100.096104.

57 Diekmann, J., Adams, K.L., Klunder, G.L. et al. (2011). Portable microcoil NMR detection coupled to capillary electrophoresis. *Analytical Chemistry* 83 (4): 1328–1335. doi: 10.1021/ac102389b.

58 Gomes, B., Pollyana, D.S., Lobo, C. et al. (2017). Strong magnetoelectrolysis effect during electrochemical reaction monitored in situ by high-resolution NMR spectroscopy. *Analytica Chimica Acta* 983: 91–95. doi: 10.1016/j.aca.2017.06.008.

59 Sorte, E.G., Jilani, S., and Tong, Y.J. (2017). Methanol and ethanol electrooxidation on PtRu and PtNiCu as studied by high-resolution in situ electrochemical NMR spectroscopy with interdigitated electrodes. *Electrocatalysis* 8: 95–102. doi: 10.1007/s12678-016-0344-8.

60 Zu-Rong, N., Cui, X.-H., Cao, S.-H. et al.(2017). A novel in situ electrochemical NMR cell with a palisade gold film electrode. *AIP Advances* 7 (8): 085205. doi: 10.1063/1.4997887.

61 Da Silva, P., Gomes, B., Lobo, C. et al. (2019). Electrochemical NMR spectroscopy: Electrode construction and magnetic sample stirring. *Microchemical Journal* 146: 658–663. doi: 10.1016/j.microc.2019.01.010.

62 Swyer, I., Soong, R., Dryden, M.D.M. et al. (2016). Interfacing digital microfluidics with high-field nuclear magnetic resonance spectroscopy. *Lab on a Chip* 16 (22): 4424–4435. doi: 10.1039/C6LC01073C.

63 Swyer, I., Von Der Ecken, S., Wu, B. et al. (2019). Digital microfluidics and nuclear magnetic resonance spectroscopy for in situ diffusion measurements and reaction monitoring. *Lab on a Chip* 19: 641–653. doi: 10.1039/c8lc01214h.

64 Hilty, C., McDonnell, E.E., Granwehr, J. et al. (2005). Microfluidic gas-flow profiling using remote-detection NMR. *Proceedings of the National Academy of Sciences of the United States of America* 102 (42): 14960–14963. doi: 10.1073/pnas.0507566102.

65 Zhivonitko, V.V., Telkki, V.-V., Leppäniemi, J. et al. (2013). Remote detection NMR imaging of gas phase hydrogenation in microfluidic chips. *Lab on a Chip* 13 (8): 1554–1561. doi: 10.1039/c3lc41309h.

66 Jiménez-Martínez, R., Kennedy, D.J., Rosenbluh, M. et al. (2014). Optical hyperpolarization and NMR detection of 129Xe on a microfluidic chip. *Nature Communications* 5 (1): 3908. doi: 10.1038/ncomms4908.

67 Kennedy, D.J., Seltzer, S.J., Jiménez-Martínez, R. et al. (2017). An optimized microfabricated platform for the optical generation and detection of hyperpolarized 129Xe. *Scientific Reports* 7 (1): 43994. doi: 10.1038/srep43994.

68 Kurhanewicz, J., Vigneron, D.B., Ardenkjaer-Larsen, J. et al. (2019). Hyperpolarized 13C MRI: Path to clinical translation in oncology. *Neoplasia* 21 (1): 1–16. doi: 10.1016/j.neo.2018.09.006.

69 Jeong, S., Eskandari, R., Park, S. et al. (2017). Real-time quantitative analysis of metabolic flux in live cells using a hyper-polarized micromagnetic resonance spectrometer. *Science Advances* 3 (6): e1700341. doi: 10.1126/sciadv.1700341.

70 Mompéan, M., Sánchez-Donoso, R.M., De La Hoz, A. et al. (2018). Pushing nuclear magnetic resonance sensitivity limits with microfluidics and photo-chemically induced dynamic nuclear polarization. *Nature Communications* 9 (1): 108. doi: 10.1038/s41467-017-02575-0.

71 Uberrück, T., Adams, M., Granwehr, J. et al.(2020). A compact X-Band ODNP spectrometer towards hyperpolarized 1H spectroscopy. *Journal of Magnetic Resonance* 314: 106724. doi: 10.1016/j.jmr.2020.106724.

72 Kiss, S. (2019). Overhauser DNP probes for compact magnetic resonance. PhD thesis. University of Freiburg, Germany.

2

Ceramic Coils for MR Microscopy

Marine A.C. Moussu[1,2], *Redha Abdeddaim*[2], *Stanislav Glybovski*[3], *Stefan Enoch*[2], *and Luisa Ciobanu*[4]

[1]*Multiwave Imaging, Marseille, France*
[2]*Aix-Marseille Université, CNRS, Centrale Marseille, Institut Fresnel, Marseille, France*
[3]*ITMO University, Saint Petersburg, Russia*
[4]*CEA, DRF, JOLIOT, NeuroSpin, Université Paris-Saclay, Gif-sur-Yvette, France*

2.1 Introduction

Conventional metallic radiofrequency (RF) coils are made of conductive materials, most often copper, and are designed to resonate at the Larmor frequency thanks to their geometry and carefully disposed lumped elements. The latter impose the coil's intrinsic noise contributions and are responsible for power losses in the acquisition chain. In magnetic resonance microscopy (MRM), the achievable signal-to-noise ratio (SNR) is significantly degraded by the interaction between the conductive biological sample and the electric field induced by the RF coil. The standard volumetric probe is the solenoid coil, a winding of copper wire closely holding the recipient containing the sample. When driven by current, the coil induces a strong B_1 field in the sample, along the solenoid axis, and, simultaneously, an excessive conservative electric field that limits the SNR by reducing the loaded quality factor.

An alternative coil design, both in terms of geometry and material, is proposed with ceramic probes. High-permittivity dielectric resonators support a variety of specific field distributions at frequencies defined by the geometry and the constitutive material properties. Each eigenmode of such components is defined by its eigenfrequency and the corresponding field distribution. Developed over the past 10 years, the new generation of RF coils exploits the modal distribution of these high-permittivity dielectric resonators, typically with cylindrical shape.

For MRM applications, the dimensions of the resonators are typically chosen to match the required field of view, and the material is selected with the specific purpose of reducing the probe's intrinsic losses while generating a strong magnetic field in the imaged volume. More precisely, the magnetic field configuration created with dielectric resonators inside a sample can be similar to that of a solenoid probe. However, for dielectric probes the conservative electric field is removed, which helps to increase the loaded quality factor for the same field of view. The most important challenge remains to limit the noise from the probe: material losses of the dielectric need to be extremely low, which is only possible to achieve using special kinds of ceramics. In practice, the magnetic field is strongly correlated to the dielectric permittivity of the material: the higher the permittivity, the better the field confinement within the sample volume. Ceramic materials based on calcium and barium/strontium titanates, with their high permittivity values (~100) and possible low loss factors, are therefore good candidates to build dielectric probes.

So far, two eigenmodes of cylindrical resonators have been exploited for MR coils: the first transverse electric (TE) mode named $TE_{01\delta}$ and the first hybrid transverse mode, $HEM_{11\delta}$. Both modes can be used to replace conventional volume probes. In particular, the first one mimics the

Magnetic Resonance Microscopy: Instrumentation and Applications in Engineering, Life Science and Energy Research, First Edition. Edited by Sabina Haber-Pohlmeier, Bernhard Blümich, and Luisa Ciobanu.

solenoid field, while the second one resembles a birdcage field. For both modes, the volume where the generated magnetic field is the strongest is also where the electric field is the weakest. This feature of modal distributions is one reason why dielectric probes have the potential to outperform conventional coils, which, on the contrary, are limited by a systematic coexistence of magnetic and electric fields within the sample.

We start this chapter by briefly discussing the state of the art of ceramic probes in MR applications. Next, we describe several theoretical tools developed to help design MR probes with optimal performance. Finally, MRM experiments with ceramic probes are detailed and discussed.

2.2 State of the Art

Take-home message: Dielectric resonators have been used in microwave engineering for decades. More recently, probes for electron paramagnetic resonance (EPR) and magnetic resonance imaging (MRI) applications at several scales have been designed as well. In EPR, several theoretical tools to quantify the transmit efficiency have been developed. The engineering of high-permittivity, low-loss materials has significantly contributed to the development of these probes.

Dielectric resonators have been used in microwave engineering for decades, since they serve as miniaturized circuit components like antennas and filters, which is of great interest for telecommunication applications [1–6]. More recently, probes for MR applications at several scales have been designed as well.

In EPR, dielectric resonators are used instead of metallic cavities since they have reduced dimensions for the same resonance frequency and much higher quality factors [7,8]. They are built with ferroelectric materials and used at microwave frequencies. Several theoretical tools have been developed to quantify the transmit efficiency of ceramic probes [7,9,10].

Prototypes of dielectric resonators probes for MRI have been proposed at several B_0 field strengths and for various applications, as listed in Table 2.1. Dielectric resonators can be used in transmit/receive mode or as transmitters only. In most cases, the constitutive dielectric material is a high- to very high-permittivity (relative permittivity $\epsilon_r \geq 100$) ceramic based on calcium and barium/strontium titanates. One exception in this review is the probe for wrist clinical imaging made of water ($\epsilon_r \sim 80$).

The engineering of high-permittivity, low-loss dielectric materials has contributed significantly to the development of these probes. The high permittivity helps to confine the electromagnetic field within the resonator, while the low losses inside the material limit noise during the signal acquisition. The development of ferroelectric ceramic materials with perovskite crystalline structures has enabled the reachable permittivity to be increased while keeping the losses low [11,12].

2.3 Modeling and Design Guidelines

Take-home message: It is possible to model the first TE mode of cylindrical dielectric resonators, at the cost of manually writing the mode's field expression and numerically solving a set of equations.

Conventional MR coils, like the solenoid, have been extensively studied from a theoretical point of view and several design tools are now readily available to optimize their performance. So far, theoretical approaches have been proposed to describe the resonant modes of dielectric resonators, but rarely in the context of their use in MRI. This implies that the noise contributions of the ceramic coil need to be identified and quantified in order to estimate the achievable SNR.

Table 2.1 Literature review of dielectric probes for MRI.

	Application	$B_0(T)$	Frequency (MHz)	$\in_r$	Imaging zone (height x diameter)	Type of probe	Dielectric material	Exploited distribution	Tuning technique
[13]	Human breast	3	128	1000	15 cm × 10 cm	Volume	$(BaSr)TiO_3$ + Mg	$TE_{01\delta}$ (5 coupled disks)	Interdisk gap adjustment
[14]	Wrist	7	298	80	10 cm × 10 cm	Volume	Water	$HEM_{11\delta}$	Capacitors (coupling loops electronic network)
[15]	Cardiac (torso)	7	298	165		Surface	$(BaSr)TiO_3$	$TE_{01\delta}$ (array of 8 resonators)	
[16]	Microscopy	14.1	600	156	24 mm × 4.8 mm	Volume	$CaTiO_3$	$TE_{01\delta}$	Overlap with copper foils
[17]	Microscopy	14.1	600	323	4 mm × 3.8 mm	Volume	$(BaSr)TiO_3$	2 coupled $TE_{01\delta}$ modes	Overlap with copper strips
[18]	Microscopy	17.2	730	536	10 mm × 5.6 mm	Volume	$(BaSr)TiO_3$ + Mg	$TE_{01\delta}$	Temperature
[19]	Microscopy	17.2	730	536.	10 mm × 5.6 mm twice	Volume	$(BaSr)TiO_3$ + Mg	2 coupled $TE_{01\delta}$ modes	Temperature

Provided that transmit power is enough to reach a desired flip angle (as in the case of MRM), the SNR can be evaluated as for other transceiver probes. In this section, we describe an approximate method to calculate the magnetic and electric fields of the first TE mode ($TE_{01\delta}$) of a cylindrical resonator sketched in Figure 2.1; however, the methodology used to study this particular mode can be applied to any higher-order or different mode type.

2.3.1 Dielectric Cylindrical Resonator Modes

Describing the eigenmodes of dielectric structures relies on a similar methodology as for metallic cavities. Here we focus on resonators with circular cross sections, therefore all the equations will be expressed in a cylindrical coordinates system (ρ, θ, y) with ρ the radius, θ the azimuth, and y the altitude. ρ and θ relate to the Cartesian coordinates system (x, y, z) according to $\rho = \sqrt{x^2 + z^2}$ and $\theta = \arctan z / x$. The unit vectors associated with this coordinates system are denoted $\boldsymbol{e}_\rho$, $\boldsymbol{e}_\theta$, and $\boldsymbol{e}_y$.

We start with the Helmholtz equation, in a source-free, linear, homogeneous, and isotropic medium, verified by the electric $\boldsymbol{E}$ and magnetic $\boldsymbol{H}$ fields denoted indifferently $\boldsymbol{U}$ [20, p. 16]. ω is the angular frequency, μ the medium permeability, $\epsilon = \epsilon' - \mathrm{j}\epsilon''$ the medium complex permittivity (with j the unit imaginary number), and k is the complex wavenumber such that $k^2 = \omega^2\mu\epsilon = k_y^2 + k_\rho^2$ with k_y and k_ρ the wavenumber components along the y- and the ρ-directions, respectively. The method of longitudinal components [17] uses the decomposition of the vector field $\boldsymbol{U}$ into its longitudinal $U_y\boldsymbol{e}_y$ and transverse $\boldsymbol{U}_T = \boldsymbol{U} - U_y\boldsymbol{e}_y$ components. These components are solutions of the scalar and vector Helmholtz equations, respectively.

$$\nabla^2\boldsymbol{U} + k^2\boldsymbol{U} = 0$$
$$\underset{\text{along } y}{\Rightarrow} \nabla^2 U_y + k^2 U_y = 0 \underset{(\rho,\,\theta,\,y)}{\Rightarrow} \nabla_T^2 U_y + \frac{\partial^2 U_y}{\partial y^2} + k^2 U_y = 0 \qquad (2.1)$$

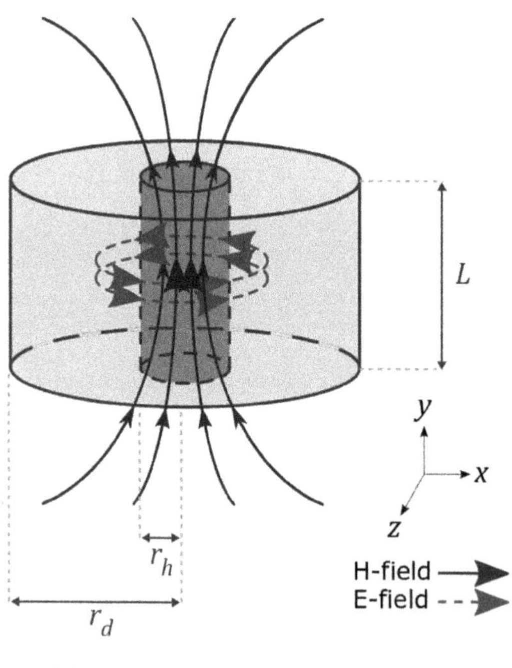

Figure 2.1 $TE_{01\delta}$ mode of a cylindrical resonator: field-line schematic and notations.

Assuming that the longitudinal component U_y can be written as the product of three functions, one for each space coordinate (in a cylindrical coordinates system in this case), the method of separation of variables allows the partial differential equation to be transformed into three independent ordinary differential equations [18]. The boundary conditions of the cavity, imposing the continuity of the tangential field components at interfaces, are then used to solve these new equations. The contributions of the axial components U_y are then separated from the transverse ones $\boldsymbol{U}_T$. The last equation for which U_y is the solution can then be solved independently in E_y, H_y, k_y, and k_ρ. The expressions for the other field components are deduced by applying the Maxwell equations, with the resulting expressions in Equation 2.2.

$$\begin{matrix} \nabla\times\boldsymbol{E}=-j\omega\mu\boldsymbol{H} \\ \nabla\times\boldsymbol{H}=+j\omega\epsilon\boldsymbol{E} \end{matrix} \Rightarrow \begin{cases} E_\rho=-\dfrac{1}{k_\rho^2}\dfrac{\partial^2 E_y}{\partial\rho\partial y}+\dfrac{j\omega\mu}{\rho k_\rho^2}\dfrac{\partial H_y}{\partial\theta} \\ E_\theta=-\dfrac{1}{\rho k_\rho^2}\dfrac{\partial^2 E_y}{\partial\theta\partial y}-\dfrac{j\omega\mu}{k_\rho^2}\dfrac{\partial H_y}{\partial\rho} \\ H_\rho=-\dfrac{1}{k_\rho^2}\dfrac{\partial^2 H_y}{\partial\rho\partial y}-\dfrac{j\omega\epsilon}{\rho k_\rho^2}\dfrac{\partial E_y}{\partial\theta} \\ H_\theta=-\dfrac{1}{\rho k_\rho^2}\dfrac{\partial^2 H_y}{\partial\theta\partial y}+\dfrac{j\omega\epsilon}{k_\rho^2}\dfrac{\partial E_y}{\partial\rho} \end{cases} \tag{2.2}$$

Here we add another simplification and consider only TE modes, meaning that the axial electric field component, E_y, is equal to zero everywhere. In this case, the mode field distribution is deduced from the solution H_y of Equation 2.1.

As an example, let us consider the TE modes of a disk resonator with radius r, height L, and its symmetry axis corresponding to the y-axis. Solving the above-mentioned problem leads to a solution inside the resonator of the form described in Equation 2.3 (magnetic field axial component inside the disk resonator for the TE modes) with A the amplitude coefficient, n an integer describing the azimuthal mode order, ϕ and ψ constant phase shifts deduced from the boundary conditions, and J_n the Bessel function of the first kind and order n.

$$H_y=A\cos(n\theta+\phi)J_n(k_\rho\rho)\cos(k_y y+\psi). \tag{2.3}$$

If the resonant structure has metallic borders, boundary conditions impose the field cancellation at the interfaces:

- In $\rho=r$, the tangential components of the magnetic field vanish, for example H_y, which quantifies the radial wavenumber:

$$H_y(\rho=r,\ \theta,y)=0\Rightarrow J_n(k_\rho r)=0\Rightarrow k_{\rho,n,m}=\frac{x_{nm}}{r}, \tag{2.4}$$

with x_{nm} the m-th zeros of the n-th order Bessel function. This boundary condition gives the radial variation order m, which is an integer.

- In $y=\pm\dfrac{L}{2}$, the magnetic field tangential components vanish, for example H_ρ, which quantifies the axial wavenumber:

$$H_\rho\left(\rho,\theta,\, y=\pm\frac{L}{2}\right)=0$$
$$\Rightarrow \sin\left(\pm k_y\frac{L}{2}+\psi\right)=0 \tag{2.5}$$
$$\Rightarrow k_{y,p}=\mp\left(\frac{2p\pi}{L}-\frac{2\psi}{L}\right),$$

with p an integer defining the axial variation order.

For a metallic cavity, the mode variations are quantified by three integers, n, m, and p, and is therefore named TE_{nmp}. It is the same principle for transverse magnetic (TM) and hybrid (HEM) mode types.

When the resonant structure has dielectric/dielectric or dielectric/air interfaces with the surrounding medium, this quantification with integer numbers is no longer valid since the electromagnetic (EM) field does not vanish at the boundaries. In such situations, the EM field outside the resonator must be explicitly expressed. An approximation consists of writing the field outside with similar variations as inside the resonator with evanescent terms, neglecting the contribution of other modes to the decaying part. This can be made part by part, as developed in [4,7] and as shown in Figure 2.2 for the disk resonator.

For example, the field expression remains unchanged within the dielectric disk, while outside it is written as in Equation 2.6, with f a decaying function depending on the computation volume boundaries, K_n the modified Bessel function of the first kind and order n, and ν the radial wavenumber in the region where the EM field is radially decaying:

$$H_y=\begin{cases} B\cos(n\theta+\phi)J_n(k_\rho\rho)f(|y-y_0|) & \text{in regions II, III} \\ C\cos(n\theta+\phi)K_n(\nu\rho)\cos(k_y y+\psi) & \text{in region VI} \end{cases}. \tag{2.6}$$

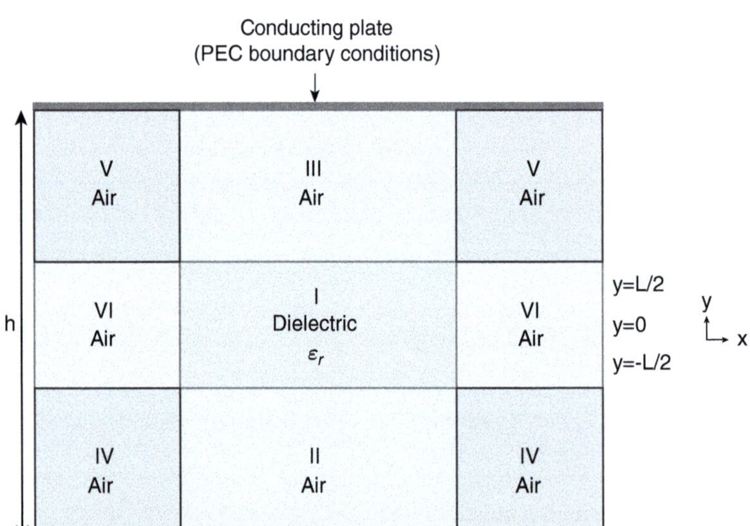

Figure 2.2 Resonant mode field estimation: part-by-part decomposition of the computation volume, case of the disk resonator.

Combining the boundary conditions at the interface air/dielectric (continuity of the tangential field component) as well as the wavenumber decomposition in each region provides a set of equations describing the modes arising from the combination of a planar dielectric waveguide with a circular dielectric waveguide.

Examples of electromagnetic field maps obtained with the same method extended to the case of a ring resonator are shown in Figure 2.3. As expected from the mode description, the electric field minimum coincides with the magnetic field maximum.

2.3.2 Power Loss Contributions in a Ceramic Probe

As ceramic probes do not require the use of an electronical circuit to tune and match at the Larmor frequency, the power losses are mainly due to the ceramic material losses and the electric field–sample interactions. Another contribution that is not considered here is that of the metallic

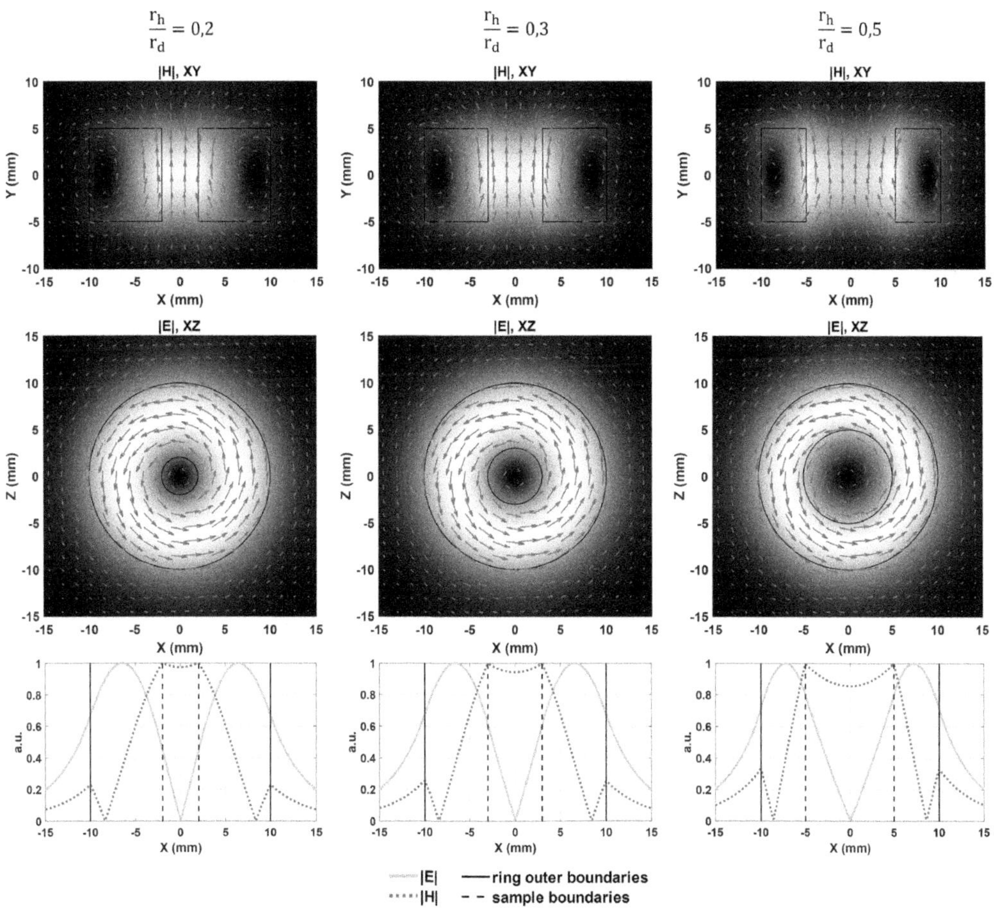

Figure 2.3 Electromagnetic field distribution of the $TE_{01\delta}$ mode of a ring resonator (relative permittivity 500, outer radius 10 mm, height 10 mm) filled with a sample (relative permittivity 50) for varying inner to outer radii ratio: magnetic field (first line) and electric field (second line) field maps and lines (gray arrows), and both fields profiles (third line). The 2D maps are plotted in grayscale with a linear value distribution. From [21].

feeding loop that is used to induce the mode's field distribution in the ceramic resonator. As it is a small, nonresonant loop, its contribution is considered insignificant.

While operating in different regimes, the loss phenomenon is the same in the ceramic material and in the sample: it is energy dissipated as heat within complex permittivity materials immersed in an electromagnetic field [19]. In practice, these power losses are expressed as the integral over the object volume V of the power loss density, which involves two local variables: the imaginary part of the material permittivity and the electric field intensity. The power losses in a material of complex permittivity $\epsilon = \epsilon' - j\epsilon''$ (with ϵ' and ϵ'' real quantities) are expressed in Equation 2.7.

$$P_{\text{loss}} = \frac{1}{2}\omega \iiint_V \epsilon'' |\boldsymbol{E}|^2 \, dv. \tag{2.7}$$

The imaginary part of the permittivity is equal to $\epsilon'' = \epsilon' \tan\delta$ with $\epsilon' = \epsilon_0 \epsilon_{\text{r}}$ the real part of the permittivity, ϵ_0 the vacuum permittivity, ϵ_{r} the material relative permittivity, and $\tan\delta$ the total loss tangent. Materials are usually described by the quantities that can be measured in experiments, the real part of the permittivity, and the total loss tangent. The latter quantifies the effects of two distinct phenomena responsible for EM power losses in the material: bound charge polarization and electric conduction when the material has a microscopic nonzero free electron density [20]. As biological samples contain a significant proportion of water, the conductivity of the sample is dominant because of dissociated ions available in the water solution, and the losses depend on the electrical conductivity σ: $P_{\text{loss}}^{\text{sample}} = \frac{1}{2} \iiint_V \sigma_{\text{sample}} |\boldsymbol{E}|^2 \, dv.$

The ceramic probe is modeled as a ceramic ring resonator (inner radius r_h, outer radius r_d, height L, relative permittivity ϵ_{r}, loss tangent $\tan\delta$) filled with a cylindrical biological sample (radius r_{h}, height L, and electrical conductivity σ_{sample}). With the theoretical insight about the $\text{TE}_{01\delta}$ mode field distribution provided in Section 2.3.1, it is possible, as detailed in [21], to develop an analytical expression for the power losses in the ceramic probe at the cost of some approximations:

- The field distribution used to express the dielectric resonator losses is that of a lossless resonator because losses in the ceramic are considered small ($\tan\delta \ll 10^{-1}$).
- The field distribution of the ring resonator is assumed equal to that of the corresponding disk without field leakages at the lateral boundaries.

With these assumptions, the power losses expression reduces to Equation 2.8 with the axial wavenumber k_y known from the mode study, H_0^{disk} the maximum magnetic field amplitude in the disk resonator, and τ a penalty coefficient accounting for the field decreasing in the sample compared with the disk field distribution, estimated from the knowledge of both the ring and disk field distributions.

$$\begin{aligned}
&P_{\text{loss}}^{\text{ring}} = \left(H_0^{\text{disk}}\right)^2 P_{\text{loss, norm}}^{\text{ring}} \\
&\text{with } P_{\text{loss, norm}}^{\text{ring}} = 2\pi \left(\frac{\omega\mu_0}{2k_\rho}\right)^2 \epsilon_0 \epsilon_{\text{r}} \omega \tan\delta \left(1 + \frac{\sin k_y L}{k_y L}\right) L \int_{r_{\text{h}}}^{r_{\text{d}}} \rho J_1^2 \left(x_{01} \frac{\rho}{r_{\text{d}}}\right) d\rho \\
&P_{\text{loss}}^{\text{sample}} = \left(H_0^{\text{disk}}\right)^2 \tau^2 P_{\text{loss, norm}}^{\text{sample}} \\
&\text{with } P_{\text{loss, norm}}^{\text{sample}} = 2\pi \left(\frac{\omega\mu_0}{2k_\rho}\right)^2 \sigma_{\text{sample}} \left(1 + \frac{\sin k_y L}{k_y L}\right) L \int_0^{r_{\text{h}}} \rho J_1^2 \left(x_{01} \frac{\rho}{r_{\text{d}}}\right) d\rho
\end{aligned} \tag{2.8}$$

The so-called Lommel's integrals, involving Bessel functions, can be found in [22,23].

2.3.3 SNR Estimation

The SNR is proportional to the transmit efficiency as follows [24]:

$$\mathrm{SNR} \sim \frac{H_0}{\sqrt{P_{\mathrm{loss}}}}, \tag{2.9}$$

with H_0 the magnetic field amplitude induced in the sample by the probe and P_{loss} the total power losses. For ceramic probes working with the $\mathrm{TE}_{01\delta}$ mode, the latter includes the ceramic resonator and the sample contributions. Assuming the magnetic field amplitude equal to that of a disk resonator in its center, weighted by the penalty coefficient τ, Equation 2.9 becomes:

$$\mathrm{SNR}_{\mathrm{ceramic}}^{\mathrm{TE}_{01\delta}} \propto \frac{\tau H_0^{\mathrm{disk}}}{\sqrt{P_{\mathrm{loss}}^{\mathrm{ring}} + P_{\mathrm{loss}}^{\mathrm{sample}}}}$$

The above equation can be simplified to:

$$\mathrm{SNR}_{\mathrm{ceramic}}^{\mathrm{TE}_{01\delta}} \propto \frac{\tau}{\sqrt{P_{\mathrm{loss,norm}}^{\mathrm{ring}} + \tau^2 P_{\mathrm{loss,norm}}^{\mathrm{sample}}}} \tag{2.10}$$

2.3.4 Mode Frequency

To be used as a coil for MR imaging, the exploited mode of the resonator must be adjusted to the working frequency. In this framework estimating the mode frequency with a simple method is very useful for predesigning the ceramic probe. Methods with low computational costs provide an approximated value of the $\mathrm{TE}_{01\delta}$ mode frequency for disk resonators. For example, a commonly used expression, derived from simulation data, is the following:

$$f_{\mathrm{MHz}}^{\mathrm{TE}_{01\delta}} = \frac{3400}{\sqrt{\epsilon_{\mathrm{r}}}\, r_{\mathrm{d,cm}}} \left(\frac{r_{\mathrm{d}}}{L} + 3.45 \right). \tag{2.11}$$

The expression in Equation 2.11 is valid with an accuracy around 2% or below in the following domain: $0.5 < \frac{r_{\mathrm{d}}}{L} < 2$ and $30 < \epsilon_{\mathrm{r}} < 50$ [1]. A diversity of alternative simplified methods can be found in the literature, some extrapolated from experimental data [25,26], some using rough approximations about the boundary conditions at the dielectric interfaces [27] or implying the numerical solving of a transcendental equation [17], and others combining the above-mentioned approaches [21].

These estimation methods are valid for disk resonators, while the dielectric probe is described as a high-permittivity ($\epsilon_{\mathrm{d}} = \epsilon_{\mathrm{r}}\epsilon_0$) ring filled with a lower permittivity ($\epsilon_{\mathrm{s}} = \epsilon_{\mathrm{r,samp}}\epsilon_0$) sample. In the specific case of the $\mathrm{TE}_{01\delta}$ mode, equating the field distribution of the probe and the disk resonators is a reasonable approximation. Indeed, between the disk and the probe, the $\mathrm{TE}_{01\delta}$ mode features are affected in proportion to the electric and magnetic energies stored in the sample volume V_{samp} [28]. The mode frequency varies from the value f_{disk} for the disk to f_{probe} for the probe according to:

$$\frac{f_{\mathrm{probe}} - f_{\mathrm{disk}}}{f_{\mathrm{disk}}} \approx \frac{-\iiint_{V_{\mathrm{samp}}} (\epsilon_{\mathrm{s}} - \epsilon_{\mathrm{d}}) |\boldsymbol{E}_{\mathrm{disk}}|^2 \, dv}{\iiint_{V_{\mathrm{samp}} \cup V_{\mathrm{ring}}} \left(\epsilon_{\mathrm{d}} |\boldsymbol{E}_{\mathrm{disk}}|^2 + \mu_0 |\boldsymbol{H}_{\mathrm{disk}}|^2 \right) dv}. \tag{2.12}$$

As the region where the sample is located corresponds to the lowest values of the E-field distribution, the frequency shift due to the sample is limited. For the $TE_{01\delta}$ mode, the ceramic probe is expected to be robust with respect to the sample loading while the permittivity contrast and the sample volume (or sample over disk radii ratio) have reasonable values. This is demonstrated in Figure 2.4.

2.3.4 Application: Design Guidelines

One way of designing a ceramic probe as depicted in Figure 2.5 consists of studying how the achievable SNR varies with the probe dimensions and/or material properties for a given sample. In the following, the required field of view constrains the ring's inner diameter and height, the

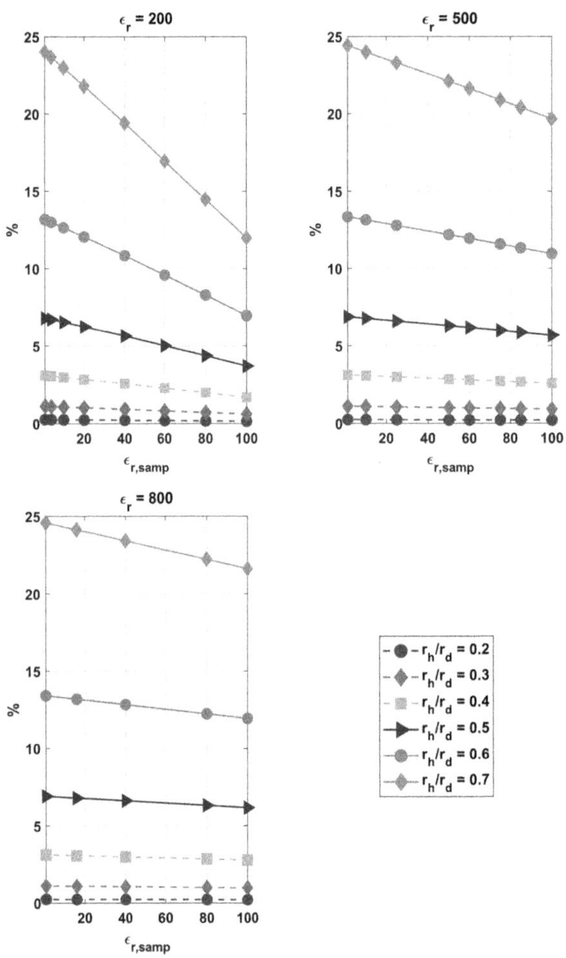

Figure 2.4 Quantification of the $TE_{01\delta}$ mode frequency shift between the disk resonator and the dielectric probe (for both, the numerical simulations results were obtained with the CST Eigenmode Solver). The probe ring has its relative permittivity equal to 200, 500, and 800. The frequency variation is plotted as a function of the sample permittivity and for several discrete values of the radii ratio. Curves in dashed lines correspond to systematic frequency shift inferior to 5%. From [21].

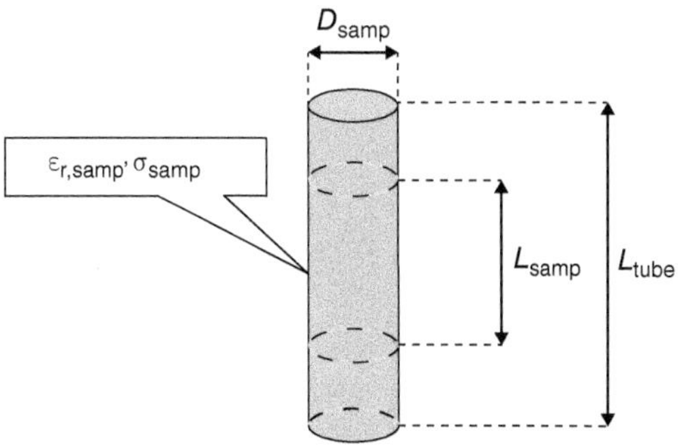

Figure 2.5 Schematics of the sample, typically contained in a water tube.

ceramic material properties optimize the SNR value, and the outer diameter and the permittivity are used for tuning the resonance at the Larmor frequency. Therefore, designing the ceramic probe consists of the following steps:

1) *Define the required field of view*: diameter D_{samp} and length L_{samp}. These two parameters define the inner radius of the dielectric ring r_{h}, and its height L, respectively. The outer radius r_{d} is kept as degree of freedom for tuning the mode at the Larmor frequency, and the material properties (permittivity ϵ_{r} and loss tangent $\tan\delta$) are used to optimize the achievable SNR.
2) For a given list of permittivity values $\overline{\epsilon_r}$, estimate the required outer radii list $\overline{r_{\text{d}}}$ for *tuning at the Larmor frequency* (N_{e} elements in each list).

Figure 2.6 is an example of such a tuning curve, with a target frequency of 730 MHz (Larmor frequency at 17 T), for a resonator with height 10 mm.

3) For a given list of loss tangent values $\overline{\tan\delta}$ (N_{t} values), estimate for each element of $\overline{\epsilon_r}$ (and the associated outer radius for tuning) the corresponding *SNR value* ($N_{\text{e}} \times N_{\text{t}}$ values).

This array of SNR values can be used as it is for optimizing the ceramic properties, or divided by the corresponding SNR of another probe for comparison. In Figure 2.7, we display the SNR gain

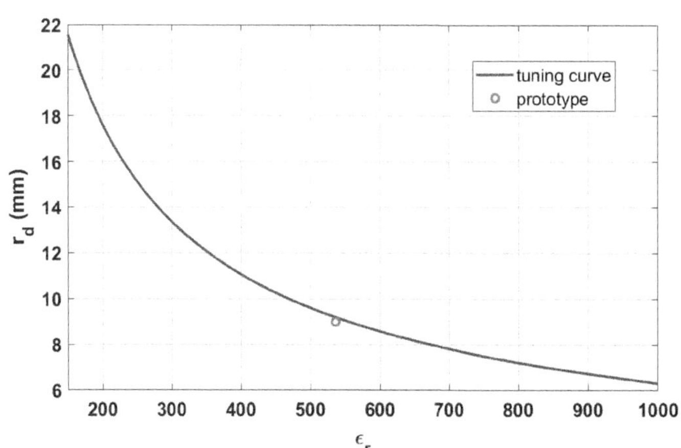

Figure 2.6 Example of tuning the first $TE_{01\delta}$ mode frequency of a disk resonator with a given height through its outer radius for varying values of the permittivity.

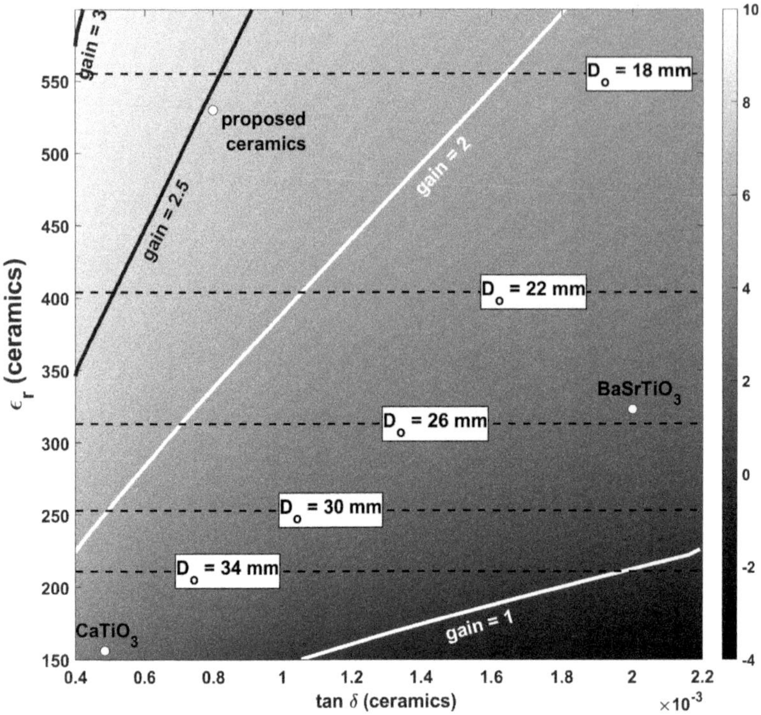

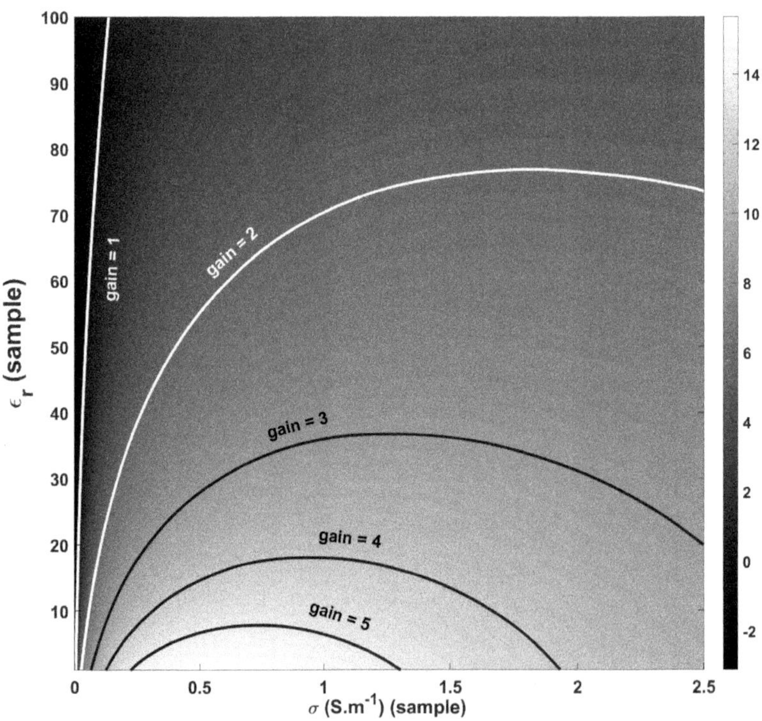

Figure 2.7 Signal-to-noise ratio (SNR) gain displayed as a function of the ceramics properties for a given sample and fixed ring height and inner diameter (top) and of the sample properties for a fixed ceramic probe design (bottom). From [30].

(Equation 2.13) over a solenoid coil with the same inner dimensions loaded with the same sample, defined as follows:

- On the left: as a function of the ceramic relative permittivity (vertical axis) and its loss tangent (horizontal axis) for given sample properties ($\epsilon_{r,samp} = 50$, $\sigma_{samp} = 1\ S/m$).
- On the right: for the proposed prototype ($\epsilon_r = 536$, $\tan\delta = 8.10^{-4}$) as a function of the sample relative permittivity and electrical conductivity.

$$\text{SNR gain} = \frac{\text{SNR}_{\text{ceramic}}}{\text{SNR}_{\text{reference}}}. \tag{2.13}$$

With the abacus that can be drawn from such calculations it is possible to design a ceramic probe working under the first TE mode with optimized properties, and to predict the SNR enhancement compared to a reference probe.

2.3.5 Validation

The accuracy of this SNR estimation model and its constitutive steps was studied for ring resonators with dimensions fitting microscopic samples, and dielectric materials with permittivity adequate for the range of frequencies of MRM. For example, the normalized power loss term in Equation 2.8 has been evaluated from the field distribution calculated in numerical simulations and with the semi-analytical method for varying electromagnetic properties of a sample for a given probe. As can be seen in Figure 2.8, the relative error between the two approaches never

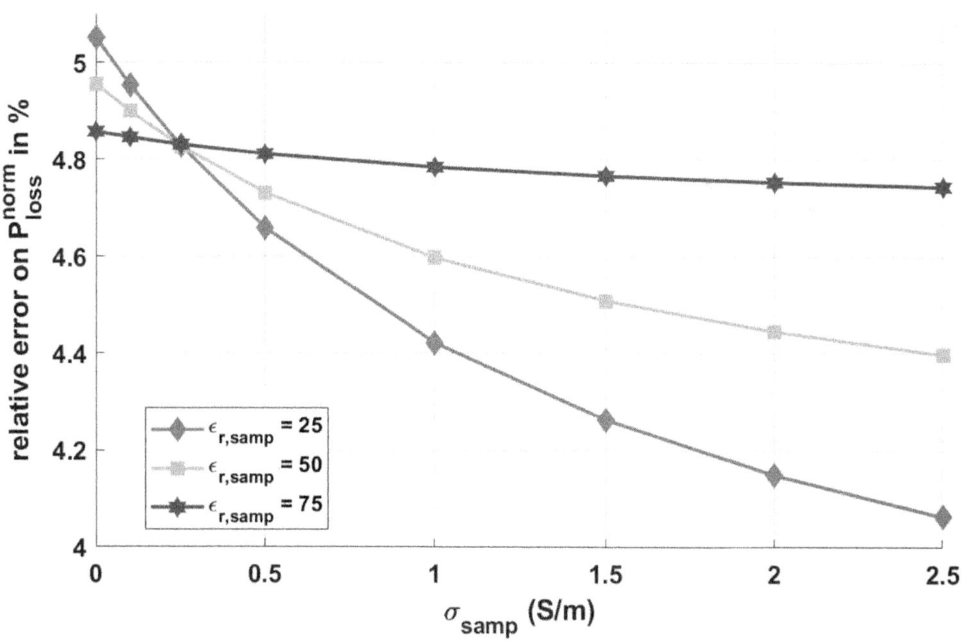

Figure 2.8 Relative error between the numerical simulations (CST Studio, Eigenmode Solver) and the semi-analytical model on the prediction of the normalized power losses term $\tau^2 P^{\text{sample}}_{\text{loss,norm}} + P^{\text{ring}}_{\text{loss,norm}}$. The error is plotted as a function of the sample conductivity and for three different sample permittivities. The ring resonator has the same properties as in [30]. The relative error is always below 5.1%. Figure reproduced with permission from [21].

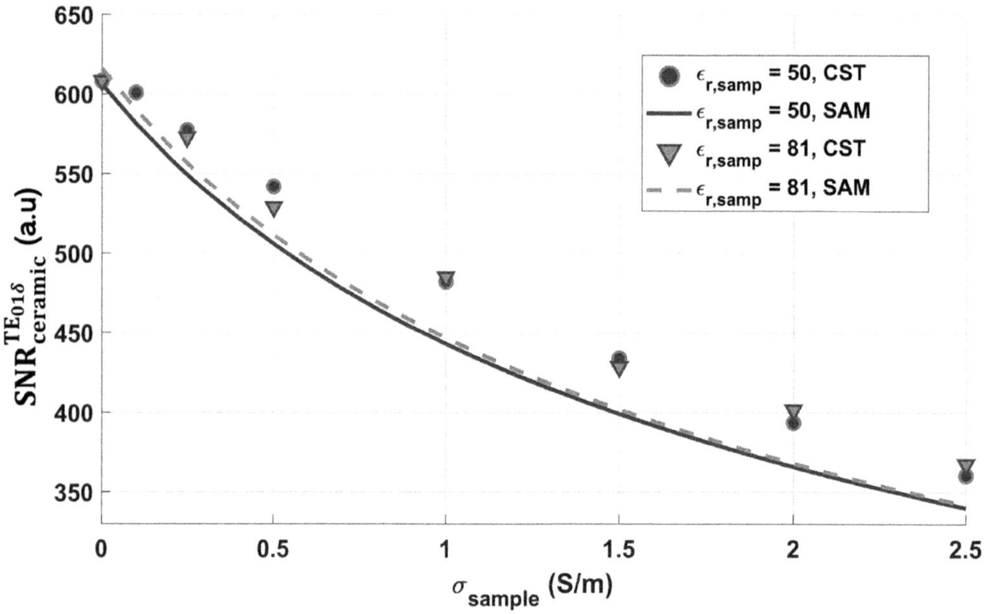

Figure 2.9 Comparison of the SNR predictions obtained with numerical simulations and with the developed semi-analytical model (SAM). The maximum relative error between the two approaches is 8%. The ring resonator has the same properties as in [30]. Data reproduced with permission from [21].

exceeds 5.1%. The SNR values predicted by numerical simulations and those obtained with the semi-analytical method are compared in Figure 2.9. The maximum relative error between the two approaches is 8% in the worst-case scenario.

2.4 MRM with Ceramic Coils

Take-home message: Several experimental proofs of concept have been made in MRM with significant SNR enhancement. Two points that need to be carefully considered are temperature stability and a tuning method that does not involve additional losses.

MR experiments involving ceramic probes have demonstrated the potential of these coils in microscopy. However, special precautions must be taken since the experimental setup differs significantly from that of conventional coils.

2.4.1 Practical Considerations and Experimental Setup

After designing the dielectric resonator to operate under a given resonant mode ($TE_{01\delta}$ or $HEM_{11\delta}$) for the required B_0 field strength and sample dimensions and properties, the ceramic material fitting the electromagnetic properties found for the resonator must be chosen. Ferroelectric materials based on oxide titanites are adequate since the final permittivity and loss tangent are adjusted through the relative proportions of each constituent [11].

To properly excite a resonant mode, the excitation source must induce an electromagnetic field that overlaps with that of the mode, as illustrated in Figure 2.10: the excitation source is in this

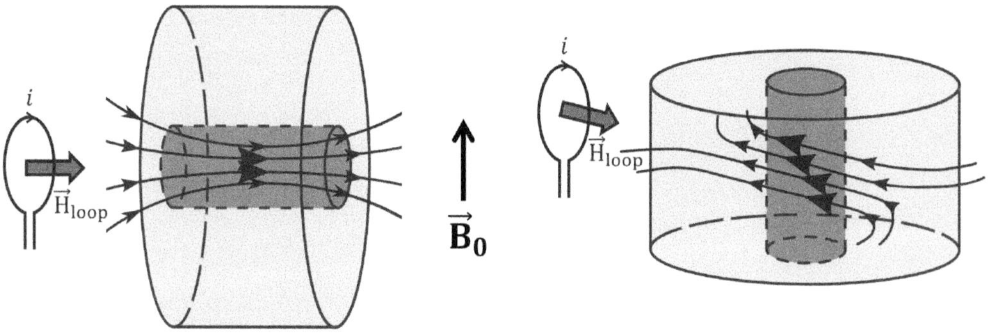

Figure 2.10 Example of excitation source: an electric current loop (magnetic dipole). Position required to excite the $TE_{01\delta}$ (left) and $HEM_{11\delta}$ (right) modes. The magnetic field lines (schematic) of each mode are represented. The static field direction is given by a thin arrow.

case equivalent to a magnetic dipole, and its position is chosen so that it is parallel to the magnetic polarization of the desired mode.

Regarding the first TE mode, analogous (in terms of field distribution) to a magnetic dipole parallel to the disk or ring axis, it is practical to use a small circular loop feed that is nonresonant at the Larmor frequency. This loop is positioned above the ceramic ring, with its axis parallel to that of the ring. In this configuration, the circulating current flowing through the loop creates an electric field distribution with the same cylindrical symmetry as that of the $TE_{01\delta}$ mode and therefore excites this mode. Higher-order modes with the same symmetry may be excited as well at their respective frequencies. Based on [29], in the case of small loops, we can maximize the quality factor of the whole probe (loop and dielectric resonator) by optimizing the distance between the loop and the resonator, or by changing the loop diameter at a given position.

In Figure 2.11, the influence of the loop position on the resonance is investigated with numerical simulations (CST Studio, Frequency Domain Solver), in the case of a disk whose first TE mode is located at 732 MHz (Eigenmode Solver of CST Studio). It was first checked that the excited mode was the $TE_{01\delta}$ mode: the magnetic field longitudinal component coincides in each configuration with the total magnetic field amplitude along the disk symmetry axis. As can be observed in Figure 2.11, the resonance frequency is slightly increased by the loop. The left column of Figure 2.11 shows that the maximum magnetic field at the center of the disk coincides with the minimum loop reflection coefficient, meaning the input power of the loop is transferred to the resonant mode of the disk. Also, but not shown in this figure, there exists an optimal loop diameter maximizing the magnetic field amplitude.

In the model proposed above (Section 2.3) for estimating the contribution of the ceramic probe to the SNR, the contribution of the feeding loop is neglected. In fact, this contribution depends on the loop's geometry and material. For the geometry in Figure 2.11 and the considered probe prototype, the probe efficiency was compared for a lossy copper loop and an ideal loop in a perfect electric conductor; the difference was less than 0.2% between the two values.

The first hybrid mode can be excited as well by the same loop positioned on its lateral side. Another potential excitation source is an electric dipole antenna, generating a circulating magnetic field around the dipole axis and positioned on the lateral side of the resonator, perpendicular to the ring axis to excite the first TE mode, and parallel for the first hybrid electromagnetic (HEM) mode.

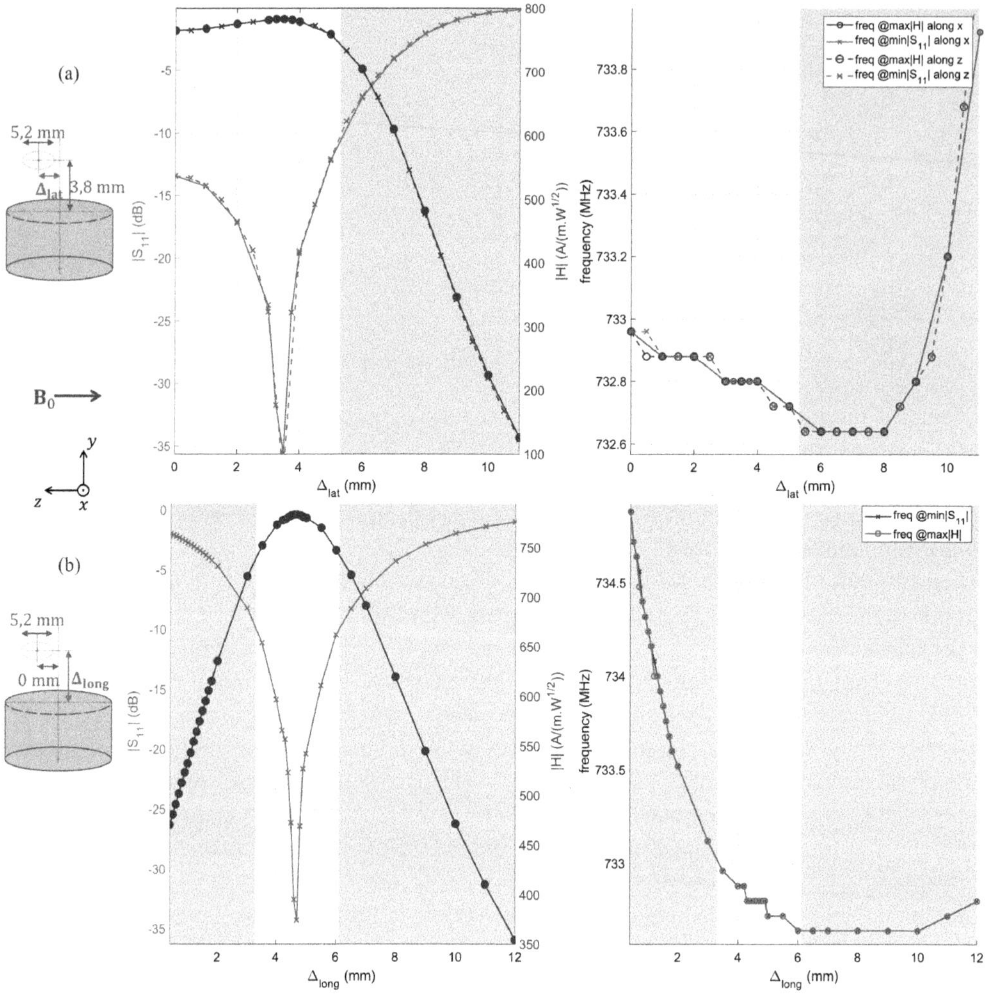

Figure 2.11 Influence on (left) the reflection coefficient S_{11} (minimum value), the magnetic field amplitude (maximum value) at the center of the resonator, and (right) the corresponding frequency of the feeding loop position relative to the dielectric disk (relative permittivity 530, loss tangent 8.10^{-4}, diameter 18 mm, height 10 mm). (a) Varying lateral position. (b) Varying longitudinal position. There is an optimal position of the feeding loop with a minimum reflection coefficient, and a maximum transmitted power to the resonant mode.

Figure 2.12 shows an example of an experimental setup exploiting a ceramic probe operating under its $TE_{01\delta}$ mode. In this setup, we use a circulating water blanket as a thermostat to homogenize and set the temperature of the ceramic probe at a given level. As depicted in Figure 2.13, the permittivity of the ceramic material and, as a result, the mode frequency, depend on its temperature, since reaching such a high permittivity value is enabled by working close to the Curie point of the ferroelectric material [7]. In [30], this dependence is used for real-time, accurate tuning of the probe to the Larmor frequency. However, in any other setup involving ceramic probes, even if another method is used for tuning, the temperature should be controlled to avoid the resonant frequency shifting away from the Larmor frequency.

Figure 2.12 Experimental setup. Fixing the resonant frequency at the Larmor frequency is enabled by thermostating the resonator at a given temperature (left). Matching is performed by sliding the excitation loop (right). With permission from [30].

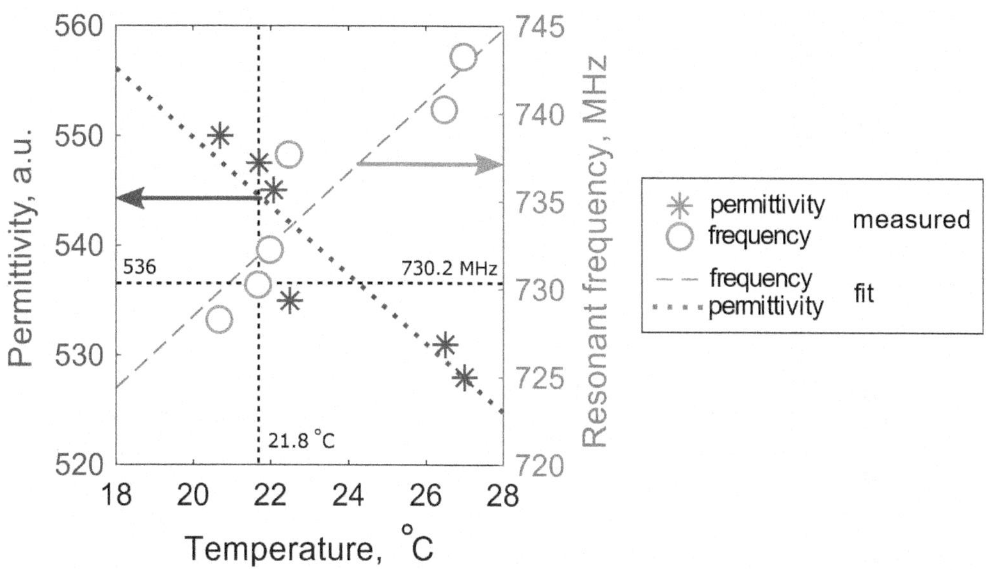

Figure 2.13 Temperature dependence of the ceramic relative permittivity and therefore of the probe resonant frequency. With permission from [30].

2.4.2 Performance

The ceramic probe designed in [30] was experimentally compared to the reference probe, which is a solenoid coil optimally designed for the same required field of view.

The measured B_1^+ field patterns in Figure 2.14 confirm that the designed ceramic probe has the same region of homogeneity as the solenoid immersed in Fluorinert™ (the Fluorinert is used to improve both B_1 and B_0 homogeneity of the solenoid [31]), and induces higher transmit field

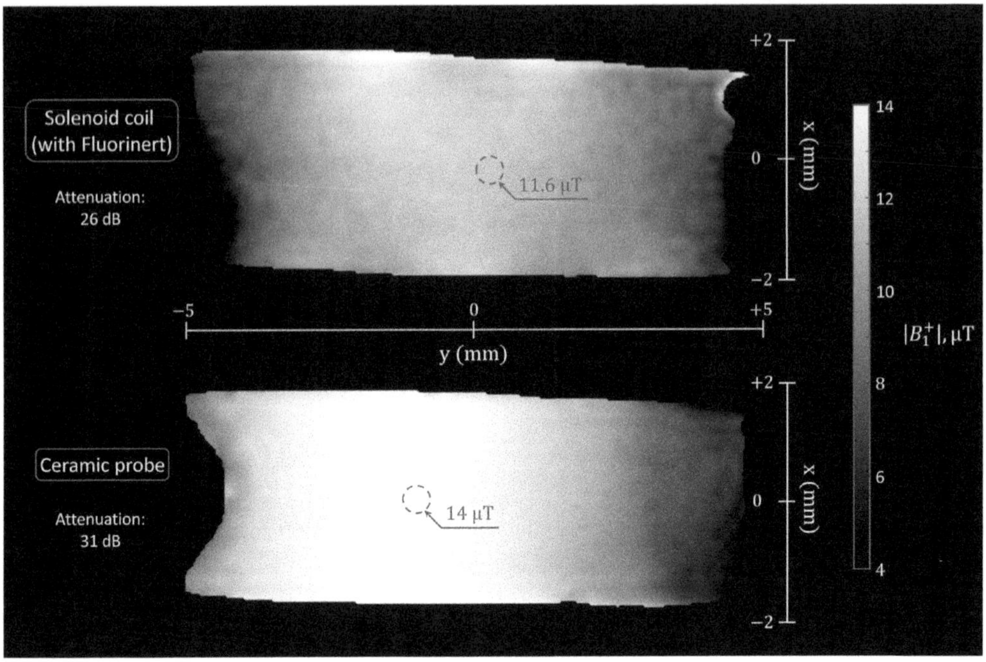

Figure 2.14 Measured transmit field pattern in a homogeneous liquid phantom for the reference optimal solenoid probe with Fluorinert (top) and for the designed ceramic (bottom), at 17.2 T. Data from [30].

intensity using less input power. Quantitatively speaking, for the same input power, the ceramic probe equivalently provides a B_1^+ field level 2.1 times higher than the solenoid immersed in Fluorinert.

The experimental SNR, measured in a tissue-mimicking liquid phantom, was also investigated for both probes in Figure 2.15 (left panel). With the same effective flip angle, the SNR was 2.2 times higher with the ceramic probe than with the solenoid immersed in Fluorinert. This has to be compared to the theoretical prediction: from Figure 2.7, for the designed ceramic probe (indicated as "proposed ceramics"), the semi-analytical method presented in Section 2.3 predicts an SNR gain of 2.5.

Better performance of the ceramic probe can be observed in Figure 2.15 (right panel): microscopy images obtained with this probe demonstrate improved quality compared with those obtained using the reference probe, since more structural details can be distinguished.

2.4.3 Dual Ceramic Coils

Since ceramic probes are based on dielectric resonators, it is possible to exploit the coupling phenomenon between such objects positioned close to each other. Figure 2.16 depicts the coupling of the $TE_{01'}$ modes of two ring resonators each holding a sample, but the principle is the same for empty rings or disks. When the resonators are placed one above the other, the electromagnetic coupling of the individual modes (at the same frequency f_0) gives rise to two coupled modes: a lower-frequency mode denoted $(++)$, with a field distribution such that the magnetic field between the resonators is enhanced; and a higher-frequency mode $(+-)$ such that the magnetic field in

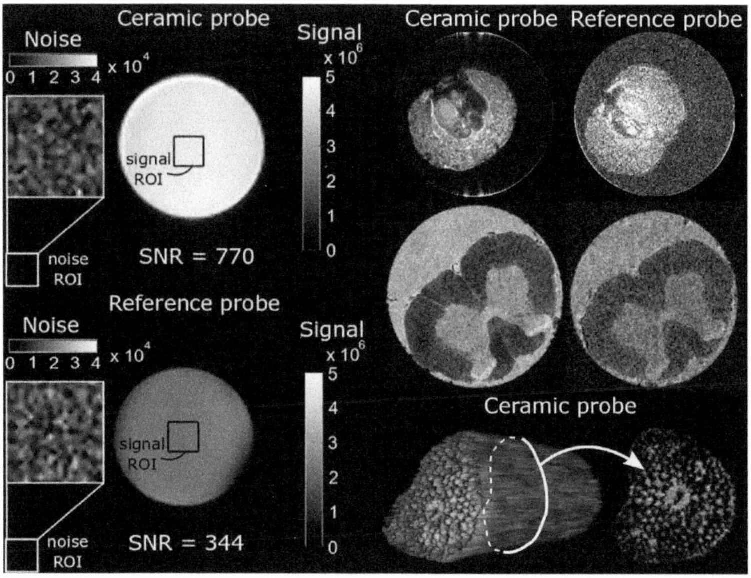

Figure 2.15 Experimental comparison of the designed ceramic probe and the reference optimal solenoid at 17 T. (Left) signal and noise maps measured in a homogeneous tissue-mimicking liquid phantom. (Right) microscopy images obtained with both probes; (first line) *Ilex aquifolium* fruit; (second line) chemically fixed rat spinal cord; (third line) 3D rendering and image slice of a plant petiole (obtained with the ceramic probe). With permission from [30].

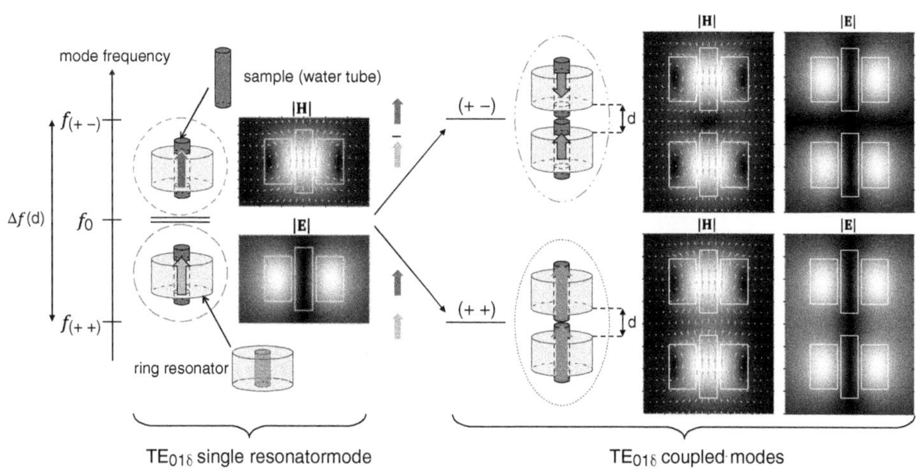

Figure 2.16 Coupling model of the first TE modes of two ring resonators. From [33].

between is decreased. In [32] the $(++)$ mode of two coupled disks is exploited to image a sample held between the disks. In [33] two samples are imaged simultaneously by using the $(++)$ mode of two ring resonators, each of them holding a sample. The corresponding magnetic field distribution is depicted in Figure 2.16 (right field maps, second line).

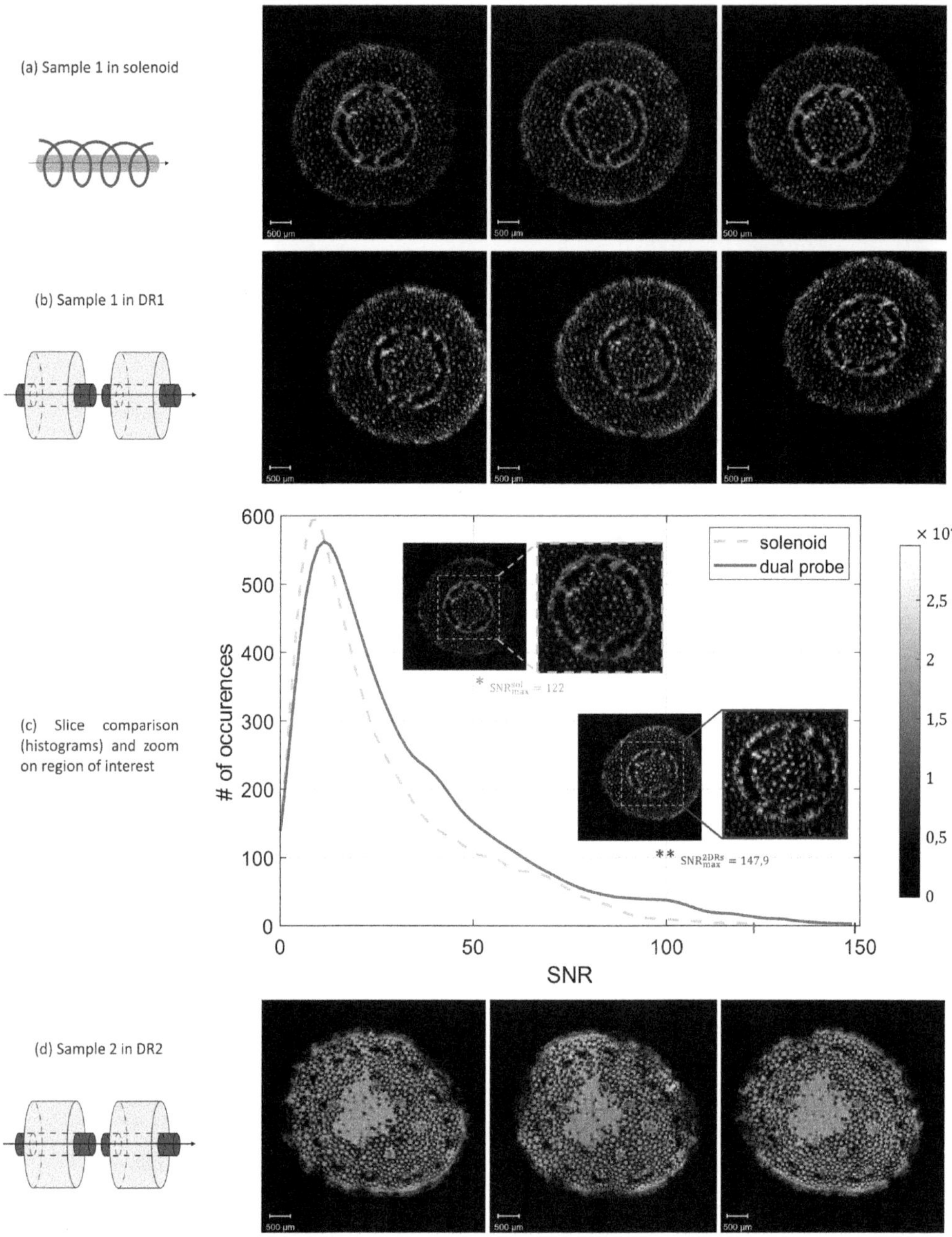

Figure 2.17 MR images of plant petioles using a solenoid coil and a ceramic probe exploiting the coupling of two dielectric ring resonators. In the latter, petioles from two different plants are imaged: sample 1 in (b) and sample 2 in (d), during the same acquisition time slot. Images of sample 1 are also acquired with the solenoid coil for comparison (a). A quantitative comparison of the two coils performance is provided in (c) with the signal-to-noise ratio (SNR) distribution of given regions of interest in sample 1. From [33].

Due to coupling, for the same input power, the SNR in each sample of the coupled ceramic probe is degraded by a factor of $\sqrt{2}$ compared with the single ceramic probe performance. However, in this study exploiting a prototype that had already overcome the reference solenoid probe by a factor of more than two, the coupled ceramic probe still provided a better SNR than the reference, with an SNR improvement of approximately 1.4, according to the theoretical modeling of the electromagnetic coupling and to numerical simulations. Experimental results from this work are shown in Figure 2.17, which displays plant petiole images obtained with the reference probe and the developed coupled ceramic probe. In the latter, two different samples are imaged, one in each dielectric resonator (DR). For sample 1, imaged with the solenoid and one of the coupled DRs, the maximum achieved SNR gain (dual ceramic probe over reference solenoid) is 1.2 (Figure 2.17c). Coupling these two resonators helps reduce the total acquisition time, since two samples can be imaged simultaneously.

2.5 Conclusion and Future Prospects

While used for years in the field of microwave engineering, ceramic probes dedicated to MRI have only recently been developed for other various applications, from clinical imaging to microscopy. The principle is to excite one specific eigenmode of a high-permittivity resonator like a disk or ring, and to exploit the stimulated mode magnetic field as a transmit and/or receive field for MRI. The particularity of the eigenmodes used is the spatial coincidence of the maximum magnetic field with low electric field levels as well as the purely magnetically induced (nonconservative) nature of the electric field in the sample. Positioning the sample in this specific region therefore limits the losses due to the electric field interaction with the conductive biological materials, while keeping B_1 values sufficiently high to perform imaging. The recent development of low-loss, high-permittivity ceramic materials has had a significant impact on the performance of dielectric probes, limiting the intrinsic losses.

Most probes proposed for MRM exploit the first TE mode of cylindrical high-permittivity resonators. Future work on developing probes exploiting the first hybrid mode would be beneficial to widen the possibilities of using ceramic probes. Another point of interest would be to develop arrays of ceramic probes to significantly reduce the acquisition time, which is a critical experimental parameter when performing MRI.

References

1 Kajfez, D. and Guillon, P. (1998). *Dielectric Resonators*, 2e. Atlanta, GA: Noble Publishing Corporation.

2 Leung, K.W., Lim, E.H., and Fang, X.S. (2012). Dielectric resonator antennas: From the basic to the aesthetic. *Proceedings of the IEEE* 100 (7): 2181–2193.

3 Long, S.A., McAllister, M.W., and Shen, L.C. (1983). The resonant cylindrical dielectric cavity antenna. *IEEE Transactions on Antennas and Propagation* 31 (3): 406–412.

4 Fiedziuszko, S.J. (1982). Dual-mode dielectric resonator loaded cavity filters. *IEEE Transactions on Microwave Theory and Techniques* 30 (9): 1311–1316.

5 Liang, X.P. and Zaki, K.A. (1993). Modeling of cylindrical dielectric resonators in rectangular waveguides and cavities. *IEEE Transactions on Microwave Theory and Techniques* 41 (12): 2174–2181.

6 Chen, S.W. (1991). Dielectric ring resonators loaded waveguide and on substrate. *IEEE Transactions on Microwave Theory and Techniques* 39 (12): 2069–2076.

7 Webb, A.G. (2011). Dielectric materials in magnetic resonance. *Concepts in Magnetic Resonance Part A* 38 (4): 148–184.

8 Golovina, I., Geifman, I., and Belous, A. (2008). New ceramic EPR resonators with high dielectric permittivity. *Journal of Magnetic Resonance* 195 (1): 52–59.

9 Mett, R.R., Sidabras, J.W., Golovina, I. et al. (2008). Dielectric microwave resonators in TE011 cavities for electron paramagnetic resonance spectroscopy. *Review of Scientific Instruments* 79 (9): 1–8.

10 Mett, R.R., Sidabras, J.W., Anderson, J.R. et al. (2019). Rutile dielectric loop-gap resonator for X-band EPR spectroscopy of small aqueous samples. *Journal of Magnetic Resonance* 307: 106585.

11 Nenasheva, E.A., Kanareykin, A.D., Kartenko, N.F. et al. (2004). Ceramics materials based on (Ba, Sr)TiO 3 solid solutions for tunable microwave devices. *Journal of Electroceramics* 13 (1–3): 235–238.

12 Nenasheva, E.A., Kartenko, N.F., Gaidamaka, I.M. et al. (2010). Low loss microwave ferroelectric ceramics for high power tunable devices. *Journal of the European Ceramic Society* 30 (2): 395–400.

13 Shchelokova, A., Ivanov, V., Mikhailovskaya, A. et al. (2020). Ceramic resonators for targeted clinical magnetic resonance imaging of the breast. *Nature Communications* 11 (1): 1–7.

14 Webb, A.G. (2012). Visualization and characterization of pure and coupled modes in water-based dielectric resonators on a human 7T scanner. *Journal of Magnetic Resonance* 216: 107–113.

15 Aussenhofer, S.A. and Webb, A.G. (2014). An eight-channel transmit/receive array of TE01 mode high permittivity ceramic resonators for human imaging at 7 T. *Journal of Magnetic Resonance* 243: 122–129.

16 Haines, K., Neuberger, T., Lanagan, M. et al. (2009). High Q calcium titanate cylindrical dielectric resonators for magnetic resonance microimaging. *Journal of Magnetic Resonance* 200 (2): 349–353.

17 Zhang, K. and Li, D. (2008). *Electromagnetic Theory for Microwaves and Optoelectronics*, 2e. New York: Springer.

18 Pozar, D.M. (2012). *Microwave Engineering*, 4e. New York: John Wiley and Sons, Inc.

19 Petit, R. (1993). *Ondes électromagnétiques en radioélectricité et en optique*. Paris: Masson.

20 Orfanidis, S.J. (2003). Electromagnetic waves and antennas. https://www.ece.rutgers.edu/~orfanidi/ewa (accessed 14 October 2021).

21 Moussu, M.A.C., Abdeddaim, R., Dubois, M. et al. (2020). A semi-analytical model of high permittivity dielectric ring resonators for magnetic resonance imaging. *IEEE Transactions on Antennas and Propagation* 68 (8): 6317–6329.

22 Abramowitz, M. and Stegun, I.A. (1972). *Handbook of Mathematical Functions with Formulas, Graphs, and Mathematical Tables*. Washington, DC: US Government Printing Office.

23 Bowman, F. (2012). *Introduction to Bessel Functions*. New York: Dover Publications Inc.

24 Minard, K.R. and Wind, R.A. (2001). Solenoidal microcoil design – Part II: Optimizing winding parameters for maximum signal-to-noise performance. *Concepts in Magnetic Resonance* 13 (3): 190–210.

25 Kishk, A.A., Glisson, A., and Junker, G.P. (2001). Bandwidth enhancement for split cylindrical dielectric resonator antennas. *Progress in Electromagnetics Research* 33: 97–118.

26 Rotaru, M. and Sykulski, J.K. Numerical investigation on compact multimode dielectric resonator antennas of very high permittivity. IET Conference Publications. 2008; 537.189–190.

27 Itoh, T. and Rudokas, R.S. (1977). New method for computing the resonant frequencies of dielectric resonators. *IEEE Transactions on Microwave Theory and Techniques* 25 (1): 52–54.

28 Harrington, R.F. (1961). *Time Harmonic Electromagnetic Fields*. New York: McGraw-Hill.

29 Guillon, P. and Garault, Y. (1982). Coupling coefficient between magnetic loop and a dielectric resonator in an evanescent waveguide. *1982 IEEE MTT-S International Microwave Symposium Digest* 392–394.

30 Moussu, M.A.C., Ciobanu, L., Kurdjumov, S. et al. (2019). Systematic analysis of the improvements in magnetic resonance microscopy with ferroelectric composite ceramics. *Advanced Materials* 31 (30): 1900912.

31 Ciobanu, L. (2017). *Microscopic Magnetic Resonance Imaging – A Practical Perspective*. Singapore: Pan Stanford Publishing.

32 Neuberger, T., Tyagi, V., Semouchkina, E. et al. (2008). Design of a ceramic dielectric resonator for NMR microimaging at 14.1 Tesla. *Concepts in Magnetic Resonance. Part B, Magnetic Resonance Engineering* 33B (2): 109–114.

33 Moussu, M.A.C., Glybovski, S., Abdeddaim, R. et al. (2020). Imaging of two samples with a single transmit/receive channel using coupled ceramic resonators for MR microscopy at 17.2 T. *NMR in Biomedicine* 33 (11): e4397.

3

Portable Brain Scanner Technology for Use in Emergency Medicine

Lawrence L. Wald[1,2,3] *and Clarissa Z. Cooley*[1,2]

[1] *Athinoula A. Martinos Center for Biomedical Imaging, Department of Radiology, Massachusetts General Hospital, Charlestown, MA, USA*
[2] *Harvard Medical School, Boston, MA, USA*
[3] *Division of Health Sciences and Technology, Harvard – Massachusetts Institute of Technology, Cambridge, MA, USA*

3.1 Where Would You Use a Portable or Small Footprint Magnetic Resonance Imager?

Portable magnetic resonance imaging (MRI) technology could expand the benefits of MRI to patients who traditionally do not have access. These recipients fall roughly into two segments: patients who live in regions with abundant MRI access but who are difficult to transport into a conventional MRI suite, and those in the large fraction of the world underserved by clinical MRI due to equipment and operational costs.

For the first group, the goal is to influence patient outcomes by bringing the scanner to vulnerable patient populations difficult to bring to a conventional MRI suite. Brain imaging is a prime target due to the supremacy of MR contrast in this important organ, and the time-sensitive need for brain evaluation in many conditions, including stroke, elevated intracranial pressure (ICP) from hemorrhage, edema, hydrocephalus, or neonatal injuries such as hypoxic-ischemic encephalopathy (HIE). Furthermore, MR is intrinsically well suited for repeated scanning scenarios (such as monitoring) if a sufficiently nonobtrusive device is available.

The vast majority of emergency departments (EDs) and intensive care units (ICUs) currently lack MRI scanners. Although these facilities are located in hospitals equipped with MRI suites, it is also not always easy to transport the patient, even within the same building. EDs rely almost exclusively on ultrasound (US) and computed tomography (CT) for tomographic imaging. While the unique imaging capabilities of MRI (e.g. for evaluation of acute stroke) have led some hospitals to install scanners within their ED, this is a rarity, existing in only a small number of large tertiary care centers [1–3]. Even fewer hospitals have an MRI scanner within their ICUs. Additionally, the COVID-19 pandemic has taught us the importance of portable imaging modalities [4], with portable chest X-ray systems by far the most utilized [5], followed by CT [6,7] and US [8]. MR has shown the potential for assessing unique neurological manifestations of COVID-19 not sufficiently evaluated by CT [9] but is little used due to the barriers to transporting these acutely ill and infectious patients. The first-choice modality for patients in such a setting would ideally be brought into the patient's room, utilized for a point-of-care (POC) scan, wheeled out, and disinfected.

The second group (access in low-income areas) is addressed mainly through cost reduction. Geographical MRI accessibility gaps have been recently reviewed in West Africa [10] and worldwide [11]. The degree to which some regions lag behind worldwide averages for the number of scanners per capita is shocking. For example, 11 African countries have no scanner. However,

Magnetic Resonance Microscopy: Instrumentation and Applications in Engineering, Life Science and Energy Research, First Edition. Edited by Sabina Haber-Pohlmeier, Bernhard Blümich, and Luisa Ciobanu.

there is also some indication that low-income countries overspend on high-end medical equipment relative to other infrastructure [12]. Together, this underscores the need to reduce the cost of the MRI equipment if we hope to serve this community. Additionally, economic accessibility gaps are not limited to low-income countries, but also rural areas of middle- to high-income countries where the density of patients is insufficient to support the cost and operation of the scanner [13].

Costs of MRI components are hard to pin down because the costs a system manufacturer faces for volume-sourced parts are proprietary and likely considerably cheaper than seen by academic researchers, who constitute a niche market. Also, in some locations import tariffs on imported equipment incurs a significant additional cost to the user, which could be avoided if the device could bypass tariffs through local manufacture [14]. Nonetheless, a recent review has attempted to outline system component costs [15]. Lowering costs by reducing component-level costs is only part of the equation. Operating expenses must be lowered, including service costs, cryogen and cryogenic equipment maintenance, and electrical consumption of both the equipment and the carefully controlled air-conditioned environment typically required [14].

3.2 Rethinking System-level Approaches

For any of these three levels of POC devices, a new system-level approach is needed. Ample industrial effort has already been put into shortening the bore of conventional high-field (1.5 T and above) superconducting magnet-based systems. After nearly four decades of whole-body scanner design, the result is still a multi-ton, nontransportable system with high power and cooling needs and a relatively large magnetic field footprint. The reason for this partly stems from trying to achieve ever-higher imaging speed and resolution. But which foundations of the existing high-field system architecture could go if that goal were relaxed? The first that comes to mind is the conventional whole-body focus of current clinical scanners. If a specific body part must be chosen, the head is an ideal target due to the importance of brain injury and disease, and because the anatomy allows for smaller bore size. Here we omit discussion of small systems for extremity imaging (knee, wrist, etc.) since small versions of these scanners have been available for some time including a mini-van mounted device [16]. Other obvious departures include lowering the static magnetic field strength (at the expense of sensitivity) and/or its homogeneity. This has multiple positive and negative implications for the system and its performance [17–19], but allows for cheaper, smaller superconducting magnets or permanent magnets where the magnetic energy density can be stored without the use of cryogens. Other departures examined include nonswitched readout gradients built into the static magnet design (saving power, cooling, and reducing acoustic noise) [20], encoding by rotation of a built-in gradient (further reducing encoding electronics) [21–27], and shrinking the imaging field of view (FOV) even further to a subset of the organ and perhaps not fully encoding all spatial dimensions.

3.3 Three Levels of POC Use

We arbitrarily divide POC MRI use cases into three levels based on their deviation from a conventional high-field scanner suite. The closest level employs modest deviations from a standard 1.5-T scanner approach and attempts to improve siting (and perhaps cost) to facilitate siting within a tight ED or ICU space. This “easy-to-site suite” scanner could utilize a standard superconducting solenoid magnet architecture, perhaps at reduced field strength, but with modifications to decrease its cost, size, and stray-field footprint. For example, the system might use a short-bore, conduction-cooled superconducting magnet to eliminate cryogens and the quench-pipe. The footprint, size, and cost can be reduced compared with conventional whole-body scanners if the magnet is sized for brain imaging and operated at mid-field (between 0.5 T and 1.0 T). This intermediate field strength can provide sensitivity and imaging contrasts similar to conventional 1.5-T scanners while increasing accessibility – a topic recently reviewed and put into historical perspective [28]. Addition of active electromagnetic interference (EMI) mitigation could further ease siting by eliminating the standard radiofrequency shielded room. Although the system retains many aspects of conventional suites including high-power electrical hookups (for conventional gradients), maintenance-prone cryogenic equipment, water cooling, and a safety exclusion zone, the potential siting benefits have motivated several commercial MRI manufacturers to initiate development of this type of device (see Figure 3.1).

The second level would be a truly portable scanner that could be pushed down the hallways of the hospital by a single staff member who brings it into the ward or even to the bedside and powers it up for POC use. This device must operate using a standard electrical outlet or perhaps battery power, the latter allowing it to be embedded in an ambulance or mobile setting. The mobile POC scanner would likely operate at low field (50–200 mT), need unconventional EMI mitigation, and must operate with substantially reduced electrical power compared with conventional systems and without water cooling or cryogenics. This class of POC devices is being actively pursued by several companies and academic groups as discussed in Section 3.3.2.

Finally, the third class is a more speculative device that extends MRI to a near "handheld" level, likely with a greatly reduced imaging capabilities, but inexpensive and small enough to be considered an MR detector or monitoring device more than a diagnostic imaging device. Such a lightweight device could reach into the bed and monitor an organ, perhaps using 1D imaging or just the MR signal itself. This rethinks the role of MRI as a tomographic imager and, as such, is the most distant from conventional MRI scanner architectures. Nonetheless, examples of this more speculative device are starting to emerge in the literature as outlined in Section 3.3.3.

3.3.1 Brain MRI in an "Easy-to-Site Suite"

A head–neck focus can still utilize conventional but "shrunk-down" superconducting magnet architectures (perhaps with asymmetric gradients) to provide the "easy-to-site suite" and

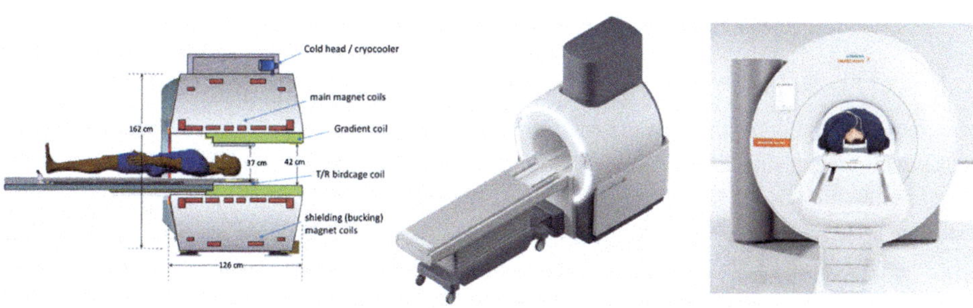

Figure 3.1 Superconducting MRI systems recently introduced to the market to provide high-quality imaging with reduced siting needs. All employ conduction-cooled magnets to eliminate the need for "quench pipes" to vent cryogenic gases. From left to right: GE (Waukesha WI) 3-T "compact head scanner," the Synaptive Medical (Toronto, ON, Canada) 0.5-T "Envry" compact scanner, and the Siemens Healthineers (Erlangen, Germany) 0.55-T "Free Max" compact 80-cm diameter patient-bore scanner (Siemens Healthcare GmbH).

possibly reduce cost. High-field superconducting head-only scanners include the Siemens Allegra 3-T clinical scanner [29] introduced in the early 2000s but no longer produced, and the more recent GE high-performance 3-T head scanner employing a conduction-cooled magnet with no cryogen vent-pipe [30]. While the magnet and gradients of these high-field head-only systems are more compact, the focus of these two systems was on performance rather than siting alone, which is only modestly simplified. There is renewed interest in making additional changes to provide easier siting of superconducting systems within an ED, ICU, or interventional suite. These approaches all leverage intermediate-field (B_0 near 0.5 T) superconducting systems with cryogen-free refrigeration systems such as a conduction-cooled 0.55-T [31] or 0.5-T scanner [32–34], both employing modern, high-performance gradient systems in a standard architecture. Other efforts are underway with an even smaller head-focused 1.5-T high-temperature superconducting magnet [35]. Figure 3.1 shows three "easy-to-site" superconducting approaches recently introduced by manufacturers. Extremely small bore size superconducting magnets have also been introduced as needed for extremity or neonatal imaging.

3.3.2 Brain MRI with a Portable Device

A portable MRI can be brought to the patient rather than bringing the patient to the MRI. Thus we define "portable" as a device that can be moved room-to-room by a single healthcare worker for use on a patient within minutes. The power and cooling infrastructure requirements must be met by a typical office or exam room infrastructure. Such portable MRI scanners equivalent to their counterparts in US, X-ray, and CT have traditionally not been available, although a commercial clinical portable MRI product has recently been introduced and tested in a clinical setting [36].

True portability seems to require both restriction to a limited body region (e.g. the head) *and* operation at low field (likely below 200 mT) and potentially employing new types of magnet architectures and encoding strategies. Here attention has turned to permanent magnets, which offer the ability to operate without external power or a cryogenic system. Modern rare-earth permanent magnets can produce head-sized B_0 field regions up to about 0.5 T with surprisingly small external field footprints. Many magnet and gradient configurations have emerged, with several prototypes being demonstrated. Notably the first fully portable clinical MRI product scanner is now Food and Drug Administration (FDA) approved for clinical care – the 64-mT Hyperfine scanner [36,37]. Figure 3.2 shows another such system under development, based on an 80-mT "Halbach bulb" rare-earth magnet configuration with a magnet weight of 24 kg [38].

3.3.3 Brain MRI as a Monitoring Device

In addition to employing permanent magnets and reducing the magnet field strength and size, we also consider more extreme changes to the scanner architecture to further reduce the scanner size toward a handheld device. This includes dropping the goal of whole-organ imaging, for example probing only a section of the body with limited image encoding (2D, 1D, or even no image encoding). But, even if detailed anatomy is not visualized, the MR data can be collected and monitored for changes that might accompany, for example, important intracranial pathology. This type of MRI is thus a patient-monitoring device in the way a pulse-oximeter or electrocardiogram (ECG) device is used at the bedside in the ED or ICU. It is new territory for MRI and pushes the technology into a more radical configuration. This extreme approach does not attempt to have the system fully encircle a body part, but is limited to a "single-sided" approach following that originally introduced for materials characterization with MR [39,40] but also considered for biological applications

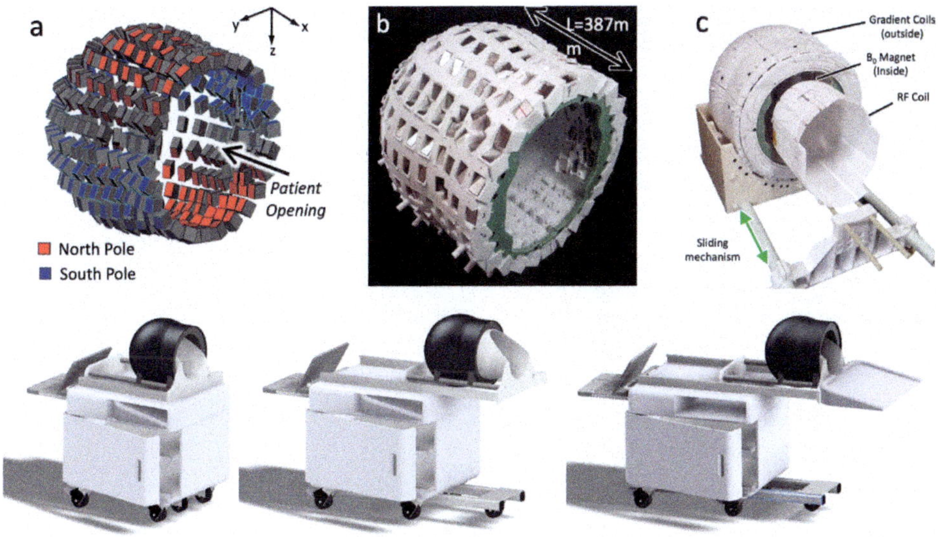

Figure 3.2 A potential portable brain magnetic resonance imaging cart based on a "Halbach bulb" permanent magnet array for bedside use in the emergency department or intensive care unit. Prototype magnet, gradient, and radiofrequency (top) and envisioned mobile cart (bottom); system approaches bed from head end and utilizes a double sliding mechanism (radiofrequency coil slides followed by magnet sliding down).

[41–43] and using remarkably inexpensive components [44]. Single-sided spectrometers have also been introduced for assessing breast tissue [45,46] and muscle hydration [47,48]. Single-sided full-imaging systems are less common, but have been demonstrated [49] and applied to burn depth [50].

Figure 3.3 shows a single-sided scanner designed for brain imaging [51]. An ultra-lightweight (<10 kg) single-sided brain scanner will have extreme inhomogeneity by medical MRI standards and require unconventional encoding strategies. But these changes are likely necessary to make the leap from a scanner where the patient's head still goes inside the device, to a more handheld scanner simply placed adjacent to the head. This step will not be without image quality reductions. For monitoring, the device must "reach" into the bed, operate adjacent to the patient (versus placing the anatomy inside the magnet), and be light and cheap enough for sustained operation as a monitoring device. An ED or ICU might benefit from such an MRI device to continuously image the brain watching for intracranial hemorrhage or changes in cerebral mass effect though monitoring a ventricular/cerebrospinal fluid left–right hemisphere asymmetry. An intracranial MR monitor could provide an early warning sign of impending herniation, particularly in patients where clinical examination is difficult (e.g. sedated patients).

3.4 Clinical Use Scenarios of "Easy-to-Site" POC, and Monitoring MR Devices

Although engineers and researchers are often content with the "build it and they will come" philosophy, clinicians, funding agencies, and hospital administrators prefer a need-driven approach. Furthermore, knowledge of the clinical use is critical to many design decisions. Here we frame the technical needs by outlining some of the clinical use scenarios envisioned for POC neurological MRI.

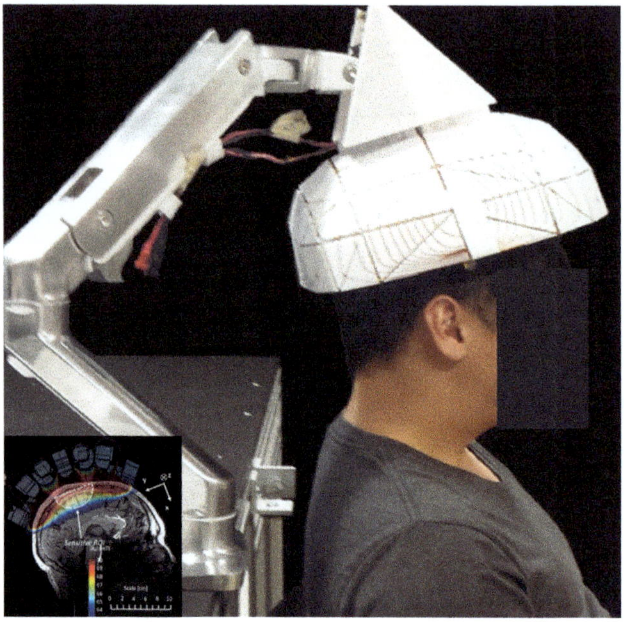

Figure 3.3 Single-sided brain magnetic resonance imaging under development for monitoring applications in the emergency department or intensive care unit. The <7 kg Halbach sphere-section magnet generates ~80 mT in a subregion of the brain.

3.4.1 ED and ICU

Neuroimaging in emergency medicine often focuses on immediately ruling out or identifying a source of increased ICP from hemorrhage, stroke, hydrocephalus (enlargement of the ventricles), or cerebral mass effect (displacement of brain tissue by blood or tumor or following trauma). This is currently done in the ED using CT scanners, which are fast, relatively cheap, and easy to site compared with MRI. The addition of POC MRI would be helpful for assessment of neurological emergencies for which CT provides a poor depiction or for patient populations sensitive to the ionizing radiation used in CT.

3.4.2 Acute Stroke Care

The use of low-field MRI in the stroke setting has recently been reviewed [52] including for resource-limited settings [53]. Portable low-field MRI has been contemplated for ruling out a hemorrhagic stroke to allow immediate reperfusion with the injection of recombinant tissue plasminogen activator (rtPA), which would exacerbate any existing cerebral hemorrhage. However, CT currently performs this job well, including portable units in EDs and even specialized CT-equipped ambulances [54–58] dispatched when stroke is suspected. Although a portable MRI would also be capable of this "rule out a bleed" role, it is worth looking to the next step of stroke care to avoid simply replacing one capable mobile imaging modality for another. Successful new intravenous thrombolytic therapies as well as widening windows for treatment have led to catheter-based thrombectomy treatments following the initial rtPA treatment [59,60]. A "hub-and-spokes" model for treatment sites has emerged for this treatment. First, patients with

suspected stroke are rushed to the nearest "spoke hospital" where they receive a CT to verify stroke and rule out hemorrhage for immediate administration of rtPA. In the few cities with CT-equipped ambulances (mobile stroke units), this treatment is received in the field. Following rtPA administration, the patient is then evaluated for transfer to a "hub" hospital for catheter-based thrombectomy [59,60]. This assessment uses CT angiography (CTA) to identify a large vessel occlusion (LVO) target in the anterior circulation and perfusion (CTP) to estimate infarct size. An infarct core size >70 cc both reduces the likely benefit of revascularization and establishes a risk for hemorrhage during the procedure. Unfortunately, the infarct size determination is difficult and imprecisely measured on CTP [61]. In contrast, diffusion MRI is the established gold standard for early infarct core volume assessment [61] but is not commonly available in EDs [62,63]. The time-critical nature of acute stroke care and the cost of transport underscore the need for improved POC MR imaging in the spoke hospital ED to determine the eligibility of the patient for transport to receive catheter-based thrombectomy.

3.4.3 Assessing Pediatric Hydrocephalus in the Developing World

A low-cost and readily accessible POC MRI could be invaluable to guide and assess surgical treatment of pediatric hydrocephalus. In hydrocephalus, the brain's ventricular system is blocked, which in turn causes swelling of the head to an abnormal size and increased cranial pressure. In Africa this is usually caused by a bacterial infection during the early months of life [64]. Untreated hydrocephalus can hamper brain development and incur a significant disease burden, including brain damage, blindness, and death [65]. Treatments include introducing a shunt to reestablish the cerebrospinal fluid (CSF) flow; however, this must be monitored and possibly replaced if clogged. More minimally invasive shuntless approaches [66] are becoming increasingly common in Africa, where follow-up assessment is difficult.

Imaging of the brain's ventricular system is needed at all stages, from diagnosis to surgical follow-ups. Fortunately, the ventricular system is a large structure that dominates routine T_1- or T_2-weighted MR images and is perhaps one of the easiest intracranial structures to image with MRI and does not require high spatial resolution or sensitivity. It is also well depicted on CT, but the focus on a pediatric population and the need for repeated follow-up imaging creates a concern about the effects of CT's ionizing radiation. This places MRI as the first choice for assessing and following this disorder. The combination of disease severity and prevalence, the available treatment, the limited spatial resolution needed, and remote distribution of the patients makes low-cost, portable MRI an ideal tool for pediatric hydrocephalus assessment [67–69]. Figure 3.4 shows a low-cost 50-mT permanent magnet-based system under development for this purpose [68,70].

3.4.3 Mass-effect Monitor in ED or ICU Setting

Similar to the hydrocephalus application, knowledge of a mass effect (compression or displacement of the brain by a pathological source) is critical in acute care because it alerts the care team to the development of significant ICP (which is life-threatening). Mass effects can arise in a wide spectrum of conditions associated with hemorrhage (e.g. traumatic brain injury, rupture of an aneurysm or vascular malformation, hemorrhagic conversion of acute stroke, and hemorrhage resulting from a postsurgical event). Other sources include alterations of CSF dynamics (e.g. obstructive and nonobstructive hydrocephalus, intracranial hypertension, CSF leaks, and other

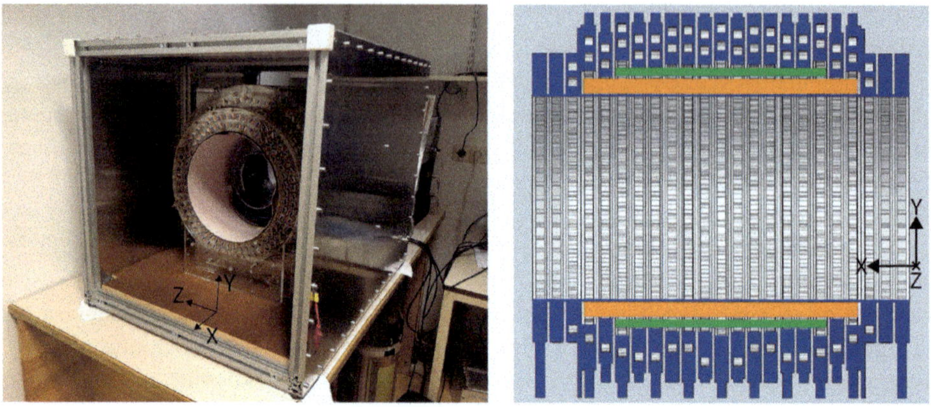

Figure 3.4 A low-cost, lightweight scanner under development for pediatric hydrocephalus monitoring. The system uses a rare-earth permanent magnet "Halbach" arrangement producing a 50-mT transverse field. Magnet bore is 27 cm diameter × 50 cm length.

disorders of CSF flow), and cerebral edema (e.g. from demyelinating disease, cerebral infections, or cytotoxic edema due to acute stroke). Finally ICP can be induced by tumors (primary brain tumors, metastases, and extra-axial masses). The presence of ICP is frequently detected by looking for a shift in mid-line structures of the brain, ventricular effacement (one ventricle bigger than the other), or sulcal effacement (CSF spaces in sulci smaller in one hemisphere).

While CT (including portable CT) is effective at identifying a mass effect, it is less useful for answering the second most important clinical question: "Is it getting worse?" Because the mass effect and thus increasing pressure could be developing for hours post injury (e.g. in the case of a hemorrhage after trauma), a single imaging time point is insufficient. Multiple time points are desirable and would ideally be acquired with a wearable monitor. CT is still relatively cumbersome to use repeatedly in a POC setting and the ionizing radiation makes it ill-suited for anything resembling continuous monitoring. Instead, compact, single-sided "MR monitors" could fill this gap. This is a potential new role for MR technology in medicine: a real-time monitor of ventricular/CSF asymmetry to provide an early warning sign of impending herniation, particularly in patients where clinical exam is difficult (e.g. sedated patients).

In a true monitoring scenario, the patient will have the MR monitor on for hours, perhaps days, likely precluding traditional architectures where the patient goes "inside" the bore (of even a small magnet). Instead, the magnet is "worn" or simply adjacent to the head. Such a device will not provide a homogeneous field, or even allow imaging over the whole head. Instead, the clinical information must be extracted from a 3D image of only a region of brain, or perhaps a 2D or 1D image. Nonetheless, this line of instrumentation can follow the extensive development effort previously applied to single-sided MR devices for spectroscopy and materials evaluation [49,71,72], and relaxometry and diffusion measurements in oil well logging [73].

3.4.4 Neonatal Intensive Care Unit (NICU)

A scanner situated in an NICU or capable of being brought to the NICU (beside incubator/isolette) could have many uses. Here we focus on one: HIE to provide a general picture of the uses and issues of MR in a NICU.

HIE and resulting hypoxic-ischemic injury (HII) is the leading cause of neonatal mortality and morbidity with an occurrence of 1 to 8 out of 1000 live births in the United States and up to 26 out of 1000 births in developing countries [74,75]. In survivors, HIE is a predictor of permanent neurodevelopmental disability. The significance of early diagnosis and intervention in HIE is profound, particularly when considering the long-term consequences in remaining years of life.

Once HIE is identified, immediate supportive care is applied to encourage brain perfusion, including intubation and mechanical ventilation for more severely affected neonates, titration of intravenous fluids, and blood pressure support [76]. Continuous electroencephalogram (EEG) monitoring is used to monitor background brain electrical activity and seizures. Expedient intervention with hypothermia neural rescue therapy within the first few hours of life is vital, improving survival rates and reducing future disabilities [77]. Thus, moderate to severe cases undergo neuroprotective interventions with therapeutic hypothermia (also known as hypothermia neural rescue therapy or "cooling") where a cooling blanket and temperature monitor maintain a body temperature decrease of ~3°C for 72 h. Successful treatment of neonatal brain injury could eliminate many decades of disease burden and healthcare needs.

As the gold standard for brain imaging, MRI is the best imaging technique for diagnosing and staging HIE, particularly with diffusion-weighted contrast [78]. CT is less suitable due to its limited soft tissue contrast and the risks of exposing neonates to ionizing radiation. Although the preferred imaging method, MRI currently requires transport of the fragile patients to a radiology suite. This introduces safety concerns, since it requires a full hand-off of patient care responsibilities and exposes the neonate to stress from transport and acoustic noise. In particular, HIE patients are often not stable enough to transfer to the scanner for several days after birth. Although safety is the number one concern, we must also consider the burden on the hospital workflow. Transfer of critical care patients requires tremendous teamwork and coordination from multiple sources. For example, a neonatologist, nurse, and respiratory therapist must accompany the infant to the scanner and remain with the patient for the duration of the study. The time required for the scan and transfer to/from the NICU is around three hours and NICU staffing coverage must be implemented during their absence. Currently, bedside cranial imaging is available with ultrasound, but provides low sensitivity for early abnormalities associated with HIE [78].

Siting a small footprint MRI in the NICU would provide tangible benefits and a bedside (portable) MRI could offer additional benefits. Figure 3.5 illustrates some existing MRI systems and components that have been developed or adapted for MRI in the NICU. The benefits of bedside MRI accrue since even the trip from one side of the NICU to the other is difficult due to the removal/replacement of leads, tubes, ventilation, and temperature maintenance equipment. Furthermore, intravenous (IV) pumps are not MR compatible so extension tubing must be placed, which can interrupt the administration of critical medications (e.g. dopamine). Even if "MR-safe," noncompatible or metallic equipment can cause susceptibility artifacts, which are more severe at higher field strengths [79]. All of these issues point to an extremely compact magnetic footprint magnet; likely a mid- to low-field magnet design with minimal fringe field.

3.5 Technological Approaches to POC and/or Portable MRI

The goal of reducing cost, siting needs, and even enabling portability has resurrected interest in low-field MRI technology, including new magnet and encoding methods. Low-field MRI, loosely defined as scanners operating between 10 mT and 500 mT, has a long history in MRI, starting

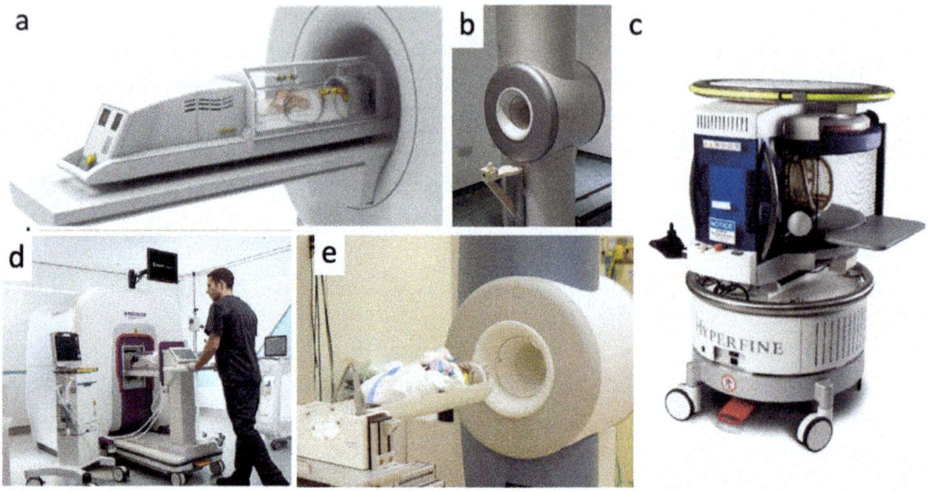

Figure 3.5 MR devices designed to support imaging in neonatal intensive care unit (NICU). (a) LMT MRI compatible incubator (LMT Medical Systems GmbH). (b) GE Healthcare investigational 3-T neonatal MR system installed at Sheffield Teaching Hospital in a dedicated NICU scan room (EMAP Publishing Limited). (c) The Hyperfine 64-mT portable MRI scanner. (d) Aspect Imaging's Embrace neonatal scanner, a dedicated 1-T permanent magnet scanner designed for installation in the NICU without a shielded room. (e) Modified 1.5-T GE orthopedic scanner adapted for the NICU at Cincinnati Children's Hospital.

with the first human imaging systems [80–83]. Although high-field superconducting magnet-equipped scanners have largely displaced them, thousands of low-field MRI systems (mainly in the 0.2–0.35-T range) are in daily operation and their clinical capabilities are well known. Additionally, commercial systems operating below 100 mT have an established clinical history; for example, the Toshiba Access 64-mT system introduced in 1991. These older generation low-field systems are often called "open MRIs" due to the two-pole-piece dipole magnet geometry used. They are still big and heavy and have most of the siting issues of their high-field superconducting cousins. This review focuses on the modern generation of low-field scanners specifically aimed at the three POC uses outlined earlier. The "low" vs. "ultralow" tradeoffs have been recently reviewed in depth [84].

In addition to low-field architectures, considerable research has also focused on ultralow field (ULF), with B_0 below 10 mT [85]. However, several barriers remain to using ULF approaches to achieve low-cost or POC and/or portable medical imaging. The sensitivity of MR as a function of field strength is well known [18], and the severe penalty at ULF requires either additional technology such as prepolarization [86], cryogenic detectors [87–90], hyperpolarized media [91], or efficient but nonstandard brain-imaging sequence schemes [92]. This additional technology can be a barrier to achieving reduced costs or portability. Additional issues arise for image encoding at ULF due to unwanted encoding by the gradient coil's concomitant field terms [93,94].

Because of the signal-to-noise ratio limitations of ULF MRI, this review focuses mainly on compact magnets with B_0 of above 50 mT or compact superconducting magnets in the 0.5–1.0-T range. We focus on magnet and system architecture and component technology; the constraints of MR physics at low field are well understood and are discussed in several recent reviews [15,18,19,84,95,96].

3.5.1 Magnet Designs

3.5.1.1 Advances in Cryogenics for Supercon Magnets

Recent advances have been made in cryogenic technology that facilitate maintenance and siting of superconducting high- and mid-field MRI systems in diverse locations. Modern cryocoolers can now remove more heat than previous generations allowing reduction and even elimination of the "bath" of liquid helium (LHe) that has typically surrounded the superconducting wire in the magnet to maintain it at 4.2 K. These cryogenic technologies and those leading up to them have been extensively reviewed [97]. In the earliest superconducting magnets, a two-cryogen system was used; LHe on the inner vessel providing a liquid bath in direct contact with the niobium titanium (NbTi) superconducting wire, and liquid nitrogen (LN) in an outer vessel to reduce heat flow to the environment. The residual heat entering the LHe vessel was countered only by the latent heat of evaporation of helium, about 20.5 kJ/kg. Thus, boiling off ~8 l of Lhe per day (~1 kg per day) compensates for a heat leak of only 0.24 W. The next step was to replace the use of nitrogen with a mechanical refrigeration system ("cold-head"). This is considerably safer, since nitrogen gas poses increased asphyxiation risk compared with helium (which floats up rather than sinking down). In addition to being more convenient, holding a radiation shields outside of the LHe vessel to 20 K and 70 K using cold-heads also decreases the heat flow compared with a similar 77 K LN cooled surface.

Superconducting MRI systems typically employ Gifford–McMahon (GM) cryocoolers, a variant of the older Stirling cycle. GM cryocoolers have a moving rotary valve in the subsystem attached to the magnet, which makes the familiar steady "washing machine" sound of a non-scanning MRI, and is also a potential source of failure and mechanical vibrations. Current cold-heads can achieve more than 1 W of cooling at 4.2 K and tens of Watts at 12 K. This is enough to not only lessen the entering heat flux, but to fully overcome it, allowing GM cryocooler technology to provide sufficient heat removal at 4 K that helium is liquefied inside the cryostat. The result is a "zero-boiloff" magnet, now the norm for new clinical MRI systems. While not a fully "dry magnet," current zero-boiloff designs eliminate the need for regular cryogen fills.

The next step is to eliminate the LHe completely, or at least reduce LHe volume to where the system can be safely operated without an emergency quench vent system (<10 l of LHe). This can save significantly on siting costs, especially in a complex hospital setting such as the ED. Elimination of most of the LHe bath requires direct "conduction cooling" of the superconductor by the cold-head (as opposed to bathing the windings in LHe). Although first envisioned as needing more exotic higher temperature superconductors such MgB_2 [98,99], "dry magnets" have recently been introduced by the major clinical MRI manufacturers using standard NbTi wire. The use of high-temperature superconductors that operate well above 4.2 K is attractive because they have an easier cooling target but is currently held back by their increased cost.

3.5.1.2 Superconducting Solenoid Designs for the Easy-to-Site Suite

Shorter superconducting solenoid magnets have clear benefits for patient acceptance and ease of siting and use in an ED setting. In contrast to the situation for cryogenic equipment, the design of superconducting MRI magnet windings is little changed over the past few decades; see Lvovsky et al. for a thorough review [97]. In standard design optimizations, the position and number of turns of discrete winding groups are optimized on the cylinder. In addition to the primary field-producing windings, counter-wound shielding windings (usually larger-diameter coils at the bore end) attempt to reduce the stray field around the magnet. The optimization seeks a short magnet on a predetermined diameter cylindrical former that achieves the target homogeneity over the imaging region (typically defined as a given diameter spherical volume (DSV) with a minimum wire cost (length of wire). The optimization either assumes a small number of winding groups (e.g. six) [100] or uses linear programming and a sparsity-promoting norm to limit the winding groups [101]. The magnet homogeneity is a relative measure (ppm) and expressing the magnet length in unitless terms (as L/DSV, where DSV is the imaging "diameter spherical volume") is also helpful. Xu et al. [102] showed that for a 1 ppm target homogeneity, the cost-optimized magnets followed a linear formula: $L_{\text{optimal}} = 1.19\ \text{DSV} + 0.77D$, where L_{optimal} and D are the magnet windings length and diameter, and DSV is the diameter of the spherical homogeneity region. Reducing the magnet length further ($<L_{\text{optimal}}$) results in skyrocketing wire costs (which are proportional to conductor volume). A similar analysis was also applied to gradient coil lengths [103].

3.5.1.3 Shorter Supercon Magnets from Relaxed Homogeneity

Because the optimization used by Xu and others is convex, it represents a true optimal solution and we must relax other constraints in order to achieve shorter magnets. In the text that follows, we replicate Xu et al.'s analysis [102] for L_{optimal} as a function of DSV and magnet diameter for a $D = 1.1$ m magnet but also examined optimized designs with homogeneity targets between 1 ppm and 10 000 ppm. Figure 3.6 shows the result of this analysis. When a length-optimized design is chosen from the knee of the cost vs. length L-curve, the cost of an optimized design can be plotted as a function of the unitless L/DSV for a given homogeneity target. Figure 3.6 shows this result, which informs the potential improvements in magnet length achievable if imaging could be performed in less homogeneous fields. Unfortunately, substantial relaxation in homogeneity is needed to get a significant reduction in magnet length. For example, relaxing the 1 ppm specification to 1000 ppm reduces the magnet length from 2.92 to 2.15 in normalized units – a 26% reduction. Thus, superconducting solenoid magnets, as they are currently envisioned, are not likely to significantly change in geometry.

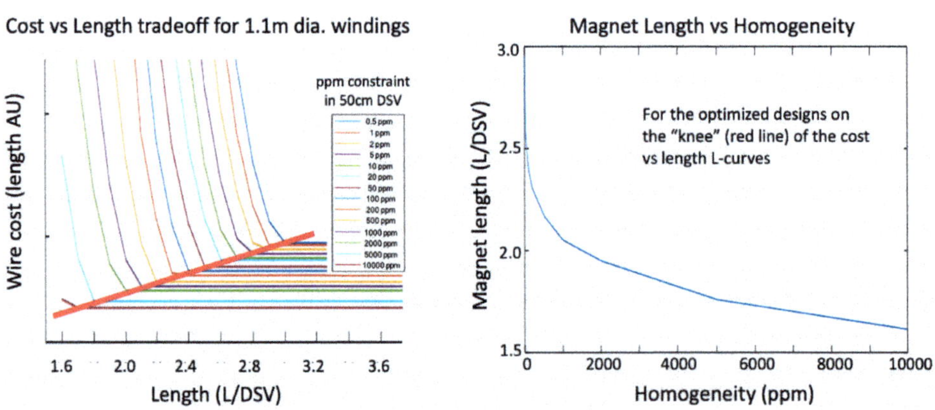

Figure 3.6 Tradeoffs in superconducting solenoid optimizations following the optimization of Xu et al.[] The wire cost is seen to rise significantly as the magnet length is shortened (here shown in units of the imaging volume diameter).The left graph shows the wire length–length tradeoff for different homogeneity constraints. Taking the points on the "knee" of these L-curves allows the plot of magnet length vs. homogeneity (right-hand graph).

3.5.1.4 Permanent Magnets for Portable MRI

Superconducting solenoids are attractive because of their lack of an external energy source, high stored magnetic field energy, and temporal stability, although with the requirement for a cryogenic subsystem. Permanent magnets offer these capabilities to some degree and do not need cryogenics. Their downside is a reduced field-generating capability; human-sized homogeneous magnets above 0.5 T require a considerable weight of material. They are also less stable, for example drifting with temperature.

Rare-earth magnets made from alloys of neodymium such as NdFeB are the strongest form of permanent magnets readily available. They are a relatively recent development, introduced in 1984 [104,105]. Formed into blocks from sintered powder and then magnetized by applying a pulsed magnetic field, the strength of the permanent magnet is measured by its remanence (B_r), which measures the magnetic field of the residual magnetization present in the absence of an external field. For NdFeB, B_r ranges from 1.0 to 1.4 T. The field of NdFeB changes with temperature, with a temperature coefficient of remanence of about −0.1% per °K (the field goes down as the temperature increases). The coercivity (H_{cl}) measures the material's resistance to demagnetization by an external field (or the field of a neighboring block). The tendency of the sintered material to corrode requires that the block be coated, usually with copper-nickel plating, or a coating of epoxy or another polymer. Because of the considerable forces between large NdFeB blocks, handling and assembling the material requires care and expertise.

The most common geometry for MRI use is a simple two-pole-piece dipole magnet. Spacing two magnetized disks with a gap (containing the imaging region) forms a uniform dipole field volume between the pole pieces. Each disk creates a dipole current pattern, and the net effect is like two stacked current loops (each with the current flowing in the same sense). Optimizing the placement of the material in the disks and/or shaping an iron pole-piece can improve homogeneity. Adding an iron yoke helps channel the flux into a closed circuit and can provide mechanical support for the two disks (which otherwise want to come together). This is the basic configuration of many commercial 0.2–0.35-T "low-field open" MRI systems, which reached a peak in popularity in the early 1990s. The "open" patient space reduced claustrophobia anxieties at a time when superconducting solenoid systems were quite long and narrow bore (55 cm diameter as opposed to the current norm of 70 cm diameter). In many ways, low-field open systems remain a reasonable choice for emergency MR, but their size and weight are comparable with 1.5-T superconducting systems, requiring a similar siting footprint. The Hyperfine 64-mT portable MRI system [36] appears to follow this geometry, as does a 0.2-T portable system mounted in a minivan for elbow imaging at baseball games [16]. The direction of the B_0 field differs from their superconducting cousins. The superconducting solenoid imposes B_0 in the "head–foot" direction, while low-field open systems usually have a posterior to anterior (vertical) B_0. It has been noted that the accompanying change in radiofrequency field geometry for low-field open geometry systems yields a reduced interaction with deep-brain stimulation leads (and lower radiofrequency heating) [106].

3.5.1.5 Halbach Arrays for Portable MRI

The standard dipole magnet described in Section 3.5.1.4, built by placing two blocks of magnetized material above and below the imaging volume can be extended by adding additional blocks around the imaging volume with intermediate magnetization angles. This extension is analogous to expanding two current loop coil dipoles in the "Helmholtz" configuration to a birdcage style structure by adding loops in a circular configuration with the currents phased at intermediate phase angles. In the permanent magnet case, the result is a magnetized annulus commonly referred to as a Halbach array shown in Figure 3.7 [107,108]. The ideal Halbach configurations have the counterintuitive property of not needing a flux return yoke because the flux is returned in the magnets themselves. In linear Halbach arrays there is no flux on one side of the structure. Similarly, the ideal cylindrical (Figure 3.7, top left) and spherical Halbach arrays have no flux outside the structure [109]. This nearly completely eliminates the stray-field footprint, as well as eliminating the heavy iron yoke associated with the open-MRI dipole magnets; two compelling practical advantages to the Halbach configurations.

The field inside an ideal cylindrical Halbach array (e.g. Figure 3.7, top left) of inner radius r_a and outer radius r_b is given by $B = B_r \ln(r_b/r_a)$, where B_r is the remanence of the material used. Thus, for a human head system with $r_b/r_a = 50$ cm/30 cm and $B_r = 1.4$ T, we can expect a field of $B_0 = 0.72$ T in this ideal structure. For a whole-body system with $r_b/r_a = 120$ cm/80 cm, $B_0 = 0.56$ T. If the cylinder length is not infinite, but equal to its outer diameter, the estimated weight of rare-earth material for these two systems would be 1500 kg and 18 000 kg – not exactly lightweight. Achieving realistic weights for POC systems necessitates reducing the field, typically to below 0.1 T for adult human imaging systems.

Practical construction necessitates breaking the continuous distribution of material into discrete blocks such as shown in Figure 3.7, coined the NMR-MANDHaLa (nuclear magnetic resonance–magnet arrangements for novel discrete Halbach layout) [39,110]. This has the advantage

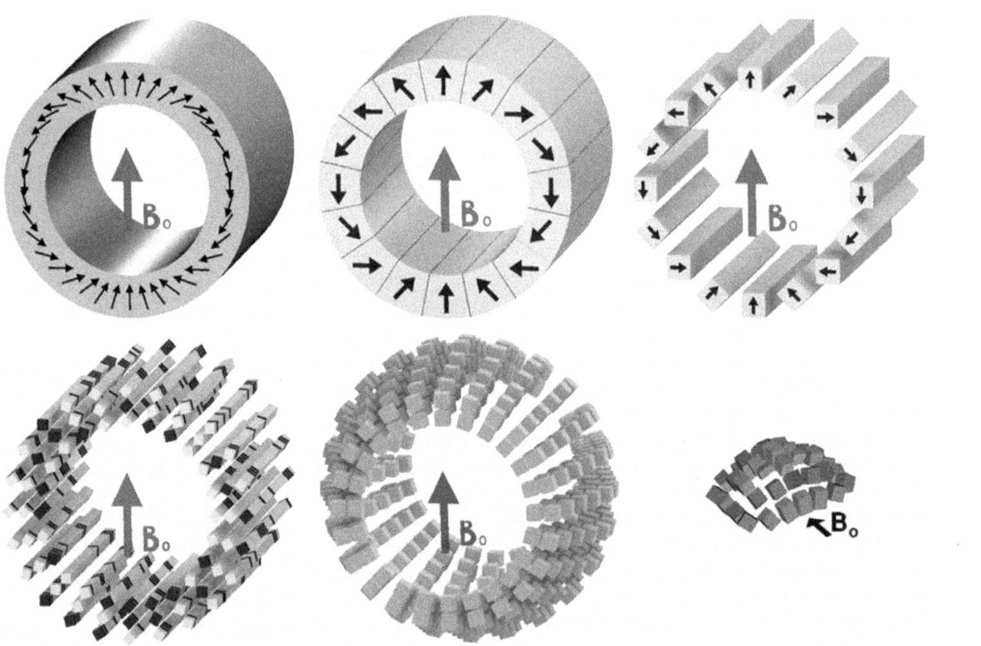

Figure 3.7 Halbach arrays of permanent magnets. Phasing the magnetization from 0 to 4π provides a magnetic dipole with a uniform transverse field inside. Higher-order modes can be obtained by varying the angle to other multiples of 2π. Top left shows the ideal continuous magnetization distribution. Top middle uses key-stone shaped pieces. Top right shows square cross-section pieces, which are readily commercially available. Bottom left shows the possibility of using multiple cubes of differing B_r, chosen to optimize a desired field pattern. The bottom middle takes better advantage of the head geometry by using a portion of a Halbach sphere. The bottom right shows a section of a Halbach sphere used in the "MR Cap" of Figure 3.3.

of using stock NdFeB magnet geometries. Multiple publications have addressed methods to hold the material in place – an important consideration considering there are significant internal forces. Typically some sort of optimization of the configuration is needed for Halbach arrays, which are discretized and of finite length. This is especially important for head magnets where the shoulders are not inside the cylinder, in which case the imaging volume is centered at the brain center, only 18 cm from the end of the magnet. The array geometry is altered to mitigate the truncation effects using some form of optimization. Altering the size or grade of the rare-earth blocks using a genetic algorithm is a popular approach [51,68,111,112] resulting in a homogeneity improvement of 10-fold for a head-sized homogeneous magnet design [111]. The block size can also be treated as a continuous variable for the optimization and later discretized [38] or the angular position can be continuously altered [113,114]. Even with careful optimization, some form of additional shimming has been needed after construction to achieve field target fidelity comparable with superconducting systems. At least two prototype Halbach cylinders, one at 50 mT [20,115] and the other at 80 mT [20,115] have been tested for human brain imaging, in addition to the 72-mT Halbach bulb design of Figure 3.2 [116]. In addition to their lightweight, compact configuration, some of these systems have employed unusual design features such as a gradient field pattern built into the magnet design to eliminate the conventional switched readout gradient [20], or a conventional uniform field design, but with the gradient coils external to the magnet [116].

3.5.1.6 Other Types of Permanent Magnet Arrays

Several additional geometries have been proposed to create uniform field volumes for MR. The approach of Manz et al. uses a variant of a Halbach sphere using a ring magnet (disk with hole), magnetized along the cylindrical axis and two simple cylinder magnets [117]. A yoked permanent magnet design was introduced for imaging cartilage in the knee at the magic angle, which requires the magnetic field to be rotated relative to the knee about two axes (the Halbach cylinder and conventional dipole are limited to a single rotation axis) [118]. The “Auberg ring” design uses radially magnetized rings on a cylinder to produce a field along the axis of the cylinder [119,120] and has seen increased interest for low-field MRI [25,121–123].

3.5.2 Other Technological Challenges

While a suitably compact magnet design constitutes one of the primary challenges to achieving the easy-to-site suite, portable brain MRI, or MR brain monitor discussed, additional technical challenges arise as the system deviates from the canonical high-field system design. Issues include trying to image in a more inhomogeneous field and trying to operate an MRI scanner without the “shielded room” Faraday cage that attenuates external EMI. As mentioned earlier, relaxing the homogeneity constraint benefits compact magnet design and eliminating the shielded room requires an alternative passive or active EMI mitigation strategy.

3.5.2.1 Image Encoding in an Inhomogeneous Field

The image acquisition strategy and/or the image reconstruction methods will likely need refinement if the homogeneity specifications of the magnet are relaxed to achieve a more desirable POC footprint. Mullins and Garwood review the signal dropout and distortion consequences of inhomogeneities of >10 kHz in conventional sequences and some acquisition approaches [124]. In conventional gradient echo acquisition sequences, simple image reconstructions require that all gradient pulses (slice-select, phase-encoding, and readout) use a gradient strength that dominates the local gradients from the inhomogeneous magnet. Thus, strong gradient fields are desirable for reducing geometric distortions, but at odds with the constraints of portable or POC use where power and cooling infrastructure might be limited. For 3D spin echo sequences, only the readout gradient must dominate since a spin echo can be arranged to refocus the spurious gradients but not the phase-encoding field.

The distortions induced in the image's coordinate system are described by the Jacobian matrix formed from the partial spatial derivatives of the B_0 field. If the magnet's B_0 field map is known, the Jacobian is fully known and, in principle, can be used to correct the distortions. But no spatial encoding occurs in locations where the magnet's static gradient is equal and opposite to the applied gradient, and the correction problem is singular. Because of this, it can be desirable to acquire the image twice, for example once with a positive readout gradient and once with the readout gradient current reversed to ensure that all locations are spatially encoded for at least one acquisition. A general approach is to use a "model-based reconstruction" where the "forward model" describes how measured data are produced given any object input to the model. Conversely, given a set of measured data, an inversion of the forward model (generally by some form of iterative search) finds the object giving the "best fit" to the data, possibly subject to some constraints or prior knowledge [125–128].

Acquisition approaches to the magnet inhomogeneity problem include relying on multiple spin echo sequences, which have a long history of use in the oil well-logging industry where NMR is performed in very inhomogeneous fields [129]. This approach was used with phase encoding and the fixed "readout" gradient inherent to an inhomogeneous magnet to image in single-sided devices [49,130] and in a Halbach cylinder with a built-in readout gradient [20,22]. Other approaches include quadratic phase-encoding approaches [131,132] and missing point steady-state free precession (MS-SSFP) methods [133].

3.5.2.2 Limited Frequency Bandwidth of Tuned Radiofrequency Coils

If the magnet is inhomogeneous, or has a built-in encoding field, the MR linewidth across the body will be larger than seen in canonical high-field scanners, which typically achieve proton linewidths of a few tens of Hertz over the brain. In addition to the encoding and reconstruction problems described earlier, the Q of the radiofrequency coil is often high enough so that the bandwidth (BW) of the coil is lower than that of the signal, diminishing detection in regions with high-frequency offsets.

On excitation, this could be compensated in the design of the radiofrequency pulse using knowledge of the coil's response to counteract the effect at the frequencies at the edge of the coil's response. Alternatively, one could simply damp the resonator with an external resistor switched in during transmit (at the expense of requiring increased radiofrequency power) [134]. But during reception, the high-Q coil will simply be inefficient at detecting spins in locations where the spin frequency exceeds the coil BW. The ability to receive across the full-imaging BW is critical, and becomes more difficult at low field. Low loss reception is important for maintaining the sensitivity of detection, so there is strong motivation to maximize Q by increasing L and reducing resistive losses in the coil. Furthermore, reduced radiofrequency losses in the tissue help keep the Q of head resonators in the 50–200 range across a wide variety of low- to mid-field strengths, since the coil BW (full width at half maximum of coil response) relates to Q through $\text{BW} = \nu_0/Q$ (where ν_0 is the Larmor frequency). As Q does not change much, in general the detection BW decreases at lower field. Assuming $Q = 100$ and an 80-mT magnet ($\nu_0 = 3.4$ MHz), the expected

BW of the coil is 34 kHz (compared with a 640-kHz BW for a $Q = 100$ coil at 1.5 T). Even for homogeneous magnets, a typical readout gradient of 10 mT/m will introduce a proton linewidth of 84 kHz over the 20-cm head. Thus, a receive coil constructed with desirable high-Q properties will introduce substantial shading across the FOV.

Increasing the receive coil Q with resistors is a poor strategy from the signal-to-noise ratio point of view since this introduces losses and thus radiofrequency noise. Lossless methods are needed to alter the frequency response of the tuned circuit. The Fano–Bode BW theory describes a general limit for the BW over which the impedance match of an MR coil can be achieved [135]. Cryoprobes at high field experience a similar issue [136]. In addition to traditional impedance matching methods, lossless manipulation of the tuned circuit frequency response can be addressed using a variant of preamplifier decoupling, typically used to reduce geometric coupling in arrays [137]. Traditional preamplifier decoupling uses a non-power-matched input impedance preamplifier to couple an extra parallel LC "trap" circuit to coil resonance. The lack of a power match is immaterial if the noise performance of the preamplifier is satisfactory with that matching condition. Because a power match is not used, it's not clear if the Fano–Bode theory applies. Although known for causing a large perturbation to the coil's frequency response, preamp decoupling at modest levels can create a minor split-resonance, which has the effect of lossless broadening the response. Figure 3.8 shows the frequency response of a low-impedance preamplifier (WanTcom Inc., Chanhassen, MN) connected to a coil tuned to 3.4 MHz. The strategy is able to reduce the coil's power response variation over the desired imaging BW (shaded region of Figure 3.8) from 4.7 dB to 0.7 dB.

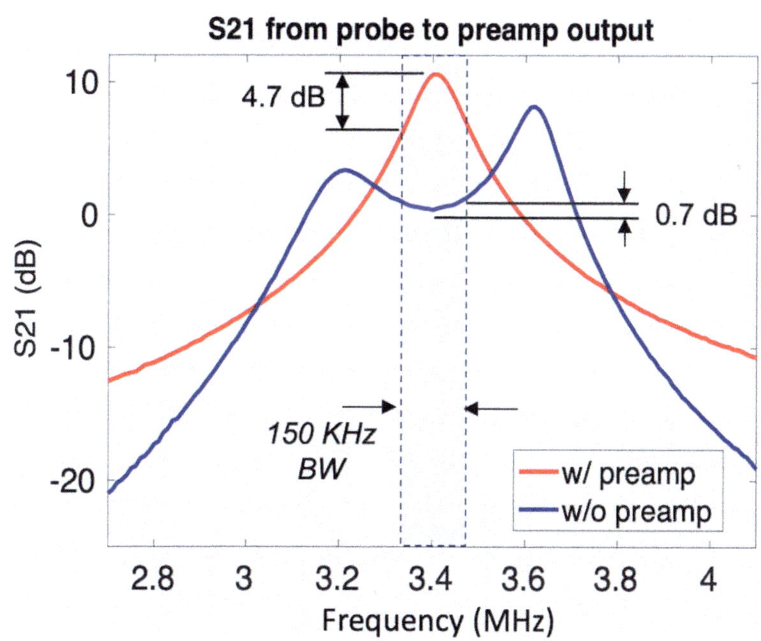

Figure 3.8 Response of a low-field magnetic resonance imaging coil with and without preamplifier decoupling. With mild preamplifier decoupling, the variation over the expected imaging bandwidth dropped from 4.7 dB to 0.7 dB.

3.5.2.3 External EMI Removal (Eliminating the Shielded Room)

Typical MRI scanners are sited inside a passively shielded conductive room (Faraday cage) to prevent external radiofrequency sources from inducing image artifacts. This >US$100 000 siting component is both costly and prevents portability. Passive shielding can be partially achieved by placing copper foil around the magnet. The passive radiofrequency shielding is generally sufficient for phantoms and objects that do not extend outside of the magnet (shielded area) but proves insufficient when a human subject is in the magnet. This is because the conductive human torso and legs act as an antenna which picks up external radiofrequency interference and pipes it into the detection coil. Imaging a cantaloupe and a living human can differ in external interference artifact by as much as 100-fold.

Additional EMI mitigation can be achieved using external interference detectors (additional radiofrequency coils outside of the magnet) and retrospective removal from the images. The interference is picked up from external coils and the artifact is removed from the image using a trained correlation model between that seen in imaging coil and the external detection coil. This approach has a broad history in electromagnetic measurements and in EEG and magnetoencephalography (MEG). Muller-Petke provides some review as well as an application to MR [138], and several methods have emerged specifically for low-field MRI systems [139,140]. Figure 3.9

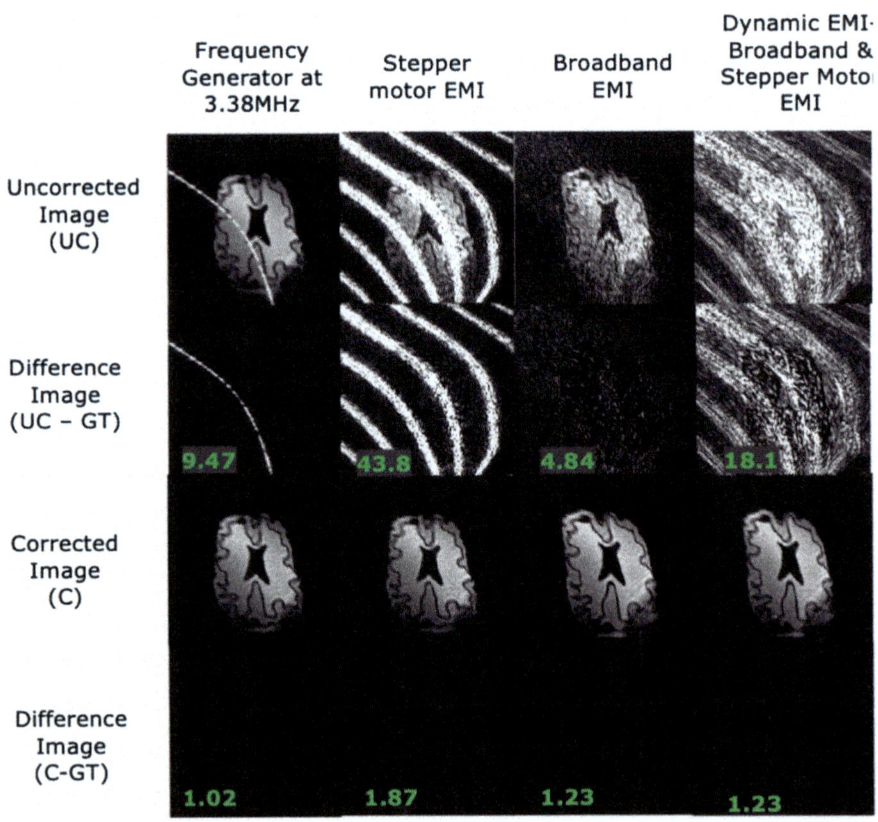

Figure 3.9 Retrospective EMI correction of a time-varying EMI using five external coils using a k-space correlation matrix approach in a low-field scanner [140]. Four electromagnetic interference (EMI) sources are shown including a dynamic source (right-hand column). The bottom row shows difference images using "Ground Truth" (GT) images obtained without the noise sources present.

shows results of an example phantom experiment in an 80-mT portable brain scanner [20,141] using retrospective EMI correction with five external detector coils [140]. Although refinements will continue, it appears that the EMI issue can be managed for POC MRI without a Faraday-shielded room.

3.6 Conclusions

Research in MRI technology has done a phenomenal job of expanding diagnostic benefit by developing acquisition techniques and instrumentation to enable MRI scanners to "see more" through improved sensitivity, spatio-temporal resolution, contrasts, and expanded clinical targets. In contrast to the clear benefits achieved in this direction, extending the *reach* of MRI by widening the range of where it can be applied has received less attention (until recently). This is especially seen for MRI technology designed to extend its use at the POC such as emergency medicine settings by reducing its cost, and siting needs such as footprint, power, cooling, venting, and shielding to the point where bedside use is possible. Despite the challenges, recent activity has advanced the technology on multiple fronts, making MRI scanners available in a wider range of settings; look for a portable MRI scanner if you are unfortunate enough to need a visit to the ED!

Acknowledgments

The authors would like to thank Jason Stockmann, Patrick McDaniel, Thomas Witzel, Mathew Rosen, and Abitha Srivnivas for their extensive technical help, and John Conklin, Sarah Bates, Farrah Mateen, and Michael Lev for their input to the clinical issues surrounding portable and POC MRI. We thank Alex Barksdale for the superconducting magnet optimizations. Research reported in this publication was supported by the National Institute of Biomedical Imaging and Bioengineering of the National Institutes of Health under award number R01EB018976.

References

1 Buller, M. and Karis, J.P. (2017). Introduction of a dedicated emergency department MR imaging scanner at the Barrow Neurological Institute. *American Journal of Neuroradiology* 38 (8): 1480–1485.
2 Redd, V., Levin, S., Toerper, M. et al. (2015). Effects of fully accessible magnetic resonance imaging in the emergency department. *Academic Emergency Medicine* 22 (6): 741–749.
3 Sanchez, Y., Yun, B.J., Prabhakar, A.M. et al. (2017). Magnetic resonance imaging utilization in an emergency department observation unit. *The Western Journal of Emergency Medicine* 18 (5): 780–784.
4 Manna, S., Voutsinas, N., Maron, S.Z. et al. (2020). Leveraging IR's adaptability during COVID-19: A multicenter single urban health system experience. *Journal of Vascular and Interventional Radiology* 31 (7): 1192–1194.
5 Jacobi, A., Chung, M., Bernheim, A. et al. (2020). Portable chest X-ray in coronavirus disease-19 (COVID-19): A pictorial review. *Clinical Imaging* 64: 35–42.
6 Chung, M., Bernheim, A., Mei, X. et al. (2020). CT imaging features of 2019 novel coronavirus (2019-nCoV). *Radiology* 295 (1): 202–207.

7 El Homsi, M., Chung, M., Bernheim, A. et al. (2020). Review of chest CT manifestations of COVID-19 infection. *European Journal of Radiology Open* 7: 100239.

8 Smith, M.J., Hayward, S.A., Innes, S.M. et al. (2020). Point-of-care lung ultrasound in patients with COVID-19 – A narrative review. *Anaesthesia* 75: 1096–1104.

9 Kandemirli, S.G., Dogan, L., Sarikaya, Z.T. et al. (2020). Brain MRI findings in patients in the intensive care unit with COVID-19 infection. *Radiology* 297: 201697.

10 Ogbole, G.I., Adeyomoye, A.O., Badu-Peprah, A. et al. (2018). Survey of magnetic resonance imaging availability in West Africa. *The Pan African Medical Journal* 30: 240.

11 Geethanath, S. and Vaughan, J.T., Jr. (2019). Accessible magnetic resonance imaging: A review. *Journal of Magnetic Resonance Imaging* 49 (7): e65–e77.

12 Lazaro, P., Beerra, A., Luengo, S. et al. (1994). Health care expenditures and expensive medical technology; the paradox of low-income countries. Annual Meeting International Society of Technology Assessment in Health Care, Baltimore USA.

13 Saslow, E. (2019). "Out here, it's just me": In the medical desert of rural America, one doctor for 11,000 square miles. *The Washington Post* 28 September.

14 panel discussion (2019). ISMRM Workshop on Accessible MRI. New Delhi, India.

15 Wald, L.L., McDaniel, P.C., Witzel, T. et al. (2020). Low-cost and portable MRI. *Journal of Magnetic Resonance Imaging* 52 (3): 686–696.

16 Nakagomi, M., Kajiwara, M., Matsuzaki, J. et al. (2019). Development of a small car-mounted magnetic resonance imaging system for human elbows using a 0.2T permanent magnet. *Journal of Magnetic Resonance* 304: 1–6.

17 Coffey, A.M., Truong, M.L., and Chekmenev, E.Y. (2013). Low-field MRI can be more sensitive than high-field MRI. *Journal of Magnetic Resonance* 237: 169–174.

18 Marques, J.P., Simonis, F.F.J., Webb, A.G., and Low-field, M.R.I. (2019). An MR physics perspective. *Journal of Magnetic Resonance Imaging* 49 (6): 1528–1542.

19 O'Reilly, T. and Webb, A. (2019). Deconstructing and reconstructing MRI hardware. *Journal of Magnetic Resonance* 306: 134–138.

20 Cooley, C.Z., McDaniel, P.C., Stockmann, J.P. et al. (2021). A portable scanner for magnetic resonance imaging of the brain. *Nature Biomedical Engineering* 5 (3): 229–239.

21 Cooley, C.Z., Stockmann, J.P., Armstrong, B.D. et al. (2014). 2D imaging in a lightweight portable MRI scanner without gradient coils. *Proceedings of the ISMRM Annual Meeting*, Milan, Italy.

22 Cooley, C.Z., Stockmann, J.P., Armstrong, B.D. et al. (2015). Two-dimensional imaging in a lightweight portable MRI scanner without gradient coils. *Magnetic Resonance in Medicine* 73 (2): 872–883.

23 Blumler, P. (2016). Proposal for a permanent magnet system with a constant gradient mechanically adjustable in direction and strength. *Concepts in Magnetic Resonance. Part B, Magnetic Resonance Engineering* 46 (1): 41–48.

24 Ren, Z.H., Maréchal, L., Luo, W. et al. (2015). Magnet array for a portable magnetic resonance imaging system. 2015 IEEE MTT-S 2015 International Microwave Workshop Series on RF and Wireless Technologies for Biomedical and Healthcare Applications (IMWS-BIO), Taipei, Taiwan.

25 Ren, Z.H., Obruchkov, S., Lu, D.W. et al. (2017). A low-field portable magnetic resonance imaging system for head imaging. *2017 Progress in Electromagnetics Research Symposium – Fall (PIERS – FALL)*, Singapore.

26 McDaniel, P.C., Cooley, C.Z., Stockmann, J.P. et al. (2019). The MR cap: A single-sided MRI system designed for potential point-of-care limited field-of-view brain imaging. *Magnetic Resonance in Medicine* in press 82 (5): 1946–1950.

27 Vogel, M.W., Giorni, A., Vegh, V. et al. (2016). Rotatable small permanent magnet array for ultra-low field nuclear magnetic resonance instrumentation: A concept study. *PLoS ONE* 11 (6): e0157040.

28 Runge, V.M. and Heverhagen, J.T. (2020). The next generation- advanced design low-field MR systems. *Siemens Magnetom Flash* Special Issue: Head-to-toe Imaging: 11–19.

29 Mohr, C. (2002). The Siemens ultra high-field program MAGNETOM Allegra and Trio 3T MR: The next dimension in clinical and research MR systems. *Siemens Magnetom Flash* 1: 21–22.

30 Foo, T., Vermilyea, M., Xu, M. et al. (2016). Dedicated high-performance, lightweight, low-cryogen compact 3.0T MRI system for advanced brain imaging. *Proceedings of the ISMRM*, Singapore.

31 Campbell-Washburn, A.E., Ramasawmy, R., Restivo, M.C. et al. (2019). Opportunities in interventional and diagnostic imaging by using high-performance low-field-strength MRI. *Radiology* 293 (2): 384–393.

32 Panther, A., Thevathasan, G., Connell, I. et al. (2019). A dedicated head-only MRI scanner for point-of-care imaging. *Proceeding of the ISMRM*, Montreal.

33 Stainsby, J., Bindseil, G., Connell, I. et al. (2019). Imaging at 0.5T with high-performance system components. *Proceeding of the ISMRM*, Montreal, 1194.

34 Stainsby, J., Harris, C., Bindseil, G. et al. High-performance diffusion imaging on a 0,5T system. *Proceedings of the ISMRM*, Montreal.

35 Vaughan, J., Wang, B., Idiyatullin, D. et al. (2016). Progress toward a portable MRI system for human brain imaging. *Proceedings of the ISMRM*, Singapore.

36 Sheth, K.N., Mazurek, M.H., Yuen, M.M. et al. (2020). Assessment of brain injury using portable, low-field magnetic resonance imaging at the bedside of critically ill patients. *JAMA Neurology* 78 (1): 41–47.

37 Welch, B., By, S., Chen, G. et al. (2020). Use environments and clinical feasibility of portable point-of-care bedside brain MRI. Proceedings of the ISMRM, virtual.

38 McDaniel, P.C., Cooley, C.Z., Stockmann, J.P. et al. (2019). Numerically optimized design for a low-cost, lightweight 86 mT whole-brain magnet. *Proceeding of the ISMRM*, Montreal, Canada.

39 Blumler, P. and Casanova, F. (2016). Hardware developments: Halbach magnet arrays. In: *Mobile NMR and MRI; Developments and Applications* (eds. M. Johns, E.O. Findjonson, S. Vogt, and A. Haber), 133–155. Cambridge, UK: Royal Society of Chemistry.

40 Brown, M.C., Verganelakis, D.A., Mallett, M.J. et al. (2004). Surface normal imaging with a hand-held NMR device. *Journal of Magnetic Resonance* 169 (2): 308–312.

41 Danieli, E. and Blumich, B. (2013). Single-sided magnetic resonance profiling in biological and materials science. *Journal of Magnetic Resonance* 229: 142–154.

42 Van Landeghem, M., Danieli, E., Perlo, J. et al. (2012). Low-gradient single-sided NMR sensor for one-shot profiling of human skin. *Journal of Magnetic Resonance* 215: 74–84.

43 Backhouse, L., Dias, M., Gorce, J.P. et al. (2004). GARField magnetic resonance profiling of the ingress of model skin-care product ingredients into human skin in vitro. *Journal of Pharmaceutical Sciences* 93 (9): 2274–2283.

44 Greer, M., Chen, C., and Mandal, S. (2019). An easily reproducible, hand-held, single-sided, MRI sensor. *Journal of Magnetic Resonance* 308: 106591.

45 Ali, T.S., Tourell, M.C., Hugo, H.J. et al. (2019). Transverse relaxation-based assessment of mammographic density and breast tissue composition by single-sided portable NMR. *Magnetic Resonance in Medicine* 82 (3): 1199–1213.

46 Tourell, M.C., Ali, T.S., Hugo, H.J. et al. (2018). T1-based sensing of mammographic density using single-sided portable NMR. *Magnetic Resonance in Medicine* 80 (3): 1243–1251.

47 Colucci, L.A., Corapi, K.M., Li, M. et al. (2019). Fluid assessment in dialysis patients by point-of-care magnetic relaxometry. *Science Translational Medicine* 11 (502).

48 Li, M., Vassiliou, C.C., Colucci, L.A. et al. (2015). (1)H nuclear magnetic resonance (NMR) as a tool to measure dehydration in mice. *NMR in Biomedicine* 28 (8): 1031–1039.

49 Perlo, J., Casanova, F., and Blumich, B. (2004). 3D imaging with a single-sided sensor: An open tomograph. *Journal of Magnetic Resonance* 166 (2): 228–235.

50 He, Z., He, W., Wu, J. et al. (2017). The novel design of a single-sided MRI probe for assessing burn depth. *Sensors* 17 (3).

51 McDaniel, P.C., Cooley, C.Z., Stockmann, J.P. et al. (2019). The MR cap: A single-sided MRI system designed for potential point-of-care limited field-of-view brain imaging. *Magnetic Resonance in Medicine* 82 (5): 1946–1960.

52 Bhat, S.S., Fernandes, T.T., Poojar, P. et al. (2020). Low-field MRI of stroke: Challenges and opportunities. *Journal of Magnetic Resonance Imaging* e27324 54 (2): 372–390

53 Okorie, C.K., Ogbole, G.I., Owolabi, M.O. et al. (2015). Role of diffusion-weighted imaging in acute stroke management using low-field magnetic resonance imaging in resource-limited settings. *The West African Journal of Radiology* 22 (2): 61–66.

54 Fassbender, K., Grotta, J.C., Walter, S. et al. (2017). Mobile stroke units for prehospital thrombolysis, triage, and beyond: Benefits and challenges. *Lancet Neurology* 16 (3): 227–237.

55 Fassbender, K., Walter, S., Grunwald, I.Q. et al. (2020). Prehospital stroke management in the thrombectomy era. *Lancet Neurology* 19 (7): 601–610.

56 Fassbender, K., Walter, S., Liu, Y. et al. (2003). "Mobile stroke unit" for hyperacute stroke treatment. *Stroke* 34 (6): e44.

57 Grunwald, I.Q., Phillips, D.J., Sexby, D. et al. (2020). Mobile stroke unit in the UK healthcare system: Avoidance of unnecessary accident and emergency admissions. *Cerebrovascular Diseases* 49 (4): 388–395.

58 Walter, S., Ragoschke-Schumm, A., Lesmeister, M. et al. (2018). Mobile stroke unit use for prehospital stroke treatment – An update. *Radiologe* 58 (Suppl. 1): 24–28.

59 Powers, W.J., Rabinstein, A.A., Ackerson, T. et al. (2019). Guidelines for the early management of patients with acute ischemic stroke: 2019 Update to the 2018 guidelines for the early management of acute ischemic stroke: A guideline for healthcare professionals from the American Heart Association/American Stroke Association. *Stroke* 50 (12): e344–e418.

60 Powers, W.J., Rabinstein, A.A., Ackerson, T. et al. (2018). Guidelines for the early management of patients with acute ischemic stroke: A guideline for healthcare professionals from the American Heart Association/American Stroke Association. *Stroke* 49 (3): e46–e110.

61 Schaefer, P.W., Souza, L., Kamalian, S. et al. (2015). Limited reliability of computed tomographic perfusion acute infarct volume measurements compared with diffusion-weighted imaging in anterior circulation stroke. *Stroke* 46 (2): 419–424.

62 Leiva-Salinas, C. and Wintermark, M. (2010). Imaging of acute ischemic stroke. *Neuroimaging Clinics of North America* 20 (4): 455–468.

63 Wintermark, M., Sanelli, P.C., Albers, G.W. et al. (2013). Imaging recommendations for acute stroke and transient ischemic attack patients: A joint statement by the American Society of Neuroradiology, the American College of Radiology and the Society of NeuroInterventional Surgery. *Journal of the American College of Radiology* 10 (11): 828–832.

64 Li, L., Padhi, A., Ranjeva, S.L. et al. (2011). Association of bacteria with hydrocephalus in Ugandan infants. *Journal of Neurosurgery: Pediatrics* 7 (1): 73–87.

65 Warf, B.C., Alkire, B.C., Bhai, S. et al. (2011). Costs and benefits of neurosurgical intervention for infant hydrocephalus in sub-Saharan Africa. *Journal of Neurosurgery: Pediatrics* 8 (5): 509–521.

66 Kulkarni, A.V., Schiff, S.J., Mbabazi-Kabachelor, E. et al. (2017). Endoscopic treatment versus shunting for infant hydrocephalus in Uganda. *The New England Journal of Medicine* 377 (25): 2456–2464.

67 Obungoloch, J., Harper, J.R., Consevage, S. et al. (2018). Design of a sustainable prepolarizing magnetic resonance imaging system for infant hydrocephalus. *MAGMA* 31 (5): 665–676.

68 O'Reilly, T., Teeuwisse, W.M., and Webb, A.G. (2019). Three-dimensional MRI in a homogenous 27cm diameter bore Halbach array magnet. *Journal of Magnetic Resonance* 307: 106578.

69 Diehl, J., Van Doesum, F., Bakker, M. et al. (2020). The embodiment of low-field MRI for the diagnosis of infant hydrocephalus in Uganda. *IEEE Global Humanitarian Technology Conference (GHTC)*, virtual.

70 O'Reilly, T., Wouter, T., Winter, L. et al. (2019). Design of a homogeneous large-bore Halbach array for low field MRI. *Proceedings of the ISMRM*, Montreal, Canada.

71 Blumich, B., Blumler, P., Eidmann, G. et al. (1998). The NMR-mouse: Construction, excitation, and applications. *Magnetic Resonance Imaging* 16 (5–6): 479–484.

72 Perlo, J., Casanova, F., and Blumich, B. (2005). Profiles with microscopic resolution by single-sided NMR. *Journal of Magnetic Resonance* 176 (1): 64–70.

73 Hurlimann, M.D. and Heaton, N.J. (2016). NMR well logging. In: *Mobile NMR and MRI: Developments and Applications* (eds. M. Johns, E.O. Findjonson, S. Vogt, and A. Haber), 11–79. Cambridge, UK: Royal Society of Chemistry.

74 Douglas-Escobar, M. and Weiss, M.D. (2015). Hypoxic-ischemic encephalopathy: A review for the clinician. *JAMA Pediatrics* 169 (4): 397–403.

75 Finer, N.N., Robertson, C.M., Richards, R.T. et al. (1981). Hypoxic-ischemic encephalopathy in term neonates: Perinatal factors and outcome. *Journal of Pediatrics* 98 (1): 112–117.

76 Oorschot, D.E., Sizemore, R.J., and Amer, A.R. (2020). Treatment of neonatal hypoxic-ischemic encephalopathy with erythropoietin alone, and erythropoietin combined with hypothermia: history, current status, and future research. *International Journal of Molecular Sciences* 21 (4): 1487.

77 Gluckman, P.D., Wyatt, J.S., Azzopardi, D. et al. (2005). Selective head cooling with mild systemic hypothermia after neonatal encephalopathy: Multicentre randomised trial. *Lancet* 365 (9460): 663–670.

78 Bano, S., Chaudhary, V., and Garga, U.C. (2017). Neonatal hypoxic-ischemic encephalopathy: A radiological review. *Journal of Pediatric Neurosciences* 12 (1): 1–6.

79 Arthurs, O.J., Edwards, A., Austin, T. et al. (2012). The challenges of neonatal magnetic resonance imaging. *Pediatric Radiology* 42 (10): 1183–1194.

80 Hinshaw, W.S., Andrew, E.R., Bottomley, P.A. et al. (1978). Internal structural mapping by nuclear magnetic resonance. *Neuroradiology* 16: 607–609.

81 Hinshaw, W.S., Bottomley, P.A., and Holland, G.N. (1977). Radiographic thin-section image of the human wrist by nuclear magnetic resonance. *Nature* 270 (5639): 722–723.

82 Edelman, R.R. (2014). The history of MR imaging as seen through the pages of radiology. *Radiology* 273 (2 Suppl.): S181–S200.

83 Feinberg, D.A., Mills, C.M., Posin, J.P. et al. (1985). Multiple spin-echo magnetic resonance imaging. *Radiology* 155 (2): 437–442.

84 Sarracanie, M. and Salameh, N. (2020). Low-field MRI: How low can we go? A fresh view on an old debate. *Frontiers in Physics* 8: 172.

85 Kraus, R.H., Espy, M.A., Magnelind, P.E. et al. (2014). *Ultra-Low Field Nuclear Magnetic Resonance, a New MRI Regime*. Oxford, UK: Oxford University Press.

86 Macovski, A. and Conolly, S. (1993). Novel approaches to low-cost MRI. *Magnetic Resonance in Medicine* 30 (2): 221–230.

87 Clarke, J.C., Hatridge, M., and Mossle, M. (2007). SQUID-detected magnetic resonance imaging in microtesla fields. *Annual Review of Biomedical Engineering* 9: 389–413.

88 Inglis, B., Buckenmaier, K., Sangiorgio, P. et al. (2013). MRI of the human brain at 130 microtesla. *Proceedings of the National Academy of Sciences of the United States of America* 110 (48): 19194–19201.

89 Lin, F.H., Vesanen, P.T., Nieminen, J.O. et al. (2013). Noise amplification in parallel whole-head ultra-low-field magnetic resonance imaging using 306 detectors. *Magnetic Resonance in Medicine* 70 (2): 595–600.

90 Savukov, I.M., Zotev, V.S., Volegov, P.L. et al. (2009). MRI with an atomic magnetometer suitable for practical imaging applications. *Journal of Magnetic Resonance* 199 (2): 188–191.

91 Tsai, L.L., Mair, R.W., Rosen, M.S. et al. (2008). An open-access, very-low-field MRI system for posture-dependent 3He human lung imaging. *Journal of Magnetic Resonance* 193 (2): 274–285.

92 Sarracanie, M., LaPierre, C.D., Salameh, N. et al. (2015). Low-cost high-performance MRI. *Scientific Reports* 5: 15177.

93 Volegov, P.L., Mosher, J.C., Espy, M.A. et al. (2005). On concomitant gradients in low-field MRI. *Journal of Magnetic Resonance* 175 (1): 103–113.

94 Nieminen, J.O. and Ilmoniemi, R.J. (2010). Solving the problem of concomitant gradients in ultra-low-field MRI. *Journal of Magnetic Resonance* 207 (2): 213–219.

95 Hayashi, N., Watanabe, Y., Masumoto, T. et al. (2004). Utilization of low-field MR scanners. *Magnetic Resonance in Medical Sciences* 3 (1): 27–38.

96 Wald, L.L. (2019). Ultimate MRI. *Journal of Magnetic Resonance* 306: 139–144.

97 Lvovsky, Y., Stautner, E.W., and Zhang, T. (2013). Novel technologies and configurations of superconducting magnets for MRI. *Superconductor Science and Technology* 26: 093001).

98 Iwasa, Y. (2017). Toward liquid-helium-free, persistent-mode MgB2 MRI magnets: FBML experience. *Superconducting Science and Technology* 30 (5): 053001.

99 Baig, T., Al Amin, A., Deissler, R.J. et al. (2017). Conceptual designs of conduction cooled MgB2 magnets for 1.5 and 3.0T full body MRI systems. *Superconductor Science and Technology* 30 (4): 043002.

100 Lugabsky, L.B. (1987). Optimal coils for producing uniform magnetic fields. *Journal of Physics E* 20 (3): 277–285.

101 Xu, H., Conolly, S., Scott, G. et al. (2000). Homogeneous magnet design using linear programming. *IEEE Transactions on Magnetics* 36 (2): 476–483.

102 Xu, H., Conolly, S.M., Scott, G.C. et al. (1999). Fundamental scaling relations for homogeneous magnets. *Proceedings of the ISMRM* 475.

103 Zhang, B., Gazdzinski, C., Chronik, B. et al. (2005). Simple design guidelines for short MRI systems. *Magnetic Resonance Part B (Magnetic Resonance Engineering)* 25B (1): 53–59.

104 Lucas, J., Lucas, P., and LeMercier, T. (2014). *Rare Earths: Science, Technology, Production and Use*, 1e, 224–225. Elsevier, Amsterdam, Netherlands.

105 Sagawa, M., Fujimura, H., Yamamoto, Y. et al. (1984). Permanent magnet materials based on the rare earth-iron-boron tetragonal compounds. *IEEE Transactions on Magnetics* 20 (5): 1584–1589.

106 Kazemivalipour, E., Bhusal, B., Vu, J. et al. (2021). Vertical open-bore MRI scanners generate significantly less radiofrequency heating around implanted leads: A study of deep brain stimulation implants in 1.2T OASIS scanners versus 1.5T horizontal systems. *Magnetic Resonance in Medicine* 86 (3): 1560–1572.

107 Halbach, K. (1979). Strong rare earth cobalt quadrupoles. *IEEE Transactions on Nuclear Science* 26 (3): 3882–3884.

108 Halbach, K. (1980). Design of permanent multipole magnets with oriented rare earth cobalt material. *Journal of Nuclear Instruments & Methods* 69: 1–10.

109 Shute, H.A., Mallison, J.C., Wilton, D.T. et al. (2000). One-sided fluxes in planar, cylindrical, and spherical magnetized structures. *IEEE Transactions on Magnetics* 36 (2): 440–451.

110 Raich, H. and Blumler, P. (2004). Design and construction of a dipolar Halbach array with a homogeneous field from identical bar magnets: NMR mandhalas. *Concepts in Magnetic Resonance* 23B (1): 16–25.

111 Cooley, C.Z., Haskell, M.W., Cauley, S.F. et al. (2018). Design of sparse Halbach magnet arrays for portable MRI using a genetic algorithm. *IEEE Transactions on Magnetics* 54 (1): 1–12.

112 Purchase, A.R., Vidarsson, L., Wachowicz, K. et al. (2021). A short and light, sparse dipolar Halbach magnet for MRI. *IEEE Access* 9: 95294–95303.

113 Choi, J.S. and Yoo, J. (2008). Design of a Halbach magnet array based on optimization techniques. *IEEE Transactions on Magnetics* 44 (10): 2361–2366.

114 Tewari, S., O'Reilly, T., and Webb, A. (2021). Improving the field homogeneity of fixed- and variable-diameter discrete Halbach magnet arrays for MRI via optimization of the angular magnetization distribution. *Journal of Magnetic Resonance* 324: 106923.

115 O'Reilly, T., Teeuwisse, W.M., De Gans, D. et al. (2021). In vivo 3D brain and extremity MRI at 50 mT using a permanent magnet Halbach array. *Magnetic Resonance in Medicine* 85 (1): 495–505.

116 Cooley, C.Z., Stockmann, J.P., and Wald, L.L. (2021). A portable brain MRI scanner based on a 72 mT, 35 kg "Halbach-bulb" magnet and external gradient coil. Proceedings of the ISMRM, virtual.

117 Manz, B., Benecke, M., and Volke, F. (2008). A simple, small and low cost permanent magnet design to produce homogeneous magnetic fields. *Journal of Magnetic Resonance* 192 (1): 131–138.

118 McGinley, J.V., Ristic, M., and Young, I.R. (2016). A permanent MRI magnet for magic angle imaging having its field parallel to the poles. *Journal of Magnetic Resonance* 271: 60–67.

119 Hugon, C., D'Amico, F., Aubert, G. et al. (2010). Design of arbitrarily homogeneous permanent magnet systems for NMR and MRI: Theory and experimental developments of a simple portable magnet. *Journal of Magnetic Resonance* 205 (1): 75–85.

120 Aubert, G. (1991). Cylindrical permanent magnet with longitudinal induced field. USA patent 5014032.

121 Ren, Z.H., Mu, W.C., and Huang, S.Y. (2019). Design and optimization of a ring-pair permanent magnet array for head imaging in a low-field portable MRI system. *IEEE Transactions on Magnetics* 55 (1): 1–8.

122 Ren, Z.H., Gong, S., and Huang, S.Y. (2019). An irregular-shaped inward-outward ring-pair magnet array with a monotonic field gradient for 2D head imaging in low-field portable MRI. *IEEE Access* 7: 48715–4872.

123 Kuang, I., Arango, N., Stockmann, J.P. et al. (2019). Equivalent-charge-based optimization of spokes and hub magnets for hand-held and classroom MR imaging. *Proceedings of the ISMRM*, Montreal, Canada.

124 Mullen, M. and Garwood, M. (2020). Contemporary approaches to high-field magnetic resonance imaging with large field inhomogeneity. *Progress in Nuclear Magnetic Resonance Spectroscopy* 120–121: 95–108.

125 Lange, K. and Carson, R. (1984). EM reconstruction algorithms for emission and transmission tomography. *Journal of Computer Assisted Tomography* 8 (2): 306–316.

126 Harshbarger, T.B. and Twieg, D.B. (1999). Iterative reconstruction of single-shot spiral MRI with off resonance. *IEEE Transactions on Medical Imaging* 18 (3): 196–205.

127 Fessler, J. (2010). Model-based image reconstruction for MRI. *IEE Signal Processing Magazine* 27 (4): 81–89.

128 Sutton, B.P., Noll, D.C., and Fessler, J.A. (2003). Fast, iterative image reconstruction for MRI in the presence of field inhomogeneities. *IEEE Transactions on Medical Imaging* 22 (2): 178–188.

129 Hurlimann, M.D. and Griffin, D.D. (2000). Spin dynamics of Carr-Purcell-Meiboom-Gill-like sequences in grossly inhomogeneous B(0) and B(1) fields and application to NMR well logging. *Journal of Magnetic Resonance* 143 (1): 120–135.

130 McDaniel, P.C., Cooley, C.Z., Stockmann, J.P. et al. (2019). The MR Cap: A single-sided MRI system designed for potential point-of-care limited field-of-view brain imaging. *Magnetic Resonance in Medicine* 82 (5): 1946–1960.

131 Ben-Eliezer, N., Shrot, Y., and Frydman, L. (2010). High-definition, single-scan 2D MRI in inhomogeneous fields using spatial encoding methods. *Magnetic Resonance Imaging* 28 (1): 77–86.

132 Snyder, A.L., Corum, C.A., Moeller, S. et al. (2014). MRI by steering resonance through space. *Magnetic Resonance in Medicine* 72 (1): 49–58.

133 Kobayashi, N., Parkinson, B., Idiyatullin, D. et al. (2021). Development and validation of 3D MP-SSFP to enable MRI in inhomogeneous magnetic fields. *Magnetic Resonance in Medicine* 85 (2): 831–844.

134 Zhen, J.Z., O'Neill, K.T., Fridjonsson E.O. et al. (2018). A resistive Q-switch for low-field NMR systems. *Journal of Magnetic Resonance* 287: 33–40.

135 Scott, G.C., Conolly, S., and Macovski, A. (1996). Low field preamp matching design for high-Q receiver coils. *Proceeding of the ISMRM*, New York.

136 Darrasse, L. and Ginefri, J.C. (2003). Perspectives with cryogenic RF probes in biomedical MRI. *Biochimie* 85 (9): 915–937.

137 Roemer, P.B., Edelstein, W.A., Hayes, C.E. et al. (1990). The NMR phased array. *Magnetic Resonance in Medicine* 16 (2): 192–225.

138 Muller-Petke, M. (2020). Non-remote reference noise cancellation-using reference data in the presence of surface-NMR signal. *Journal of Applied Geophysics* 177: 104040.

139 Rearick, T., Charvat, G., Rosen, M.S. et al. (2016). Noise suppression methods and apparatus patent 9,797,971. 10 March 2016.

140 Srinivas, S.A., Cauley, S.F., Stockmann, J.P. et al. (2021). External dynamic interference estimation and removal (EDITER) for low field MRI. *Magnetic Resonance in Medicine*. https://onlinelibrary.wiley.com/doi/10.1002/mrm.28992.

141 Dyvorne, H., Rearick, T., Poole, M. et al.(2021). Freeing MRI from its Faraday cape with interference rejection. *Proceedings of the 29th Annual Meeting of ISMRM*, 0749, Virtual.

4

Technology for Ultrahigh Field Imaging

Kamil Uğurbil

Center for Magnetic Resonance Research (CMRR), University of Minnesota, Minneapolis, MN, USA

4.1 Introduction

Developments in ultrahigh field magnets for human imaging are intricately tied to the use of magnetic resonance (MR) techniques to image alterations in human brain activity, i.e. functional brain mapping with magnetic resonance imaging (fMRI). It was self-evident from its inception that the sensitivity of the fMRI measurement had to increase to improve the detectability of the extremely small signal changes linked to brain activity at any spatial resolution and, particularly, at very high resolutions relevant for neuronal ensembles of a few thousand neurons that perform elementary computations. To date, this challenge has been and, in fact, continues to be addressed primarily by increasing the magnetic field strength of the experiment. This effort has pushed superconducting magnet technologies to achieve very high field magnets with large bore sizes, but in addition, has necessitated the development of other technologies that are integral to the MR measurement, namely gradients and radiofrequency (RF) transmission and detection technologies.

The magnetic field of 7 Tesla (7 T) using a 90-cm bore magnet large enough to accommodate humans was introduced in 1999, predominantly justified for its advantages in fMRI and to a lesser extent for MR spectroscopy [1,2]. Neither fMRI nor 7 T was of particular interest to manufacturers of clinical MRI scanners used for diagnosis in humans. As such, the instrument was developed in-house as a research and development effort by the group led by the author (CMRR), and not by the manufacturers of MRI scanners. This development was motivated by results obtained at 4 T for functional mapping in the human brain at a time when 3 T human MR scanners did not exist, and 1.5 T represented the "high-field" end of clinical scanners. One of the two independent and contemporaneous efforts [3,4] that introduced fMRI was conducted at 4 T [3]. It was clear from these very early results that 4 T provided superior functional images (Figure 4.1). While the 1.5-T data showed a diffuse area of "activation" in the general region of the human visual cortex, the higher-resolution 4 T images were detailed and followed the counters of the cortex.

This early T2*-weighted 4 T functional map (Figure 4.1) already contained indications that high magnetic fields produced stimulus-invoked signal changes in the cortical gray matter where such activation should appear. There was, however, also evidence of apparent and clearly erroneous "activation" outside the cortex in the space occupied by cerebrospinal fluid (CSF). Subsequent studies identified the latter as originating from draining veins, abundant on the pial surface [5]. Thus, while 4 T fMRI data displayed functional signals associated with

Magnetic Resonance Microscopy: Instrumentation and Applications in Engineering, Life Science and Energy Research, First Edition. Edited by Sabina Haber-Pohlmeier, Bernhard Blümich, and Luisa Ciobanu.

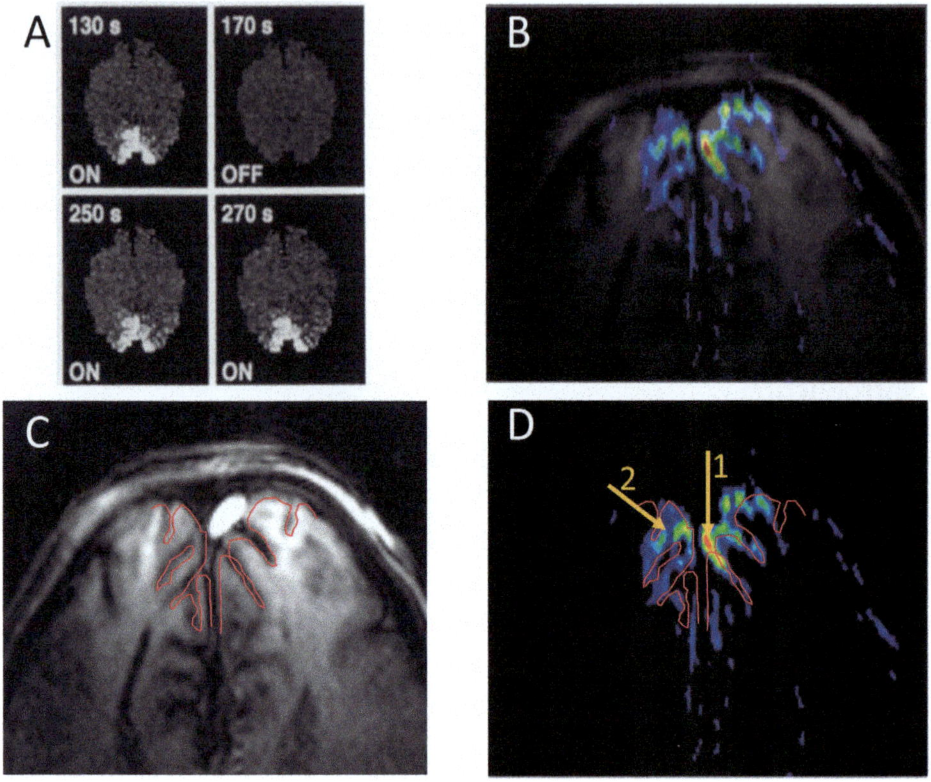

Figure 4.1 First functional images reported from (A) the Massachusetts General Hospital using a 1.5-T scanner specially built to be able to perform EPI data acquisition [4]. (B) the Center for Magnetic Resonance Research (CMRR) using a 4 T instrument [3] with the functional image in color superimposed on the anatomical image of the slice. Panel (C) shows the anatomical image only corresponding the image shown in panel B, with a red line outlining the cortical gray matter, and panel (D) displays the same image as panel B but with the red lines outlining the cortical gray matter illustrated in panel C now superimposed on this image so as to display how the functional map in color corresponds spatially to the gray matter contours. There are regions where the apparent activation is outside the cortex (e.g. indicated by the arrow labeled as 1) but there are also regions where the activity falls clearly in the gray matter (region indicated by the arrow labeled as 2).

parenchyma as well as draining veins, the latter was shown to dominate functional maps obtained at lower magnetic fields [6–9]. Concurrent efforts undertaken at the time to better understand the mechanisms underlying fMRI [10] also pointed to blood oxygenation level-dependent (BOLD) contrast improvements with magnetic fields, more so for weaker responses associated with the parenchyma as opposed to the undesirable signals arising from the draining veins. This led to efforts in our group (CMRR) to develop a 7 T system [1,2]. After exploratory initiatives that started in 1995, the decision was made to pursue a 7 T 90-cm bore magnet. The bore diameter of 90 cm was considered important because shielded gradients were not as well developed at the time; in addition, gradients were not being developed simultaneously with the magnet to optimize the designs and minimize the interactions between the two. Therefore, protection against the gradient–magnet interactions as well as flexibility for future gradient developments were sought in the large bore of the magnet. This system was assembled from parts acquired from different manufactures with system integration, multichannel receivers, and RF coils provided in house. First functional images from this system demonstrated clearly the gains anticipated from this ultrahigh [magnetic] field (UHF, defined as ≥7 T) relative to 4 T [11].

Approximately two decades after the introduction of and first publications coming from this 7 T system, today we witness an increasing interest in 7 T with a rapidly expanding user base. This development is motivated cumulatively by numerous developments. These include the introduction of commercial 7 T systems by the major manufacturers of clinical MR scanners, increased understanding of the physics of imaging the human body at the high operating frequency (~300 MHz) [12–14], gradual improvement of the commercially available instruments as a result of incorporating technologies necessary to deal with the challenges of this ultrahigh field strength, an accumulating body of neuroscientific applications demonstrating superior results for human brain studies, the successful expansion of the UHF applications to the entire human body, and an increasing number of unique clinical applications (e.g. reviews [15–23]; also see collection of articles in *Neuroimage* Special Issue [24]). These developments have further motivated initiatives to pursue even higher magnetic fields than 7 T. A 10.5-T/88-cm bore whole-body system is already operational at the CMRR, University of Minnesota. An 11.7-T/90-cm bore whole-body system is in the process of being commissioned at Neurospin (France), and a 11.7-T/68-cm bore head-only magnet is waiting to be energized at the intramural research program of the National Institutes of Health (NIH) in the USA. All of these > 10 T magnets are still based on conventional niobium–titanium superconducting wire but operate at <4°K, achieved by pumping on their liquid helium reservoir. These magnets operate close to the limit of magnetic field that this superconducting wire can support. Initiatives exist for achieving a 14 T head size system, which would be a new frontier in magnet design and development.

The afore-listed >10 T initiatives are again justified primarily based on neuroimaging, particularly functional brain imaging in the case of the 10.5 T system in CMRR. Nevertheless, they will undoubtedly produce impressive improvements for imaging other organ systems, as already demonstrated by initial results from 10.5 T [23,25], as well as for multinuclear spectroscopy and imaging with magnetic nuclei other than protons.

4.2 Neuroimaging and UHF

The reason neuroimaging has been the main justification for the push to attain high magnetic fields resides in the difficult task we confront in studying the human brain. To quote Eric Kandel, "The central challenge of science in the twenty-first century is to understand the human mind in biological terms" [26]. This central challenge also extends to clinical problems: neuropsychiatric (mental, behavioral, and neurological) disorders are the primary cause of disability (measured as disability adjusted life years [DALYs] lost by cause) in the US healthcare system. However, our knowledge of such disorders is largely descriptive and the treatments mostly palliative. To this end, national initiatives, such as the Human Connectome Project (HCP) [27,28] and the BRAIN (**B**rain **R**esearch **T**hrough **A**dvancing **I**nnovative **N**eurotechnologies) Initiative [29], have called for the development of noninvasive technologies to map brain circuits toward achieving "progress in diagnosing, treating, and potentially curing neurological and psychiatric diseases and disorders that devastate so many lives." MRI is central to these initiatives as one of, if not the most powerful noninvasive methodology available to study the human brain; increasingly this will involve ultrahigh field magnets of 7 T and inevitably even higher magnetic fields.

The inadequacy of the existing methods for studying the brain that justified the launch of the BRAIN Initiative was laid out in an article authored by the leadership of NIH [29]. In this article, the authors stated: "Human neuroimaging captures the whole brain in action, but each 1-mm^3 voxel includes at least 80,000 neurons and 4.5 million synapses. An fMRI scan, with 680,000 voxels, is capturing local changes in blood flow and oxygen consumption – but these changes are low-resolution and slow surrogates for neuronal activity." Their criticisms were not just targeting the human imaging methods, but were also aimed at traditional and canonical invasive methods such as electrophysiology, pointing out its shortcomings with the words, "Neural circuits involve at least 106 cells in a complex, recursive network, but neurophysiology has been based classically on single-cell recordings or, more recently, on small ensembles of cells." In response to this challenge, the strategic plan titled "BRAIN2025: A scientific Vision," developed and submitted to the Director of NIH [30,31] by the 14-member BRAIN Initiative Working Group, was based on the fact that brain function is mediated by hierarchical local and long-range circuits organized at multiple spatial scales, and bridging and spanning these multiple scales of organization is essential for understanding brain function and ultimately dysfunction. Consequently, the report called for noninvasive, MRI-based brain mapping methods "to measure the wiring diagram and functional activity of the human brain at multiple scales – neuronal ensembles, circuits, and larger scale networks ('circuits of circuits')" and identified the need for technological developments with MRI to achieve "submillimeter spatial resolution descriptions of neuronal activity, functional and structural connectivity, and network analysis in the human brain through advances in instrumentation, data acquisition and analysis techniques" [30], establishing even a specific resolution limit of ≤0.1 microliter (μl) voxel volume to be achieved in the first phase of the BRAIN Initiative.

The 1-μl voxel volume, which was considered inadequate in the announcement of the BRAIN Initiative, is in fact a challenge for contemporary MR techniques that report on brain function and connectivity. The most advanced *whole-brain* MR data on connectivity has been produced by the HCP, which achieved significant advances in spatial and temporal sampling of fMRI and diffusion imaging (dMRI) signals [32,33], the latter employed for tractography-based structural connectivity in the brain. Yet, the 1200-subject, 3-T resting state fMRI (r-fMRI), task/stimulus-based fMRI (t/s-fMRI), and dMRI data collected in the HCP have 2 and 1.25 mm isotropic resolution for fMRI and dMRI, respectively, corresponding to 8 and 2 μl voxel volumes. The more advanced, 180-subject, 7 T HCP data accomplished higher resolutions of 1.6 mm isotropic (~4 μl volume) and 1.05 mm isotropic (~1.2 μl volume) for r-fMRI and dMRI, respectively [32–34]. In other efforts, with field-of-view (FOV) restriction, 7-T t/s-fMRI in the human brain has reached nominal voxel volumes of ~0.5–1 μl to permit the detection of activity in cortical layers (e.g. [35–43]) and columns (e.g. [36, 43–46]), and other fine-scale organizations (e.g. [47–51], see reviews [52,53]), albeit with marginal functional contrast-to-noise ratio (fCNR) and spatial resolution, adequate for demonstrating feasibility but not necessarily for easy use as a routine neuroscientific research tool. These accomplishments, as noteworthy and important as they are, nevertheless fall far short of the BRAIN Initiative target of whole-brain studies with of 0.1 μl or smaller voxel volumes [30]. Achieving such a goal will involve numerous complementary technologies but will certainly involve 7 T and even higher magnetic fields.

4.3 UHF and fMRI

The ≤0.1 μl goal set for the MR methods by the BRAIN Initiative is not an arbitrary target. At this spatial resolution, the number of neurons detected in an MR voxel reduces from hundreds of thousands in a typical 3 T study to a few thousand neurons typical of neuronal ensembles of the mammalian cortex that perform elementary computations, which are spatially clustered and repeated numerous times in each cortical area. A canonical example of these clusters is the orientation columns discovered by Hubel and Wiesel [54–56] for which a Nobel Prize was awarded in 1981. These organizations exist in three dimensions, showing clustered and repeating patches on the cortical surface but are also subject to a rich laminar architecture of neural circuitry across the cortex, which is ~2–3 mm thick in the human brain and is divided into six layers based on neuronal cell types. Long before the BRAIN Initiative was launched, however, efforts to reach functional imaging with MR at this submillimeter organizational scale were initiated [57–61]. These prescient efforts ultimately led to the introduction of 7 T and ultimately the mapping of orientation columns in the human brain for the first time [45]. In a way, the fMRI efforts pursued already between ~1996 and ~2000 anticipated the BRAIN Initiative launched in 2013 and ultrahigh magnetic field of 7 T was at the center of these efforts.

However, submillimeter resolution fMRI to image at the level of cortical columns and layers is not the only reason for pursuing UHF fMRI. Even at conventional, supra-millimeter resolutions employed routinely for functional mapping at 3 T, UHF can provide significant gains in being able to detect functional maps and alterations of such maps in their topography and/or response magnitudes with cognitive processes [52]. This is particularly important for building computational models of brain function and evaluating their predictive power. For example, in a study conducted at both 3 and 7 T using $1.5 \times 1.5 \times 1.5\ mm^3$ (3.4 μl voxel volume) resolution and aimed at evaluating computational models of natural sound representation in the human brain, prediction accuracy of

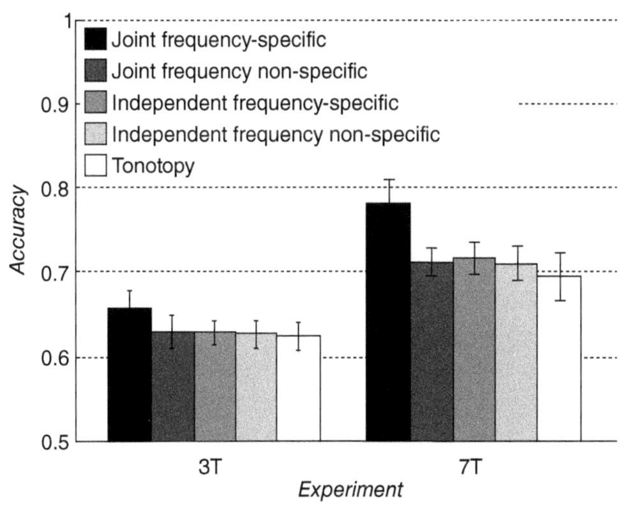

Figure 4.2 Prediction accuracy for five models of how natural sounds are represented in the human brain as measured in both 3 T and 7 T fMRI experiments acquired with $1.5 \times 1.5 \times 1.5\ mm^3$ resolution. All models did better at 7 T. The joint frequency-specific MTF-based model showed significantly better prediction accuracy than all other models, particularly at 7 T. Accuracies are normalized between 0 and 1. Chance level is 0.5. Reproduced from [62].

all models considered was significantly higher at 7 T (Figure 4.2). The 7-T data permitted the identification of the most accurate model with less ambiguity; remarkably, the best performing model was able to generalize to stimuli not used for parameter estimation only at 7 T [62]. Other examples of cognitive tasks studied with supra-millimeter resolution fMRI where a field-dependent comparison was performed include a mental maze solving task performed at 4 and 7 T, with 7.3 and 5.2 μl voxel volumes, respectively, demonstrating that 7 T [63] provided much finer detail of the directional tuning in the superior parietal lobule (SPL), as compared with those obtained at 4 T [64]. A more recent work, comparing a computational cognitive neuroscience paradigm at 3 T and 7 T using a resolution of $1.98 \times 1.98 \times 2.0\ mm^3$ (7.8 μl voxel volume) at both field strengths, found that the model predictive power on a numerosity task was on average four times higher at 7 T than at 3 T [65]. These are just a few examples of a growing number of similar observations demonstrating superiority of UHF fMRI data in tackling cognitive neuroscience questions.

A different type of study carried in the HCP examined functional connectivity with whole-brain, supra-millimeter resolution resting state fMRI; carefully conducted measurements performed on 24 individuals scanned both at 7 T and 3 T demonstrated that the contrast-to-noise ratio was higher at 7 T, permitting robust connectivity measures where none was detectable at 3 T [66]; this was the case despite the use of a higher (1.6 mm isotropic) resolution at 7 T, versus 2 mm isotropic at 3 T. This resolution difference translates into a factor 2 smaller voxel volume at 7 T, and consequently a factor 2 lower image signal-to-noise ratio (SNR).

Thus, whether one is performing fMRI with conventional supra-millimeter resolutions typical of contemporary studies carried out at 3 T, or pushing into the submillimeter resolution mesoscopic scale, 7 T has proven to be significantly advantageous relative to 3 T in functional mapping in the human brain. These UHF advantages originate from several important gains: (i) higher BOLD contrast (e.g. [10,67–69]); (ii) higher fCNR and image SNR; (iii) improved spatial fidelity of functional mapping signals to neuronal activity (see review [70]); and (iv) significantly improved parallel imaging for higher accelerations [71–74] for more accurate temporal sampling of the fMRI time series [75].

The most commonly used functional imaging approach relies on T_2*-weighted gradient echo (GE) echo planar imaging (EPI) readouts. In such a sequence, functional mapping signals predominantly originate from the BOLD mechanism, particularly at high magnetic fields [10]; this mechanism relies on the magnetic susceptibility difference between intravascular space containing deoxyhemoglobin and surrounding tissue and, as such, is amplified with magnetic field magnitude. Models of the extravascular BOLD effect predict supralinear gains for microvasculature ($\lesssim$20 μm diameter blood vessels) and linear gains for larger blood vessels containing deoxyhemoglobin (e.g. [10,67–69]) with increasing magnetic fields. These early BOLD models are largely accurate, albeit incomplete with respect to identifying the sources of the fMRI signals. Some of the unaccounted for contributions, such as intravascular BOLD [76–78] and inflow effects associated with slice-based image acquisitions [79–81], form the basis for the successful use of lower field magnets, such as 1.5 T, for functional mapping; however, these effects are highly nonspecific to the regions of altered neuronal activity that elicit them, and, when acquired at high resolution, look like functional "angiography" depicting extracerebral blood vessels rather than parenchymal neuronal activation [6]. Nevertheless, a more comprehensive understanding of the origins of fMRI signals ([78] and references therein) has also reconfirmed our initial expectations that ultrahigh fields provide major advantages for BOLD contrast particularly for the microvasculature.

Ultimately the success of functional mapping, whether the aim is to improve spatial resolution or to minimize the errors at a given spatial resolution, is not defined by just the magnitude of the BOLD effect but by fCNR. For task- or stimulus-based fMRI studies, fCNR can be cast as $(\Delta S / S)(S / N)$ where $(\Delta S / S)$ is the stimulus- or task-induced fractional or percentage signal change (i.e. the BOLD effect) and (S / N) represents SNR measured as the mean signal divided by the "noise" in the form of standard deviation of the signal fluctuations in a detrended fMRI time series during baseline periods (also referred to as temporal SNR (tSNR)). However, the nature and origin of what constitutes "noise" in an fMRI time series is a complex question. Thermal noise associated with the MR detection process [82,83], arising from the electronics and/or the interaction of the RF coil with the sample, is an important noise source in fMRI; this is a "white" or "random" noise source and would be zero-mean Gaussian distributed. Most contemporary MR acquisitions, including those employed for fMRI, use parallel imaging (PI) to accelerate acquisitions. The use of PI introduces a spatially nonuniform amplification of the "thermal" noise by the g-factor [84]. The conditions under which thermal noise, with or without g-factor amplification, becomes dominant in an fMRI time series depend on the static magnetic field strength, the voxel volume, and image repetition time (TR) used in the experiment, becoming more prominent at higher resolutions (i.e. smaller voxel volumes), short TRs, and/or lower magnetic fields [85,86]. It is the *dominant* contribution at ~0.5 μl voxel volumes (e.g. ~0.8 mm isotropic dimensions) typically employed in high-resolution 7 T fMRI studies; it remains dominant at 7 T up to ~10 μl voxel volumes, gradually plateauing beyond that [85,86].

In the limit of high-resolution and/or rapidly acquired data, when thermal noise dominates temporal signal fluctuations [85–87] in an fMRI time series, fCNR is directly proportional to the product of neuronal activity-induced fractional signal change ($\Delta S/S$) and intrinsic image SNR [70]. In other words, tSNR is dominated by image SNR. ($\Delta S/S$) was modeled as a function of B_0, in good agreement with experimental data [78]. The field dependence of image SNR is complex in the UHF regime (discussed in greater detail in the next section) but was reported to increase as ${B_0}^{1.65}$ between 3 T and 9.4 T in the human head for a 32-channel receive array [88], consistent with a greater than twofold SNR gain reported between 7 and 10.5 T in the torso of human-size pigs with array coils [89]. Thus, the anticipated fCNR–B_0 relationship for fMRI can be calculated using the ${B_0}^{1.65}$ dependence on image SNR and previously modeled ($\Delta S/S$), and displays dramatic gains with increasing B_0 [70,78,90–92] (Figure 4.3).

Physiological processes of respiration and cardiac pulsation (e.g. [93–99]) also constitute a source of "noise" in an fMRI time series that degrades the tSNR. In addition, for task/stimulus fMRI, the spatially correlated spontaneous fluctuations ascribed to functional networks in rs-fMRI [100] represent yet another source of "noise" in an fMRI time series [101]. Unlike thermal noise, however, all of these noise sources are nonwhite in nature [93–96] (i.e. they are structured) and are clearly detectable in the Fourier transform of the fMRI time series when the SNR is sufficiently high and the sampling rate is sufficiently fast (e.g. [102]). The fluctuations induced by these sources are proportional to signal magnitude [86,87,103–105]. As such, they become dominant only when a voxel's signal (which is proportional to voxel volume) is large compared with instrumental thermal noise, as encountered, for example, when combinations of low spatial resolutions, high flip angles, long TRs, and high magnetic fields [85,86,106,107] are used. Thus, under certain conditions, UHF fMRI can suffer more from the structured noise in the fMRI time series relative to lower magnetic fields, blunting any field-dependent gains for functional mapping [85,86]. However, there are numerous methods for suppressing these structured noise components (e.g. review [94]) and their use must be considered when operating in a regime where they are the dominant contribution to the "noise" in an fMRI time series.

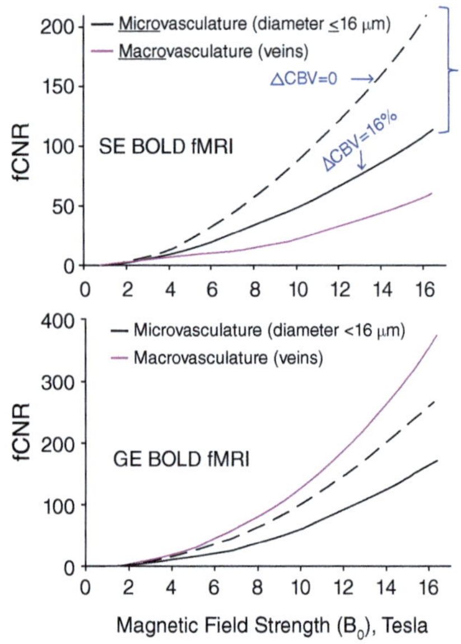

Figure 4.3 Functional contrast-to-noise ratio (fCNR) as a function of B_0 (normalized to 1.5 T). ΔCBV = change in blood volume, calculated assuming either 0 or 16% change for microvasculature (black) (arterioles, capillaries, and venules) and ΔCBV = 0 for macro-vasculature. Adapted from [70], and [3,78].

One of the major advantages of ultrahigh fields is increased spatial accuracy of fMRI signals. The most commonly used approach in fMRI applications, i.e. T_2*-weighted GE EPI-based acquisitions, suffers from spatial inaccuracies in functional mapping because of large draining vein contributions, as previously mentioned. Often one is concerned with the large blood vessels on the pial surface of the brain. This major confound was realized and documented early in the history of fMRI [5,7,8,108] and was recently further reemphasized [37]. GE fMRI studies would also have a strong contribution from ~50 to ~100 μm diameter intracortical veins that run perpendicular to the cortical surface and drain the different layers. These are the veins that were visualized in the original BOLD imaging work conducted in rats [109]; they can also be seen in very high resolution human brain images with T_2* or phase contrast (e.g. [110–112]). These intracortical small veins can be limiting to laminar-specific studies [113–115] although they would certainly produce signals more accurate than the draining pial veins.

The contribution of these large vein effects in BOLD-based functional imaging increasingly becomes a major source of error as fine-grained structures are targeted with high-resolution mapping (e.g. see [70]). Therefore, either they have to be suppressed in the acquisition stage by using alternative functional contrast encoding such as spin echo BOLD rather than gradient echo BOLD (review [70]), or eliminated with post-acquisition data processing strategies [116,117]. Irrespective of the strategy employed, however, the data must contain parenchymal functional mapping signals so that when the large vein contributions are suppressed, there is enough signal left associated with the parenchyma that can be used for functional mapping of the desired functional feature. In this respect, UHF provides definitive advantages (Figure 4.3) as exemplified by the SE BOLD-based functional mapping of orientation domains and ocular dominance columns [45,46]. This UHF advantage will continue to increase beyond 7 T and provides justification for pursuing very high magnetic fields like 10.5 T and 11.7 T.

4.4 SNR, Increasing Magnetic Fields, and Dense Arrays

One of the primary motivations for going to ultrahigh magnetic fields despite the large escalation in cost associated with building such magnets is the field-dependent increases in SNR. Such SNR improvements have been experimentally demonstrated (e.g. [118,119]) as previously mentioned. However, field dependence of SNR is complex, especially at UHF where the traveling wave characteristics of the RF produce substantial $\boldsymbol{B}_1$ inhomogeneities characterized by the sample size and geometry and RF coil layout [12]. Even without this UHF complexity, however, field dependence of SNR has had a complex history.

Initial efforts to develop MRI [120,121] were carried out at magnetic fields of ~0.05 to 0.35 T (e.g. [122,123]), likely because of the cost and difficulty of building high-field magnets that can accommodate humans, and the difficulties anticipated in achieving the penetration depth of the RF frequency needed to excite and detect the water protons in the conductive human body. Early theoretical calculations suggested that "frequency of operation of the spectrometer with human samples should be less than about 10 MHz [~0.24 T]" [124]. These early general concerns, though not the exact frequency limits that were prescribed, persisted even at the time 7 T was being introduced. However, catalyzed largely by the effort undertaken to exploit 7 T and the effort to understand the physics of working at ~300 MHz and beyond, today we have a much better understanding of SNR as a function of field strength.

Electromagnetic modeling studies predict spatially nonuniform (much higher in the periphery than in the center in a human head-like object) and spatially dependent linear to supralinear increases in ultimate intrinsic SNR (uiSNR) [74,125,126]. Capturing this uiSNR advantage requires high-density array coils composed of multiple distributed small receive elements for signal reception [127–129]. Modeling studies predict that for unaccelerated imaging, ~32 receive coils capture most of the uiSNR in the center of the human head; at the periphery, the uiSNR is intrinsically much higher and approaching it requires going significantly beyond 32 elements [129]; such high channel count arrays perform better at UHF due to increased coupling to the sample even though the coil dimensions become smaller [129]. Thus, the experimentally measured SNR will result from the combination of magnetic field effects and the RF coil employed to measure them. For example, in the first measurement exploring SNR at 7 T relative to a lower magnetic field (4 T), a single-channel volume coil was employed at both field strengths [130]. This arrangement provided supralinear gains in SNR in the center of the brain but significantly less than linear gains in the periphery. This pattern was demonstrated to be simply a result of the destructive interference of the $\boldsymbol{B}_1$ fields produced by the different current-carrying elements of the coil and not intrinsic to the magnetic field [13]. Measurements performed at 3, 7, and 9.4 T with a 32-channel receive array reported a relatively uniform $\boldsymbol{B}_0^{1.65}$ dependence over the human head [118]; this is also likely a consequence of combining coil- and field-dependent effects. In this case, most of the uiSNR in the center of the brain is expected to be captured, while the coil fails to do the same in the periphery with only 32 channels; a larger number of coils in the array would likely boost the peripheral SNR gains with increasing magnetic field in principle; however, as the number of coils in the array covering the same surface area increases, coil dimensions decrease and such small coils do not perform as well at low magnetic fields.

Currently, the highest magnetic field starting to be employed for human imaging is 10.5 T [25,131], and progress at this field strength is advanced enough to provide preliminary hints at what kinds of SNR gains may be feasible in this new >10 T regime in the human head. Table 4.1 shows preliminary SNR measurements at 10.5 T relative to 7 T in a "phantom" mimicking human head electrical

Table 4.1 Preliminary data on the ratio of SNR at 10.5 T to SNR at 7 T measured in a phantom mimicking human head size and electrical properties using 32- and 64-channel receive array coils at 7 T and 10.5 T, respectively.

d, depth (mm)	Peripheral VOI SNR	Central VOI SNR	Overall average SNR
10	7.70	2.55	3.45
20	5.42	2.23	
30	4.43	2.14	

properties and approximate dimensions. Relative SNR gains are reported in a peripheral and a central volume of interest (VOI) integrated over the entire sample, demarcated with a boundary defined by the parameter *d* designating the distance from the surface of the phantom. The comparison is between an industry standard 32-channel array at 7 T versus a 64-channel array at 10.5 T. In the center of this phantom, there is an SNR gain of ~2, consistent with the experimentally determined SNR $\sim B_0^{1.65}$ dependence previously reported [118]. On the other hand, in the periphery, where the uiSNR is higher, significantly larger gains are observed depending on the depth (i.e. the thickness) of the peripheral volume of interest. Preliminary measurements with a 128-channel array suggest that another factor 2 gain may be achieved in the periphery [132] though there are certainly challenges in achieving such high count arrays without compromising performance. These peripheral gains come from the increase in the field magnitude *as well as* the ability to capture this uiSNR better by going to higher channel counts at these very high magnetic fields.[1]

High resolution requires both high SNR and high acceleration factors achieved by PI with minimal g-factor penalties; the latter is critical for high-resolution coverage of the whole brain or large parts of the brain and is employed to significantly reduce $\boldsymbol{k}$-space sampling [84,133,134] and/or the number of slices needed to cover a given volume [32,135–137]. As previously mentioned, there is a strong interplay between image SNR, B_0, and PI performance. Both SNR and PI performance improve significantly with increasing B_0 [71–74]; they also improve with increasing number of coil elements in a receiver array and this gain itself is more pronounced at higher B_0 [129]. PI is an indispensable technology for high-resolution imaging enabled by the higher SNR of ultrahigh fields. Therefore, in realistic acquisition schemes, one must confront not the SNR in *unaccelerated* images, but SNR with g-factor noise amplification in *accelerated* images; fortuitously, the g-factor is diminished at UHF [72], and/or by using larger number of elements in the array. Thus, again, synergistic use of high magnetic fields and a large number of receive coils can improve experimentally achievable SNR everywhere in the sample. This was demonstrated recently for 7 T [128] (Figure 4.4), and additional gains would be predicted for magnetic fields >10 T. Therefore, it is natural to contemplate even higher channel count receive arrays, for example 96 or 128 channels, at very high magnetic fields such as 10.5 T and beyond.

1 Data on the human head at 10.5 T are not yet available because 10.5 T operates under a US Food and Drug administration (FDA) Investigational Device Exemption (IDE) and all RF coils must have rigorous safety testing and approval by the FDA prior to use in the human head.

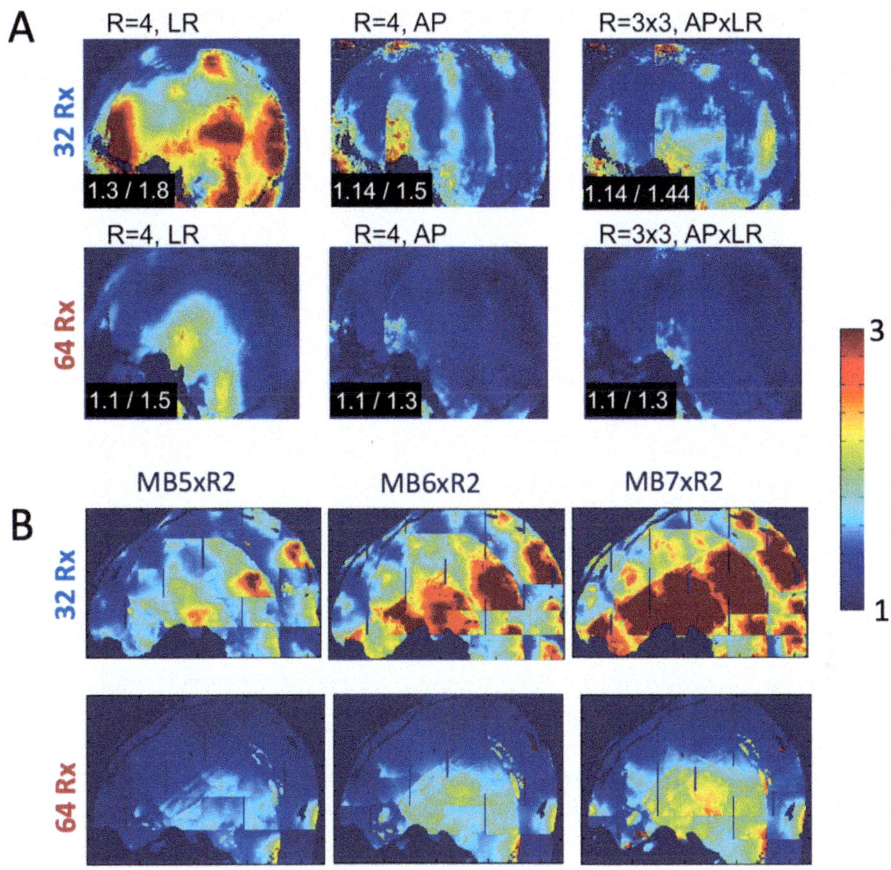

Figure 4.4 Noise amplification due to parallel imaging (g-factor) for two different arrays, one with 64 and other with 32 receive channels (Rx). Data obtained at 7 T. g-actors are presented as a maximum intensity projection onto a sagittal slice, for combinations of slice and phase encode acceleration given as MB and R factors, respectively. Panel A shows one-dimensional fourfold acceleration (R = 4) with the reduction applied in the left–right (LR) or anterior–posterior (AP) dimension and a two-dimensional 3 × 3 reduction. High g-factor (i.e. red colors mean high noise amplification). The numbers given in the lower left corner of each image report the mean and maximum g-factor values. Panel B displays accelerations for a simultaneous multislice acquisition with phase encode undersampling. For example, MB5xR2 implies 10-fold two-dimensional acceleration achieved with simultaneous excitation and imaging of five slices with twofold undersampling in the phase encode dimension. [128] Kamil Uğurbil et al. (2019), figure 05 & 06[p.08]/with permission from John Wiley & Sons, Inc.

4.5 High Performance Gradients

High performance gradients with high maximal gradient strength and high slew rates are also critical technologies for UHF MRI. High performance gradients are usually discussed in the context of diffusion imaging for tractography (dMRI), though fMRI also benefits from gradient performance because of the impact on $\boldsymbol{k}$-space encoding with fast techniques like EPI and spiral trajectories.

In dMRI, diffusion weighting occurs typically during a spin echo with echo time TE with gradient pulses embedded in the intervals surrounding the refocusing pulse(s). During the echo time TE, SNR is lost according to e^{-TE/T_2}. Because of this loss, and the fact that T_2 gets shorter with higher magnetic fields (e.g. [78]), there is a general belief in the dMRI community that dMRI is best performed at 3 T; logical extension of this argument would suggest that 1.5 T would be even better. However, this is a completely erroneous argument for the same reason 3 T is better than 1.5 T and that no one is advocating for conducting dMRI studies at 1.5 T. Essentially, at sufficiently short *TE*s, the losses due to the shorter T_2 of the higher magnetic field can be substantially less than the intrinsic SNR gain available from the higher magnetic field strength.

Figure 4.5 demonstrates a simple calculation of relative SNR as a function of *TE* for 10.5 T versus 3 T as a function of echo time *TE*; in other words, it displays plots of $(Ce^{TE(1/T_2^{10.5T}-1/T_2^{3}T)})$ where C is the ratio of SNR at 10.5 T versus 3 T at $TE = 0$, and $T_2^{10.5T}$ and T_2^{3T} represent the T_2 at 10.5 T and 3 T, respectively. In this figure, C was taken either as (10.5/3) or as $(10.5/3)^{1.65}$ in accordance with the ${B_0}^{1.65}$ dependence previously reported [118]; as discussed in Section 4.4, based on data presented in Table 4.1, a ${B_0}^{1.65}$ gain is expected in a large center volume of the head with much larger gains in the periphery coming from field dependent gains and the use of high-density receive arrays. A plot for $C = (10.5/3)$, representing a linear gain in SNR with field magnitude, also demonstrate that even according to this conservative estimate, which has been considered in the early days of UHF, and may be applicable for uiSNR (though not necessarily the experimentally realizable SNR) in the periphery, the higher magnetic field has advantages with *TE*s less than ~100 ms.

In a real diffusion experiment, *TE* will be determined by the number of refocusing pulses used and the highest *b* value employed in diffusion encoding and not by the magnetic field of the experiment. In the past, double refocused sequences were employed to suppress the deleterious consequences of eddy currents induced during diffusion encoding, leading to long *TEs*. However, postacquisition correction of eddy currents with a single refocused dMRI sequence was introduced in the HCP and is now the preferred method of diffusion encoding [139,140]. The HCP 3 T dMRI protocol employed a maximum *b* value of 3000 s/mm^2 [139]. Using a gradient set with maximal gradient (G_{max}) of 70 mT/m, and 200 slew rate (Siemens SC72 gradient used for 7 T scanners), the *TE* for diffusion encoding with $b = 3000$ s/mm^2 was 75 ms [141]. Therefore, even using such a "standard" gradient at 10.5 T would give an approximately fourfold SNR advantage to 10.5 T relative to 3 T.

There is no impediment to achieving even shorter *TEs* with higher performance gradients in ultrahigh field systems, for example by using gradients with performance specifications that match the 3 T "connectome" scanner [142] or better. Such high performance gradients can be "head gradients" that can provide high G_{max} as well as high maximal slew rates, without being limited severely by peripheral nerve stimulation encountered with whole-body gradients and without needing amplifiers beyond what is currently available for driving existing whole-body gradients. Head gradients are also advantageous with respect to gradient magnet interactions. Such head gradients are gaining recent attention and popularity. A head gradient for a compact 3 T scanner was reported to achieve 80 mT/m G_{max} and 700 T/m/s slew rate using commonly available 1 MVA drivers; significant improvements in EPI and dMRI were demonstrated with these gradients as would be expected [143]. Another head gradient effort reported dMRI data at 3 T using $G_{max} = 100$ mT/m and 600 mT/m maximum slew rate, achieving 19 ms echo time for $b = 1000$ s/mm^2 [144]. Head gradients with such performance can easily be developed and used at ultrahigh magnetic fields as well. A Siemens head gradient with 200 mT/m G_{max}, 700 T/m/s slew rate performance based on 2200 V/1200 A drivers has recently been installed on the 7 T system at the University of California–Berkeley. In collaboration with Stanford University (Professor Brian Rutt), we are in the process of building a head gradient for the 10.5 T system in the CMRR that will achieve 175 mT/m G_{max} and 900 T/m/s maximum slew rate with 950 A, 2250 V drivers. With such gradient sets, the echo times will get shorter, and gains in dMRI will increase for 7 T and higher magnetic fields relative to 3 T equipped with similar gradients. For example, with the 19 ms *TE* achieved for $b = 1000$ s/mm^2, 10.5 T would provide approximately sevenfold SNR gain over 3 T (Figure 4.5).

Thus, all else being equal, when the *TE*s are kept short so that the plots in Figure 4.5 remain above the relative SNR ratio of 1, very high magnetic fields such as 10.5 T will retain an advantage for dMRI. However, one could argue that all else is not equal. Likely one will need to use PI and undersample the ***k***-space in the phase encode dimension to keep the echo train length short at the high magnetic field, while it is possible to get away with full ***k***-space sampling at 3 T. In the HCP, for example, 3 T dMRI data were acquired without ***k***-space undersampling [139] but factor 3 undersampling (i.e. acceleration) in the phase encode dimension was used in 7 T dMRI acquisitions [145]. Although this would lead to an SNR loss by square root of the acceleration factor, it would also allow a shorter *TE*, making up for some of the SNR loss due to PI undersampling. However, with high slew rate gradients EPI encoding will also improve, thus shortening the echotrain length and decreasing the need for the use of extensive ***k***-space undersampling along the phase encode direction.

Even without using these high performance gradients, superior 7 T dMRI results were demonstrated in the HCP [145,146] that had a 3 T and a 7 T component, and also in post HCP 7 T dMRI data acquired using the same three *b* shells employed in the HCP dMRI 3 T protocol [141]. With the introduction of high performance gradients into UHF systems, the gains in UHF will outstrip those at lower magnetic fields and the best dMRI scanners will ultimately be operating at very high magnetic fields.

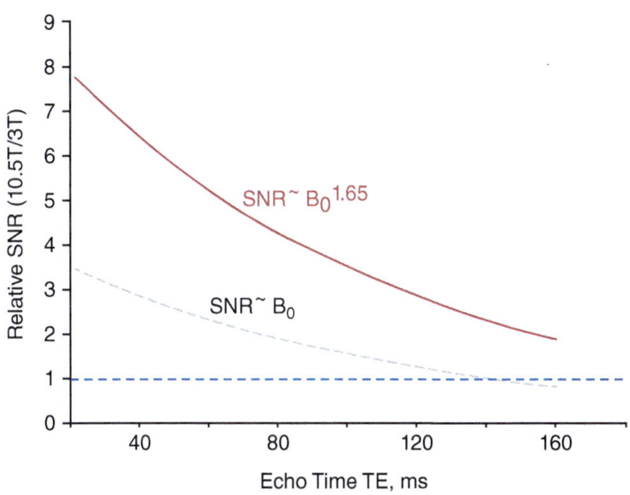

Figure 4.5 Signal-to-noise ratio (SNR) at 10.5 T versus 3 T during a decay of the traverse magnetization with their respective T_2s. The 10.5 T T_2 was estimated to be 38 ms from T_2 data compiled as a function of field strength in [78]. The T_2 at 3 T was reported to be between 55 and 69 ms ([138] and references therein); for the calculation presented in this figure, it was assumed to be the average of this range, i.e. 62 ms. The relative SNR at TE = 0 represents the intrinsic higher SNR available at the higher magnetic field. This field-dependent SNR gain was taken as $B_0^{1.65}$ (red curve) (see discussion in the paper) or to increase linearly with field magnitude (gray dashed line).

4.6 Imaging the Human Torso at UHF

One of the most significant challenges faced at UHFs is the transmit $\boldsymbol{B}_1$ ($\boldsymbol{B}_1^+$) field heterogeneities, which scale in complexity with the size of the imaging region of interest. Such complex $\boldsymbol{B}_1$ behavior was clearly visible in early efforts at 4 T [147], at a time when high-end clinical MRI scanners operated at 1.5 T. Originally, these inhomogeneities were often ascribed to and described as "dielectric resonances." Ultimately, however, this RF phenomenon was characterized not as a dielectric "resonance" per se but as arising from attenuated traveling wave behavior displayed in the electrical milieu of the human body when RF wavelength becomes shorter than the dimensions of the object to be imaged [12]. In this regime, significant $\boldsymbol{B}_1^+$ distortions occur due to spatial phase variation and amplitude modulation caused by the interference of RF traveling waves determined by the tissue electromagnetic properties, object size and geometry, and the coil configuration [148,149].

In the human torso, the $\boldsymbol{B}_1^+$ behavior becomes even more complex because of the large body dimension relative to the wavelength (approximately 11 cm at 7 T). The ability to manipulate the transmit RF using multichannel transmit with channel-specific independent control [112,150,151] is ultimately necessary to overcome this problem. The multichannel approach, particularly combined with local transmit coil arrays placed in close proximity to the torso, introduced the ability to maximize and/or homogenize the $\boldsymbol{B}_1^+$ over a predetermined region of interest (ROI), enabling imaging of targeted organs in the human torso at 7 T for the first time [151–158]. Since then, multichannel parallel transmit techniques and local multichannel transmit and receive array coils have been adapted and utilized for human torso studies by numerous groups (e.g. [159–171]).

The gains already achieved for imaging the human body at 7 T will undoubtedly continue to grow as the installed base of the latest generation 7 T scanners increase. The technologies developed for 7 T also permit imaging the torso at much higher magnetic fields. Local transmitters close to the surface of the body have been the preferred approach for 7 T imaging in the human torso. Following this approach, a 10-channel transceiver dipole array [172] for 10.5 T was straightforward to model in electromagnetic simulations, which were subsequently validated experimentally in phantoms, leading to approval of this RF array for human imaging by the US Food and Drug Administration. Using this and similar RF coils, human imaging in the torso at 447 MHz [23,25,173] was possible as an additional component to an ongoing 10.5-T safety study [174].

The primary hurdles faced in such an effort are the transmit $\boldsymbol{B}_1$ inhomogeneities and operation within safe power deposition (i.e. specific absorption ratio [SAR]) limits. The feasibility of overcoming these hurdles was demonstrated using electromagnetic simulations and experimental data obtained with a human torso-shaped phantom, comparing the 10-channel 10.5-T transceiver dipole array with a similar antenna array configuration at 7.0 T [172]. It was shown that combining a two-spoke pulse design with the 10.5 T dipole array provided comparable (when imaging kidneys or heart) or even better (when imaging prostate) RF performance than single-spoke designs at 7 T with respect to $\boldsymbol{B}_1$ inhomogeneities and peak SAR. This work also showed that SNR gains were greater than twofold at 10.5 T compared with 7 T [172].

Examples of 10.5 T imaging in the human torso have been published [23,25], demonstrating that volumetric imaging with excellent results is possible in the human torso even at 10.5 T. A collection of images at the level of the human hip and of the heart are reproduced in Figure 4.6.

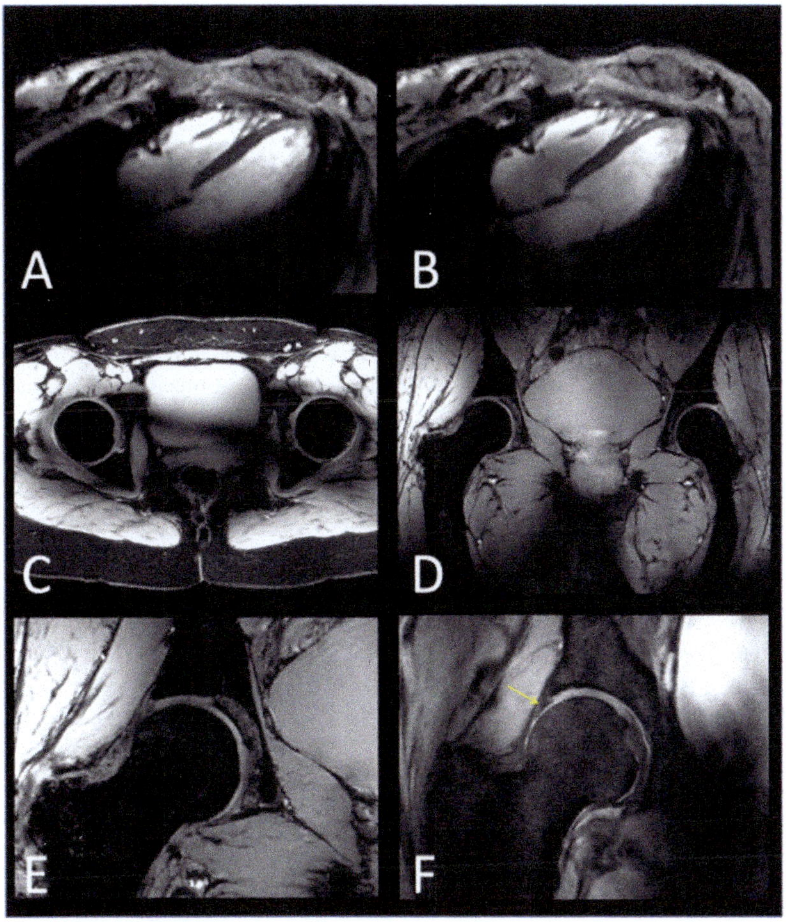

Figure 4.6 Initial images obtained from the human torso at 10.5 T using a 10-channel transmit and receive dipole array. Top row, 10.5-T cardiac images obtained with (A) phase-only B_1 shim and (B) 1spoke pTx RF pulses, retrospectively gated gradient-recalled echo cine acquisition in the four-chamber view. Lower two rows, anatomic hip imaging: (C) An axial multislice 2D gradient echo acquisition obtained with fat saturation. (D) 3D coronal MEDIC acquisition. (E) Zoomed version of the right femoral head from the MEDIC (D) acquisition. (F) A proton-density (PD) weighted turbo spin echo (TSE) acquisition showing the expected contrast between the labrum and cartilage (yellow arrow). [25] He et al. (2020), figure 04 & 08[p.007 & 010]/with permission from John Wiley & Sons, Inc.

Imaging of the human torso at high magnetic fields was considered almost an impossible task, a concept that was cemented by the early, virtually unrecognizable images of the torso first reported at 4 T [147]. Following the launch of 7 T, however, critically important strides have been achieved in our understanding of the physics of RF interactions in the human body, which in turn has led to solutions for overcoming the difficulties encountered at such ultrahigh magnetic fields. As a result, the original poor image quality at 4 T has progressed to the high-quality and high-resolution images obtained at 7 T, and now at 10.5 T in the human torso. Despite these tremendous advances, however, work remains to further improve SNR and contrasts in these UHF images, and fully capitalize on the potential advantages UHF has to offer.

4.7 Conclusions

The need to expand the information content and the spatial resolution of MR images to biologically relevant scales has propelled the development of magnets operating at 7 T and higher field strengths with large enough bores to accommodate humans. At these high magnetic fields, imaging in the human body encountered severe challenges due to the RF wavelengths becoming shorter than the dimensions of the object being imaged, thus departing from the comfortable near-field operating regime of MR systems used in clinical practice. However, numerous technological and methodological advances have demonstrated that these challenges can be overcome, leading to previously unattainable results in the human brain as well as in other organ systems. As a result, we witness today a new ultrahigh field effort aimed at magnetic fields >10 T for whole-body imaging. There is still much work to accomplish toward optimization of technologies and image acquisition strategies for these >10-T systems; however, most of what still needs to be accomplished is well understood and attainable with time. Inevitably, this effort will also produce new techniques as well as complementary image reconstruction and image analysis methods. Together, these developments will certainly usher in a new era of biomedical discovery.

Acknowledgments

The work reported in this article from the author's laboratory was supported by NIH grants P41 EB027061, P41 EB015894, U01 EB025144, P30 NS076408, and the HCP (U54 MH091657)

References

1 Uğurbil, K. (2012). The road to functional imaging and ultrahigh fields. *Neuroimage* 62 (2): 726–735.
2 Uğurbil, K. (2018). Imaging at ultrahigh magnetic fields: History, challenges, and solutions. *Neuroimage* 168: 7–32.
3 Ogawa, S., Tank, D.W., Menon, R. et al. (1992). Intrinsic signal changes accompanying sensory stimulation: Functional brain mapping with magnetic resonance imaging. *Proceedings of the National Academy of Sciences of the United States of America* 89 (13): 5951–5955.
4 Kwong, K.K., Belliveau, J.W., Chesler, D.A. et al. (1992). Dynamic magnetic resonance imaging of human brain activity during primary sensory stimulation. *Proceedings of the National Academy of Sciences of the United States of America* 89 (12): 5675–5679.
5 Menon, R.S., Ogawa, S., Tank, D.W. et al. (1993). 4 Tesla gradient recalled echo characteristics of photic stimulation-induced signal changes in the human primary visual cortex. *Magnetic Resonance in Medicine* 30 (3): 380–386.
6 Segebarth, C., Belle, V., Delon, C. et al. (1994). Functional MRI of the human brain: Predominance of signals from extracerebral veins. *Neuroreport* 5 (7): 813–816.
7 Lai, S., Hopkins, A.L., Haacke, E.M. et al. (1993). Identification of vascular structures as a major source of signal contrast in high resolution 2D and 3D functional activation imaging of the motor cortex at 1.5T: Preliminary results. *Magnetic Resonance in Medicine* 30 (3): 387–392.

8 Haacke, E.M., Hopkins, A., Lai, S. et al. (1994). 2D and 3D high resolution gradient echo functional imaging of the brain: Venous contributions to signal in motor cortex studies [published erratum appears in *NMR Biomed* 1994 Dec;7(8):374]. *NMR in Biomedicine* 7 (1–2): 54–62.

9 Hoogenraad, F.G., Hofman, M.B., Pouwels, P.J. et al. (1999). Sub-millimeter fMRI at 1.5 Tesla: Correlation of high resolution with low resolution measurements. *Journal of Magnetic Resonance Imaging* 9 (3): 475–482.

10 Ogawa, S., Menon, R.S., Tank, D.W. et al. (1993). Functional brain mapping by blood oxygenation level-dependent contrast magnetic resonance imaging. A comparison of signal characteristics with a biophysical model. *Biophysics Journal [Comparative Study]* 64 (3): 803–812.

11 Yacoub, E., Shmuel, A., Pfeuffer, J. et al. (2001). Imaging brain function in humans at 7 Tesla. *Magnetic Resonance in Medicine: Official Journal of the Society of Magnetic Resonance in Medicine/Society of Magnetic Resonance in Medicine* 45 (4): 588–594.

12 Yang, Q.X., Wang, J., Zhang, X. et al. (2002). Analysis of wave behavior in lossy dielectric samples at high field. *Magnetic Resonance in Medicine* 47 (5): 982–989.

13 Van de Moortele, P.F., Akgun, C., Adriany, G. et al. (2005). B(1) destructive interferences and spatial phase patterns at 7 T with a head transceiver array coil. *Magnetic Resonance in Medicine* 54 (6): 1503–1518.

14 Uğurbil, K. (2014). Magnetic resonance imaging at ultrahigh fields. *IEEE Transactions on Bio-medical Engineering* 61 (5): 1364–1379.

15 Umutlu, L., Ladd, M.E., Forsting, M. et al. (2014). 7 Tesla MR imaging: Opportunities and challenges. *RöFo* 186 (2): 121–129.

16 Trattnig, S., Bogner, W., Gruber, S. et al. (2016). Clinical applications at ultrahigh field (7T). Where does it make the difference? *NMR in Biomedicine* 29 (9): 1316–1334.

17 Kraff, O. and Quick, H.H. (2017). 7T: Physics, safety, and potential clinical applications. *Journal of Magnetic Resonance Imaging* 46 (6): 1573–1589.

18 Niendorf, T., Sodickson, D.K., Krombach, G.A. et al. (2010). Toward cardiovascular MRI at 7 T: Clinical needs, technical solutions and research promises. *European Radiology* 20 (12): 2806–2816.

19 Niendorf, T., Paul, K., Oezerdem, C. et al. (2016). W(h)ither human cardiac and body magnetic resonance at ultrahigh fields? Technical advances, practical considerations, applications, and clinical opportunities. *NMR in Biomedicine* 29 (9): 1173–1197.

20 Stab, D., Al Najjar, A., O'Brien, K. et al. (2019). Cardiac magnetic resonance imaging at 7 Tesla. *JoVE* (143). doi:10.3791/55853.

21 De Boer, A., Hoogduin, J.M., Blankestijn, P.J. et al. (2016). 7 T renal MRI: Challenges and promises. *MAGMA* 29 (3): 417–433.

22 Erturk, M.A., Li, X., Van de Moortele, P.F. et al. (2019). Evolution of UHF body imaging in the human torso at 7T: Technology, applications, and future directions. *Topics in Magnetic Resonance Imaging* 28 (3): 101–124.

23 Uğurbil, K., Van de Moortele, P.F., Grant, A. et al. (2021). Progress in imaging the human torso at the ultrahigh fields of 7 and 10.5 T. *Magnetic Resonance Imaging Clinics of North America* 29 (1): e1–e19.

24 Polimeni, J.R. and Uludag, K. (2018). Neuroimaging with ultra-high field MRI: Present and future. *Neuroimage* 168: 1–6.

25 He, X., Erturk, M.A., Grant, A. et al. (2020). First in-vivo human imaging at 10.5T: Imaging the body at 447 MHz. *Magnetic Resonance in Medicine* 84 (1): 289–303.

26 Kandel, E.R. (2012). *The Age of Insight: The Quest to Understand the Unconscious in Art, Mind, and Brain, from Vienna 1900 to the Present*. London: Random House.

27 Van Essen, D.C. and Uğurbil, K. (2012). The future of the human connectome. *Neuroimage* 62 (2): 1299–1310.

28 Van Essen, D.C., Uğurbil, K., Auerbach, E. et al. (2012). The Human Connectome Project: A data acquisition perspective. *Neuroimage* 62 (4): 2222–2231.

29 Insel, T.R., Landis, S.C., and Collins, F.S. (2013). Research priorities. The NIH BRAIN Initiative. *Science* 340 (6133): 687–688.

30 BRAIN Initiative Working Group (2014). BRAIN 2025: A scientific vision. http://braininitiative.nih.gov/pdf/BRAIN2025_508C.pdf (accessed 14 October 2021).

31 Jorgenson, L.A., Newsome, W.T., Anderson, D.J. et al. (2015). The BRAIN Initiative: Developing technology to catalyse neuroscience discovery. *Philosophical Transactions of the Royal Society of London. Series B, Biological Sciences* 370 (1668).

32 Uğurbil, K., Xu, J., Auerbach, E.J. et al. (2013). Pushing spatial and temporal resolution for functional and diffusion MRI in the Human Connectome Project. *Neuroimage* 80: 80–104.

33 Glasser, M.F., Smith, S.M., Marcus, D.S. et al. (2016). The Human Connectome Project's neuroimaging approach. *Nature Neuroscience* 19 (9): 1175–1187.

34 Van Essen, D.C., Smith, S.M., Barch, D.M. et al. (2013). The WU-Minn Human Connectome Project: An overview. *Neuroimage* 80: 62–79.

35 Yacoub, E., Duong, T.Q., Van De Moortele, P.F. et al. (2003). Spin-echo fMRI in humans using high spatial resolutions and high magnetic fields. *Magnetic Resonance in Medicine* 49 (4): 655–664.

36 Zimmermann, J., Goebel, R., De Martino, F. et al. (2011). Mapping the organization of axis of motion selective features in human area MT using high-field fMRI. *PLoS ONE* 6 (12): e28716.

37 Polimeni, J.R., Fischl, B., Greve, D.N. et al. (2010). Laminar analysis of 7T BOLD using an imposed spatial activation pattern in human V1. *Neuroimage* 52 (4). 1334–1346.

38 Koopmans, P.J., Barth, M., Orzada, S. et al. (2011). Multi-echo fMRI of the cortical laminae in humans at 7 T. *Neuroimage* 56 (3): 1276–1285.

39 Siero, J.C., Petridou, N., Hoogduin, H. et al. (2011). Cortical depth-dependent temporal dynamics of the BOLD response in the human brain. *Journal of Cerebral Blood Flow and Metabolism* 31 (10): 1999–2008.

40 De Martino, F., Zimmermann, J., Muckli, L. et al. (2013). Cortical depth dependent functional responses in humans at 7T: Improved specificity with 3D GRASE. *PLoS ONE* 8 (3): e60514.

41 Olman, C.A., Harel, N., Feinberg, D.A. et al. (2012). Layer-specific fMRI reflects different neuronal computations at different depths in human V1. *PLoS ONE* 7 (3): e32536.

42 Muckli, L., De Martino, F., Vizioli, L. et al. (2015). Contextual feedback to superficial layers of V1. *Current Biology* 25: 2690–2695.

43 De Martino, F., Moerel, M., Uğurbil, K. et al. (2015). Frequency preference and attention effects across cortical depths in the human primary auditory cortex. *Proceedings of the National Academy of Sciences of the United States of America* 112 (52): 16036–16041.

44 Nasr, S., Polimeni, J.R., and Tootell, R.B. (2016). Interdigitated color- and disparity-selective columns within human visual cortical areas V2 and V3. *Journal of Neuroscience* 36 (6): 1841–1857.

45 Yacoub, E., Harel, N., and Uğurbil, K. (2008). High-field fMRI unveils orientation columns in humans. *Proceedings of the National Academy of Sciences of the United States of America* 105 (30): 10607–10612.

46 Yacoub, E., Shmuel, A., Logothetis, N. et al. (2007). Robust detection of ocular dominance columns in humans using Hahn Spin Echo BOLD functional MRI at 7 Tesla. *Neuroimage* 37 (4): 1161–1177.

47 Stringer, E.A., Chen, L.M., Friedman, R.M. et al. (2011). Differentiation of somatosensory cortices by high-resolution fMRI at 7 T. *Neuroimage* 54 (2): 1012–1020.

48 Sanchez-Panchuelo, R.M., Francis, S., Bowtell, R. et al. (2010). Mapping human somatosensory cortex in individual subjects with 7T functional MRI. *Journal of Neurophysiology* 103 (5): 2544–2556.

49 Siero, J.C., Hermes, D., Hoogduin, H. et al. (2014). BOLD matches neuronal activity at the mm scale: A combined 7T fMRI and ECoG study in human sensorimotor cortex. *Neuroimage* 101: 177–184.

50 Moerel, M., De Martino, F., Uğurbil, K. et al. (2015). Processing of frequency and location in human subcortical auditory structures. *Scientific Reports* 5: 17048.

51 De Martino, F., Moerel, M., van de Moortele, P.F. et al. (2013). Spatial organization of frequency preference and selectivity in the human inferior colliculus. *Nature Communications* 4: 1386.

52 De Martino, F., Yacoub, E., Kemper, V. et al. (2018). The impact of ultra-high field MRI on cognitive and computational neuroimaging. *Neuroimage* 168: 366–382.

53 Dumoulin, S.O., Fracasso, A., van der Zwaag, W. et al. (2018). Ultra-high field MRI: Advancing systems neuroscience towards mesoscopic human brain function. *Neuroimage* 168: 345–357.

54 Hubel, D.H. and Wiesel, T.N. (1968). Receptive field and functional architecture of monkey striate cortex. *Journal of Physiology (London)* 195: 215–243.

55 Hubel, D.H. and Wiesel, T.N. (1977). Functional architecture of macaque monkey visual cortex. *Proceedings of the Royal Society London* B198 (1–59).

56 Hubel, D.H., Wiesel, T.N., and Stryker, M.P. (1978). Anatomical demonstration of orientation columns in macaque monkey. *Journal of Comparative Neurology* 177 (3): 361–380.

57 Menon, R.S., Ogawa, S., and Uğurbil, K. (1996). Mapping ocular dominance columns in human V1 using fMRI. *NeuroImage* 3: S357.

58 Kim, D.S., Duong, T.Q., and Kim, S.G. (2000). High-resolution mapping of iso-orientation columns by fMRI. *Nature Neuroscience* 3 (2): 164–169.

59 Cheng, K., Waggoner, R.A., and Tanaka, K. (2001). Human ocular dominance columns as revealed by high-field functional magnetic resonance imaging. *Neuron* 32 (2): 359–374.

60 Duong, T.Q., Kim, D.S., Uğurbil, K. et al. (2001). Localized cerebral blood flow response at submillimeter columnar resolution. *Proceedings of the National Academy of Sciences of the United States of America* 98 (19): 10904–10909.

61 Goodyear, B.G. and Menon, R.S. (2001). Brief visual stimulation allows mapping of ocular dominance in visual cortex using fMRI. *Human Brain Mapping* 14 (4): 210–217.

62 Santoro, R., Moerel, M., De Martino, F. et al. (2014). Encoding of natural sounds at multiple spectral and temporal resolutions in the human auditory cortex. *PLoS Computational Biology* 10 (1): e1003412.

63 Jerde, T.A., Lewis, S.M., Goerke, U. et al. (2008). Ultra-high field parallel imaging of the superior parietal lobule during mental maze solving. *Experimental Brain Research* 187 (4): 551–561.

64 Gourtzelidis, P., Tzagarakis, C., Lewis, S.M. et al. (2005). Mental maze solving: Directional fMRI tuning and population coding in the superior parietal lobule. *Experimental Brain Research* 165 (3): 273–282.

65 Cai, Y., Hofstetter, S., Van Der Zwaag, W. et al. (2021). Individualized cognitive neuroscience needs 7T: Comparing numerosity maps at 3T and 7T MRI. *Neuroimage* 237: 118184.

66 Vu, A.T., Jamison, K., Glasser, M.F. et al. (2017). Tradeoffs in pushing the spatial resolution of fMRI for the 7T Human Connectome Project. *Neuroimage* 154: 23–32.

67 Fujita, N. (2001). Extravascular contribution of blood oxygenation level-dependent signal changes: A numerical analysis based on a vascular network model. *Magnetic Resonance in Medicine* 46 (4): 723–734.

68 Boxerman, J.L., Hamberg, L.M., Rosen, B.R. et al. (1995). MR contrast due to intravascular magnetic susceptibility perturbations. *Magnetic Resonance in Medicine* 34: 555–556.

69 Kennan, R.P., Zhong, J., and Gore, J.C. (1994). Intravascular susceptibility contrast mechanisms in tissue. *Magnetic Resonance in Medicine* 31: 9–31.

70 Uğurbil, K. (2016). What is feasible with imaging human brain function and connectivity using functional magnetic resonance imaging. *Philosophical Transactions of the Royal Society of London. Series B, Biological Sciences* 371 (1705): 20150361.

71 Wiesinger, F., Van De Moortele, P.F., Adriany, G. et al. (2004). Parallel imaging performance as a function of field strength – an experimental investigation using electrodynamic scaling. *Magnetic Resonance in Medicine* 52 (5): 953–964.

72 Wiesinger, F., Van de Moortele, P.F., Adriany, G. et al. (2006). Potential and feasibility of parallel MRI at high field. *NMR in Biomedicine* 19 (3): 368–378.

73 Ohliger, M.A., Grant, A.K., and Sodickson, D.K. (2003). Ultimate intrinsic signal-to-noise ratio for parallel MRI: Electromagnetic field considerations. *Magnetic Resonance in Medicine* 50 (5): 1018–1030.

74 Wiesinger, F., Boesiger, P., and Pruessmann, K.P. (2004). Electrodynamics and ultimate SNR in parallel MR imaging. *Magnetic Resonance in Medicine* 52 (2): 376–390.

75 Smith, S.M., Miller, K.L., Moeller, S. et al. (2012). Temporally-independent functional modes of spontaneous brain activity. *Proceedings of the National Academy of Sciences of the United States of America* 109 (8): 3131–3136.

76 Duong, T.Q., Yacoub, E., Adriany, G. et al. (2003). Microvascular BOLD contribution at 4 and 7 T in the human brain: Gradient-echo and spin-echo fMRI with suppression of blood effects. *Magnetic Resonance in Medicine* 49 (6): 1019–1027.

77 van Zijl, P.C., Eleff, S.M., Ulatowski, J.A. et al. (1998). Quantitative assessment of blood flow, blood volume and blood oxygenation effects in functional magnetic resonance imaging [see comments]. *Nature Medicine* 4 (2): 159–167.

78 Uludag, K., Muller-Bierl, B., and Uğurbil, K. (2009). An integrative model for neuronal activity-induced signal changes for gradient and spin echo functional imaging. *Neuroimage* 48 (1): 150–165.

79 Frahm, J., Merboldt, K.-D., Hanicke, W. et al. (1994). Brain or vein-oxygenation or flow? On signal physiology in functional MRI of human brain activation. *NMR in Biomedicine* 7 (1/2): 45–53.

80 Segebarth, C., Belle, V., Delon, C. et al. (1994). Functional MRI of the human brain: Predominance of signals from extracerebral veins. *NeuroReport* 5: 813–816.

81 Glover, G.H., Lemieux, S.K., Drangova, M. et al. (1996). Decomposition of inflow and blood oxygen level-dependent (BOLD) effects with dual-echo spiral gradient-recalled echo (GRE) fMRI. *Magnetic Resonance in Medicine* 35 (3): 299–308.

82 Hoult, D.I. and Richards, R.E. (1976). The signal-to-noise ratio of the nuclear magnetic resonance phenomenon. *Journal of Magnetic Resonance* 24 (71): 71–85.

83 Edelstein, W.A., Glover, G.H., Hardy, C.J. et al. (1986). The intrinsic signal-to-noise ratio in NMR imaging. *Magnetic Resonance in Medicine* 3 (4): 604–618.

84 Pruessmann, K.P., Weiger, M., Scheidegger, M.B. et al. (1999). SENSE: Sensitivity encoding for fast MRI. *Magnetic Resonance in Medicine* 42 (5): 952–962.

85 Triantafyllou, C., Hoge, R.D., Krueger, G. et al. (2005). Comparison of physiological noise at 1.5 T, 3 T and 7 T and optimization of fMRI acquisition parameters. *Neuroimage* 26 (1): 243–250.

86 Triantafyllou, C., Polimeni, J.R., and Wald, L.L. (2011). Physiological noise and signal-to-noise ratio in fMRI with multi-channel array coils. *Neuroimage* 55 (2): 597–606.

87 Yacoub, E., Van De Moortele, P.F., Shmuel, A. et al. (2005). Signal and noise characteristics of Hahn SE and GE BOLD fMRI at 7 T in humans. *Neuroimage* 24 (3): 738–750.

88 Pohmann, R., Speck, O., and Scheffler, K. (2015). Signal-to-noise ratio and MR tissue parameters in human brain imaging at 3, 7, and 9.4 Tesla using current receive coil arrays. *Magnetic Resonance in Medicine* 75: 801–809.

89 Erturk, M.A., Wu, X., Eryaman, Y. et al. (2016). Toward imaging the body at 10.5 Tesla. *Magnetic Resonance in Medicine* 77: 434–443.

90 Uludag, K. and Blinder, P. (2018). Linking brain vascular physiology to hemodynamic response in ultra-high field MRI. *Neuroimage* 168: 279–295.

91 Uludag, K. and Uğurbil, K. (2015). Physiology and physics of the fMRI signal. In: *fMRI: From Nuclear Spins to Brain Function* (ed. K. Uludag, K. Uğurbil, and L. Berliner), 163–214. New York: Springer.

92 Uğurbil, K. (2021). Ultrahigh field and ultrahigh resolution fMRI. *Current Opinions in Biomedical Engineering* 18.

93 Kay, K.N., Rokem, A., Winawer, J. et al. (2013). GLMdenoise: A fast, automated technique for denoising task-based fMRI data. *Frontiers in Neuroscience* 7: 247.

94 Caballero-Gaudes, C. and Reynolds, R.C. (2017). Methods for cleaning the BOLD fMRI signal. *Neuroimage* 154: 128–149.

95 Lund, T.E., Madsen, K.H., Sidaros, K. et al. (2006). Non-white noise in fMRI: Does modelling have an impact? *Neuroimage* 29 (1): 54–66.

96 Murphy, K., Birn, R.M., and Bandettini, P.A. (2013). Resting-state fMRI confounds and cleanup. *Neuroimage* 80: 349–359.

97 Hu, X. and Kim, S.G. (1994). Reduction of signal fluctuation in functional MRI using navigator echoes. *Magnetic Resonance in Medicine* 31 (5): 495–503.

98 Bianciardi, M., Fukunaga, M., van Gelderen, P. et al. (2009). Sources of functional magnetic resonance imaging signal fluctuations in the human brain at rest: A 7 T study. *Magnetic Resonance Imaging* 27 (8): 1019–1029.

99 Bianciardi, M., van Gelderen, P., Duyn, J.H. et al. (2009). Making the most of fMRI at 7 T by suppressing spontaneous signal fluctuations. *Neuroimage* 44 (2): 448–454.

100 Smith, S.M., Fox, P.T., Miller, K.L. et al. (2009). Correspondence of the brain's functional architecture during activation and rest. *Proceedings of the National Academy of Sciences of the United States of America* 106 (31): 13040–13045.

101 Fox, M.D., Snyder, A.Z., Zacks, J.M. et al. (2006). Coherent spontaneous activity accounts for trial-to-trial variability in human evoked brain responses. *Nature Neuroscience* 9 (1): 23–25.

102 Xu, J., Moeller, S., Auerbach, E.J. et al. (2013). Evaluation of slice accelerations using multiband echo planar imaging at 3 T. *Neuroimage* 83: 991–1001.

103 Kruger, G. and Glover, G.H. (2001). Physiological noise in oxygenation-sensitive magnetic resonance imaging. *Magnetic Resonance in Medicine* 46 (4): 631–637.
104 Kruger, G., Kastrup, A., and Glover, G.H. (2001). Neuroimaging at 1.5 T and 3.0 T: Comparison of oxygenation-sensitive magnetic resonance imaging. *Magnetic Resonance in Medicine* 45 (4): 595–604.
105 Hyde, J.S., Biswal, B.B., and Jesmanowicz, A. (2001). High-resolution fMRI using multislice partial k-space GR-EPI with cubic voxels. *Magnetic Resonance in Medicine* 46 (1): 114–125.
106 Bodurka, J., Ye, F., Petridou, N. et al. (2007). Mapping the MRI voxel volume in which thermal noise matches physiological noise–implications for fMRI. *Neuroimage* 34 (2): 542–549.
107 Gonzalez-Castillo, J., Roopchansingh, V., Bandettini, P.A. et al. (2011). Physiological noise effects on the flip angle selection in BOLD fMRI. *Neuroimage* 54 (4): 2764–2778.
108 Kim, S.G., Hendrich, K., Hu, X. et al. (1994). Potential pitfalls of functional MRI using conventional gradient-recalled echo techniques. *NMR in Biomedicine* 7 (1–2): 69–74.
109 Ogawa, S. and Lee, T.M. (1990). Magnetic resonance imaging of blood vessels at high fields: In vivo and in vitro measurements and image simulation. *Magnetic Resonance in Medicine* 16: 9–18.
110 Budde, J., Shajan, G., Hoffmann, J. et al. (2011). Human imaging at 9.4 T using T(2) *-, phase-, and susceptibility-weighted contrast. *Magnetic Resonance in Medicine* 65 (2): 544–550.
111 Budde, J., Shajan, G., Scheffler, K. et al. (2013). Ultra-high resolution imaging of the human brain using acquisition-weighted imaging at 9.4T. *Neuroimage* 86: 592–598.
112 Vaughan, T., DelaBarre, L., Snyder, C. et al. (2006). 9.4T human MRI: Preliminary results. *Magnetic Resonance in Medicine* 56 (6): 1274–1282.
113 Goense, J.B. and Logothetis, N.K. (2006). Laminar specificity in monkey V1 using high-resolution SE-fMRI. *Magnetic Resonance Imaging* 24 (4): 381–392.
114 Harel, N., Lin, J., Moeller, S. et al. (2006). Combined imaging-histological study of cortical laminar specificity of fMRI signals. *Neuroimage* 29 (3): 879–887.
115 Yu, X., Glen, D., Wang, S. et al. (2012). Direct imaging of macrovascular and microvascular contributions to BOLD fMRI in layers IV-V of the rat whisker-barrel cortex. *Neuroimage* 59 (2): 1451–1460.
116 Kay, K., Jamison, K.W., Zhang, R.Y. et al. (2020). A temporal decomposition method for identifying venous effects in task-based fMRI. *Nature Methods* 17 (10): 1033–1039.
117 Menon, R.S. (2002). Postacquisition suppression of large-vessel BOLD signals in high-resolution fMRI. *Magnetic Resonance in Medicine* 47 (1): 1–9.
118 Pohmann, R., Speck, O., and Scheffler, K. (2016). Signal-to-noise ratio and MR tissue parameters in human brain imaging at 3, 7, and 9.4 Tesla using current receive coil arrays. *Magnetic Resonance in Medicine* 75 (2): 801–809.
119 Erturk, M.A., Wu, X., Eryaman, Y. et al. (2017). Toward imaging the body at 10.5 Tesla. *Magnetic Resonance in Medicine* 77 (1): 434–443.
120 Lauterbur, P.C. (1974). Magnetic resonance zeugmatography. *Pure and Applied Chemistry* 40: 149–157.
121 Lauterbur, P.C. (1973). Image formation by induced local interaction: Examples employing nuclear magnetic resonance. *Nature* 241: 190–191.
122 Smith, F.W., Mallard, J.R., Reid, A. et al. (1981). Nuclear magnetic resonance tomographic imaging in liver disease. *Lancet* 1 (8227): 963–966.
123 Crooks, L., Arakawa, M., Hoenninger, J. et al. (1982). Nuclear magnetic resonance whole-body imager operating at 3.5 KGauss. *Radiology* 143 (1): 169–174.

124 Hoult, D.I. and Lauterbur, P.C. (1979). The sensitivity of the zeugmatographic experiment involving human samples. *Journal of Magnetic Resonance* 34: 425–433.

125 Ocali, O. and Atalar, E. (1998). Ultimate intrinsic signal-to-noise ratio in MRI. *Magnetic Resonance in Medicine* 39 (3): 462–473.

126 Guerin, B., Villena, J.F., Polimeridis, A.G. et al. (2016). The ultimate signal-to-noise ratio in realistic body models. *Magnetic Resonance in Medicine* 78: 1969–1980.

127 Keil, B. and Wald, L.L. (2013). Massively parallel MRI detector arrays. *Journal of Magnetic Resonance* 229: 75–89.

128 Uğurbil, K., Auerbach, E., Moeller, S. et al. (2019). Brain imaging with improved acceleration and SNR at 7 Tesla obtained with 64-channel receive array. *Magnetic Resonance in Medicine* 82 (1): 495–509.

129 Vaidya, M.V., Sodickson, D.K., and Lattanzi, R. (2014). Approaching ultimate intrinsic SNR in a uniform spherical sample with finite arrays of loop coils. *Concepts in Magnetic Resonance Part B, Magnetic Resonance Engineering* 44 (3): 53–65.

130 Vaughan, J.T., Garwood, M., Collins, C.M. et al. (2001). 7T vs. 4T: RF power, homogeneity, and signal-to-noise comparison in head images. *Magnetic Resonance in Medicine* 46 (1): 24–30.

131 Sadeghi-Tarakameh, A., DelaBarre, L., Lagore, R.L. et al. (2020). In vivo human head MRI at 10.5T: A radiofrequency safety study and preliminary imaging results. *Magnetic Resonance in Medicine* 84 (1): 484–496.

132 Lagore, R.L., Jungst, S., Radder, J. et al. (2021). A 128-channel receive array for 10.5T human head imaging. *Proceedings of the International Society for Magnetic Resonance in Medicine(Annual Meeting Abstracts)* 0177.

133 Griswold, M.A., Jakob, P.M., Heidemann, R.M. et al. (2002). Generalized autocalibrating partially parallel acquisitions (GRAPPA). *Magnetic Resonance in Medicine* 47 (6): 1202–1210.

134 Sodickson, D.K. and Manning, W.J. (1997). Simultaneous acquisition of spatial harmonics (SMASH): Fast imaging with radiofrequency coil arrays. *Magnetic Resonance in Medicine* 38 (4): 591–603.

135 Larkman, D.J., Hajnal, J.V., Herlihy, A.H. et al. (2001). Use of multicoil arrays for separation of signal from multiple slices simultaneously excited. *Journal of Magnetic Resonance Imaging* 13 (2): 313–317.

136 Moeller, S., Yacoub, E., Olman, C.A. et al. (2010). Multiband multislice GE-EPI at 7 Tesla, with 16-fold acceleration using partial parallel imaging with application to high spatial and temporal whole-brain fMRI. *Magnetic Resonance in Medicine* 63 (5): 1144–1153.

137 Setsompop, K., Gagoski, B.A., Polimeni, J.R. et al. (2012). Blipped-controlled aliasing in parallel imaging for simultaneous multislice echo planar imaging with reduced g-factor penalty. *Magnetic Resonance in Medicine* 67 (5): 1210–1224.

138 Stanisz, G.J., Odrobina, E.E., Pun, J. et al. (2005). T1, T2 relaxation and magnetization transfer in tissue at 3T. *Magnetic Resonance in Medicine* 54 (3): 507–512.

139 Sotiropoulos, S.N., Jbabdi, S., Xu, J. et al. (2013). Advances in diffusion MRI acquisition and processing in the Human Connectome Project. *Neuroimage* 80: 125–143.

140 Andersson, J.L. and Sotiropoulos, S.N. (2015). Non-parametric representation and prediction of single- and multi-shell diffusion-weighted MRI data using Gaussian processes. *Neuroimage* 122: 166–176.

141 Gulban, O.F., De Martino, F., Vu, A.T. et al. (2018). Cortical fibers orientation mapping using in-vivo whole brain 7T diffusion MRI. *Neuroimage*. 178: 104–118.

142 Setsompop, K., Kimmlingen, R., Eberlein, E. et al. (2013). Pushing the limits of in vivo diffusion MRI for the Human Connectome Project. *Neuroimage* 80: 220–233.

143 Foo, T.K.F., Laskaris, E., Vermilyea, M. et al. (2018). Lightweight, compact, and high-performance 3T MR system for imaging the brain and extremities. *Magnetic Resonance in Medicine* 80 (5): 2232–2245.

144 Wilm, B.J., Hennel, F., Roesler, M.B. et al. (2020). Minimizing the echo time in diffusion imaging using spiral readouts and a head gradient system. *Magnetic Resonance in Medicine* 84 (6): 3117–3127.

145 Vu, A.T., Auerbach, E., Lenglet, C. et al. (2015). High resolution whole brain diffusion imaging at 7T for the Human Connectome Project. *Neuroimage* 122: 318–331.

146 Sotiropoulos, S.N., Hernandez-Fernandez, M., Vu, A.T. et al. (2016). Fusion in diffusion MRI for improved fibre orientation estimation: An application to the 3T and 7T data of the Human Connectome Project. *Neuroimage* 134: 396–409.

147 Barfuss, H., Fischer, H., Hentschel, D. et al. (1990). In vivo magnetic resonance imaging and spectroscopy of humans with a 4 T whole-body magnet. *NMR in Biomedicine* 3 (1): 31–45.

148 Glover, G.H., Hayes, C.E., Pelc, N.J. et al. (1985). Comparison of linear and circular-polarization for magnetic-resonance imaging. *Journal of Magnetic Resonance* 64 (2): 255–270.

149 Bottomley, P.A. and Andrew, E.R. (1978). RF magnetic field penetration, phase shift and power dissipation in biological tissue: Implications for NMR imaging. *Physics in Medicine and Biology* 23 (4): 630–643.

150 Adriany, G., Van De Moortele, P.F., Wiesinger, F. et al. (2005). Transmit and receive transmission line arrays for 7 Tesla parallel imaging. *Magnetic Resonance in Medicine* 53 (2): 434–445.

151 Vaughan, J.T., Snyder, C.J., DelaBarre, L.J. et al. (2009). Whole-body imaging at 7T: Preliminary results. *Magnetic Resonance in Medicine* 61 (1): 244–248.

152 Snyder, C.J., DelaBarre, L., Metzger, G.J. et al. (2009). Initial results of cardiac imaging at 7 Tesla. *Magnetic Resonance in Medicine* 61 (3): 517–524.

153 Metzger, G.J., Snyder, C., Akgun, C. et al. (2008). Local B1+ shimming for prostate imaging with transceiver arrays at 7T based on subject-dependent transmit phase measurements. *Magnetic Resonance in Medicine* 59 (2): 396–409.

154 Metzger, G.J., Auerbach, E.J., Akgun, C. et al. (2013). Dynamically applied B1+ shimming solutions for non-contrast enhanced renal angiography at 7.0 Tesla. *Magnetic Resonance in Medicine* 69 (1): 114–126.

155 Suttie, J.J., Delabarre, L., Pitcher, A. et al. (2012). 7 Tesla (T) human cardiovascular magnetic resonance imaging using FLASH and SSFP to assess cardiac function: Validation against 1.5 T and 3 T. *NMR in Biomedicine* 25 (1): 27–34.

156 Snyder, C.J., Delabarre, L., Moeller, S. et al. (2012). Comparison between eight- and sixteen-channel TEM transceive arrays for body imaging at 7 T. *Magnetic Resonance in Medicine* 67 (4): 954–964.

157 Schmitter, S., Delabarre, L., Wu, X. et al. (2013). Cardiac imaging at 7 Tesla: Single- and two-spoke radiofrequency pulse design with 16-channel parallel excitation. *Magnetic Resonance in Medicine* 70 (5): 1210–1219.

158 Ellermann, J., Goerke, U., Morgan, P. et al. (2012). Simultaneous bilateral hip joint imaging at 7 Tesla using fast transmit B(1) shimming methods and multichannel transmission – A feasibility study. *NMR in Biomedicine* 25 (10): 1202–1208.

159 Grams, A.E., Kraff, O., Umutlu, L. et al. (2011). MRI of the lumbar spine at 7 Tesla in healthy volunteers and a patient with congenital malformations. *Skeletal Radiology* 41: 509–514.

160 Orzada, S., Maderwald, S., Poser, B.A. et al. (2010). RF excitation using time interleaved acquisition of modes (TIAMO) to address B1 inhomogeneity in high-field MRI. *Magnetic Resonance in Medicine* 64 (2): 327–333.

161 Umutlu, L., Kraff, O., Orzada, S. et al. (2011). Dynamic contrast-enhanced renal MRI at 7 Tesla: Preliminary results. *Investigative Radiology* 46 (7): 425–433.

162 Umutlu, L., Orzada, S., Kinner, S. et al. (2011). Renal imaging at 7 Tesla: Preliminary results. *European Radiology* 21 (4): 841–849.

163 Frauenrath, T., Hezel, F., Heinrichs, U. et al. (2009). Feasibility of cardiac gating free of interference with electro-magnetic fields at 1.5 Tesla, 3.0 Tesla and 7.0 Tesla using an MR-stethoscope. *Investigative Radiology* 44 (9): 539–547.

164 Frauenrath, T., Hezel, F., Renz, W. et al. (2010). Acoustic cardiac triggering: A practical solution for synchronization and gating of cardiovascular magnetic resonance at 7 Tesla. *Journal of Cardiovascular Magnetic Resonance* 12: 67.

165 von Knobelsdorff-Brenkenhoff, F., Frauenrath, T., Prothmann, M. et al. (2010). Cardiac chamber quantification using magnetic resonance imaging at 7 Tesla – A pilot study. *European Radiology* 20 (12): 2844–2852.

166 van Elderen, S.G., Versluis, M.J., Webb, A.G. et al. (2009). Initial results on in vivo human coronary MR angiography at 7 T. *Magnetic Resonance in Medicine* 62 (6): 1379–1384.

167 von Knobelsdorff-Brenkenhoff, F., Tkachenko, V., Winter, L. et al. (2013). Assessment of the right ventricle with cardiovascular magnetic resonance at 7 Tesla. *Journal of Cardiovascular Magnetic Resonance* 15: 23.

168 Graessl, A., Renz, W., Hezel, F. et al. (2013). Modular 32-channel transceiver coil array for cardiac MRI at 7.0T. *Magnetic Resonance in Medicine* 72: 276–290.

169 Niendorf, T., Graessl, A., Thalhammer, C. et al. (2013). Progress and promises of human cardiac magnetic resonance at ultrahigh fields: A physics perspective. *Journal of Magnetic Resonance* 229: 208–222.

170 Maas, M.C., Vos, E.K., Lagemaat, M.W. et al. (2013). Feasibility of T-weighted turbo spin echo imaging of the human prostate at 7 Tesla. *Magnetic Resonance in Medicine* 71: 1711–1719.

171 Raaijmakers, A.J., Ipek, O., Klomp, D.W. et al. (2011). Design of a radiative surface coil array element at 7 T: The single-side adapted dipole antenna. *Magnetic Resonance in Medicine* 66 (5): 1488–1497.

172 Erturk, M.A., Wu, X.P., Eryaman, Y. et al. (2017). Toward imaging the body at 10.5 Tesla. *Magnetic Resonance in Medicine* 77 (1): 434–443.

173 Sadeghi-Tarakameh, A., Jungst, S., Wu, Z.X. et al. (editors) (2020). Imaging the spine at 10.5T. Americas Committee for Treatment and Research in Multiple Sclerosis Forum. West Palm Beach, FL.

174 Grant, A., Metzger, G.J., Van de Moortele, P.F. et al. (2020). 10.5 T MRI static field effects on human cognitive. *Vestibular, and Physiological Functional Magnetic Resonance Imaging* 73: 163–176.

5

Sweep Imaging with Fourier Transformation (SWIFT)

Djaudat Idiyatullin and Michael Garwood

Center for Magnetic Resonance Research and Department of Radiology, University of Minnesota Medical School, Minneapolis, MN, USA

5.1 Introduction

A radically new method to accomplish magnetic resonance imaging (MRI), called sweep imaging with Fourier transformation (SWIFT), was proposed in 2006 [1]. SWIFT stands alone from any other MRI techniques due to its unique image acquisition protocol (pulse sequence), which involves exciting spins with a frequency-swept pulse and simultaneously acquiring the spins' signals. This type of frequency-swept technique itself is the oldest nuclear magnetic resonance (NMR) method, which was used routinely in spectroscopy until the late 1970s when it began to be replaced by the pulsed radiofrequency (RF) technique with Fourier transformation. The advantages of the pulsed technique for spectroscopy became obvious at that time, due to scientific and technological progress, most notably the fast Fourier transformation (FFT) [2], increased availability of computers, and motivating publications by Richard Ernst and others [3–6]. The architectures of modern MRI scanners that have emerged as a result are based on pulsed FT-NMR [7]. Now, the almost obsolete sweep technique is again attracting interest due to its ability to overcome certain demands and limitations of modern MRI, such as unavoidable B_1 inhomogeneity (large RF coils) and the consequent need for RF power efficiency. One of the revived ideas from the past frequency-swept NMR technique is the advance of adiabatic pulses [8–12], which nowadays have replaced some of the conventional excitation and refocusing pulses in many MRI applications. Another example is the SWIFT method, which enables MRI investigations of objects with ultrashort spin–spin relaxation times [13]. The SWIFT technique has been developed in continuous competition with two other short T_2-sensitive MRI methods applicable for ex vivo and in vivo imaging, which are the ultrashort TE (UTE) method [14–16] and the free induction decay (FID)-projection imaging method, also called zero TE (ZTE) [17–21]. These radial imaging methods hold much potential for MR microscopy, because high-quality images of almost any liquid or solid-like material can be acquired relatively quickly. The SWIFT sequence has flexibility that allows limitations of other sequences to be overcome, such as the slow encoding rate during the ramped readout gradient in UTE and the high peak power required for RF excitation in ZTE to achieve high bandwidth, both of which have negative consequences for short T_2 imaging. This chapter presents an introduction to SWIFT, along with its variations and applications.

Magnetic Resonance Microscopy: Instrumentation and Applications in Engineering, Life Science and Energy Research, First Edition. Edited by Sabina Haber-Pohlmeier, Bernhard Blümich, and Luisa Ciobanu.

5.1.1 Sweep Excitation

Signals of nuclear spins are usually generated by a magnetic field that oscillates at a constant RF. Alternatively, signals can be generated by a pulse in which the frequency is swept across the resonance frequencies of interest. A swept-frequency pulse can be expressed as an explicit function of both time and frequency. In the case of a pulse having constant RF amplitude with a linear frequency sweep, the excitation function $x(t)$ is represented by the chirp function $c(t)$ [22]:

$$x(t) = c(t) = e^{(-i\pi a t^2)}, \tag{5.1}$$

where $a = \dfrac{\omega}{2\pi t}$ is the sweep rate in s^{-2}. A useful property of the chirp function, ignoring edge effects, is that the Fourier transform of a chirp function is itself a chirp function in the frequency domain:

$$F\{c(t)\} = \sqrt{i/a}\ c(t)^{-1} = C(\omega). \tag{5.2}$$

For small flip angles, whereby the time invariant spin system can be considered a linear system, the spin response, $r(t)$, is the convolution ($\oplus$) of the FID, $h(t)$, and the excitation function, $x(t)$ [23]:

$$r(t) = h(t) \oplus x(t). \tag{5.3}$$

According to the Fourier theorem, the spectral response in this case is a product of the spin spectrum, $H(\omega)$, and the excitation profile, $X(\omega)$:

$$R(\omega) = H(\omega) X(\omega). \tag{5.4}$$

The uniqueness of chirp excitation is that the response to such excitation has both the time and the frequency domain features, depending on the relationship between the sweep rates and transverse relaxation times (T_2) of the spins. The oldest continuous wave (CW) method uses slow passage of the resonance lines, which means the time used to pass through a resonance linewidth is smaller than its T_2, i.e. $aT_2^2 \ll 1$. On such a slow time scale, $h(t)$ can be represented by a delta function, which is nonzero only at the time of passing the resonance, i.e. representing a spectrum $H(\omega)$ multiplied by $x(t)$. With phase-sensitive detection used in the CW method, which is the equivalent to multiplication by $c(t)^{-1}$ or $C(\omega)$, the responding spin signal, according to Equation 5.3, will directly yield the spectrum $H(\omega)$ [24].

Another extreme limit is the excitation with infinitely high sweep rate, i.e. $a \sim \infty$, which we call pulsed NMR. In this case, $x(t)$ is a delta function, and according to Equation 5.3, the time response will be exactly the free induction decay $h(t)$, i.e. pure time domain signal.

With intermediate sweep rates, the response represents a mix of frequency and time domain signals. Throughout much of the decades-long history of NMR, the intermediate regime was known for its distorted spectra in CW methods [25] or twisted phase of the off-resonance signals excited by longer pulses in pulsed FT methods. Later, it was recognized that an undistorted spectrum of the spins, $H(\omega)$, can be extracted in the intermediate sweep regime also, by one of the schemes of deconvolution (correlation): by considering the response to chirp excitation as a frequency-domain signal [24,26]:

$$H(\omega) = F\{F^{-1}\{r(t)C(\omega)\}c(t)^{-1}\}, \tag{5.5}$$

or as a signal in the time domain:

$$H(\omega) = F\{r(t)\}C(\omega). \tag{5.6}$$

The first of these two cases is mostly used for retrieving undistorted spectra collected with the Rapid-Scan method [24,26], in which the phase-sensitive receiver is locked to the transmitted RF signal and the acquired signal is proportional to $r(t)C(\omega)$. The second case is convenient to apply to the data collected directly as $r(t)$ with fixed receiver frequency. In general, Equations 5.5 and 5.6 are equivalent due to properties of chirp transformation [22]. However, the second method can be generalized to any type of excitation function, as:

$$H(\omega) = \frac{F\{h(t) \oplus x(t)\}}{F\{x(t)\}} = \frac{R(\omega)}{X(\omega)}, \tag{5.7}$$

and could be used for unscrambling the signal phase after a random excitation in stochastic NMR [27,28] as well as in SWIFT utilizing mostly nonlinear (nonchirp) sweep excitation.

Figure 5.1 illustrates the features of the scanning methods with increasing sweep rate. The scan times change from minutes to milliseconds with the RF power increasing from milliwatts to kilowatts, and at the same time, the response signal is moving from the frequency to the time domain. SWIFT considers the response signal in the time domain and occupies the region between Rapid-Scan and Pulse-FT methods. At times, the line separating these techniques is arbitrary. As will be shown, SWIFT can be converted to different forms and in some cases it is difficult to differentiate, for example, continuous SWIFT from Rapid-Scan, gradient-modulated SWIFT from UTE, or multi-band SWIFT from the regular Pulse-FT method.

5.1.2 Tailored Excitation Profile

In order to shorten the scan time, MRI requires very fast excitation, and at the same time, the frequency excitation profile must be flat to avoid spatially varying image contrast. Using a simple chirp pulse in this case is not optimal. Equations 5.1 and 5.2 do not account for truncation effects, but the

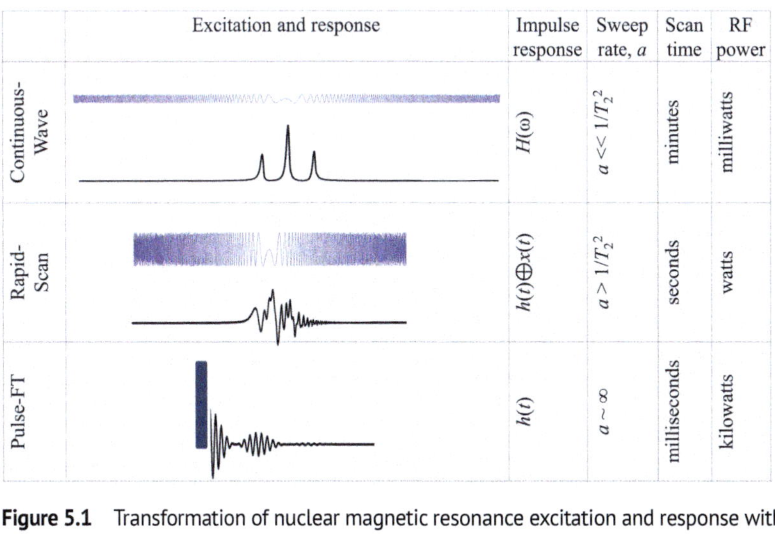

Figure 5.1 Transformation of nuclear magnetic resonance excitation and response with increasing sweep rate.

contribution of these effects increases when shortening the pulse length and this inevitably destroys the excitation profile. Truncation effects can be minimized by optimizing the amplitude modulation (AM) function [9]. Let us consider the frequency-modulated (FM) HSn pulses [29], which originally were developed for the purpose of accomplishing adiabatic inversion (i.e. adiabatic full passage, AFP pulses) with reduced peak amplitude as compared with the original hyperbolic secant (HS) pulse [8]. The RF amplitude-modulated function $m_a(t)$ is chosen to be a stretched HS function:

$$m_a(t) = \operatorname{sech}\left(\beta\left(\frac{2t}{T_p}-1\right)^n\right), \tag{5.8}$$

where n is a dimensionless shape factor (typically a positive integer), β is a dimensionless truncation factor (usually $\beta \approx 5.3$), and T_p is the pulse length (i.e. $0 \leq t \leq T_p$). For $n \to \infty$ (also for $\beta \to 1$), the function $m_a(t)$ becomes a rectangle, which describes the cases of chirp and also square pulses. The frequency modulation, $m_f(t)$, is written as:

$$m_f(t) = \left(\frac{\int_0^t m_a^2(\tau)d\tau}{\int_0^{T_p} m_a^2(\tau)d\tau} - \frac{1}{2}\right). \tag{5.9}$$

Most modern NMR instruments execute FM pulses by modulating the pulse phase $\varphi(t)$, rather than the frequency:

$$\varphi(t) = 2\pi b_w \int_0^t m_f(t)d\tau, \tag{5.10}$$

where b_w is the excitation bandwidth in Hz. Thus, the amplitude- and frequency-modulated version of the RF pulse is described by the function:

$$x(t) = \omega_1 m_a(t) e^{-i\varphi(t)}, \tag{5.11}$$

where $\omega_1 = \gamma B_1$ is the maximum amplitude of the RF field in frequency units. Figure 5.2 presents the AM and FM functions (on the left) and excitation profiles for the HS1 and HS4 pulses in

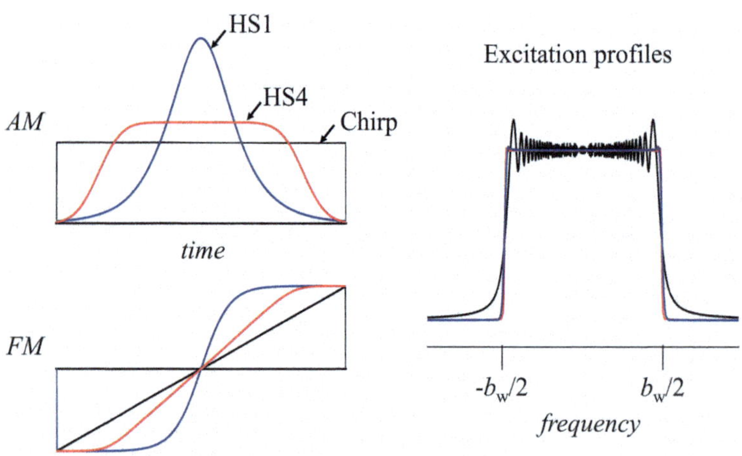

Figure 5.2 Comparison of different AM and FM pulses including HS1, HS4, and chirp (left) and their excitation profiles (right).

comparison with the chirp pulse. The profile of the chirp excitation is highly distorted due to the abrupt leading and trailing edges of the pulse, whereas the profile of HSn excitation is relatively flat and closely approximates a rectangular shape.

As was shown for small flip angles (up to 30°), the excitation flip angle for any HSn pulses can be calculated by the semi-phenomenological formula [30]:

$$\theta = \omega_1 k \frac{\sqrt{R}}{b_{\mathrm{w}}}, \tag{5.12}$$

where R is the dimensionless pulse length–bandwidth product, $R = T_{\mathrm{p}} b_{\mathrm{w}}$, and the pulse shape coefficient $k = \beta^{-\frac{1}{2n}}$. In the case of a square pulse, for example, the desirable excitation bandwidth (matched to the field of view, FOV), with <5% falloff at the edges of the bandwidth, is equal to one-third of the inverse length of the pulse [9,30], $b_{\mathrm{w}} = 1/(3T_{\mathrm{p}})$, and

$$\theta = \frac{\omega_1}{3b_{\mathrm{w}}}. \tag{5.13}$$

Thus, unlike the simple AM pulses, the FM pulses have an ability to independently regulate the flip angle by changing the pulse length and/or excitation bandwidth as an argument.

5.2 The Original Gapped SWIFT Technique and Its Components

In its most basic form, SWIFT executes a swept RF excitation while nearly simultaneously acquiring the spins' signal. To encode spatial information, this process is performed in the presence of a field gradient. The orientation of this gradient changes from one swept RF pulse to the next, until a full set of the n projections (views) has been acquired (Figure 5.3). In SWIFT,

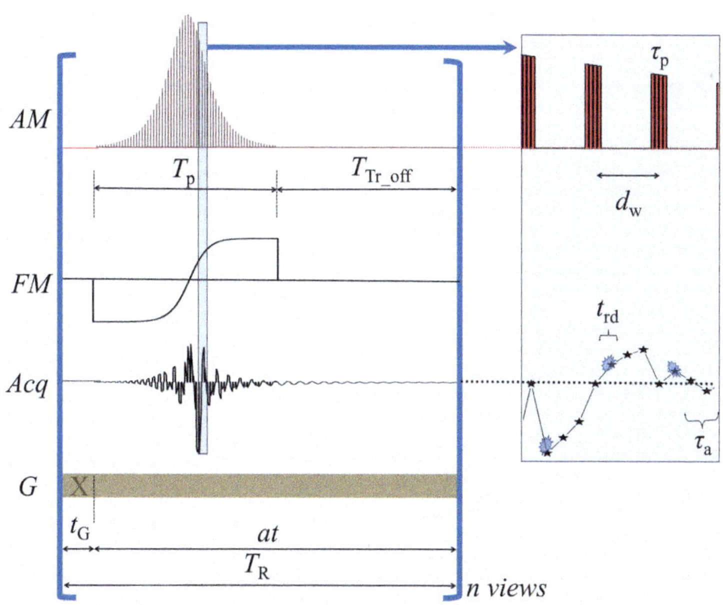

Figure 5.3 Schematic of original SWIFT sequence. Typical parameters values: acquisition oversampling = 4, d_c= 0.25, radiofrequency (RF) fraction = T_p / at = 0.25, RF pulse oversampling = 16.

spins are excited sequentially with the average sweep rate, $a = b_w / T_p$, satisfying these conditions: (i) *rapid passage* [5], $aT_2^2 \gg 1$, and (ii) *linear region*, $a / \omega_2^2 \gg 1$. In original SWIFT, HSn pulses are typically used; however, any other AM and FM pulses satisfying those two conditions are applicable. Of note, the difference between the *linear region* and the *adiabatic region* lies in the fact that the second condition specifies $a / \omega_2^2 \ll 1$. Accordingly, the conversion of a pulse from the adiabatic to the linear region requires reducing the RF amplitude or increasing the sweep rate, while other pulse parameters remain constant. As such, any adiabatic rapid passage pulse can be used in the nonadiabatic *rapid passage, linear region* to rotate the magnetization by an angle <90°, as needed for excitation in SWIFT.

In the original (gapped) version of SWIFT, a time-shared excitation and acquisition scheme is used. In this case, the RF transmitter must be gated "on" and "off" at a rate that is at least as large as the bandwidth, b_w, of the used FM pulse. Technically, the data sampling is not interrupted, but rather the receiver is gated synchronously with the transmitter to block the unwanted signal during the transmitter "on" time, τ_p, and during the coil ring-down time, t_{rd}. The dwell time of data sampling is $d_w = T_p / N$, where N is the number of gaps, which is usually lower than the length–bandwidth product $N \le R$. The transmitter duty cycle is equal to $d_c = \tau_p / d_w$ and an acquisition duty cycle during the gapped pulse is equal to $d_a = \tau_a / d_w$, where $\tau_a = d_w - t_{rd} - \tau_p$. In some cases, it is advantageous to increase the effective acquisition duty cycle by continuing data acquisition after the gapped FM pulse [31], during which time the RF transmitter amplifier remains "off."

In the original SWIFT technique, a deconvolution is done by using a correlation method (Equation 5.7). In addition to this robust and simple method, an algebraic model-based reconstruction method has been proposed [31]. Because model-based reconstruction is computationally more time consuming, it is less used than the standard FT method, which works well for original SWIFT. However, an algebraic reconstruction method is practical and useful when it is needed to recover missed k-space points for ZTE or Multi-Band-SWIFT sequences (see Section 5.4.2).

5.2.1 View Ordering and Silent MRI

SWIFT uses three-dimensional radial center-out *k*-space (kooshball) sampling, where the orientation of the readout gradient is typically updated in small increments. In this way, a low sound pressure level [32,33] and minimal eddy currents are produced in the scanner [13]. In fact, ear protection is unnecessary in a SWIFT-only MRI session. The terminus of the radial spokes can be isotropically distributed on a sphere (Figure 5.4) as a modified generalized spiral set [34] with Euler angles:

$$\theta_k = \arccos\left(\frac{k-1}{n-0.5} - 1\right), \qquad \varphi_k = \theta_k \sqrt{\pi n}, \tag{5.14}$$

where $k = \{1, n\}$ is a counter for the numbers n of views.

When using a randomized view order optimized for acquisition with higher temporal resolution [33,35] or in the case of magnetization preparation (MP) modules embedded between the SWIFT readouts (see Section 5.3.1), the acoustic noise increases modestly; however, the SWIFT

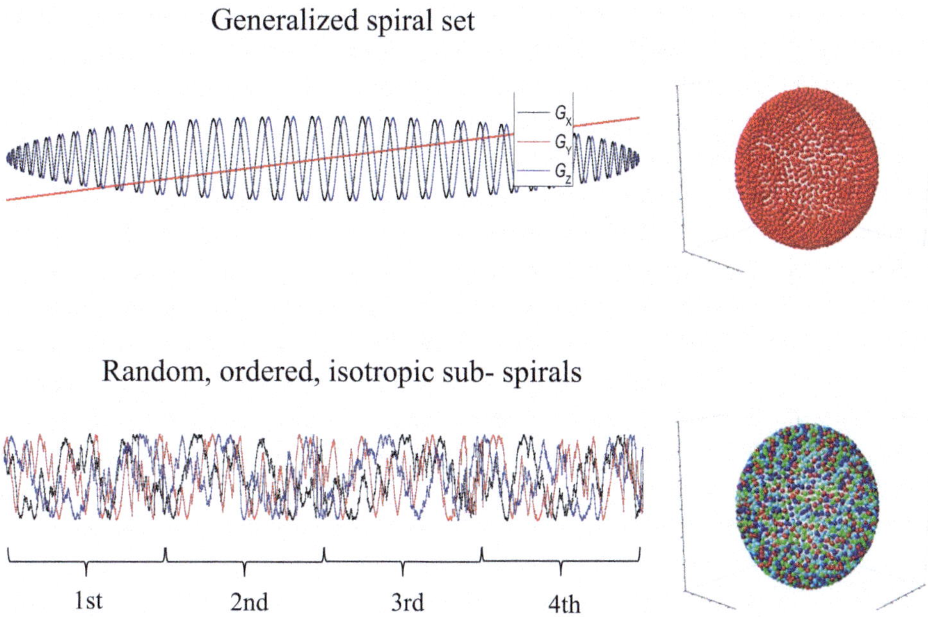

Figure 5.4 The x, y, and z gradient components for continuous spiral (top) and randomized (bottom) view ordering obtained by randomization long continuous spirals to maintain an isotropic distribution of gradient terminus on the surface of the sphere (right) for each sub-spiral.

sequence remains much quieter than the standard gradient echo (GRE) based sequence with similarly embedded MP modules, by as much as 40–50 dB (see Figure 5.5).

Linogram sampling is a "semi-Cartesian" *k*-space sampling method, where the sampling pattern is a cubic grid [36]. SWIFT acquisition with linogram sampling and Hermitian extrapolation in the reconstruction can be used to remove off-resonance blurring, which is common in radial MRI sequences. With linogram sampling, the off-resonance effects are manifest as displacement/distortion of images, similar to what occurs in Cartesian sampling [37,38].

5.2.2 SWIFT-specific Artifact and Its Correction

The AM and FM functions of RF pulses must be discretized for execution by the scanner. When executing the original SWIFT sequence, gaps must also be introduced in the pulse to accommodate signal acquisition. These modulations to the RF pulse functions can both generate sidebands that contaminate and reduce the flatness of the baseband excitation profile, as well as introduce systematic noise in the acquired data [30]. Additional undesirable modifications of the pulse functions can come from the scanners' hardware, such as amplitude nonlinearity and phase instability of the RF amplifier. All of these effects lead to a "bullseye" artifact in a radial SWIFT image, which degrades image quality. This artifact is reduced by pulse oversampling (Figure 5.6a), which pushes the excitation sidebands away from the acquisition window [30]. A further cancellation of this artifact can be achieved by a gap cycling scheme applied from view to view (Figure 5.6b) [39]. In addition to these solutions, an effective postprocessing method for RF distortion correction uses the average projection [40], as described in detail in the text that follows.

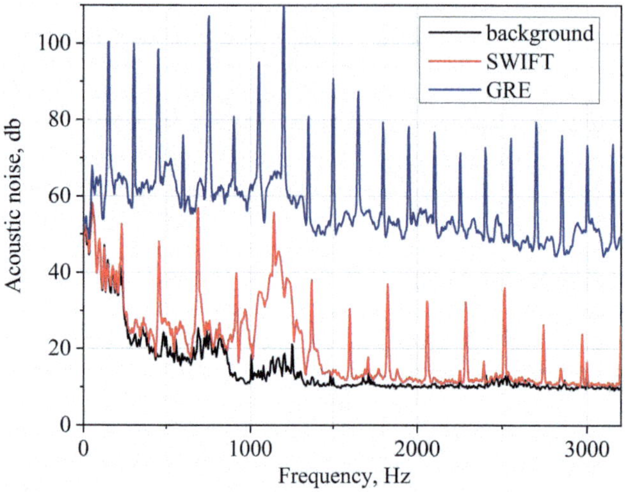

Figure 5.5 Plot showing acoustic noise measurements during gradient echo (GRE) and SWIFT acquisitions, both 3D and with embedded inversion-recovery magnetization preparation (MP) modules as well as background noise in the room. Courtesy of Mathias Nittka, Siemens 2010.

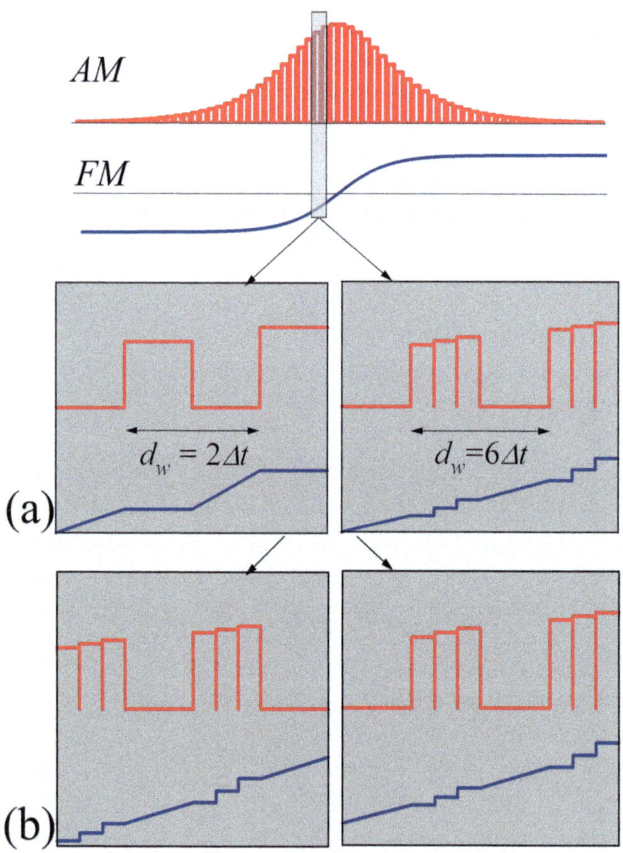

Figure 5.6 Two methods for reducing bullseye artifacts in the gapped SWIFT sequence: (a) pulse oversampling [30] and (b) gap cycling [39].

Let us assume that RF pulse perturbation is an additive and real pulse function $x(t)$ deviating from the expected pulse function, $x_{\text{exp}}(t)$, by an error Δx, i.e. $x(t) = x_{\text{exp}}(t) + \Delta x$. Correlation with the expected pulse function in the frequency domain yields the spectrum H_{calc}, which is different from spectrum H of the spin system:

$$H_{\text{calc}} = \left(1 + \Delta X / X_{\text{exp}}\right) H. \tag{5.15}$$

For radial acquisitions, the persistent perturbation of RF pulse can be estimated from the data themselves. The premise for doing this is that the average projection from all views of an object is smooth, and that the error is consistent for all views:

$$\overline{H_{\text{calc}}} = \sum\nolimits_{\text{all views}} H_{\text{calc}}, \tag{5.16}$$

the corrected projection can then be determined from division by the average projection ($\overline{H_{\text{calc}}}$) and multiplication by the low pass filtered averaged projection ($\overline{\overline{H_{\text{calc}}}} = LP(\overline{H_{\text{calc}}})$):

$$H = H_{\text{calc}} \overline{\overline{H_{\text{calc}}}} / \overline{\overline{H_{\text{calc}}}}. \tag{5.17}$$

The effect of the bullseye and its correction is illustrated in Figure 5.7.

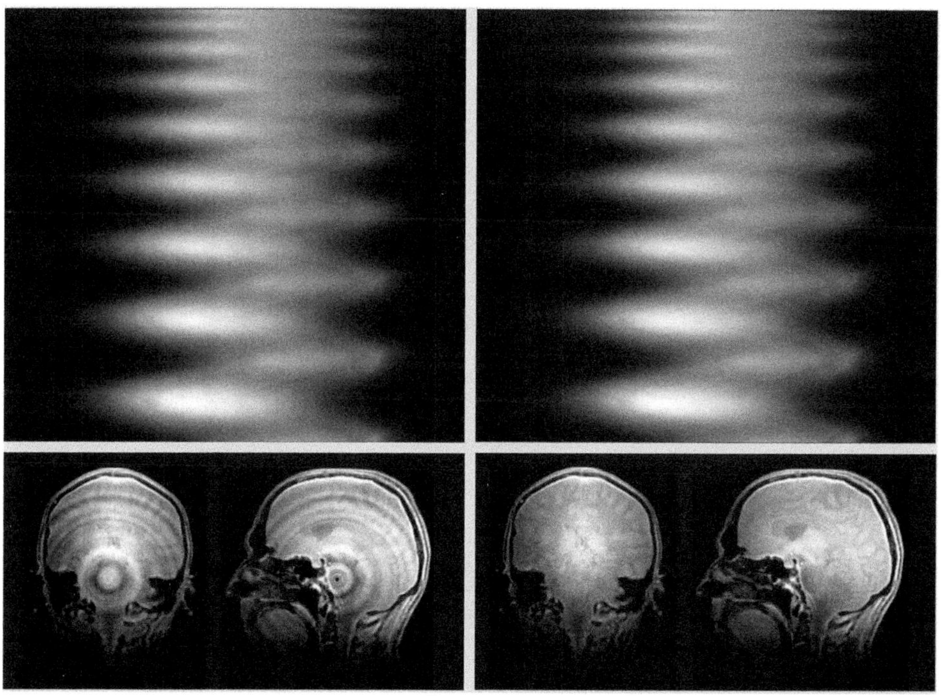

Figure 5.7 The sinograms and images before (left) and after (right) correcting the bullseye effect. [40] Moeller et al. (2008)/with permission from ISMRM.

5.3 SWIFT Utilization

The SWIFT method is insensitive to T_2 differences except in cases of extremely short T_2 ($< d_w$). The contrast of SWIFT images depends on a steady state established by repeated excitation of the spin system. Usually image acquisition starts after performing a set of dummy scans for a duration of about T_1 to establish a steady state. Under these conditions, the signal intensity depends on T_1 and the equilibrium magnetization M_0 as [6]:

$$S(t) = M_0 \frac{1 - E_1}{1 - E_1 \cos(\theta)} \sin(\theta), \tag{5.18}$$

where $E_1 = \exp(-T_1 / T_R)$. For a fixed T_R, the maximum signal intensity is attained with optimum flip angle θ_E (Ernst angle) satisfying the equation $E_1 = \cos(\theta_E)$.

The image contrast is dominated mainly by proton density when the excitation flip angle is small and by T_1-weighting when it is large relative to the Ernst angle. Due to the latter, SWIFT is well suited for variable flip angle (VFA)-based T_1 quantification [41–43].

Noteworthy, the SWIFT VFA allows a true T_1 contrast without the inevitable T_2 contamination that comes with any GRE-based method. Figure 5.8 illustrates the effect of changing the image contrast from proton density to true T_1 contrast with increasing excitation flip angle. The example in Figure 5.9 demonstrates a mixed T_1 and proton density contrast allowing crack detection of 20 μm width in teeth [44].

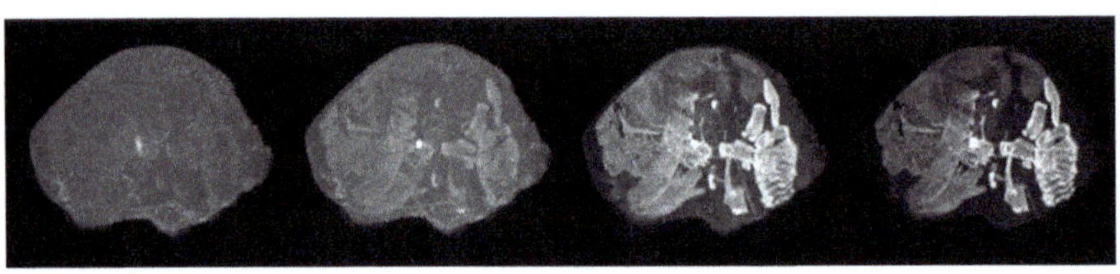

Figure 5.8 Maximum intensity projections of SWIFT images of mouse fetus with different flip angles (from left to right: 5, 10, 15, and 20°), 9.4 T, scan time 7.5 min, T_R=2.3 ms, b_w=125 kHz, field of view = 20^3 mm^3, matrix 256^3. [41] Zhang et al. (2012)/with permission from ISMRM.

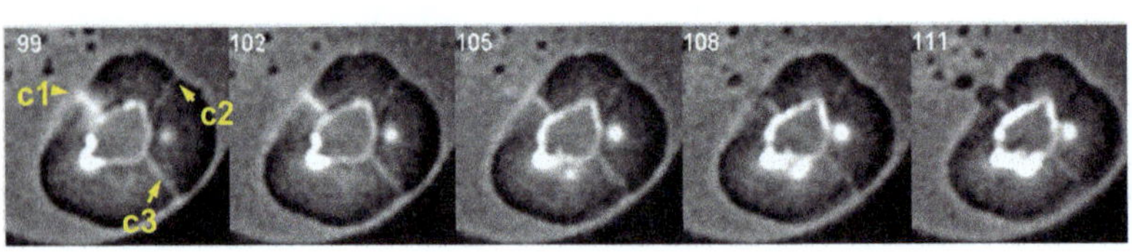

Figure 5.9 Selected slices of ex vivo SWIFT images of cracked teeth collected with an in vivo setup of a whole-body scanner at 4 T. Three cracks, labeled c1, c2, and c3, were measured by micro-computed tomography to have 50, 30, and 40 μm width, respectively (at slice 106). The crack widths decrease from left to right and about 20 μm at slice 111 for crack c2 and at slice 120 for cracks c1 and c3 (not shown). The field of view was 120^3 mm^3 and scan time 3.5 min. Images reconstructed to a nominal resolution of 270 μm isotropic voxel size. [42] Idiyatullin et al. (2016), figure 04[p.16]/The British Institute of Radiology.

5.3.1 MP Schemes

The steady-state magnetization of SWIFT can be manipulated to attain almost any type of a contrast (T_1, T_2, $T_{1\rho}$, $T_{2\rho}$, *diffusion*, etc.) by using MP modules repetitively incorporated between groups of low flip angle SWIFT readouts. To avoid a disturbance of the steady state, the time between modules must be less than T_1. Usually, incorporated MP modules increase the scan time by 10–20%. Figure 5.10 presents two examples of MP modules designed to obtain T_1 and T_2 contrasts [32]. At the beginning and at the end of each module, the readout gradient is ramping down and up, respectively, and additional crusher gradients are optionally applied to avoid interference from transverse coherences.

Obviously, the contrasts obtained by using MP modules are, to some extent, mixed with T_1 contrast. Let us consider an MP module that is embedded into the sequence after each k-th SWIFT readout, providing a decay of longitudinal magnetization based on some other (than T_1) targeted mechanism, for example as: $E_{MP} = \exp(-T_{MP} / kT_{other}$, where T_{other} is the characteristic relaxation time under that mechanism. The exact description of the signal evolution in this case is complicated due to relaxation during the readout [45,46]. However, for average signal values and by neglecting T_1 relaxation during the MP block, the signal acquired during the n-th readout, for $n \gg k$ can be approximated with:

$$S_n = \left(S_s - S_{ss}\right)E_{eff}^n + S_{ss}, \quad (5.19)$$

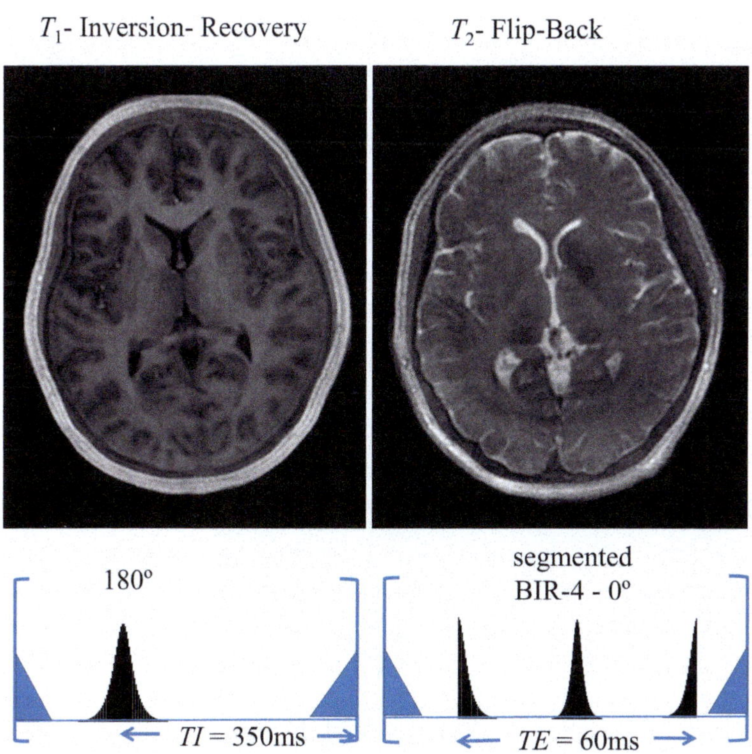

Figure 5.10 SWIFT images (top) and schematic of magnetization preparation modules (bottom) used to obtain T_1 (left) and T_2 (right) contrasts. [32] Chamberlain et al. (2011)/with permission from ISMRM.

where S_s is the starting signal amplitude, S_{ss} is the signal at steady state:

$$S_{ss} = M_0 \sin(\theta)(1 - E_1)/(1 - E_{eff}), \tag{5.20}$$

and E_{eff} describes an effective relaxation:

$$E_{eff} = E_1 E_{MP} \cos(\theta). \tag{5.21}$$

Here, the three types of experiments can be considered: inversion – recovery with $S_s = -M_0 \sin(\theta)$, saturation – recovery with $S_s = 0$, and without any preparations starting with equilibrium magnetization $S_s = M_0 \sin(\theta)$. Thus, T_{other} can be evaluated from fitting the effective decay by using Equation 5.19 or from S_{ss} values collected for different T_{MP}, with Equation 5.20.

Examples of SWIFT using MP modules include, but are not limited to: an inversion pulse followed by a recovery period to produce T_1-weighting [32,47]; a magnetization rotation to produce adiabatic $T_{1\rho}$-weighting [45,46] or relaxation along a fictitious field (RAFF) [48]; diffusion weighting produced by applying field gradients during the TE/2 delays of a spin echo module [49], or a frequency-selective pulse to saturate one or more parts of the targeted spectra (Figure 5.11) [50–54].

5.3.1.1 Look-Locker Sequence

The transition time between equilibrium and steady state can be exploited to accelerate the measurements with the so-called Look-Locker approach [55]. The fast view-shared data acquisition, saturation recovery Look-Locker scheme was implemented in the 3D SWIFT sequence to generate 3D T_1 maps [56]. The method is applicable for mapping relaxation in objects with ultrashort T_2^* and is less sensitive to B_1 inhomogeneity than the VFA method.

5.3.2 Spectroscopic SWIFT

Beyond the 3D spatial information of objects, magnetic resonance spectroscopic imaging (MRSI) yields a fourth (spectroscopic) dimension that can provide valuable information for biomedical and materials science research. In many cases, spectroscopic properties of objects can be more informative than just the relaxation parameters (T_1 and T_2) of MRI. The applications of MRSI are broad, ranging from mapping magnetic field inhomogeneity to imaging distributions of metabolites. Nowadays, MRSI techniques are mostly based on multidimensional Fourier methods utilizing one of the many available phase-encoding schemes with or without slice selection [57–59]. As such, these conventional MRSI methods are restricted to spectroscopy of soft tissue and

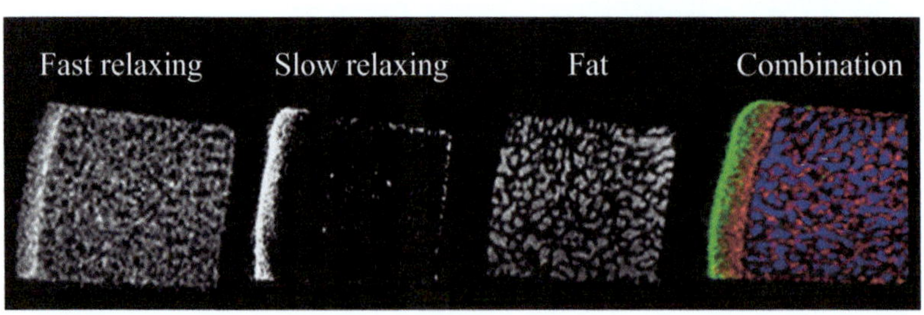

Figure 5.11 Images of different spin pools and color-coded composite image of an osteochondral pig specimen: fast relaxing (red), slow relaxing (green), fat (blue). [45] Zhang et al. (2013)/with permission from ISMRM.

objects that have long relaxation times. These methods tend to fail when applied to objects having relaxation times comparable with or less than the duration of delays needed for phase encoding and/or slice selection. For spectroscopic imaging of objects having short relaxation times, T_2 or T_2^*, an ultrashort *TE* spectroscopic imaging (UTESI) technique based on projection imaging with variable TE was developed [60–63]. Nevertheless, even with the UTESI method, signals in hard tissues such as cortical bone or lung parenchyma having T_2 or T_2^* values under 300–400 μs will considerably decay during RF excitation and gradient ramping time.

The spectroscopic SWIFT (S-SWIFT) method is based on the old idea to reveal both the spatial distributions and the intrinsic frequency spectra of NMR signals from a set of projections obtained in the presence of spatial gradients of varying strength and orientation [64]. Instead of collecting the FIDs following hard RF pulses, as done in the original method, this technique uses SWIFT, which is less demanding of RF field amplitude and, therefore, is more applicable to human studies [65].

To reconstruct a 4D image, a set of projections with orientations isotropically distributed on the surface of "4D sphere" was collected. This was accomplished by using a set of shrinking spatial spheres with accordingly decreased acquisition bandwidth (Figure 5.12) [66]. The 4D

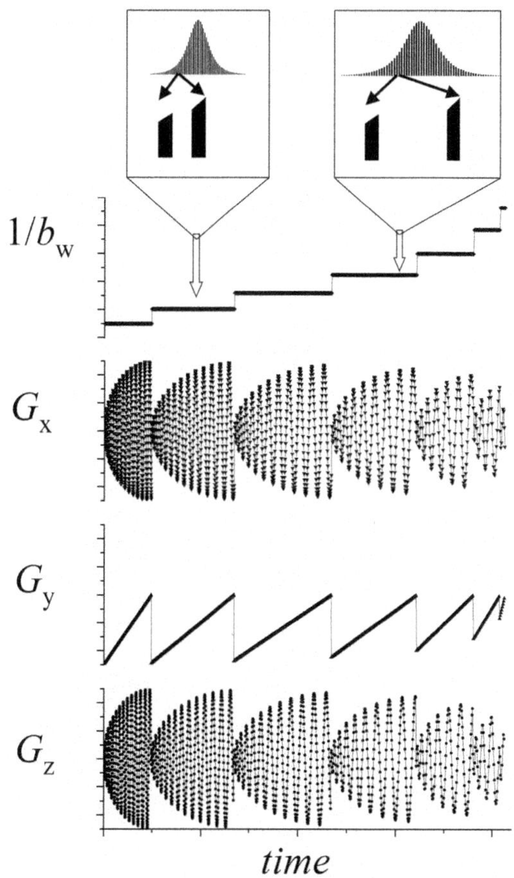

Figure 5.12 Gradients used for the 4D spectroscopic SWIFT. The figure describes shrinking spatial hemispheres in time. The opposite hemispheres are also acquired using inverted gradient directions. Zhang et al. (2016). Characterizing iron oxide nanoparticles using 4D spectroscopic SWIFT. ISMRM Annual Scientific Meeting & Exhibition, Singapore.

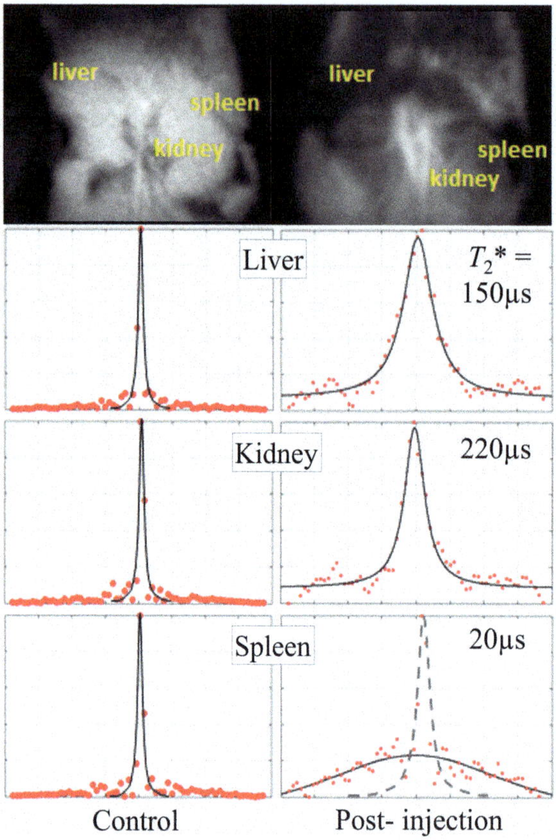

Figure 5.13 The slices of 4D spectroscopic images at the water frequency of a mouse body (top) and spectra of representative regions of interest on different organs with the fitted Lorentzians before (left) and after one-week IONPs injection (right). Parameters: b_w = 125 kHz, bandwidth of the frequency dimension Ω = 15 kHz with 234 Hz resolution, total number of projections 41 792, acquisition time 12 min. [66] Zhang et al. (2016)/with permission from ISMRM.

spectroscopic SWIFT technique was applied to characterize the concentration of injected iron oxide magnetic nanoparticles (IONPs) in different organs of mice in vivo [66]. The apparent line broadening effect was observed in the postinjection case for all organs (Figure 5.13). The measured T_2^* values of organs with deposited IONPs are in the range of 20–200 μs, which is difficult for conventional imaging because it creates signal voids in the images. Therefore, the application of this method is expected to be useful for ultrashort T_2 mapping, fat–water separation, and B_0/susceptibility mapping.

5.4 SWIFT Variations

5.4.1 Gradient-modulated SWIFT

The RF power deposition of FM pulses applied in the presence of field gradients can be significantly reduced by modulating the amplitude of those gradients. The gradient-modulated offset-independent adiabaticity (GOIA) approach [67] as well as the variable-rate selective excitation (VERSE) [68] can be applied to modify the RF pattern in order to achieve the desired power reduction and at the same time to keep a flat excitation profile. The schematic of gradient-modulated SWIFT (GM-SWIFT) using the GOIA approach [67] is presented in Figure 5.14. The time-dependent gradient in this case replicates the time-dependent amplitude $m_a(t)$ of the used RF pulse:

$$G(t) = G_{max}\left(1 - g_m m_a(t)\right), \tag{5.22}$$

where g_m is the gradient modulation factor and G_{max} is the maximum amplitude of the readout gradient. To keep a flat excitation profile in the presence of the variable gradient the frequency modulation $m_f(t)$ of the RF pulse is recalculated under this constraint:

$$\frac{d}{dt}\left(\frac{b_w m_f(t)}{G(t)}\right) = \frac{2\pi\left(\omega_1 m_a(t)\right)^2}{KG(t)}, \tag{5.23}$$

where the adiabaticity factor K is held constant over the range of the pulse time and targeted excitation frequency.

An implementation of this approach in SWIFT is not trivial because the gradient modulation influences both the excitation and the phase evolution of the spins, which results in nonequidistant k-space sampling. As shown previously [69] the correlation procedure Equation 5.7 can still be done; however, in this case, the transformation of the response and the pulse function to the

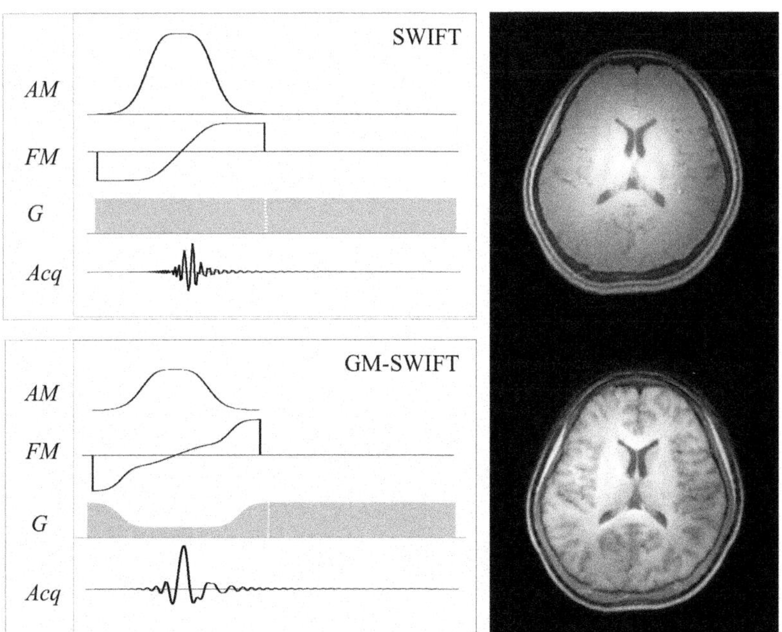

Figure 5.14 Schematic presentation of sequences (left) and human brain images collected with original SWIFT (top) and gradient-modulated SWIFT (GM-SWIFT) (bottom) using same excitation radiofrequency power. [69] Zhang et al. (2016), figure 07[p.543]/with permission from John Wiley & Sons, Inc.

frequency domain must be done nonuniformly. Mathematically this transformation can be described as:

$$R(\omega)=\int_0^{\infty} r(t)\exp\left(-i\omega\int_0^t G(\tau)d\tau\right)G(t)dt,$$
$$X(\omega)=\int_0^{\infty} x(t)\exp\left(-i\omega\int_0^t G(\tau)d\tau\right)dt. \tag{5.24}$$

The GM-SWIFT offers additional flexibility to the SWIFT sequence. By manipulating the shape of gradients, GM-SWIFT maximizes the efficiency of RF power and allows optimization of RF power/SAR, scanning time, and image quality. For the case shown in Figure 5.14 with $g_m = 0.8$, GM-SWIFT reduces the RF power/SAR by about 80% while keeping a similar flip angle [69]. Due to the reduced effective excitation bandwidth, the sensitivity of the GM-SWIFT decreases to extremely short T_2 relative to a standard SWIFT sequence. For some applications this might be considered to be a benefit, for example, decreasing background signal originating from coil holders constructed from ultrashort T_2 (under 100 μs) materials. One example of such application is imaging of lung parenchyma having T_2 of about 300–400 μs. For example, as was shown in [70], GM-SWIFT has high enough sensitivity to detect lung proton density changes due to gravity, different phases of respiration, and accumulation of extravascular water in the rat with acute respiratory distress syndrome lungs (Figure 5.15). The GM-SWIFT is some what similar to the UTE sequence using ramp-up gradient during a signal acquisition [71].

5.4.2 Multi-band SWIFT

Gapped excitation with duty cycle d_c and periodicity $d_w = 1/b_w$ creates sidebands in the frequency domain with periodicity b_w and amplitudes given by:

$$A_m = \mathrm{sinc}(md_c), \tag{5.25}$$

where m is the sideband order [30]. The concept underlying multi-band SWIFT (MB-SWIFT) involves using many sidebands, and as such, distinguishes it from original SWIFT, which uses only the baseband to excite the full field of view (Figure 5.16) [72].

MB-SWIFT treats the sidebands as sub-voxel excitations. As compared with the original SWIFT pulse, it uses fewer gaps N_G, which greatly reduces the transmitter duty cycle d_c but keeps the pulse duration approximately similar. In addition, the method uses a higher number of sampled data points N_{OS} per each gap. The amplitude σ of the frequency modulation in MB-SWIFT is variable in the range $0 \le \sigma \le b_w/2$, which is different from the original SWIFT, where σ is usually fixed and equal to $b_w/2$. In the simplest version of MB-SWIFT, b_w is matched to the linear width of the voxel (in frequency units). As such, the number N of discrete frequencies resolved is equal to the number of sidebands covering the linear size of the FOV, i.e. $N = N_{OS}$, and the total excitation bandwidth is equal to $N_{OS}b_w$. The magnitude spectrum of the gapped excitation profile is shown in Figure 5.16, right. The case presented in the figure is $\sigma < b_w/2$, whereby the profile looks like a comb. The abscissa is frequency unit normalized to b_w, and the integer part of f/b_w is equal to the corresponding sideband order m. For steady-state imaging, it is important to keep the excitation (in this case amplitudes A_m of the sidebands) as uniform as possible

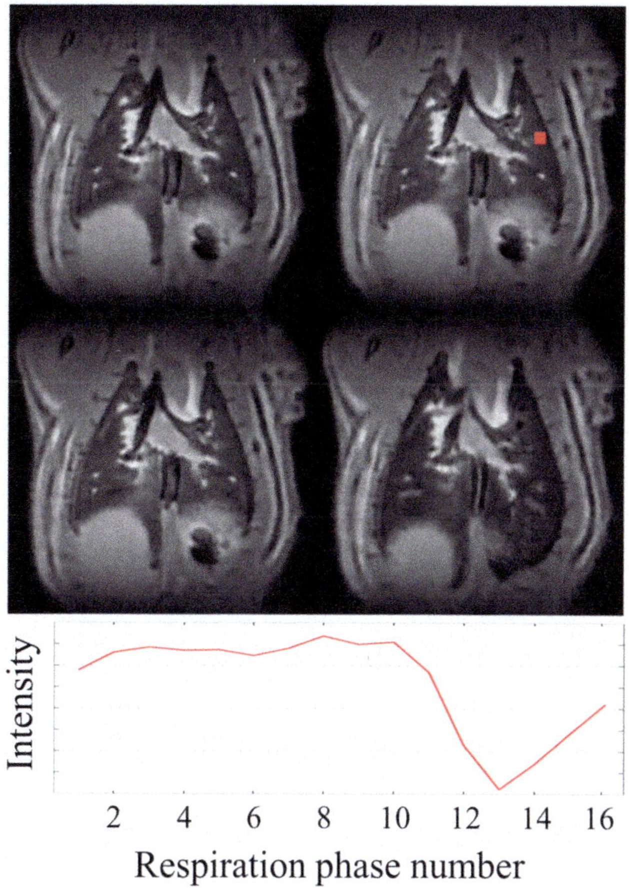

Figure 5.15 3D cine magnetic resonance imaging of rat lung with gradient-modulated SWIFT (GM-SWIFT) and DC navigator retrospective gating (top) and lung's proton density changes during respiration phases (bottom). [70] Kobayashi et al. (2015), figure 04[p.941718-5]/SPIE.

throughout the FOV ($N_{OS}b_W$) [13]. If a full bandwidth is chosen to be equal to the inverse length $\tau_p = 1/(N_{OS}b_W)$ of the pulse element, the used maximum sideband order according to Equation 5.25 is equal to $m_{max} = int\left(1/(2d_c)\right)$. The width of the excited bands (or strips) is equal to or smaller than the linear size of the voxel and can be altered according to the relationship between σ and b_W. Due to the relationship between acquisition time window and resolution [73], the width of the strips cannot be smaller than $1/T_p$, where $T_p = N_G / b_W$ is the total pulse length. Thus, the excited fraction p of the voxel can be estimated as $p = \left(2\sigma(N_G - 1)/b_W + 1\right)/N_G$. Therefore, the smallest excited fraction is simply $p = 1/N_G$ when $\sigma = 0$, and the voxel is fully excited if $p = 1$ for $b_W = 2\sigma$. After acquisition, the response signal is Fourier transformed and correlated with the pulse function in the frequency domain (see Equation 5.7) similarly to the original SWIFT. Afterwards, the signal is inverse Fourier transformed and chopped after the first N_{OS} samples (or after $N_{sr}N_{OS}$ samples in the case of the super-resolution scheme, see the text that follows). This downsampling procedure yields the desired spoke in k-space, which in the

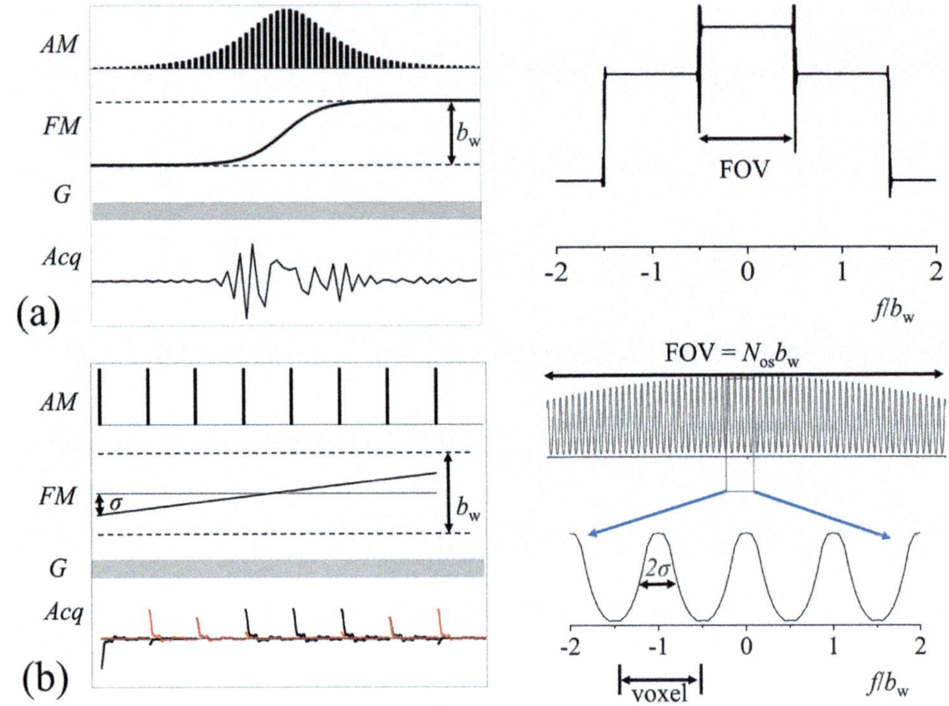

Figure 5.16 Schematics depicting pulse sequences (on left) and magnitude spectra of excitation profiles (on right) of original SWIFT with HS1 (a) and multi-band SWIFT (MB-SWIFT) with chirp (b) excitation pulses. Adapted from [72].

frequency (projection) domain represents the amplitudes of excited bands (strips), i.e. in the simplest case, one point per strip along the projection. In an actual experiment, the few first samples during and after each of the pulse elements are contaminated by coil ring-down, and thus, are discarded and then predicted based on the remaining "good" data points [31,74] or measured in a separate experiment, as done with PETRA (pointwise encoding time reduction with radial acquisition) methods [20].

MB-SWIFT allows experiments with highly increased excitation and acquisition bandwidths, which is a huge benefit for short T_2 imaging, such as, for example, material imaging (Figure 5.17), as well as dental and mandibular imaging (Figure 5.18) [75]. MB-SWIFT allows functional images free of artifacts from susceptibility or dynamic distribution of the B_0 field, and provides contrast for measuring fMRI-based brain activity in close proximity to implanted leads, which is not possible with conventionally used gradient echo and spin echo – echo planar imaging fMRI techniques (Figure 5.19) [33,76]. As compared with original SWIFT, MB-SWIFT can generate improved images in situations requiring a high acquisition bandwidth to minimize distortions from nonuniform B_0 and/or chemical shift. A demonstration of the latter can be seen in Figure 5.20, which shows images of tibia bone containing a metallic implant, acquired without fat saturation [77].

Sub-voxel excitation in the case of MB-SWIFT contains additional spatial information that can be used for resolution enhancement ($N > N_{OS}$), without any increase of the readout time, by applying a "super-resolution" approach that has similarities with a previously proposed method

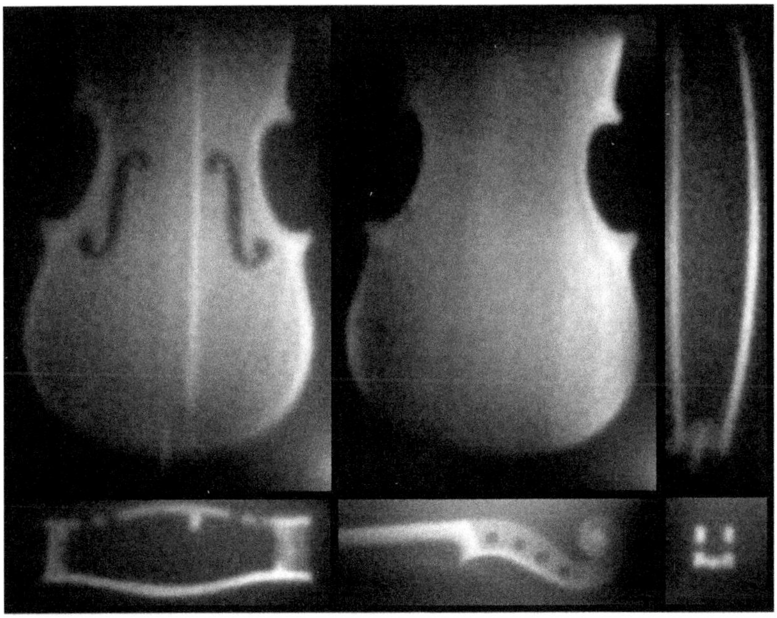

Figure 5.17 The multi-band SWIFT (MB-SWIFT) images of a modern violin without varnish. Flip angle 6°, $N_{OS}b_w$ = 250 kHz, volume transverse electromagnetic coil. Courtesy of Shalom Michaeli.

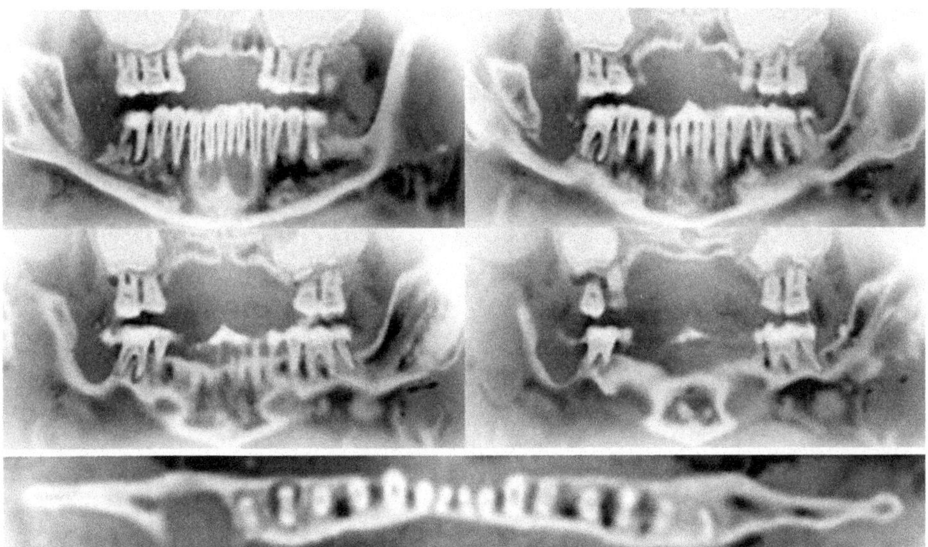

Figure 5.18 Preoperative assessment of mandibular invasion with multi-band SWIFT (MB-SWIFT) (inverted grayscale). Subject with oral cancer abutting the mandible was scanned prior to surgery to assess bone invasion preoperatively. N_{sr} = 2, $N_{OS}b_w$ = 250 kHz. [75] Idiyatullin et al. (2014)/with permission from ISMRM.

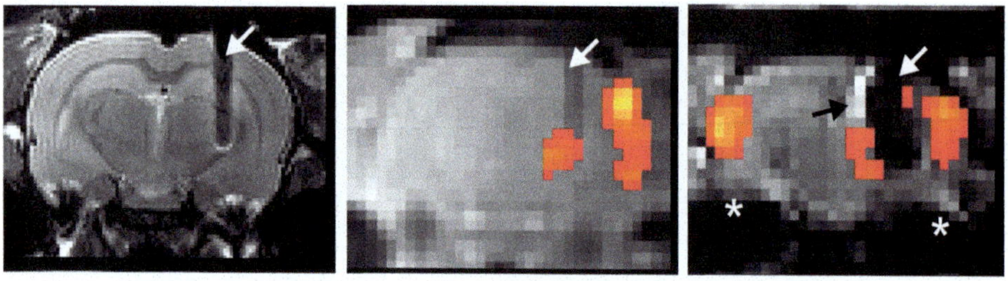

Figure 5.19 Functional response of the rat brain to deep brain stimulation of the ventral posteromedial nucleus. From left to right: high-resolution anatomical fast spin echo image; activation maps overlaid on multi-band SWIFT (MB-SWIFT) and spin echo-planar imaging images. The white arrow points to the artifact due to the implanted electrode; the black arrow and white asterisk point to a signal pile-up artifact and magnetic susceptibility artifact due to the tissue/air interface, respectively. [76] Lehto et al. (2017), [p.18]/with permission from Elsevier.

[78]. In the case of MB-SWIFT, the idea can be realized in at least two ways [72]: (i) by increasing the number of strips per voxel, and (ii) by increasing the number of samples per strip. The first way uses narrow-banded excitation ($\sigma = 0$) in separate scans with subsequent shifting of the carrier frequency. Combining these projections yields a projection with enhanced resolution. The excitation of more than one strip per b_{W} is also possible during one readout by incrementally shifting the excitation frequency (similarly to that done in a Frank pulse) [79], followed by

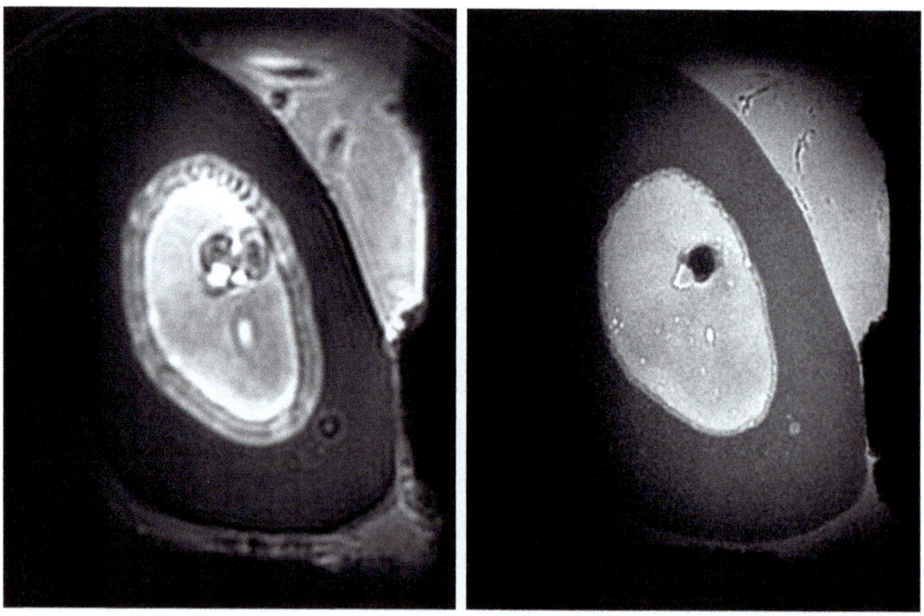

Figure 5.20 Tibia bone images with brass implant in bone marrow by original SWIFT ($b_{\mathrm{W}} = 100$ kHz) (left) and multi-band SWIFT (MB-SWIFT) with high bandwidth ($N_{\mathrm{OS}} b_{\mathrm{W}} = 385$ kHz, $N_{\mathrm{sr}} = 4$, $N_{\mathrm{G}} = 4$) (right). 9.4 T, nominal resolution 156 μm. [77] Idiyatullin et al (2014)/with permission from ISMRM.

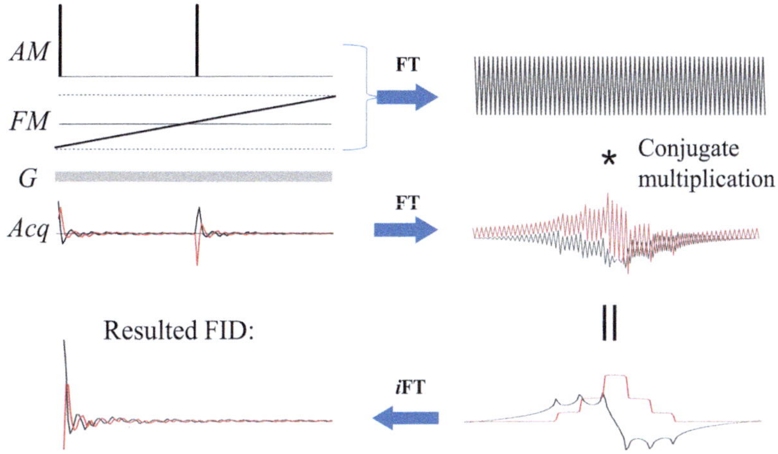

Figure 5.21 The schematic explains a super-resolution approach twice increasing resolution $N_{sr} = 2$ for the simplest version of multi-band SWIFT (MB-SWIFT) with $N_G = 2$ on an example of a stepped object. Adapted from [72].

correlation and forward FT using the first $N = N_{sr}N_{OS}$ data points, where N_{sr} is the number of recovered resolution points in this case equal to the number of frequency steps [72].

The second way, which looks more practically attractive, uses whole-voxel excitation and increased samples per voxel. It is straightforward to apply with quadratic phase modulation inside of each voxel by choosing $\sigma = b_w / 2$. The FIDs with $N_{sr}N_{OS}$ number of points are obtained after FT and correlation. A scheme showing the flow of such procedure for the simplest case is presented in Figure 5.21.

Thus, there are many different schemes by which super-resolution can be used to recover nominal resolution. To avoid a "bullseye" artifact in the image, it may be necessary to choose a scheme that yields the flattest excitation profile after the points have been combined. Alternatively, consecutive excitations can be averaged out by means of frequency or gap cycling [39]. The main advantage of such MB-SWIFT implementations, as compared with the standard ZTE sequence of similar resolution, is the ability to have more pulse elements per projection, which allows increased flip angles for a given RF amplitude setting. MB-SWIFT allows essentially increased excitation and acquisition bandwidths relative to the original SWIFT for the same hardware switching parameters and requires less peak amplitude of the RF field (or a greater flip angle at the same peak amplitude) as compared with ZTE [72]. As such, MB-SWIFT provides a bridge between SWIFT and ZTE sequences.

5.4.3 Continuous SWIFT

The large difference between levels of transmit (Tx) and receive (Rx) NMR signals requires high isolation between Tx and Rx channels. In modern NMR scanners, isolation is usually achieved by separating these events "in time" (i.e. acquisition after pulsed excitation) or more rarely "in frequency" (i.e. using different frequencies for Tx and Rx) [80,81]. In both cases, there is unavoidable compromise between the efficiency of excitation on one side and the optimal signal acquisition on the other. A method to achieve truly simultaneous transmission and reception (STAR) at the same frequency, with nearly 100% efficiency on both sides, has yet to be developed. Interestingly, in the first CW NMR experiments, Bloch already showed that, by careful orthogonal adjustment of the Tx and Rx coils, sufficient isolation could be reached to permit spectral data collection in a STAR mode. Obviously, adapting the classical CW experiment for MRI is impractical due to its time-consuming manner of sweeping frequency. Accelerating the excitation requires increasing the Tx power, which inevitably causes increased "leakage" of the Tx signal into the Rx channel and thus requires greater effort in isolating the channels. The first attempt at STAR MRI was a SWIFT sequence performed in continuous mode (C-SWIFT) with chirp excitation, which requires the smallest RF amplitude per bandwidth [82]. In those experiments, C-SWIFT employed a standard quadrature coil setup with a hybrid to interface with Tx and Rx channels (Figure 5.22). It was found that the level of leakage signal could be highly minimized by slightly mismatching the coil tuning to reach inter-compensation between leaked and reflected RF signals. The leakage $A(\omega)$ in this scheme has at least two sources: $A(\omega) \approx A_{\text{hh}}(\omega) + A_{\text{ch}}(\omega)$. The first term A_{hh} is due to direct coupling between transmitter and receiver via connectors inside the hybrid, which depends on the quality of the hybrid. The second term A_{ch} appears due to a mismatch of impedances between the probe coil and the hybrid. Considering that the hybrid's bandwidth is much broader than that of the RF coil, one can conclude that the frequency dependence of the leakage is mostly determined by the $A_{\text{ch}}(\omega)$ term [83]:

$$A_{\text{ch}}(\omega) \propto R_{\text{probe}} \frac{1 + i\Delta\omega\tau}{1 + \Delta\omega^2\tau^2} - R_{\text{h}}, \tag{5.26}$$

where $\Delta\omega = \omega - \omega_0$, ω_0 is the resonance frequency, $\tau = 2Q/\omega_0$ is the ring-down time of the coil with quality factor Q, and R_{h} and R_{probe} are, respectively, the hybrid and probe resistances, which are usually (by matching or by design) about 50 Ω. For the case of $\Delta\omega\tau \ll 1$, which usually holds in practice, Equation 5.26 predicts almost linear off-resonance dependence of the leakage

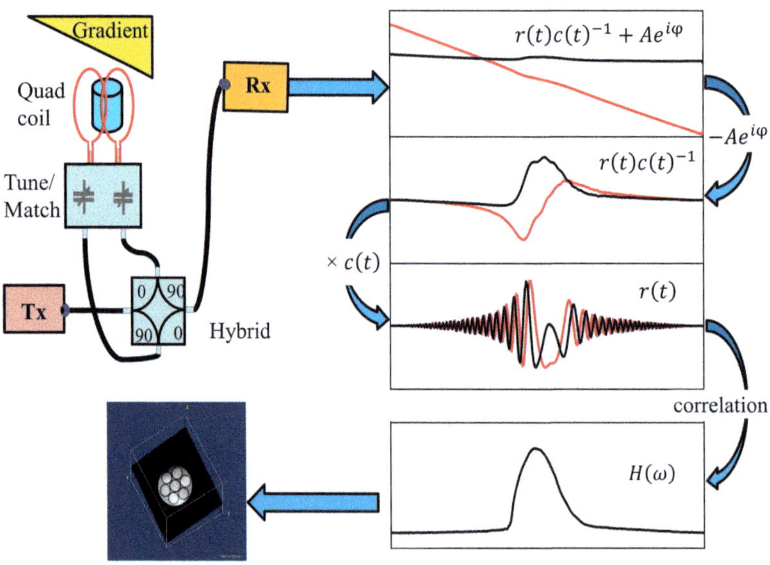

Figure 5.22 Graphical presentation of coil connection and data processing in continuous SWIFT (C-SWIFT). [82] Idiyatullin et al. (2012), figure 05[p.30]/with permission from Elsevier.

(see the imaginary component of the raw data (red) in Figure 5.22). With the leakage minimized at the resonance frequency, the magnitude of leakage signal as a function of resonance offset is v-shaped. The main challenge in implementing C-SWIFT is maintaining the leakage level lower than the threshold of the linear dynamic range of the receiver's preamplifier. Obviously, this becomes more difficult to accomplish with increased sweep range (bandwidth). In practice, however, this effect can be compensated by introducing some linear phase shift synchronized with the sweeping frequency. Additionally, the leakage amplitude depends on the excitation RF power. With chirp excitation at the optimal Ernst condition, in the case where the acquisition bandwidth is equal to the excitation bandwidth, the RF power, and consequently the leakage amplitude, will follow the relation: $A \propto \sqrt{a}/(\gamma Q)$, where γ is gyromagnetic ratio and $a = b_{\mathrm{w}}/T_{\mathrm{p}}$ is sweep rate [30]. Thereby complications increase with the square root of the sweep rate and inversely with the coil quality.

In the continuous mode the acquired raw signal, S, consists of the mixture of the spin response signal ($r(t)$) with the "leakage" of the transmitted chirp signal:

$$S(t) = \left(r(t) + Ae^{i\varphi}c(t)\right)c(t)^{-1} = r(t)c(t)^{-1} + Ae^{i\varphi}. \tag{5.27}$$

Here the multiplication by $c(t)^{-1}$ describes phase-sensitive detection with the receiver frequency locked to the excitation frequency (the FM frame) [11], and A and φ (usually $\varphi \approx 90°$) are the relative amplitude and phase of the transmitter leakage signal. In the FM frame, the leakage signal is a smooth function of time and can be subtracted using some type of model-based fit using Equation 5.26. After subtraction, the multiplication by $c(t)^{-1}$ transfers the data back from the FM frame to the regular rotating frame. At this point, the data become similar to the data acquired with original SWIFT and further processing continues in the same way (Figure 5.22).

As an example, Figure 5.23 presents a C-SWIFT image collected without any transmitter amplifier and the hybrid directly connected to the output of the console [82].

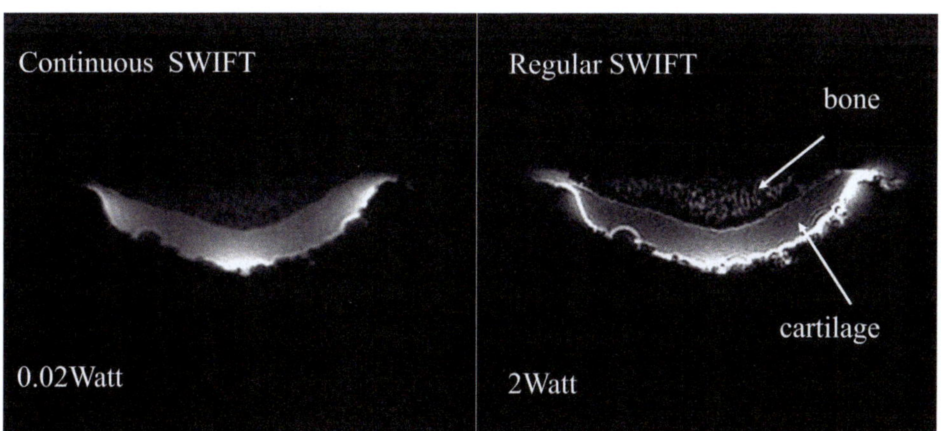

Figure 5.23 Images of a human total knee arthroplasty sample collected with C-SWIFT (left) and original SWIFT (right) at 9.4 T. The continuous SWIFT (C-SWIFT) acquisition was done without a transmitter amplifier using parameters: b_{w} = 71 kHz, 128 000 views, 7 min. [82] Idiyatullin et al. (2012), figure 06[p.30]/with permission from Elsevier.

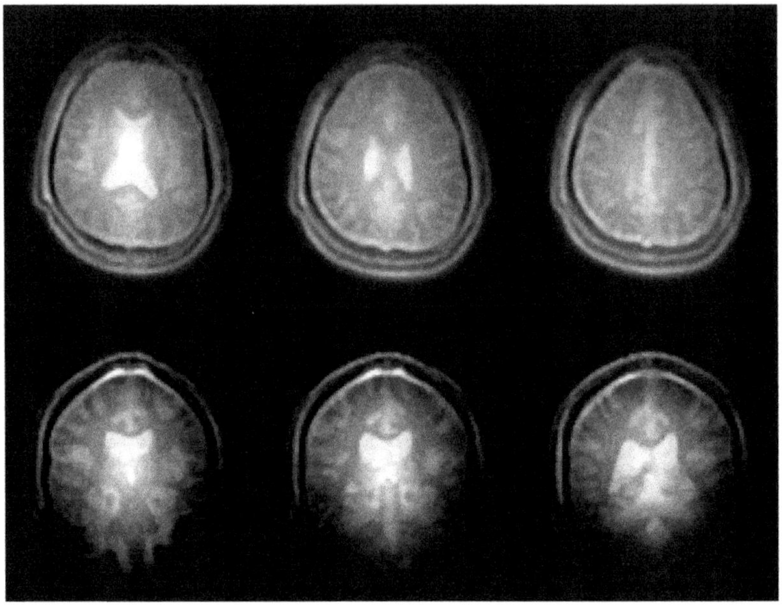

Figure 5.24 The first in vivo human head images with continuous SWIFT (C-SWIFT) at 4 T. Acquisition parameters: Tx power 50 mW, b_W = 32.5 kHz, 128 000 views, 256 complex points per view, isotropic resolution of 1.7 mm, and scan time 10 min. [86] Sohn et al. (2016), figure 05[p.05]/with permission from John Wiley & Sons, Inc.

The Tx/Rx isolation might be done in different ways, for example by using a nonmagnetic circulator [84] or with destructive interference decoupling using extra Tx coils [85]. However, when imaging living objects, motion is a main limitation of these direct methods, since it modulates the transmitter leakage signal by changing of the coil loading. To solve this problem, active leakage cancellation systems were proposed that adjust the amplitude and phase of the sampled Tx signal and add it back to Rx to cancel the leakage [86,87]. The performance of such a system can be appreciated from in vivo C-SWIFT images collected using the ultra-low RF peak power of 50 mW (Figure 5.24) [86].

Due to the complete absence of "dead time," C-SWIFT has the potential to extend applications of MRI and spectroscopy in studies of spin systems having extremely fast relaxation or broad chemical shift distributions beyond the range of existing MRI sequences. Another advantage of this technique is extremely low amplitude of RF field used for the spin excitation. The curves in Figure 5.25 present the calculated values of RF amplitudes used to rotate a proton spin system by the "Ernst angle" in a spoiled steady-state experiment with similar conditions for all classes of SWIFT and ZTE sequences [88]. Note, that C-SWIFT, in comparison with pulsed MRI, requires about 100 times (40 db) lower amplitude of the RF pulse or even less, depending on the R (pulse length–bandwidth product) values. However, to utilize this method routinely and to take advantage of all above-mentioned benefits, more development efforts are still needed.

5.5 Applications and Outlook

Presently, fully functioning SWIFT sequences have been implemented only on research scanners. The main limitation for clinical scanners arises from the limited Tx/Rx switching capabilities of

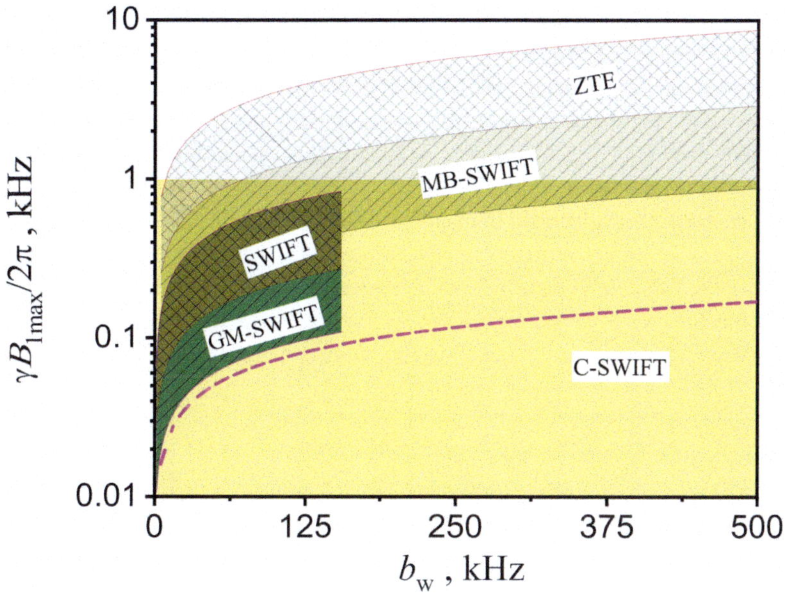

Figure 5.25 Calculated radiofrequency (RF) amplitudes needed for excitation at the Ernst angle for T_1= 1 s, T_R = $at + t_G$, gradient switching time t_G = 150 μs, N = 256 number of points. The RF amplitude reachable in head coil is ~1 kHz. [88].

available coils, which in practice prevent implementation at sufficiently high bandwidths. However, as a proof of principle, the SWIFT sequence was successfully implemented on a Siemens clinical scanner [89,90].

The applications of SWIFT are extremely broad. It was shown that SWIFT can successfully be used for the following applications: dental imaging [44,91–95], visualizing lung parenchyma [96,97], bone density quantification [98], detecting brain calcification [99] and dysmyelination [100], generating exchange-induced contrast [101], imaging traumatic brain injury [102], tracking iron oxide nanoparticles [42,53,56,66,103–105] and paramagnetic chemical exchange saturation transfer (PARACEST) contrast agents [51], high-spatial-temporal-resolution dynamic contrast-enhanced breast imaging [35], identifying prostate to bone tumor [52], detecting mandibular invasion by carcinoma [106], brain fMRI [33,76], analysis of osteochondral tissue [43], and craniofacial imaging [41].

The novel SWIFT technique was developed mostly for MRI studies of human and animals, and as such, has yet to be fully tapped for magnetic resonance microscopy, where we expect that, once unleashed, this swept-frequency NMR technique will be a highly valuable tool.

Acknowledgments

The SWIFT related work described herein was funded by National Institutes of Health grants P41 EB015894 and P41 EB027061.

References

1 Idiyatullin, D., Corum, C., Park, J.-Y. et al. (2006). Fast and quiet MRI using a swept radiofrequency. *Journal of Magnetic Resonance* 181 (2): 342–349.

2 Cooley, J.W. and Tukey, J.W. (1965). An algorithm for the machine calculation of complex Fourier series. *Mathematics of Computation* 19: 297–307.

3 Ernst, R.R. (1965). Sensitivity enhancement in magnetic resonance. I. Analysis of the method of time averaging. *Review of Scientific Instruments* 36 (12): 1689–1695.

4 Ernst, R.R. and Anderson, W.A. (1965). Sensitivity enhancement in magnetic resonance. II. Investigation of intermediate passage conditions. *Review of Scientific Instruments* 36 (12): 1696–1706.

5 Ernst, R.R. (1966). Sensitivity enhancement in magnetic resonance. *Advances in Magnetic Resonance* 2: 1–135.

6 Ernst, R.R. and Anderson, W.A. (1966). Application of Fourier transform spectroscopy to magnetic resonance. *Review of Scientific Instruments* 37: 93–102.

7 Lauterbur, P.C. (1973). Image formation by induced local interactions: Examples employing nuclear magnetic resonance. *Nature* 242: 190–191.

8 Silver, M.S., Joseph, R.I., and Hoult, D.I. (1985). Selective spin inversion in nuclear magnetic resonance and coherent optics through an exact solution of the Bloch-Riccati equation. *Physical Review. A* 31 (4): 2753–2755.

9 Kunz, D. (1986). Use of frequency-modulated radiofrequency pulses in MR imaging experiments. *Magnetic Resonance in Medicine* 3 (3): 377–384.

10 Pipe, J. (1995). Spatial encoding and reconstruction in MRI with quadratic phase profiles. *Magnetic Resonance in Medicine* 33: 24–33.

11 Garwood, M. and DelaBarre, L. (2001). The return of the frequency sweep: Designing adiabatic pulses for contemporary NMR. *Journal of Magnetic Resonance* 153: 155–177.

12 Bernstein, M.A., King, K.F., and Zho, X.J. (2004). *Handbook of MRI Pulse Sequences*. San Diego, CA: Elsevier Academic Press.

13 Garwood, M., Idiyatullin, D., Corum, C.A. et al. (2012). Capturing signals from fast-relaxing spins with frequency-swept MRI: SWIFT. In: *Encyclopedia of Magnetic Resonance*. Hoboken, NJ: John Wiley & Sons, Ltd.

14 Bergin, C.J., Pauly, J.M., and Macovski, A. (1991). Lung parenchyma: Projection reconstruction MR imaging. *Radiology* 179 (3): 777–781.

15 Gatehouse, P.D. and Bydder, G.M. (2003). Magnetic resonance imaging of short T_2 components in tissue. *Clinical Radiology* 58 (1): 1–19.

16 Robson, M.D. and Bydder, G.M. (2006). Clinical ultrashort echo time imaging of bone and other connective tissues. *NMR in Biomedicine* 19 (7): 765–780.

17 Hafner, S. (1994). Fast imaging in liquids and solids with the Back-projection Low Angle ShoT (BLAST) technique. *Magnetic Resonance Imaging* 12 (7): 1047–1051.

18 Madio, D.P. and Lowe, I.J. (1995). Ultra-fast imaging using low flip angles and FIDs. *Magnetic Resonance in Medicine* 34 (4): 525–529.

19 Wu, Y., Dai, G., Ackerman, J.L. et al. (2007). Water- and fat-suppressed proton projection MRI (WASPI) of rat femur bone. *Magnetic Resonance in Medicine* 57 (3): 554–567.

20 Grodzki, D.M., Jakob, P.M., and Heismann, B. (2012). Ultrashort echo time imaging using pointwise encoding time reduction with radial acquisition (PETRA). *Magnetic Resonance in Medicine* 67 (2): 510–518.

21 Weiger, M. and Pruessmann, K.P. (2012). MRI with zero echo time. In: *Encyclopedia of Magnetic Resonance*. Hoboken, NJ: John Wiley & Sons, Ltd.

22 Barrett, H.H. and Myers, K. (2004). *Foundations of Image Science*. Hoboken, NJ: John Wiley & Sons, Ltd. 1540.

23 Shaw, D. (1984). *Fourier Transform NMR Spectroscopy*, 2e. Amsterdam: Elsevier. 344.

24 Gupta, R.K., Ferretti, J.A., and Becker, E.D. (1974). Rapid scan Fourier transform NMR spectroscopy. *Journal of Magnetic Resonance* 13: 275–290.

25 Jacobsohn, B.A. and Wangsness, R.K. (1948). Shapes of nuclear induction signals. *Physical Review* 73: 942–946.

26 Dadok, J. and Sprecher, R.F. (1974). Correlation NMR spectroscopy. *Journal of Magnetic Resonance* 13: 243–248.

27 Ernst, R.R. (1970). Magnetic resonance with stochastic excitation. *Journal of Magnetic Resonance* 3: 10–27.

28 Kaiser, R. (1970). Coherent spectrometry with noise signals. *Journal of Magnetic Resonance* 3: 28–43.

29 Tannus, A. and Garwood, M. (1996). Improved performance of frequency-swept pulses using offset-independent adiabaticity. *Journal of Magnetic Resonance* A 120 (1): 133–137.

30 Idiyatullin, D., Corum, C., Moeller, S. et al. (2008). Gapped pulses for frequency-swept MRI. *Journal of Magnetic Resonance* 193 (2): 267–273.

31 Weiger, M., Hennel, F., and Pruessmann, K.P. (2010). Sweep MRI with algebraic reconstruction. *Magnetic Resonance in Medicine* 64 (6): 1685–1695.

32 Chamberlain, R., Moeller, S., Corum, C., et al. (2011). Quiet T1- and T2-weighted brain imaging using SWIFT. ISMRM 19th Annual Meeting, Montreal, Canada (7–11 May).

33 Paasonen, J., Laakso, H., Pirttimäki, T. et al. (2020). Multi-band SWIFT enables quiet and artefact-free EEG-fMRI and awake fMRI studies in rat. *NeuroImage* 206: 116338.

34 Saff, E.B. and Kuijlaars, A.B.J. (1997). Distributing many points on a sphere. *Mathematical Intelligencer* 19 (1): 5.

35 Corum, C.A., Benson, J.C., Idiyatullin, D. et al. (2015). High-spatial- and high-temporal-resolution dynamic contrast-enhanced MR breast imaging with sweep imaging with Fourier transformation: A pilot study. *Radiology* 274: 540–547.

36 Edholm, P.R. and Herman, G.T. (1987). Linograms in image reconstruction from projections. *IEEE Transactions on Medical Imaging* 6 (4): 301–307.

37 Kobayashi, N., Idiyatullin, D., Corum, C., et al. (2012). Reconstruction of 3D radial MRI with linogram sampling. 20th Meeting of ISMRM, Melbourne, Australia (5–11 May).

38 Kobayashi, N., Idiyatullin, D., Corum, C., et al. Off-resonance blurring tolerant image reconstruction of 3D radial MRI with linogram sampling. 23th Meeting of ISMRM, Toronto, Ontario, Canada (30 May–5 June).

39 Corum, C.A., Idiyatullin, D., Snyder, C.J. et al. (2015). Gap cycling for SWIFT. *Magnetic Resonance in Medicine* 73 (2): 677–682.

40 Moeller, S., Corum, C., Idiyatullin, D., et al. (2008). Correction of RF pulse distortions, with application in radial imaging using SWIFT. ISMRM 16th Scientific Meeting & Exhibition, Toronto, Ontario, Canada (3–9 May).

41 Zhang, J., Idiyatullin, D., Salazar, V.L., et al. (2012). Hyperintense signal from craniofacial bones in SWIFT images of fetal mouse. ISMRM 20th Scientific Meeting & Exhibition, Melbourne, Australia (5–11 May).

42 Wang, L., Corum, C.A., Idiyatullin, D. et al. (2013). T1 estimation for aqueous iron oxide nanoparticle suspensions using a variable flip angle SWIFT sequence. *Magnetic Resonance in Medicine* 70 (2): 341–347.
43 Nissi, M.J., Lehto, L.J., Corum, C.A. et al. (2015). Measurement of T1 relaxation time of osteochondral specimens using VFA-SWIFT. *Magnetic Resonance in Medicine* 74 (1): 175–184.
44 Idiyatullin, D., Garwood, M., Gaalaas, L. et al. (2016). Role of MRI for detecting micro cracks in teeth. *Dentomaxillofacial Radiology* 45 (7): 20160150.
45 Zhang, J., Nissi, M., Idiyatullin, D., et al. (2013). SWIFT with magnetization preparation: signal partitioning and 3D measurement of adiabatic T1rho in osteochondral specimen. ISMRM 21th Annual Meeting, Salt Lake City, Utah, USA (20–26 April).
46 Zhang, J., Nissi, M.J., Idiyatullin, D. et al. (2016). Capturing fast relaxing spins with SWIFT adiabatic rotating frame spin-lattice relaxation ($T1\rho$) mapping. *NMR in Biomedicine* 29 (4): 420–430.
47 Corum, C., Idiyatullin, D., Moeller, S., et al. (2009). Adiabatic T2 weighted MP-SWIFT in brain at 4 T. ISMRM 17th Annual Scientific Meeting & Exhibition, Honolulu, Hawaii, USA (18–24 April).
48 Liimatainen, T., Idiyatullin, D., Zhang, J., et al. (2014) .Maximizing tissue contrast in MRI using RAFFn preparation with SWIFT and spin echo readouts. ISMRM Annual Scientific Meeting & Exhibition, Milan, Italy (10–16 May).
49 Corum, C., Idiyatullin, D., Hutter, D., et al. (2014). DW-MP-SWIFT for high spatial resolution diffusion weighted breast MRI. ISMRM Annual Scientific Meeting & Exhibition, Milan, Italy (10–16 May).
50 Ellermann, J., Idiyatullin, D., Corum, C., et al. (2009). The ultrastructure of bone using selective saturation in SWIFT at 9.4 T. ISMRM 17th Scientific Meeting & Exhibition, Honolulu, Hawaii, USA (18–24 April).
51 Soesbe, T.C., Togao, O., Takahashi, M. et al. (2012). SWIFT-CEST: A new MRI method to overcome T2 shortening caused by PARACEST contrast agents. *Magnetic Resonance in Medicine* 68 (3): 816–821.
52 Luhach, I., Idiyatullin, D., Lynch, C.C. et al. (2013). Rapid ex vivo imaging of PAIII prostate to bone tumor with SWIFT-MRI. *Magnetic Resonance in Medicine* 32 (3): 858–863.
53 Magnitsky, S., Pickup, S., Garwood, M. et al. (2018). Imaging of a high concentration of iron labeled cells with positive contrast in a rat knee. *Magnetic Resonance in Medicine* 81 (3): 1947–1954.
54 Magnitsky, S., Garwood, M., and Idiyatullin, D. (2018). Selective saturation method for improved detection of therapeutic cells ISMRM Annual Scientific Meeting & Exhibition, Paris, France (16–21 June).
55 Look, D.C. and Locker, D.R. (1970). Time saving in measurement of NMR and EPR relaxation times. *The Review of Scientific Instruments* 41 (2): 250–251.
56 Zhang, J., Chamberlain, R., Etheridge, M. et al. (2014). Quantifying iron-oxide nanoparticles at high concentration based on longitudinal relaxation using a three-dimensional SWIFT look-locker sequence. *Magnetic Resonance in Medicine* 71 (6): 1982–1988.
57 Brown, T.R., Kincaid, B.M., and Ugurbil, K. (1982). NMR chemical shift imaging in three dimensions. *Proceedings of the National Academy of Sciences of the United States of America* 79 (11): 3523–3526.
58 Maudsley, A.A., Hilal, S.K., Perman, W.H. et al. (1983). Spatially resolved high resolution spectroscopy by "four-dimensional" NMR. *Journal of Magnetic Resonance* 51 (1): 147–152.

59 Cho, Z.H., Jones, J.P., and Singh, M. (1993). *Foundations of Medical Imaging*. New York, NY: John Wiley & Sons, Inc. 586.

60 Gold, G., Pauly, J., Macovski, A. et al. (1995). MR spectroscopic imaging of collagen: Tendons and knee menisci. *Magnetic Resonance in Medicine* 34: 647–654.

61 Gold, G.E., Pauly, J.M., Leung, A.N. et al. (2002). Short echo time MR spectroscopic imaging of the lung parenchyma. *Journal of Magnetic Resonance Imaging : JMRI* 15 (6): 679–684.

62 Robson, M.D., Tyler, D.J., and Neubauer, S. (2005). Ultrashort TE chemical shift imaging (UTE-CSI). *Magnetic Resonance in Medicine* 53 (2): 267–274.

63 Du, J., Takahashi, A.M., and Chung, C.B. (2009). Ultrashort TE spectroscopic imaging (UTESI): Application to the imaging of short T2 relaxation tissues in the musculoskeletal system. *Journal of Magnetic Resonance Imaging* 29 (2): 412–421.

64 Bernardo, M.L., Lauterbur, P.C., and Hedges, L.K. (1985). Experimental example of NMR spectroscopic imaging by projection reconstruction involving an intrinsic frequency dimension. *Journal of Magnetic Resonance* 61 (1): 168–174.

65 Idiyatullin, D., Corum, C., Moeller, S., et al. (2009). Spectroscopic SWIFT. ISMRM 17th Annual Scientific Meeting & Exhibition, Honolulu, Hawaii, USA (18–24 April).

66 Zhang, J., Garwood, M., and Idiyatullin, D. (2016). Characterizing iron oxide nanoparticles using 4D spectroscopic SWIFT. ISMRM Annual Scientific Meeting & Exhibition, Singapore (7–13 May).

67 Tannus, A. and Garwood, M. (1997). Adiabatic pulses. *NMR in Biomedicine* 10: 423–434.

68 Conolly, S., Nishimura, D., Macovski, A. et al. (1988). Variable-rate selective excitation. *Journal of Magnetic Resonance* 78 (3): 440–458.

69 Zhang, J., Idiyatullin, D., Corum, C.A. et al. (2016). Gradient-modulated SWIFT. *Magnetic Resonance in Medicine* 75 (2): 537–546.

70 Kobayashi, N., Lei, J., Utecht, L., et al. (2015). 3D cine magnetic resonance imaging of rat lung ARDS using gradient-modulated SWIFT with retrospective respiratory gating. *Proceedings of SPIE – The International Society for Optical Engineering* 9417: 18.

71 Garwood, M. (2013). MRI of fast-relaxing spins. *Journal of Magnetic Resonance* 229 (0): 49–54.

72 Idiyatullin, D., Corum, C.A., and Garwood, M. (2015). Multi-band-SWIFT. *Journal of Magnetic Resonance* 251: 19–25.

73 Folland, G. and Sitaram, A. (1997). The uncertainty principle: A mathematical survey. *Journal of Fourier Analysis and Applications* 3 (3): 207–238.

74 Kuethe, D.O., Caprihan, A., Lowe, I.J. et al. (1999). Transforming NMR data despite missing points. *Journal of Magnetic Resonance* 139 (1): 18–25.

75 Idiyatullin, D., Corum, C., Garwood, M., et al. (2014). Preoperative assessment of mandibular invasion with SWIFT. ISMRM Annual Scientific Meeting & Exhibition, Milan, Italy (10–16 May).

76 Lehto, L.J., Idiyatullin, D., Zhang, J. et al. (2017). MB-SWIFT functional MRI during deep brain stimulation in rats. *NeuroImage* 159: 443–448.

77 Idiyatullin, D., Corum, C., and Garwood, M. (2014). Multi-band-SWIFT. ISMRM Annual Scientific Meeting & Exhibition, Milan, Italy (10–16 May.

78 Ropele, S., Ebner, F., Fazekas, F. et al. (2010). Super-resolution MRI using microscopic spatial modulation of magnetization. *Magnetic Resonance in Medicine* 64 (6): 1671–1675.

79 Blümich, B., Gong, Q., Byrne, E. et al. (2009). NMR with excitation modulated by Frank sequences. *Journal of Magnetic Resonance* 199 (1): 18–24.

80 Anderson, W.A. (1962). Applications of modulation techniques to high resolution nuclear magnetic resonance spectrometers. *The Review of Scientific Instruments* 33 (11): 1160–1166.

81 Brunner, D.O., Pavan, M., Dietrich, B., et al. (2011). Sideband excitation for concurrent RF transmission and reception. ISMRM Annual Scientific Meeting & Exhibition, Montreal, Quebec, Canada (7–13 May).

82 Idiyatullin, D., Suddarth, S., Corum, C.A. et al. (2012). Continuous SWIFT. *Journal of Magnetic Resonance* 220 (0): 26–31.

83 de Graaf, R.A. (1998). *In Vivo NMR Spectroscopy: Principles and Techniques*. New York: John Wiley & Sons. 449.

84 Yuksel, H., DelaBarre, L., Idiyatullin, D., et al. (2020). A non-magnetic staggered commutation based circulator to achieve time-efficient simultaneous transmit and receive (STAR) MRI. ISMRM 28th Annual Scientific Meeting & Exhibition, Sydney, NSW, Australia (8–14 August).

85 Özen, A., Bock, M., and Atalar, E. (2015). Active decoupling of RF coils using a transmit array system. *Magnetic Resonance Materials in Physics, Biology and Medicine* 28: 565–576.

86 Sohn, S.-M., Vaughan, J.T., Lagore, R.L. et al. (2016). In vivo MR imaging with simultaneous RF transmission and reception. *Magnetic Resonance in Medicine* 76 (6): 1932–1938.

87 Özen, A.C., Atalar, E., Korvink, J.G. et al. (2018). In vivo MRI with concurrent excitation and acquisition using automated active analog cancellation. *Scientific Reports* 8 (1): 10631.

88 Idiyatullin, D. (2015). SWIFT, Sunrise Educational Course, UTE: Applications & advances. ISMRM Annual Scientific Meeting & Exhibition, Toronto, Ontario, Canada (30 May – 5 June).

89 Valette, J., Moeller, S., Idiyatullin, D., et al. (2009). Implementation of SWIFT on a Siemens clinical scanner. ISMRM 17th Annual Scientific Meeting & Exhibition, Honolulu, Hawaii, USA (18–24 April).

90 Moeller, S., Kobayashi, N., Adriany, G., et al. (2017). Progress in adapting SWIFT to a clinical scanner. ISMRM Annual Scientific Meeting & Exhibition, Honolulu, Hawaii, USA (22–27 April).

91 Idiyatullin, D., Corum, C., Moeller, S. et al. (2011). Dental magnetic resonance imaging: Making the invisible visible. *Journal of Endodontics* 37 (6): 745–752.

92 Nixdorf, D., Idiyatullin, D., and Leon-Salazar, V. (2011). Detecting early carious lesions. IADR/AADR General Session & Exhibition, San Diego, California, USA (16–19 March).

93 Sun, Y., Brauckmann, O., Nixdorf, D.R. et al. (2016). Imaging human teeth by phosphorus magnetic resonance with nuclear Overhauser enhancement. *Scientific Reports* 6: 30756.

94 Gaalaas, L.R., Schuurmans, T., Kim, M. et al. (2019). MRI: Promising dental applications that suggest clinical utility. *Oral Surgery, Oral Medicine, Oral Pathology and Oral Radiology* 127 (1): e36.

95 Schuurmans, T.J., Nixdorf, D.R., Idiyatullin, D.S. et al. (2019). Accuracy and reliability of root crack and fracture detection in teeth using magnetic resonance imaging 45: 750–755. *Journal of Endodontics*

96 Corum, C.A., Idiyatullin, D., Moeller, S., et al. (2010). Lung imaging in the mouse with SWIFT. ISMRM Annual Scientific Meeting & Exhibition, Stockholm, Sweden (1–7 May).

97 Kobayashi, N., Idiyatullin, D., Corum, C. et al. (2015). SWIFT MRI enhances detection of breast cancer metastasis to the lung. *Magnetic Resonance in Medicine: Official Journal of the Society of Magnetic Resonance in Medicine/Society of Magnetic Resonance in Medicine* 73 (5): 1812–1819.

98 Surowiec, R.K., Ram, S., Idiyatullin, D. et al. (2020). In vivo quantitative imaging biomarkers of bone quality and mineral density using multi-band-SWIFT magnetic resonance imaging. *Bone* 143: 115615.

99 Lehto, L.J., Sierra, A., Corum, C.A. et al. (2012). Detection of calcifications in vivo and ex vivo after brain injury in rat using SWIFT. *NeuroImage* 61 (4): 761–772.

100 Satzer, D., DiBartolomeo, C., Ritchie, M.M. et al. (2015). Assessment of dysmyelination with RAFFn MRI: Application to murine MPS I. *PLoS ONE* 10: 2.

101 Liimatainen, T., Hakkarainen, H., Mangia, S. et al. (2015). MRI contrasts in high rank rotating frames. *Magnetic Resonance in Medicine* 73 (1): 254–262.

102 Lehto, L.J., Corum, C.A., Idiyatullin, D., et al. (2012).Iron imaging in traumatic brain injury rat model using SWIFT. ISMRM 20th Scientific Meeting & Exhibition, Melbourne, Australia (5–11 May).

103 Zhou, R., Idiyatullin, D., Moeller, S. et al. (2010). SWIFT detection of SPIO-labeled stem cells grafted in the myocardium. *Magnetic Resonance in Medicine* 63 (5): 1154–1161.

104 Zhang, J., Ring, H.L., Hurley, K.R. et al. (2016). Quantification and biodistribution of iron oxide nanoparticles in the primary clearance organs of mice using T1 contrast for heating. *Magnetic Resonance in Medicine* 78: 702–712.

105 Magnitsky, S., Garwood, M., and Idiyatullin, D. (2019). In vivo monitoring of mesenchymal stem cells in a rat knee with tissue saturation SWIFT. ISMRM 27th Annual Scientific Meeting & Exhibition, Montreal, Canada (11–16 May).

106 Kendi, A.T.K., Khariwala, S.S., Zhang, J. et al. (2011). Transformation in mandibular imaging with sweep imaging with Fourier transform magnetic resonance imaging. *Archives of Otolaryngology – Head & Neck Surgery* 137 (9): 916–919.

6

Methods Based on Solution Flow, Improved Detection, and Hyperpolarization for Enhanced Magnetic Resonance

Patrick Berthault and Gaspard Huber

Université Paris-Saclay, CNRS, CEA, Nanosciences et Innovation pour les Matériaux, la Biomédecine et l'Energie, Gif-sur-Yvette, France

6.1 Introduction

Compared with other analytical [1] or imaging [2] techniques, nuclear magnetic resonance (NMR) is a modality with an inherently low sensitivity, regardless of which spectroscopy (magnetic resonance spectroscopy, MRS) or imaging (magnetic resonance imaging, MRI) is considered. Therefore, it is scarcely used to monitor rare or fugacious events. For instance, real operando NMR is limited to only few domains, such as long-term analysis of the electrolyte degradation in batteries [3,4] or monitoring dynamic changes in the water concentration of fruit and stems [5]. To catch molecular systems in fast evolution, it is thus important to look for instrumentation and methods to increase either the mass sensitivity or the concentration sensitivity.

At a given magnetic field and sample temperature, instrumental ways to achieve NMR sensitivity gain can be grouped into three categories: (i) spin flow, (ii) cryo- and/or micro-detection, and (iii) hyperpolarization.

In the first category, developed since the 1980s, the idea is to constantly replace the nuclear spins saturated after radiofrequency excitation by new ones in the detection region. This is basically the principle of the white blood or time-of-flight (TOF) MRI angiography techniques, in which contrast between the static tissues and the flowing blood inside the vessels is enhanced by the spin motion [6]. Also, for analytical applications employing slowly relaxing nuclei such as ^{13}C, closed-loop circuits are employed in which the solution is constantly renewed in the MRS detection area [7].

The second category deals with the optimization of the detection coil. Here we will only rely on the description of the classical induction-based methods to detect nuclear spins. It deals with reduction of electronic noise through temperature lowering in cryocoils and high-superconducting coils, on the one hand, and, on the other hand, the reduction of the size of coils in order to increase the sensitivity for mass-limited samples [8].

The third category, hyperpolarization, which is the transient boost of nuclear polarization through transfer from a more ordered system, is more and more widely used. The main methods are optical pumping of noble gases (spin-exchange optical pumping, SEOP, and metastability-exchange optical pumping, MEOP), parahydrogen-induced polarization (PHIP), and dynamic nuclear polarization (DNP). Pros and cons of these techniques have been reviewed [9].

Some recent solutions aiming at combining two or three of these categories to reach even higher NMR sensitivity for the study of systems in operation are the subject of this chapter. After

Magnetic Resonance Microscopy: Instrumentation and Applications in Engineering, Life Science and Energy Research, First Edition. Edited by Sabina Haber-Pohlmeier, Bernhard Blümich, and Luisa Ciobanu.

briefly recalling the theoretical and practical aspects of these methods, we focus on the MRI methods designed to optimally manage the out-of-equilibrium nature of hyperpolarization. Then we describe the instrumental developments using both flow and coil optimization, and those introducing hyperpolarization in complement to one of these two methods or both. Next, methods and setups for efficient dissolution of hyperpolarized species are reviewed. Finally, we consider what could be an ideal system encompassing hyperpolarization, flow, and optimized detection (remaining however on conventional NMR induction). We will illustrate the text with some figures arising from work performed in our laboratory.

6.2 Basics of the Methods to Increase NMR Sensitivity

6.2.1 Spin Hyperpolarization

The main advantage of NMR, i.e. its nonperturbative and innocuous character due to the low energies involved, also leads to its main drawback, its lack of sensitivity. For a spin ½, the nuclear polarization at thermodynamic equilibrium (normalized population difference of the low-energy level to the high-energy level) is given by the Boltzmann distribution law:

$$P = \frac{n_\downarrow - n_\uparrow}{n_\downarrow + n_\uparrow} = \tanh \frac{\gamma \hbar B_0}{2kT}, \tag{6.1}$$

where γ is the gyromagnetic ratio and B_0 the magnetic field. This means that at room temperature, even in a magnetic field of 11.7 T, the proton nuclear polarization is only ~4×10^{-5} (24 999 spins on the high-energy level, 25 001 on the other one). But nuclear spins have an advantage: they relax slowly (much slower than the electrons or the photons in any case), and therefore they can receive polarization from other – more ordered – systems.

Let us describe now the three main hyperpolarization methods, keeping in mind that there exist many others (Figure 6.1). The interested reader could for instance consult references [10] and [11] for a more complete description of them.

As electrons are fermions, they also obey the Boltzmann law. But due to a much higher ratio of their charge over their mass, their gyromagnetic ratio is ~700 times higher than that of protons. Therefore, in the DNP experiments using electrons as a source of nuclear polarization a theoretical gain of 700 for the proton NMR signal can be expected. The work of Ardenkjær-Larsen and co-workers has even demonstrated a much higher gain of a factor 10 000 for dissolution-DNP [21]. The method consists of forming glassy matrices containing nuclei of interest and stable radicals. Nuclear spins are first polarized by means of microwave irradiation at the electron spin transition frequency at low temperatures (≈1 K) and then the polarized sample is dissolved rapidly using a pressurized heated solvent and quickly transferred to the NMR spectrometer for data acquisition.

Ultimate gains of polarization could be obtained by transfer from photons. As they are bosons, they obey the Bose–Einstein distribution law, and it is well known that a photon beam can be fully circularly polarized (meaning that all photons have a spin +1 or −1). This is the basis of the optical pumping process. However, the situation is not so simple. Every nucleus in every molecule cannot be polarized in this way. First of all, the Zeeman levels responsible for the nuclear spin transitions are not isolated, and only particular atomic transitions can be excited by light. Second, for efficient optical pumping, an energy level not directly reached by the optical transition and which does not relax (or de-excite) too fast is required. Third, the relaxation of the target

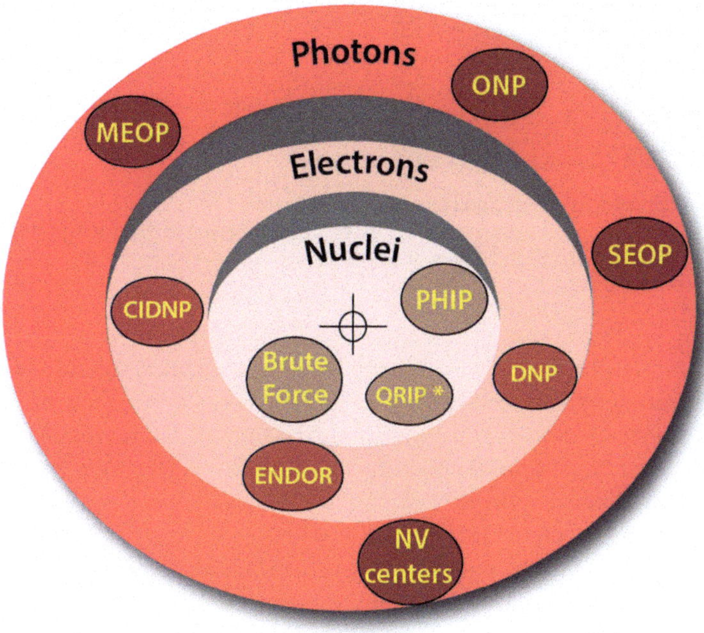

Figure 6.1 Main hyperpolarization methods. For details on the different techniques, see for instance the following references: optical nuclear polarization (ONP) [12], SEOP [13], MEOP [14], nitrogen vacancy (NV) centers [15], DNP [16], chemically induced DNP (CIDNP) (and photo-CIDNP) [17], electron-nuclear double resonance (ENDOR) [18], quantum rotor induced polarization (QRIP) (* in fact correlations between rotational and nuclear spin states) [19], PHIP [10], and brute force [20].

nucleus has to be considered. Taking all these into consideration, it follows that the most widely used optical pumping deals with noble gases of spin ½: ^{129}Xe and ^{3}He. For the former, SEOP is the most commonly used method [13]. In SEOP, a circularly polarized light beam colinear to a magnetic field induces specific electronic transitions of alkali metal atoms, resulting in an excess of population of one of the two electronic levels of the alkali metal ground state. Then, during collisions, spin-exchange takes place between the alkali electron spins and the nuclear spins of the noble gas, resulting in its hyperpolarization. Although it is a quadrupolar nucleus thus experiencing faster relaxation, ^{83}Kr can also be hyperpolarized in this way. For ^{3}He, as it is a light atom with strict selection rules, the metastable 2 $^{3}S_1$ level (a singlet state) can be populated by a radiofrequency discharge. Then optical pumping occurs between the $^{3}S_1$ and $^{3}P_J$ sub-levels. In this method, which does not require the use of an alkali metal, nuclear polarization is transferred to ^{3}He atoms in the ground state via metastability-exchange collisions: this is the principle of MEOP. Its quantum yield is excellent, but only low quantities of polarized nuclei per unit time are produced [22].

Saying that an ensemble is completely polarized is equivalent to defining it as being in a pure state. In the case of parahydrogen, the molecules are in a pure singlet state, for which no NMR transition is possible. In the PHIP methods [23], the overpopulation of some nuclear spin states of protons is achieved through pairwise addition of the parahydrogen nuclei on an unsaturated chemical compound, rendering them magnetically inequivalent. This hyperpolarization is then transferred to other nuclei mainly via scalar coupling. More recently, a parahydrogen-based

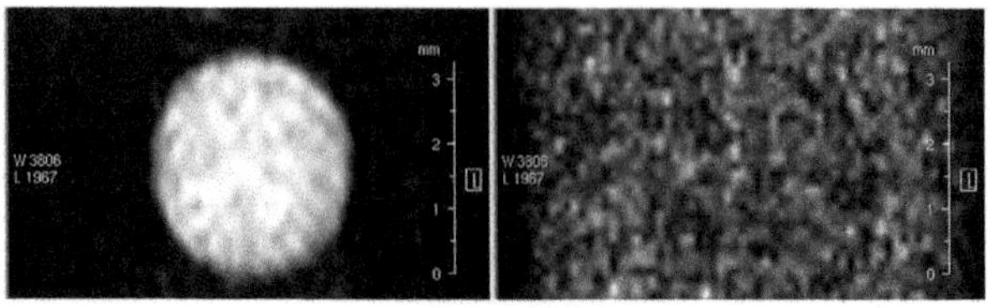

Figure 6.2 Axial ^{1}H echo planar imaging of a solution containing pyridine (43 mM) and Crabtree's catalyst (2 mM) in a 5-mm NMR tube equipped with a J. Young's valve. (Left) hyperpolarized through the dissolution of parahydrogen. (Right) 40 s later. Slice: 4 mm; acquired points: 64 × 64 (theoretical resolution: 125 μm/point). Experiment time: 1 s.

method to hyperpolarize substrates in solution without modifying their chemical structure was proposed. The signal amplification by reversible exchange (SABRE) method was shown to be particularly useful to hyperpolarize molecules with heteroaromatic rings such as pyridine [24]. Not only MRS but also MRI may be performed using the SABRE method (Figure 6.2). Very recently, the method has been extended in order to hyperpolarize, in an aprotic solvent, virtually any molecule carrying an exchangeable proton [25].

In all these hyperpolarization methods, the magnetic field value is highly important, both at the production place and for the transport and storage. For instance, considering hyperpolarization via parahydrogen in the "irreversible" hydrogenative method, the magnetic field value and the process of transport to the NMR detection region will have an influence on the populated energy levels and therefore on the resulting ^{1}H spectra (basis of the PASADENA and ALTADENA methods, see for instance C.R. Bowers' chapter in reference [26] for a detailed explanation). In the same vein, in the SABRE approach, the NMR methods to transfer polarization to the substrate of interest will not be the same at low and high magnetic fields [27]. Therefore, in the conception and building of a system integrating hyperpolarization and NMR (micro)detection, these parameters will have to be taken into account.

6.2.2 Microcoils and Cryocoils

Without consideration of the sample part, i.e. without taking into account the spin relaxation and susceptibility effects, the signal-to-noise ratio (SNR) of an NMR experiment is given by the following equation, initially expressed by Hoult and Richards [28] and reformulated here:

$$\mathrm{SNR} = \frac{k_0 \dfrac{B_1}{i} V_\mathrm{s} N \gamma \hbar^2 I(I+1) \dfrac{\omega_0^2}{3\sqrt{2} k T_\mathrm{s}}}{\sqrt{4(R_\mathrm{c} + R_\mathrm{s}) k T_\mathrm{c} \Delta f}}. \tag{6.2}$$

In this equation, k_0 is a constant that accounts for the receiving coil geometry, V_S is the sample volume, N the number of spins per unit volume, γ the gyromagnetic ratio, I the spin angular momentum, ω_0 the Larmor frequency, T_s and T_c the sample and coil temperatures, R_s and R_c the sample and coil resistances, and Δf the receiver bandwidth.

The term B_1/i, the magnetic field per unit current, can be maximized by considering the Biot and Savart law. In the case of a single coil turn of radius r traversed by a current i, the law giving the magnetic field per unit current at its center is expressed as:

$$\frac{B_1}{i} = \frac{\mu_0}{2r}, \tag{6.3}$$

where μ_0 is the magnetic permeability of the free space. To get an idea of the order of magnitude, an electric current of 1 mA in a loop of 1 mm diameter would generate a field in the microtesla range.

As the reciprocity principle states that the signal received from a sample by a coil is proportional to the magnetic field that would be created in the sample if unit current were passed through the coil, the smaller the loop radius, the more sensitive the detection. As stated by Webb [8], the sensitivity gain by reduction of the coil diameter has to be tempered by the decrease of the sample volume (term V_s in Equation 6.2). In conclusion, for mass-limited samples and a given coil volume, the best solution for optimal detection is when the space between the coil and the sample is as small as possible (highest filling factor).

Another important term in Equation 6.2 is the coil temperature at the denominator, which explains the massive advent of cryogenic probes designed to reduce the thermal noise. Here, the thermal noise in the detection pathway, including the receiver coil, cables, and preamplifier, is reduced by decreasing the temperature. However, since the sample remains at, or near room temperature, a thermal insulation is needed between the sample and the coil. This results in a reduction of the filling factor that decreases the signal. Overall, at the expense of a technologically more complex apparatus, the sensitivity gain of a cryo-detection system may reach a factor of 3 to 4 [29]. The resistance R_c of the coil and preamplifier can also be decreased using radiofrequency coils made of high-temperature superconducting materials [30,31].

Various geometries of NMR coils have been proposed in the past. The advantages of each of them have been recently reviewed [32]. The widely used saddle coil geometry in commercial probes is not suitable for a down-scaling. Compared with planar microcoils, solenoid coils offer higher SNR, but are more tedious to manufacture. Stripline coils [33] are very promising as they produce high and homogeneous B_1 field, a key parameter of the Hoult and Richards's equation, and show good scalability.

6.2.3 Solution Flow

All ways to increase NMR sensitivity, i.e. intending to maximize the SNR in a given period, are faced with the same issue: most of the time spent in a classical pulse sequence is constituted of waiting delays. As between two experiments the interscan delay is designed to let the excited nuclear spins return to equilibrium – or close to equilibrium – through longitudinal relaxation, it can become prohibitively large. This makes the NMR observation of some nuclei, particularly those with spin half and low gyromagnetic ratio, very cumbersome. This is the case for instances of ^{89}Y or ^{183}W in symmetrical environments (^{89}Y T_1 is larger than 600 s for YCl_3 in D_2O) [34], but also for less exotic nuclei, such as some quaternary ^{13}C and some ^{15}N. For instance, the ^{13}C T_1 in urea is ~40 s at room temperature, and the ^{15}N T_1 values in acetonitrile and benzonitrile are on the order of a few minutes [35].

Obviously, such an issue may compromise not only the sensitivity of the NMR experiment, but also its capability to reproduce quantitatively the relative amounts of molecules in the sample, if the interscan delay is too short. To overcome such difficulties, for a long time it has been proposed to create a solution flow in a closed-loop circuit and to limit the NMR detection area to only a part of this loop [7,36]. The use of paramagnetic species immobilized in another part of the circuit (outside the NMR detection area), designed to shorten the relaxation times of the nuclei of interest has also been suggested [37].

6.3 MRI Methods Dedicated to Hyperpolarization

Hyperpolarized species can also be used to largely improve the spatial resolution in NMR images. In every MRI experiment, for a given field-of-view in one dimension, the intrinsic spatial resolution (in Hz/point) is given by the inverse of the highest value of the wave vector:

$$k_{max} = \gamma G_{max}\ t_{max}. \tag{6.4}$$

Here G_{max} and t_{max} are the amplitude and duration of the gradient giving rise to the most distant point from the center of the **k**-space in this dimension, respectively.

The same rule applies for localized spectroscopy. For given excitation and/or inversion pulses, the size of each side of the voxel is also determined by $\gamma\ G_{max}\ t_{max}$. Here the maximal application time of the highest gradient in combination with the frequency selective pulse will give the thinner slice in the considered dimension.

But obviously other factors come to degrade this "nominal" spatial resolution, such as diffusion, transverse relaxation, or variation of the resonance frequency (due for instance to local magnetic susceptibility effects). It is also important to consider that diminishing the voxel size will decrease the SNR, and it is for this last point that hyperpolarized species may become interesting.

Dealing with localized spectroscopy, first the gyromagnetic ratio is important. For instance, considering that $\gamma(^{1}H) = 3.61\ \gamma(^{129}Xe)$, the volume of the xenon voxel will be 47 times larger than that of the proton voxel for the same gradient strength.

Diffusion and relaxation should also be taken into account, but since we are focusing here on dissolved hyperpolarized species the self-diffusion coefficient values of which are close to those of the solvent molecules, they are of less influence than in the gas phase. Finally, the transient nature of the hyperpolarization imposes certain constraints on the recording of the NMR experiment. Preference will be given to pulse sequences that acquire the center of the **k**-space first.

For Cartesian sampling, this means using the centric encoding mode with alternation of the phase direction ($0, +\Delta G, -\Delta G, +2\Delta G, -2\Delta G$ ...) or sampling only half the **k**-space. Note that according to the lifetime of the hyperpolarization with respect to the experiment time, a blurring may occur due to the decrease of the magnetization before the phase gradient from **k**-line to **k**-line in the phase dimension. To make the signal from hyperpolarization even more profitable, the transverse magnetization after each acquisition can be recycled (the condition here is that $T_2^* \gg$ TR). The steady-state free precession (SSFP) experiments, which include the true fast imaging and steady precession (TrueFISP) sequence, are used to image hyperpolarized species and were particularly developed for ^{13}C [38]. The principle of such a sequence is based on the "rewinding" of the phase gradients in order to compensate for the displacement in the phase dimension of the **k**-space.

Non-Cartesian imaging methods like ultrashort echo time (UTE), zero echo time (ZTE), or spiral could be preferred to Cartesian ones, but they may be more difficult to implement as prior knowledge of the **k**-space trajectories created by gradient shapes is necessary for data reconstruction. For the sake of sensitivity, on some systems the ^{1}H channel can be used to acquire these trajectories.

Dealing with localized spectroscopy, classical sequences such as point resolved spectroscopy (PRESS) or stimulated echo acquisition mode (STEAM) use many radiofrequency pulses for slice selection, a process rarely suitable for hyperpolarized species for the above-mentioned reasons. However, in certain circumstances, it is suitable to benefit from chemical exchange to perform fast, simple, and sensitive spectroscopic imaging experiments. This is the case in the field of ^{129}Xe MRI-based pH sensors [39,40]. In this approach, cage-molecules reversibly encapsulate hyperpolarized xenon, and give it specific spectral signatures at frequencies very different from that of free dissolved xenon, which resonates at ~200 ppm. A constant in–out exchange of xenon occurs with these systems. In a proof-of-concept experiment (Figure 6.3), it was shown that the simultaneous use of hyperpolarized xenon in two such xenon hosts possessing ionic groups enabled sensitive and bias-free measurement of solution pH. With the pulse sequence depicted in Figure 6.3b, free xenon is never excited, which enables us to keep xenon hyperpolarization for several minutes, according to its long T_1 value in D_2O. The xenon in–out exchange, slow with respect to the frequency difference between free and caged xenon, but fast with respect to the acquisition time, is used here to constantly replenish the small reservoir (caged xenon) in hyperpolarization.

For detection of small amounts of analytes in such exchange regimes, it can also be advantageous to use indirect methods belonging to the chemical exchange saturation transfer (CEST) family. In 2006, the HyperCEST approach was proposed by Schröder et al. [41]. In this method, the radiofrequency saturation applied at the frequency of the small

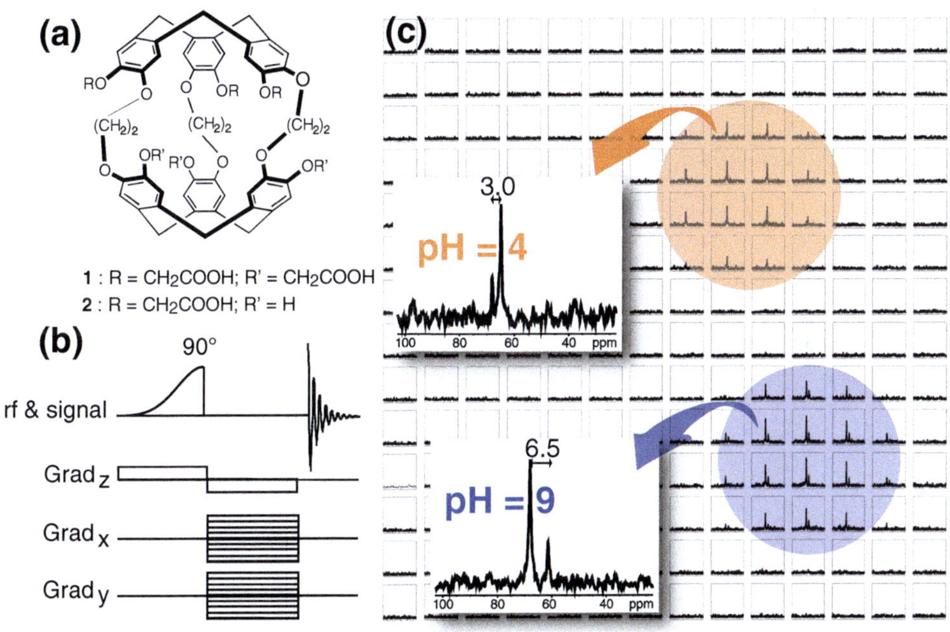

Figure 6.3 ^{129}Xe magnetic resonance spectroscopic imaging for pH sensing. Two tubes of 2.3 mm inner diameter contain a mixture of two cryptophanes (the xenon hosts) at pH 4 (orange) and pH 9 (purple), at concentrations between 150 and 250 μM. The experiment time was 53 s. (a) Structure of the two cryptophanes; (b) pulse sequence, benefiting from in–out xenon exchange. The half-Gaussian soft pulse was centered on the Xe@cryptophane region, at ~60 ppm; (c) ^{129}Xe spectroscopic axial image with intrinsic spatial resolution of 500 μm per pixel. The chemical shift splitting gives access to the pH. Adapted from [40] with permission from Wiley.

magnetization reservoir (caged xenon) leads to a decrease of the major signal, which is detected (Figure 6.4a). This enabled detection of picomolar substrate concentrations. The recording of a *Z*-spectrum (in which the main reservoir signal intensity is displayed as a function of the saturation frequency), however, requires a very stable level of original hyperpolarization, which is not easy to obtain. For hyperpolarized species, it is more efficient to perform a spectrum-per-spectrum averaging than a point-per-point averaging. In this case, ultrafast *Z*-spectroscopy [42] can be advantageous (Figure 6.4b). In such a method, which has also been applied to DNP-hyperpolarized species, saturation is applied concomitantly with a pulsed field gradient. It amounts to saturate spins contained in a slice of the sample. Then, after the read pulse, detection follows in the presence of a second gradient, enabling the recording of the *Z*-polarization. By repeating the scan with and without saturation gradient, the *Z*-spectrum appears with high sensitivity.

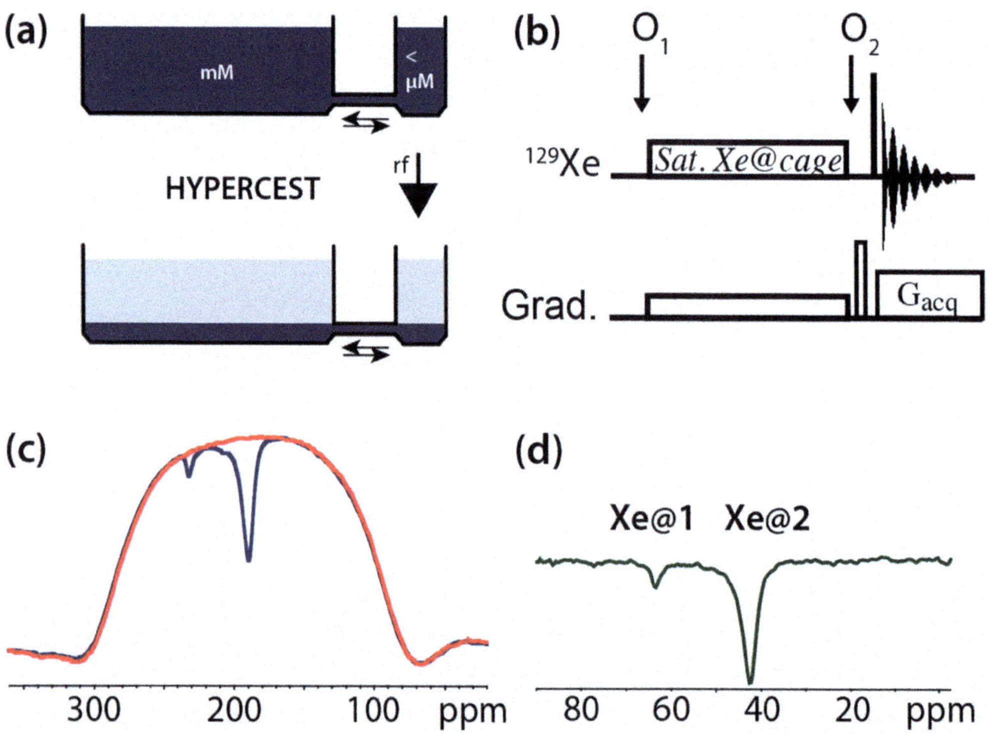

Figure 6.4 (a) Principle of the (hyper)CEST spectroscopy. Saturation at the small pool frequency (xenon caged in cryptophanes for instance, signal ~60 ppm) will also decrease the large pool (free xenon, signal ~200 ppm) if the chemical exchange kinetics is appropriate. Adapted from [43] (b) Pulse sequence used for ultrafast Z-spectroscopy. In each even scan, the minor xenon pool frequency is saturated at the O_1 frequency in the presence of a field gradient. Then the read pulse is applied at the resonance frequency of dissolved xenon O_2 (major pool). Detection is made in the presence of a second gradient. (c) In the absence of radiofrequency saturation the resulting ^{129}Xe spectrum displays the density distribution of xenon along the tube (red spectrum), whereas with saturation it reveals the presence of two xenon-binding sites Xe@1 and Xe@2 (blue spectrum). (d) The difference between the two spectra provides the caged xenon signals. Adapted with permission from [42]. Copyright (2013) American Chemical Society.

6.4 Combining Methods to Gain Sensitivity

6.4.1 Flow and Micro-detection

A reduction of the size of the NMR detection coil acts as an increase of the mass sensitivity. However, the sample preparation, i.e. the protocol for positioning the sample in the detection area while preserving its integrity and purity, becomes increasingly tricky when the sample size decreases. In some examples, the sample was fixed with epoxy glues [44], embedded into an agarose gel [45], or confined to a particular area of a larger volume. But more often, a stopped-flow system was used to position the sample precisely [46]. For example, a micro-fluidics system was conceived to position and analyze tiny samples. Here, individual worms and ova of typical dimensions around 0.1 mm were analyzed by ^{1}H NMR. The high sensitivity, of 2.5×10^{13} spin/Hz$^{1/2}$, allowed the detection of the most concentrated family of metabolites in the organisms [47].

On-flow analyses, in which the solution to be analyzed continuously flows in a channel that crosses the detection area during the analysis, offer the advantage that solutions may be hyphenated with other techniques. Decreasing the size of both the fluidic system and the coil has several advantages in addition to an increase in mass sensitivity. Lower quantities of reactants and solvents are needed, a useful feature for optimizing chemical reactions. This is particularly advantageous when harmful chemicals are involved. Also, it reduces the time of homogeneous mixing of the reactants to values less than a millisecond, if the chamber is adequately designed [48]. This avoids the NMR analysis of an inhomogeneous solution that would not only broaden the signal by the local inhomogeneities resulting from the different magnetic susceptibilities of each fluid, but also would not account for the kinetics of the reaction that indeed occurs at the molecular scale. More valuable kinetic data and faster kinetics are therefore reachable. Finally, pressure and temperature are more easily controlled. This is particularly important for highly endo- or exothermic reactions. A successful combination of flow and reduced dimensions of the mixing chamber and the detection coil to probe fast chemical reactions with the observation of intermediates has been exemplified [33]. Nevertheless, the flux mode has the disadvantage of broadening the NMR signals, simply because part of the solution leaves the detection coil during the time required for signal detection. Moreover, when the flow rate is high, which is often required to probe fast kinetics, and when the reaction chamber is not submitted to the same magnetic field as the analysis chamber, the spin ensemble may not have the time to reach the Boltzmann equilibrium. Thus, as soon as longitudinal relaxation times differ, the peak integrals are no longer proportional to concentration. To counter this, a method based on the use of the Bloch equations in a fluid dynamics simulation has recently been proposed to quantitatively predict the transport of magnetization for any cell geometry and flow rates [49]. Nevertheless, the method requires the knowledge of longitudinal relaxation data. Placing several microcoils along a capillary in which a reaction occurs has also been proposed to minimize the amount of needed reactants [50]. Remote detection is another elegant method to take advantage of the flow of the spins to be analyzed. Here, two coils are individually optimized. The spins are excited in the first coil, then stored as longitudinal magnetization as they flow toward a second coil used for detection [51]. The coupling of microcoils and flow finds applications, e.g. in reaction monitoring and process control [48], but also in monitoring the metabolic answer in a living animal. For example, the fluctuation of lactate during neuronal simulation of a rat, using a homemade solenoidal microcoil, placed in the isocenter of an MRI scanner, has been demonstrated [52]. Here, the flow mode is used to perform a microdialysis, thus separating unwanted signals belonging to macromolecules or lipids from the signals of the small metabolites of interest, in the present case lactate.

As already mentioned, another setup for on-flow analyses is to use a closed circuit for the solution, which has the advantage of requiring a smaller sample volume than in the open mode. In a recent development, the closed-circuit mode was combined with microcoil detection in a simple and disposable device made by 3D printing [53]. It used the principle of a bubble mini-pump to induce circulation of the solution of interest in a small closed-loop circuit [54]. A significant gain of sensitivity was demonstrated when a solution of ^{13}C-enriched urea was analyzed by ^{13}C NMR spectroscopy compared with the static mode, i.e. when the solution does not circulate. Two versions of this setup were proposed. In the first one (Version 1 in Figure 6.5), the insert containing the fluidic channels and the NMR microcoil is electrically plugged onto a micro-imaging probe basis [53]. Two advantages are retained with this version: (i) it enables both spectroscopic and imaging experiments, and (ii) nuclei that are blinded in commercial probe heads can now be observed. Think for instance of ^{3}He or ^{205}Tl, with resonance frequencies higher than that of ^{31}P but lower than that of ^{19}F, a region usually not covered by commercial probe heads. However, such a setup has the inconvenience that it is only adapted to a given commercial probe head basis (in the present case the Bruker micro-5 micro-imaging probe). Therefore, a second version (Version 2 in Figure 6.5) that uses inductive coupling between its microcoil and the coil of every NMR probe head has been proposed [55]. As for the open-circuit mode, a stopped-flow version results in a higher sensitivity than the on-flow alternative. When flow is abruptly stopped between each scan, all excited spins are detected during the full acquisition time.

In these two versions, a programmable syringe pump pushes the carrier gas into the inlet channel (1) reaching the solution in the lower part of the insert. The bubbles that are created are drawn upward, which induces the flow of the solution (2). A solenoid microcoil (3) detects a part of the solution. Specifically, in Version 2 of the insert (dubbed "Wifi-NMRS"), a brass or plastic rod attached to the insert and ended by a vernier (4) ensures perfect axial alignment of the micro-solenoid with the coil (5) of the host probe head (6) to maximize the coupling between them. In these two setups, other inlet pipes can be added (7), for instance for introduction of reactants in the solution circuit. An advantage of these devices is that the solution circuit remains always in a magnetic field close or equal to that of the NMR detection area. This is important for quantitative results, as exemplified in Figure 6.6.

6.4.2 Cryocoils and Flow

The technologies needed for cryoprobes on one hand and flow on the other hand are a priori compatible. Thus, combining them should afford a significant advantage in term of sensitivity gain. Despite this, only a few achievements have been published. Schmidt et al. [56] used a hyphenated technique in which 20 compounds of comparable chemical structure at micromolar concentration were separated by liquid chromatography. Then, while a small fraction was used for mass spectrometry analysis, which requires much less material than NMR due to its higher sensitivity, the large majority of the fractions were collected for concentration by solid-phase extraction and then injected in stopped-flow mode to the NMR cell of a commercial 5-mm cryoprobe. ^{1}H spectra and correlation spectroscopy (COSY) experiments, acquired within minutes to hours, allowed the identification of the majority of compounds.

Recently, another realization took advantage of flow to obtain a temporal resolution on the order of a few seconds [57]. Here, the reactants were introduced into a fixed-bed chromatographic reactor. The stationary phase was made of a catalytical bed that induced both the chemical reaction and the chromatographic separation. The solutes were detected in flow by using a commercial

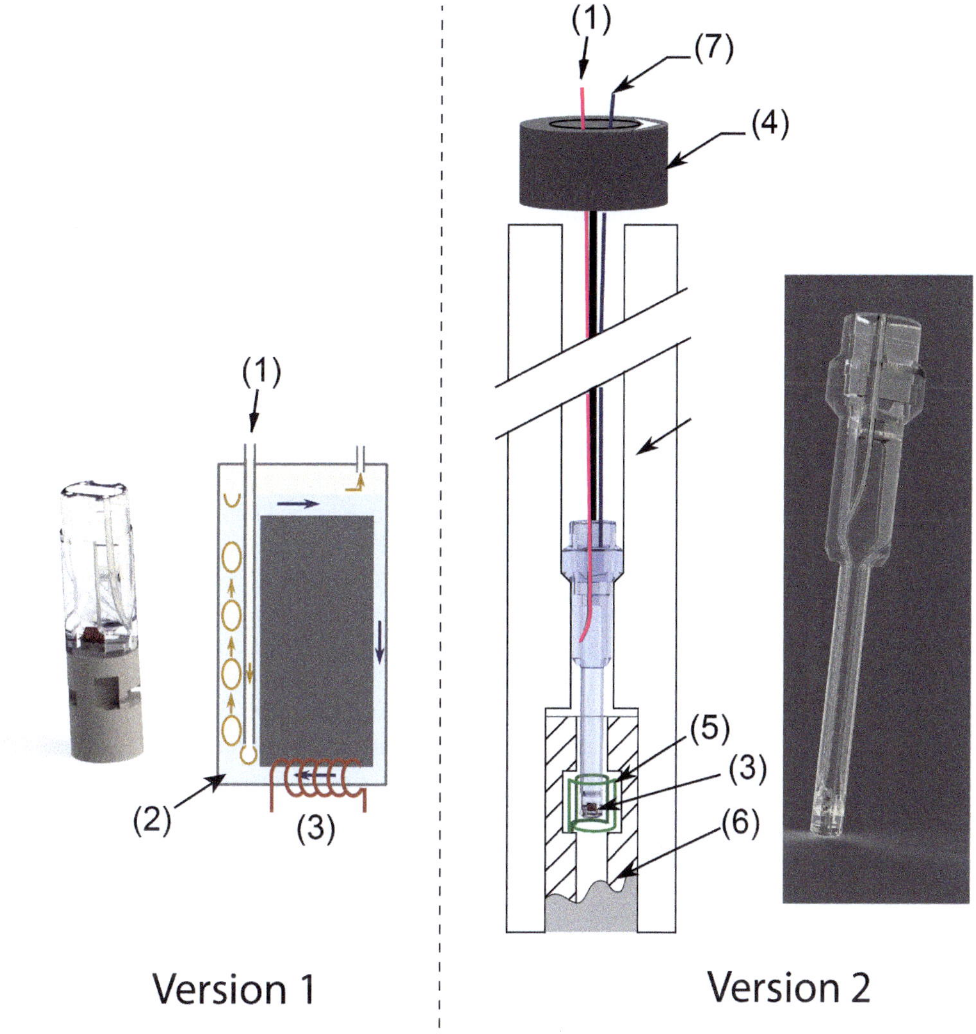

Figure 6.5 3D-printed nuclear magnetic resonance inserts combining flow of solution in a closed-loop circuit and micro-detection. (1) gas inlet, (2) solution of interest, (3) microcoil for NMR detection, (4) vernier placed on top of the magnet to manually optimize the inductive coupling, (5) coil of the host probe, (6) host probe head, (7) second inlet tube for potential addition of other reactants. Reproduced from [53] and [55] with permission from American Chemical Society.

cryoprobe equipped with a homemade tube of 1 mm inner diameter. The tube diameter was reduced in order to avoid large distribution of the solute residence time values. However, it resulted in a lower filling factor of the cryoprobe, therefore decreasing the sensitivity of the analysis. The tube diameter had, therefore, to be compromised.

6.4.3 Hyperpolarization and Micro-detection

Several reasons other than the increase in mass sensitivity alone may encourage the combination of spin hyperpolarization and micro-detection (see chapter 11 in reference [58]). This can facilitate the conception of integrated systems encompassing hyperpolarization and NMR. In the case of a precious sample, the reduction of the quantity required for NMR analysis may also be mandatory. In addition, the use of small samples can be beneficial for microwave penetration in DNP experiments, or light penetration in photo-CIDNP ones. Finally, strong coupling between the sample magnetization and the detection system may provide a further gain in sensitivity.

6.4.3.1 Toward Integrated Mini-devices for Hyperpolarization?

Hyperpolarized species can be produced either in the direct mode or in the batch mode, both methods having their pros and cons. Their "on-flow" production enables them to minimize relaxation issues, while being more difficult to set up: issues related to the lack of space and to the pollution of the sample (trace of alkali metal in the SEOP method, for instance) may be encountered.

Regardless of the chosen production mode and the hyperpolarization method, the setup is usually bulky, not transportable, and can hardly be integrated into a microfluidic platform that would contain NMR micro-detection. Very few studies have focused on reducing the size of these devices.

Jiménez-Martínez et al. designed an on-chip setup for xenon optical pumping that consisted of four chambers 1-mm thick and several millimeters wide connected by microchannels [59]. A flow of xenon and quenching gas (such as nitrogen) passed from chamber to chamber. Each chamber also contained the alkali metal (rubidium). The second chamber received circularly polarized photons from the pump laser, while the third chamber received photons from the probe laser. A transverse DC field in the pump and probe chambers enabled the characterization of the ^{129}Xe polarization as a function of the total gas flow rate, alkali atomic density, and pumping power. In this microtesla field, the NMR polarization of ^{129}Xe exceeded 0.5% at a flow rate of 5 ml/s.

Recently, it has been also shown how the use of microcoils and flows brings decisive advantages to the development of the CIDNP technique [60]. In particular, the fact that the volume to

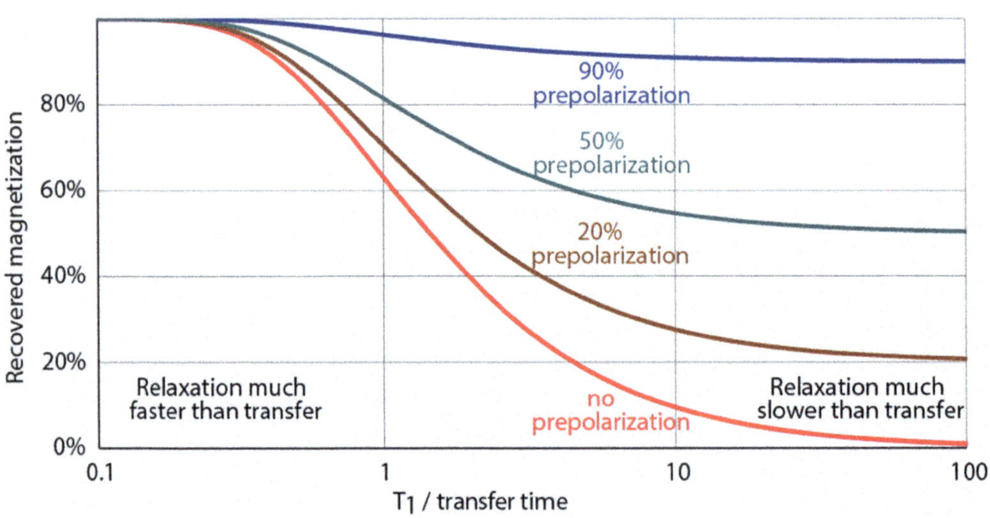

Figure 6.6 Amount of final magnetization as a function of the ratio T_1 over transfer time (sudden approximation) for a system where the sample is transported from the outside to the inside of the nuclear magnetic resonance detection area. As in the monitoring of synthesis intermediates the transfer time has to be as short as possible, the difference between the four curves shows the advantage of prepolarizing the sample.

be irradiated is only 1 μl renders the use of a high-power laser for illumination obsolete. A sub-picomole detection limit has been demonstrated.

6.4.3.2 A Further Sensitivity Gain in the Nonlinear Regime

It has been known for some years that intense magnetizations such as those created by optical pumping [61], parahydrogen [62], or DNP [63] can lead to remarkable nonlinear effects (i.e. where the recorded signal is not directly proportional to magnetization), such as radio amplification by stimulated emission of radiation (RASER) or spectral clustering.

These effects are maximized when the coupling between the magnetization and the detection system is optimized. In 2009, Desvaux et al. [64] performed the following experiment. In a 5-mm NMR tube, a micro-solenoid immersed in a solution was inductively coupled to the ^{129}Xe channel of the NMR probe head. Xenon, previously hyperpolarized via SEOP in batch mode, was introduced into the solution. It could be detected in a few minutes by classical NMR detection, but also by spin noise detection, i.e. without using radiofrequency excitation (see Figure 6.7). Remarkably, the spin noise detection favored the spins inside the microcoil. A sensitivity enhancement by a factor ≈50 was achieved by resorting to this inductively coupled microcoil. The phenomenon that amplifies the spin noise detection is the radiation damping characteristic rate, which can be expressed as [65]:

$$\lambda_r = \frac{\mu_0}{2}\eta Q \gamma M_0. \tag{6.5}$$

Here η and Q are the filling factor and the quality factor of the probe, respectively; and γ and M_0 are the gyromagnetic ratio and the magnetization of the nuclear species of interest. This formula shows that both factors inherent to the detection system (filling factor and quality factor) and to the hyperpolarization level (through the M_0 term) are important to enhance the detection sensitivity. The microcoils, designed to exhibit high filling and quality factors, are thus perfectly suited to enhance these effects. Even for classical inductive detection, this nonlinear regime [65] (the Bloch equations are no longer linear) ensures that the resulting gain is higher than the simple factor of the gains due to hyperpolarization and to (micro-)detection. This may, however, be bought at the price of a loss of quantitative character if precautions are not taken.

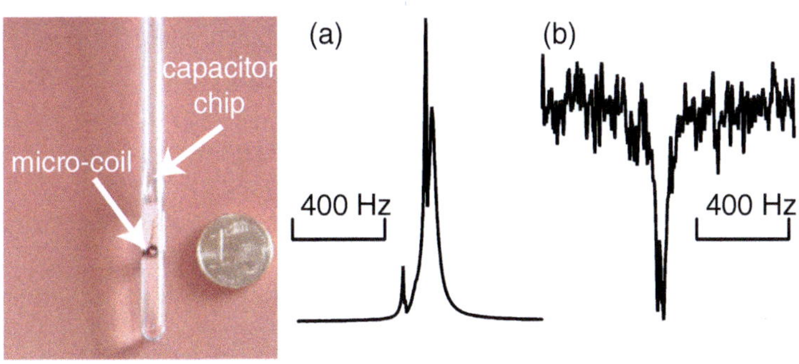

Figure 6.7 Laser-polarized ^{129}Xe spectra of a 5-mm nuclear magnetic resonance tube filled with a cyclohexane solution containing 225 mM ^{129}Xe. A microcoil is immersed in this solution. From left to right, the small peak corresponds to xenon in the hollow plug handling the microcoil, the narrow signal to the bulk phase, and the wider signal to xenon inside the microcoil. (a) After a small flip angle pulse. (b) Noise power spectrum. Adapted from [64] with permission from Wiley.

6.5 Efficient Dissolution of Hyperpolarized Species

Practical considerations on the use of hyperpolarized dissolved species in NMR have recently been reviewed [66]. Dealing with the use of hyperpolarized gases (or precursors of hyperpolarization for parahydrogen), the conception of devices integrating both the mixing chamber and the detection part inside the cavity of the superconducting magnet may face several issues [58].

First, these gases are poorly soluble in aqueous solvents, and in spectroscopy manual shaking of the NMR tube is often the best way to speed up their dissolution. The situation is worse for hydrogen, which is soluble only at the level of 0.8 mM/bar at 298 K [67], than for xenon or krypton, soluble at ~4.5 and 2.6 mM/bar, respectively, at the same temperature. This may lead to a lack of reproducibility, and it obliges one to consider the volume of hyperpolarized gas on top of the solution in the equations of state.

The use of hollow membranes with small pore sizes and high hydrophobic surface area is an elegant way to efficiently dissolve hyperpolarized species such as xenon and parahydrogen in solutions, under continuous flow conditions [66]. Baumer et al. introduced the concept using membranes made from polypropylene and with an average pore size of only 40 nm, to dissolve hyperpolarized xenon in blood with no observable loss of polarization [68]. Furthermore, these membranes were shown to be stable in other solvents such as dimethyl sulfoxide (DMSO) and alcohols. Subsequently, the same hollow fiber membranes were successfully used to dissolve parahydrogen in an aqueous solvent [69].

6.5.1 Use of Dissolved Noble Gases

The device presented in Section 6.4.1 was initially intended to introduce controlled quantities of optically polarized xenon into the solution of interest (Figure 6.8) [54]. The specifications that led to the conception of this 3D-printed device were (i) synchronizing the xenon introduction with the NMR pulse sequence, (ii) avoiding the requirement for large volumes of solution, and (iii) avoiding the presence of bubbles in the NMR detection area that would degrade the field homogeneity. In this device, laser-polarized xenon, produced in the batch mode using a home-built SEOP polarizer, is separated from helium and nitrogen through cryo-condensation and stored in a glass coil on top of the NMR magnet. At this location, a field of 0.01 T is sufficient to prevent fast depolarization of the noble gas. In this setup, the role of xenon is double. Not only does it enrich the solution in hyperpolarization, but also it serves as a carrier gas to induce circulation of the solution in the closed loop. The design of the insert ensures that no bubble is present in the detection region, as proved by the ^{129}Xe spectrum displayed in Figure 6.8c.

Such a device presents several advantages, for example, being compatible with the use of hollow fiber membranes [70], and enabling both spectroscopy and imaging experiments with the same setup. However, as the initial setup was constrained by its electrical plugging to the basis of a given type of micro-imaging probe head, other versions using inductive coupling with the coil of every commercial probe head such as those presented in Section 6.4.1 have been built following the same principle [55].

6.5.2 Use of Dissolved Parahydrogen

Parahydrogen can also be suitable for flow applications and NMR detection by means of microcoils. Devices based on this association can reach an excellent sensitivity. A successful

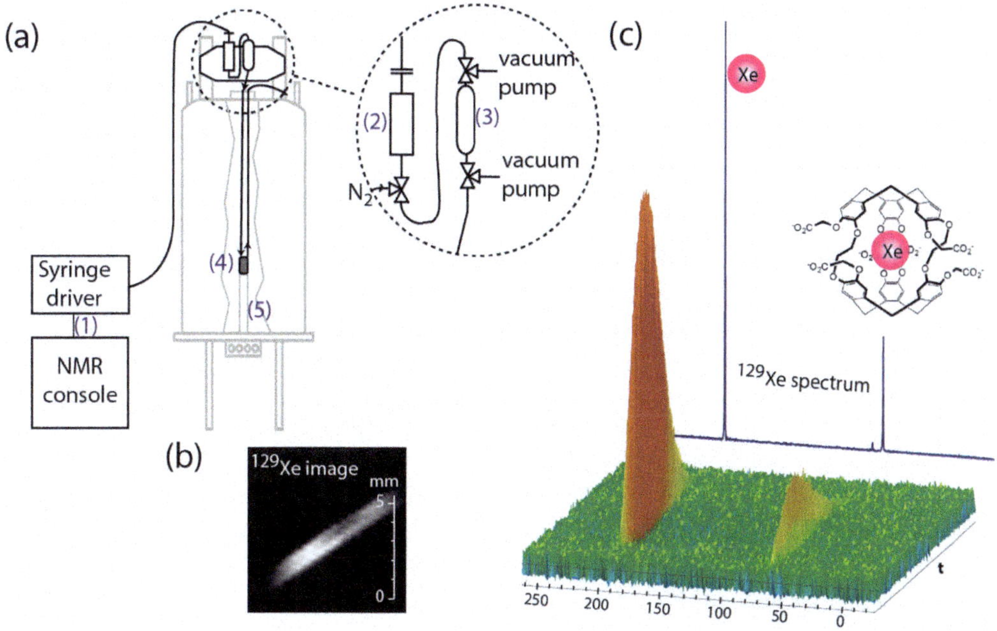

Figure 6.8 (a) Principle of the device delivering hyperpolarized xenon (produced in the batch mode) into the 3D-printed nuclear magnetic resonance (NMR) inserts of Figure 6.5. A programmable syringe pump driven by the NMR console (1) pushes in a controlled way a buffer gas such as nitrogen (2), which itself pushes the hyperpolarized noble gas contained into the reservoir (3) toward the NMR insert (4) plugged onto the NMR probehead (5). (b) ^{129}Xe axial image of the solution inside the microcoil. (c) ^{129}Xe NMR spectra obtained with this setup from an aqueous solution of cryptophane, as a function of time. The signal at ~70 ppm corresponds to xenon in the cage-molecule, the signal near 200 ppm to xenon free in solution. The absence of signal at 0 ppm indicates that there is no bubble in the detection region. Adapted from [54] with permission from the Royal Society of Chemistry.

example concerns a system in which parahydrogen is introduced into a solution by means of a membrane included in a microfluidic chip [71]. The solution contains both the substrate to be hydrogenated and a suitable catalyst. Thanks to a detection volume of only 2.5 μl, picomole amounts of the substrate can be detected. Another device, based on high-field SABRE was recently proposed [72]. The head of an NMR probe designed for proton resonance at 500 MHz had been modified so that parahydrogen can diffuse in a liquid containing substrates responsive to SABRE and an appropriate catalyst. The liquid then flows through an NMR microcoil that serves both to excite spins by single excitation pulses, and detect them. Thanks to the presence of a co-substrate, the measurement of the concentration of substrates in the micromolar range was possible, based on a methodology developed by Eshuis et al. [73]. However modest enhancement factors were observed. SABRE is indeed known to be much less efficient at high field than at lower field, being optimal around 0.01 T for protons. However, associating the proof of hardware concepts of this article and the recently proposed methods based on SABRE and efficient at high field [27], one would obtain a system combining sensitivity and dosage performances that could result in interesting applications in health or food analysis applications, for example.

6.5.3 Dissolution-DNP

The detrimental effects associated with obtaining highly polarized samples via dissolution-DNP lie in the inherent paramagnetism linked to the presence of dissolved radicals. Several solutions, based mainly on the use of chemical scavengers to transform them into diamagnetic species [74] or using ultraviolet-generated, labile radicals [75], have been proposed.

6.6 Current Status and Perspectives

It is highly likely that the use of systems integrating hyperpolarization, solution flow, and optimized detection will be expanded in the near future to exceed the current capabilities of liquid-state NMR. Indeed, while hyperpolarization is mandatory to reach low detection thresholds, it could be of great interest to produce the hyperpolarized species very close to the place where they are detected, for several reasons: (i) avoiding any human handling leads to a higher reproducibility, (ii) the time between the production of these transient polarizations and their detection can be reduced, and (iii) synchronization between the injection of the hyperpolarized substance and the triggering of the pulse sequence can more easily be achieved. Thus in an ideal system provided to the chemist and aimed at quickly analyzing a mixture of solutes, the previously hyperpolarized solution would flow into an optimized detection area.

However, several limitations can be anticipated, mainly inherent to the out-of-equilibrium nature of hyperpolarized species in such a system. First, except maybe with DNP, not all nuclei of all molecules can be hyperpolarized. Second, the quantitative character of NMR may be lost, and ways have to be found to calibrate the amplitude of the response to the number of spins present.

Also, the issues that may be encountered when trying to integrate the polarizer and/or the mixing chamber inside or close to the NMR spectrometer depend on the hyperpolarization method.

For DNP, a difficulty may lie in the necessity to give access both to the microwave and the radiofrequency field at the same volume, without altering their individual performances (for a more detailed discussion, refer to chapter 11 of reference [58]). Thus, the most often encountered systems use the principle of sample shuttling. To this end either a device with two magnetic centers is used [76], or a magnetic tunnel designed to minimize relaxation during the sample transfer is built [77]. Conversely to what was described in Section 6.4.1 for nonhyperpolarized species, here the faster the transfer, the higher the resulting polarization and thus the NMR signal.

For optically pumped noble gases, it is also important to keep them in (or very close to) the high magnetic field of an NMR spectrometer. Indeed, while optical pumping only requires a moderate magnetic field (in order to split the electron spin levels in SEOP, for instance), the relaxation of a laser-polarized noble gas batch is accelerated when it diffuses in magnetic field gradients [13]. The best solution would be to place the optical cell inside the magnet bore, provided that the magnetic field gradients are not too high. Here anyway the main difficulty lies in the arrangement of the photon beam. While there exist optical fibers that can support high photon flux (keep in mind that the efficiency of the optical pumping is related to the number of photons that cross the optical cell, per unit time), they cannot maintain the light polarization. At their extremity a mixture of linear s and p polarization will be present, and, except if the magnetic field is high enough to allow discrimination of the transitions through their frequency alone, a beam splitter cube and two quarterwave plates (one for each linear polarization) are required, which can become cumbersome. If the NMR system offers enough space, a solution with microfabricated chips can be proposed [59].

For PHIP, parahydrogen may be produced outside the magnet and delivered via tubing with slow isomerization, therefore keeping almost intact its capacity to induce hyperpolarization. However, as soon as parahydrogen is in contact with an appropriate catalyst in solution, the hyperpolarized signal decreases quickly at a rate defined by the longitudinal relaxation of the involved nuclei. Therefore, a system is needed in which parahydrogen is quickly introduced in the solution at or near the region of homogeneous field in the core of the magnet, and the NMR spectra can quickly be recorded. An alternative is to use heterogeneous catalysis, in which the solution containing dissolved hydrogen flows through a solid-phase catalyst. To do so, the device should fit within the available volume inside the NMR probe, in the core of the magnet, paying special attention to the parahydrogen dissolution process. The device proposed by Eills et al. [71] resolves two important challenges: accelerating parahydrogen dissolution by the use of a membrane, and increasing the pressure to increase the concentration of the dissolved gas. It uses a microcoil detector. In this article, the PASADENA (Parahydrogen and Synthesis Allow Dramatically Enhanced Nuclear Alignment) hyperpolarization method gave intense signals, albeit in antiphase, and was limited to molecules comprising a multiple bond. Based on this kind of device, it would be profitable to apply other parahydrogen-based methodologies prone to produce high polarization levels at high magnetic field. Some molecules, such as pyruvate, may be produced in a hyperpolarized state based on the side arm methodology [78]. Also using the device with a signal amplification by reversible exchange (SABRE) methodology would extent the applicability to a large number of molecules. Various efficient SABRE techniques at high field have been developed, of which one is particularly promising when molecules with natural isotope abundance have to be quantitatively analyzed [79]. When coupled to a standard addition method, it gives accurate and reasonably precise concentrations of solute, even at micromolar concentrations. It would also be highly profitable to provide in phase signals, for instance using the selective excitation of polarization using PASADENA (SEPP) concept [80]. Then, by adding to the NMR sequence an exchange spectroscopy (EXSY) pulse scheme, chemical shifts of the molecules free in solution would be recovered. This would open the route toward the use of SABRE-Relay at high field – a method developed so far only at low magnetic field, but applicable to the large panel of molecules comprising at least one exchangeable proton [25].

Finally, the use of cryocoils or high-temperature superconducting coils represents a gain of sensitivity that does not alter the other properties of NMR. Research to optimize instrumentation in combining micro and cryo aspects of the detection coil, as well as combining hyperpolarization methods with such devices, will probably become popular even with an apparently modest sensitivity gain. The success of probes equipped with cryocoils results from the absence of additional constraints for the final user, which are neither instrumental nor a compromise with the previously cited qualities of NMR.

Acknowledgments

Support from the French Ministry of Research (project 17-LCV2-0002-01 LabCom DESIR, project ANR-19-CE19-0024 PHOENIX and project ANR-20-CE30-0021 HELPING) is acknowledged. Dr. V. Gervais (I2BC) is warmly thanked for reading the manuscript carefully.

References

1 Zhu, Y. and Fang, Q. (2013). Analytical detection techniques for droplet microfluidics – A review. *Analytica Chimica Acta* 787: 24–35.

2 Shin, T.-H., Choi, Y., Kim, S. et al. (2015). Recent advances in magnetic nanoparticle-based multi-modal imaging. *Chemical Society Reviews* 44: 4501–4516.

3 Grey, C.P. and Tarascon, J.M. (2017). Sustainability and *in situ* monitoring in battery development. *Nature Materials* 16: 45–56.

4 Ilott, A.J., Mohammadi, M., Chang, H.J. et al. (2016). Real-time 3D imaging of microstructure growth in battery cells using indirect MRI. *Proceedings of the National Academy of Sciences of the United States of America* 113: 10779–10784.

5 Windt, C.W. and Blümler, P. (2015). A portable NMR sensor to measure dynamic changes in the amount of water in living stems or fruit and its potential to measure sap flow. *Tree Physiology* 35: 366–375.

6 Hartung, M.P., Grist, T.M., and François, C.J. (2011). Magnetic resonance angiography: Current status and future directions. *Journal of Cardiovascular Magnetic Resonance* 13: 19.

7 Sudmeier, J.L., Günther, U.L., Albert, K. et al. (1996). Sensitivity optimization in continuous-flow FTNMR. *Journal of Magnetic Resonance A* 118: 145–156.

8 Webb, A.G. (2013). Radiofrequency microcoils for magnetic resonance imaging and spectroscopy. *Journal of Magnetic Resonance* 229: 55–66.

9 Lee, J.H., Okuno, Y., and Cavagnero, S. (2014). Sensitivity enhancement in solution NMR: Emerging ideas and new frontiers. *Journal of Magnetic Resonance* 241: 18–31.

10 Hövener, J.-B., Pravdivtsev, A.N., Kidd, B. et al. (2018). Parahydrogen-based hyperpolarization for biomedicine. *Angewandte Chemie International Edition* 57: 11140–11162.

11 Green, R.A., Adams, R.W., Duckett, S.B. et al. (2012). The theory and practice of hyperpolarization in magnetic resonance using *para*hydrogen. *Progress in Nuclear Magnetic Resonance Spectroscopy* 67: 1–48.

12 Buntkowsky, G., Hoffmann, W., Kupka, T. et al. (1998). Application of optical nuclear polarization enhanced ^{13}C NMR. *Journal of Physical Chemistry A* 102: 5794–5801.

13 Walker, T.G. and Happer, W. (1997). Spin-exchange optical pumping of noble-gas nuclei. *Reviews of Modern Physics* 69: 629–642.

14 Batz, M., Nacher, P.-J., and Tastevin, G. (2011). Fundamentals of metastability exchange optical pumping in helium. *Journal of Physics: Conference Series* 292: 012002.

15 Suter, D. and Jelezko, F. (2017). Single-spin magnetic resonance in the nitrogen-vacancy center of diamond. *Progress in Nuclear Magnetic Resonance Spectroscopy* 98–99: 50–62.

16 Maly, T., Debelouchina, G.T., Bajaj, V.S. et al. (2008). Dynamic nuclear polarization at high magnetic fields. *Journal of Chemical Physics* 128: 052211.

17 Goez, M. (2009). Photo-CIDNP spectroscopy. In: *Annual Reports on NMR Spectroscopy* (ed. G.A. Webb), 66: 77–147. Amsterdam: Elsevier.

18 Hoffman, B.M. (2003). Electron-nuclear double resonance spectroscopy (and electron spin-echo envelope modulation spectroscopy) in bioinorganic chemistry. *Proceedings of the National Academy of Sciences of the United States of America* 100: 3575–3578.

19 Meier, B., Dumez, J.-N., Stevanato, G. et al. (2013). Long-lived nuclear spin states in methyl groups and quantum-rotor-induced polarization. *Journal of the American Chemical Society* 135: 18746–18749.

20 Hirsch, M.L., Kalechofsky, N., Belzer, A. et al. (2015). Brute-force hyperpolarization for NMR and MRI. *Journal of the American Chemical Society* 137: 8428–8434.

21 Ardenkjær-Larsen, J.H., Fridlund, B., Gram, A. et al. (2003). Increase in signal-to-noise ratio of > 10,000 times in liquid-state NMR. *Proceedings of the National Academy of Sciences of the United States of America* 100: 10158–10163.

22 Gentile, T.R., Nacher, P.J., Saam, B. et al. (2017). Optically polarized ^{3}He. *Reviews of Modern Physics* 89.

23 Kovtunov, K.V., Pokochueva, E.V., Salnikov, O.G. et al. (2018). Hyperpolarized NMR spectroscopy: D-DNP, PHIP, and SABRE techniques. *Chemistry, an Asian Journal* 13: 1857–1871.

24 Adams, R.W., Aguilar, J.A., Atkinson, K.D. et al. (2009). Reversible interactions with parahydrogen enhance NMR sensitivity by polarization transfer. *Science* 323: 1708–1711.

25 Iali, W., Rayner, P.J., and Duckett, S.B. (2018). Using parahydrogen to hyperpolarize amines, amides, carboxylic acids, alcohols, phosphates, and carbonates. *Science Advances* 4: eaao6250.

26 Bowers, C.R. (2002). Sensitivity enhancement utilizing parahydrogen. In: *Encyclopedia of Nuclear Magnetic Resonance* Vol. 9. (eds. D.M. Grant and R.K. Harris), 750–769. Chichester, UK: Wiley.

27 Barskiy, D.A., Knecht, S., Yurkovskaya, A.V. et al. (2019). SABRE: Chemical kinetics and spin dynamics of the formation of hyperpolarization. *Progress in Nuclear Magnetic Resonance Spectroscopy* 114–115: 33–70.

28 Hoult, D.I. and Richards, R.E. (1976). The signal-to-noise ratio of the nuclear magnetic resonance experiment. *Journal of Magnetic Resonance* 24: 71–85.

29 Kovacs, H., Moskau, D., and Spraul, M. (2005). Cryogenically cooled probes – A leap in NMR technology. *Progress in Nuclear Magnetic Resonance Spectroscopy* 46: 131–155.

30 Brey, W.W., Edison, A.S., Nast, R.E. et al. (2006). Design, construction, and validation of a 1-mm triple-resonance high-temperature-superconducting probe for NMR. *Journal of Magnetic Resonance* 179: 290–293.

31 Laistler, E., Poirier-Quinot, M., Lambert, S.A. et al. (2015). In vivo MR imaging of the human skin at subnanoliter resolution using a superconducting surface coil at 1.5 Tesla. *Journal of Magnetic Resonance Imaging* 41: 496–504.

32 Gomez, M.V. (2017). de la Hoz A. NMR reaction monitoring in flow synthesis. *Beilstein Journal of Organic Chemistry* 13: 285–300.

33 Oosthoek-de Vries, A.J., Nieuwland, P.J., Bart, J. et al. (2019). Inline reaction monitoring of amine-catalyzed acetylation of benzyl alcohol using a microfluidic stripline nuclear magnetic resonance setup. *Journal of the American Chemical Society* 141: 5369–5380.

34 Merritt, M.E., Harrison, C., Kovacs, Z. et al. (2007). Hyperpolarized ^{89}Y offers the potential of direct imaging of metal ions in biological systems by magnetic resonance. *Journal of the American Chemical Society* 129: 12942–12943.

35 Colell, J.F.P., Logan, A.W.J., Zhou, Z. et al. (2017). Generalizing, extending, and maximizing nitrogen-15 hyperpolarization induced by parahydrogen in reversible exchange. *Journal of Physical Chemistry C* 121: 6626–6634.

36 Laude, D.A., Jr, Lee, R.W.K., and Wilkins, C.L. (1984). Signal enhancement of long-relaxing ^{13}C nuclei by flow NMR. *Journal of Magnetic Resonance* 60: 453–459.

37 Zhang, Y. and Laude, D.A., Jr (1990). Immobilized free-radical substrates for magnetization of carbon-13 nuclei in flow NMR measurements. *Journal of Magnetic Resonance* 87: 46–55.

38 Golman, K., Olsson, L.E., Axelsson, O. et al. (2003). Molecular imaging using hyperpolarized ^{13}C. *The British Journal of Radiology* 76: S118–S127.

39 Berthault, P., Huber, G., and Desvaux, H. (2009). Biosensing using laser-polarized xenon NMR/MRI. *Progress in Nuclear Magnetic Resonance Spectroscopy* 55: 35–60.

40 Léonce, E., Dognon, J.-P., Pitrat, D. et al. (2018). Accurate pH sensing using hyperpolarized ^{129}Xe NMR spectroscopy. *Chemistry: A European Journal* 24: 6534–6537.

41 Schröder, L., Lowery, T.J., Hilty, C. et al. (2006). Molecular imaging using a targeted magnetic resonance hyperpolarized biosensor. *Science* 314: 446–449.

42 Boutin, C., Léonce, E., Brotin, T. et al. (2013). Ultrafast Z-spectroscopy for ^{129}Xe NMR-based sensors. *Journal of Physical Chemistry Letters* 4: 4172–4176.

43 Zhang, S., Merritt, M., Woessner, D.E. et al. (2003). PARACEST agents: Modulating MRI contrast via water proton exchange. *Accounts of Chemical Research* 36: 783–790.

44 Peck, T.L., Magin, R.L., and Lauterbur, P.C. (1995). Design and analysis of microcoils for NMR microscopy. *Journal of Magnetic Resonance B* 108: 114–124.

45 Grisi, M., Vincent, F., Volpe, B. et al. (2017). NMR spectroscopy of single sub-nL ova with inductive ultra-compact single-chip probes. *Scientific Reports* 7: 44670.

46 Seeber, D.A., Cooper, R.L., Ciobanu, L. et al. (2001). Design and testing of high sensitivity microreceiver coil apparatus for nuclear magnetic resonance and imaging. *Review of Scientific Instruments* 72: 2171–2179.

47 Montinaro, E., Grisi, M., Letizia, M.C. et al. (2018). 3D printed microchannels for sub-nL NMR spectroscopy. *PLoS ONE* 13: e0192780.

48 Liu, Y. and Jiang, X. (2017). Why microfluidics? Merits and trends in chemical synthesis. *Lab on a Chip* 17: 3960–3978.

49 Friebel, A., Specht, T., von Harbou, E. et al. (2020). Prediction of flow effects in quantitative NMR measurements. *Journal of Magnetic Resonance* 312: 106683.

50 Ciobanu, L., Jayawickrama, D.A., Zhang, X. et al. (2003). Measuring reaction kinetics by using multiple microcoil NMR spectroscopy. *Angewandte Chemie International Edition* 42: 4669–4672.

51 McDonnell, E.E., Han, S., Hilty, C. et al. (2005). NMR analysis on microfluidic devices by remote detection. *Analytical Chemistry* 77: 8109–8114.

52 Crémillieux, Y., Dumont, U., Mazuel, L. et al. (2019). Online quantification of lactate concentration in microdialysate during cerebral activation using ^{1}H-MRS and sensitive NMR microcoil. *Frontiers in Cellular Neuroscience* 13: 89.

53 Carret, G., Berthelot, T., and Berthault, P. (2017). Enhancing NMR of nonrelaxing species using a controlled flow motion and a miniaturized circuit. *Analytical Chemistry* 89: 2995–3000.

54 Causier, A., Carret, G., Boutin, C. et al. (2015). 3D-printed system optimizing dissolution of hyperpolarized gaseous species for micro-sized NMR. *Lab on a Chip* 15: 2049–2054.

55 Carret, G., Berthelot, T., and Berthault, P. (2018). Inductive coupling and flow for increased NMR sensitivity. *Analytical Chemistry* 90: 11169–11173.

56 Schmidt, S., Piechotta, C., Godejohann, M. et al. (2010). Characterisation of commercially available linear alkylbenzenesulfonates by LC-SPE-NMR/MS (liquid chromatography-solid phase extraction-nuclear magnetic resonance spectroscopy-mass spectroscopy). *Talanta* 82: 143–150.

57 Brächer, A., Kreußer, L.M., Qamar, S. et al. (2018). Application of quantitative inline NMR spectroscopy for investigation of a fixed-bed chromatographic reactor process. *Chemical Engineering Journal* 336: 518–530.

58 Anders, J. and Korvink, J.G. (eds.) (2018). *Micro and Nano Scale NMR: Technologies and Systems*. Weinheim, Germany: Wiley-VCH.

59 Jiménez-Martínez, R., Kennedy, D.J., Rosenbluh, M. et al. (2014). Optical hyperpolarization and NMR detection of ^{129}Xe on a microfluidic chip. *Nature Communications* 5: 3908.

60 Mompeán, M., Sánchez-Donoso, R.M., De La Hoz, A. et al. (2018). Pushing nuclear magnetic resonance sensitivity limits with microfluidics and photo-chemically induced dynamic nuclear polarization. *Nature Communications* 9: 108.

61 Marion, D.J.-Y., Berthault, P., and Desvaux, H. (2009). Spectral and temporal features of multiple spontaneous NMR-maser emissions. *The European Physical Journal D* 51: 357–367.

62 Suefke, M., Lehmkuhl, S., Liebisch, A. et al. (2017). Para-hydrogen raser delivers sub-millihertz resolution in nuclear magnetic resonance. *Nature Physics* 13: 568–572.

63 Chen, H.-Y., Lee, Y., Bowen, S. et al. (2011). Spontaneous emission of NMR signals in hyperpolarized proton spin systems. *Journal of Magnetic Resonance* 208: 204–209.

64 Desvaux, H., Marion, D.J.Y., Huber, G. et al. (2009). Nuclear spin-noise spectra of hyperpolarized systems. *Angewandte Chemie International Edition* 48: 4341–4343.

65 Desvaux, H. (2013). Non-linear liquid-state NMR. *Progress in Nuclear Magnetic Resonance Spectroscopy* 70: 50–71.

66 Berthault, P., Boutin, C., Martineau-Corcos, C. et al. (2020). Use of dissolved hyperpolarized species in NMR: Practical considerations. *Progress in Nuclear Magnetic Resonance Spectroscopy* 118–119: 74–90.

67 Purwanto, D.R., Deshpande, R. M., Chaudhari, R.V. et al. (1996). Solubility of hydrogen, carbon monoxide, and 1-octene in various solvents and solvent mixtures. *Journal of Chemical & Engineering Data* 41: 1414–1417.

68 Baumer, D., Brunner, E., Blümler, P. et al. (2006). NMR spectroscopy of laser-polarized ^{129}Xe under continuous flow: A method to study aqueous solutions of biomolecules. *Angewandte Chemie International Edition* 45: 7282–7284.

69 Roth, M., Kindervater, P., Raich, H.-P. et al. (2010). Continuous ^{1}H and ^{13}C signal enhancement in NMR spectroscopy and MRI using parahydrogen and hollow-fiber membranes. *Angewandte Chemie International Edition* 49: 8358–8362.

70 Simkins, J.W., Schuhmann, S., Guthausen, G. et al. (2020). Characterization of biofilm distribution in hollow fiber membranes using compressed sensing magnetic resonance imaging. *Journal of Membrane Science* 594: 117437.

71 Eills, J., Hale, W., Sharma, M. et al. (2019). High-resolution nuclear magnetic resonance spectroscopy with picomole sensitivity by hyperpolarization on a chip. *Journal of the American Chemical Society* 141: 9955–9963.

72 Bordonali, L., Nordin, N., Fuhrer, E. et al., (2019). Parahydrogen based NMR hyperpolarisation goes micro: An alveolus for small molecule chemosensing. *Lab on a Chip* 19: 503–512.

73 Eshuis, N., Hermkens, N., van Weerdenburg, B.J.A. et al. (2014). Toward nanomolar detection by NMR through SABRE hyperpolarization. *Journal of the American Chemical Society* 136: 2695–2698.

74 Miéville, P., Ahuja, P., Sarkar, R. et al. (2010). Scavenging free radicals to preserve enhancement and extend relaxation times in NMR using dynamic nuclear polarization. *Angewandte Chemie International Edition* 49: 6182–6185.

75 Pinon, A.C., Capozzi, A., and Ardenkjær-Larsen, J.H. (2020). Hyperpolarized water through dissolution dynamic nuclear polarization with UV-generated radicals. *Communications Chemistry* 3: 57.

76 Leggett, J., Hunter, R., Granwehr, J. et al. (2010). A dedicated spectrometer for dissolution DNP NMR spectroscopy. *Physical Chemistry Chemical Physics* 12: 5883–5892.

77 Milani, J., Vuichoud, B., Bornet, A. et al. (2015). A magnetic tunnel to shelter hyperpolarized fluids. *Review of Scientific Instruments* 86: 024101.

78 Reineri, F., Boi, T., and Aime, S. (2015). Parahydrogen induced polarization of ^{13}C carboxylate resonance in acetate and pyruvate. *Nature Communications* 6: 5858.

79 Eshuis, N., Aspers, R.L.E.G., van Weerdenburg, B.J.A. et al. (2015). 2D NMR trace analysis by continuous hyperpolarization at high magnetic field. *Angewandte Chemie International Edition* 54: 14527–14530.

80 Eshuis, N., Aspers, R.L.E.G., vanWeerdenburg, B.J.A. et al. (2016). Determination of long-range scalar 1H–1H coupling constants responsible for polarization transfer in SABRE. *Journal of Magnetic Resonance* 265: 59–66.

7

Advances and Adventures with Mobile NMR

Bernhard Blümich, Denis Jaschtschuk, and Christian Rehorn

Institut für Technische und Makromolekulare Chemie, RWTH Aachen University, Aachen, Germany

7.1 Introduction

The conception of mobile nuclear magnetic resonance (NMR) was nourished by the need of the petrochemical industry to characterize oil wells and optimize their yield [1,2]. Even today, well-logging NMR is still the dominant application of mobile NMR [3]. Although it is nothing less than an adventure for those who partake in it for the first time, well logging is not covered in the following text as it is a rather mature and established technique [1,3,4]. Instead the focus is placed on compact stray-field NMR sensors, which are the small, single-sided descendants of the inside-out NMR logging tools. They were explored first at the Southwest Research Institute, San Antonio, Texas, from the 1980s to the 1990s [5]. Similar to the well-logging tools, they can be employed to measure relaxation and translational diffusion of nuclear spins in the inhomogeneous stray fields of a magnet and the radiofrequency coil to assess physical properties of materials; however, they can be far more compact so that the lighter ones can easily be carried and deployed by a single person. Through their portability, these sensors enable studies of an object of interest at its site, be it inside a factory, a hospital or a museum, at an excavation site, or out in the wilderness. Their small size puts them in line with other compact NMR instruments such as tabletop relaxometers and the even younger generation of tabletop spectrometers [6,7], but in contrast to those, compact, mobile NMR opens up a new class of exciting studies. Some of these studies are addressed in the following.

7.2 Compact Stray-field NMR Sensors

Compact NMR instruments rely on permanent magnets because they are smaller than electromagnets for field strengths in the range of 2 T and below. There are two basic designs of compact stray-field NMR sensors, the sweet-spot sensor (Figure 7.1a) and the flat-slice sensor (Figure 7.1b). They differ in the shape and size of the sensitive region outside the magnet, from which the NMR signal is collected. The position as well as the size and the shape of this volume are defined by the excitation frequency, the direction of the magnetic field and the gradient vector at each point of the stray field, the pulse sequence in general, the pulse duration in particular, and the geometry of the radiofrequency coil [8,9]. In first approximation, the bandwidth of the excitation defines the Larmor-frequency window of the magnetization components in the stray field, which

Magnetic Resonance Microscopy: Instrumentation and Applications in Engineering, Life Science and Energy Research, First Edition. Edited by Sabina Haber-Pohlmeier, Bernhard Blümich, and Luisa Ciobanu.

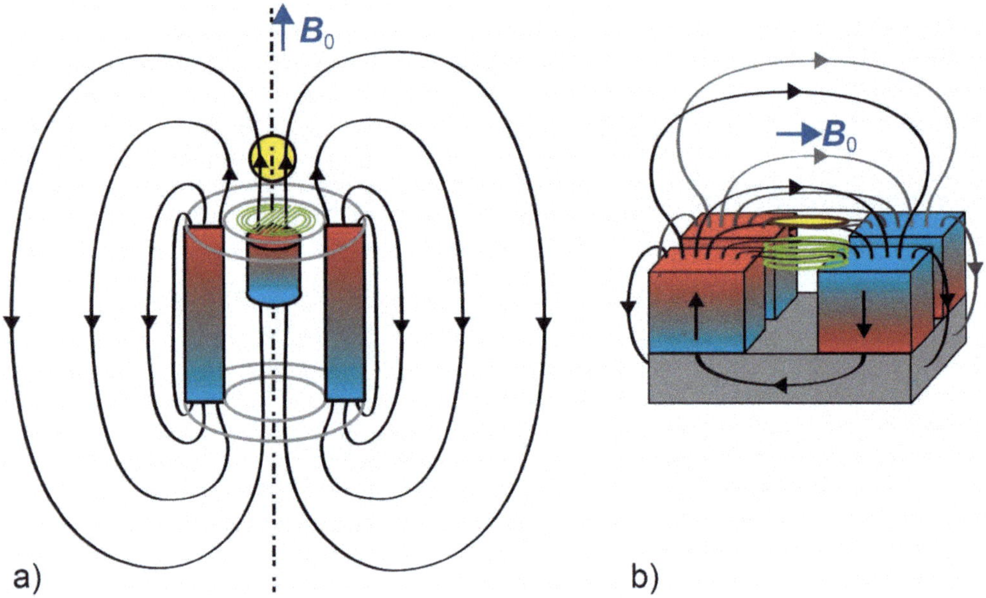

Figure 7.1 Stray-field sensors. Sweet-spot sensors (a) differ from flat-slice sensors (b) in that they have a larger sensitive volume (yellow), a smaller stray-field gradient and a lower stray-field magnitude. Depending on the direction of the magnetic stray field B_0, the geometry of the coil (green) differs.

can be excited and detected. The larger the field gradient is, the narrower is the sensitive volume in the gradient direction, a situation similar to slice selection in magnetic resonance imaging (MRI) [10,11].

Many applications do not require high depth resolution. These can be addressed with sweet-spot sensors like the barrel magnet [12] (Figure 7.1a) and the NMR-MOLE (NMR MObile Lateral Explorer) [13]. Sweet-spot sensors are designed with attention to a low stray-field gradient so that the volume in which Larmor frequencies can be addressed with a short excitation pulse is large. Typically, a low gradient accompanies a low stray field. The main advantage of a large sensitive volume is that sample inhomogeneities like local variations of cross-link density in technical elastomers [4,6,14,15] and fluctuations in the content of stone aggregate in asphalt [16] are averaged out better. Another advantage is a large depth of access, although it comes at the price of larger diameter radiofrequency coils operated at low frequency. Both attributes are unfavorable for short dead time, so the use of such sensors is largely confined to studies of soft matter with long transverse relaxation times. In that respect, sensors like the original NMR-MOUSE (NMR Mobile Universal Surface Explorer) [17] perform best at short distances, because they collect signal from a thin sensitive volume, benefiting from higher stray-field gradients and stronger magnetic fields. The main advantage of high-gradient sensors is that the sensitive volume is thin. It can be flattened into a planar sensitive slice at a fixed distance from the magnet surface (Figure 7.1b), so that the sensor can be employed to measure depth profiles in a point-by-point fashion either by shifting it between measurements away from the object or by conventional MRI frequency encoding, for example, with a spatial resolution as high as 3 μm for the Profile NMR-MOUSE [18]. It depends on the particular sensor, and it may be debated, which type of sensor, sweet-spot or flat-slice, produces the better signal-to-noise ratio [19,20], but in general one has the choice between a larger sensitive volume at lower field or a smaller sensitive volume at higher field. The main challenge with depth profiling is to properly align the sensitive slice with a typical diameter of 10–40 mm with the stratigraphy of the object. For example, to ensure spatial resolution better than 0.2 mm, the misalignment angle between the sensitive slice and the object layers needs to be <0.5° for a 10-mm diameter sensitive slice characteristic of an NMR-MOUSE with 5–10 mm depth of access (Figure 7.2) [21].

The spatial variation of the stray field is determined by the arrangement of magnet elements that make up the stray-field sensor [20]. Shim magnets can be incorporated to shape the profile of the stray field. For example, the axial position of the magnet in the center of the barrel magnet (Figure 7.1a) determines the location of the sweet spot, i.e. the depth of access of the sensor. Similarly, a version of the NMR-MOUSE has been modified with sets of adjustable shim magnets to generate a 2-mm-thick sensitive slice with a homogeneous gradient so that a depth profile over the entire thickness of the slice can be acquired in a single shot in terms of the Fourier transform of the echo response following the principle of frequency encoding in MRI [22]. A simpler way of shaping the stray field is to introduce grooves on the magnet surface or gaps between the blocks

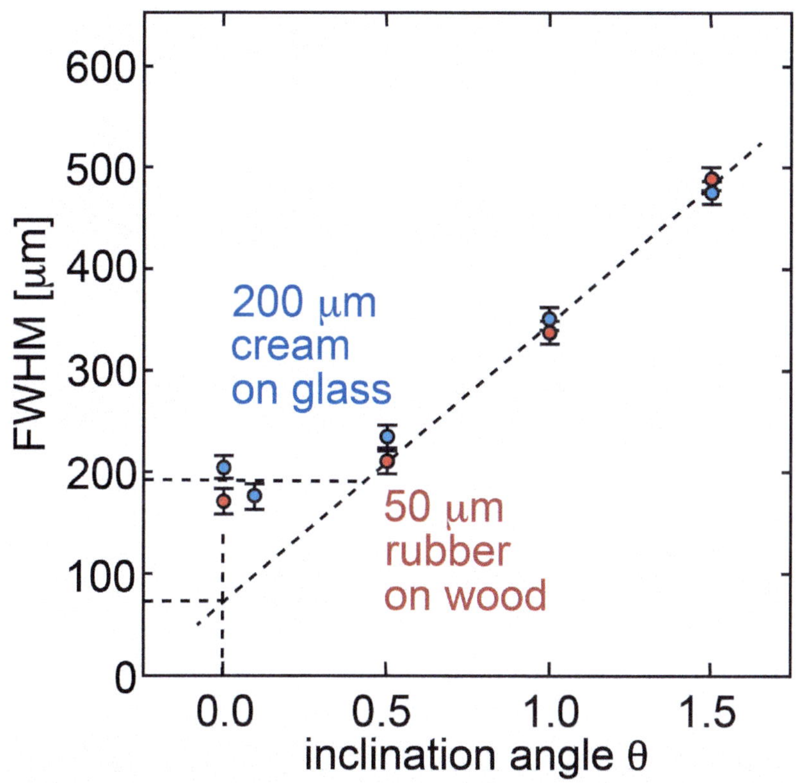

Figure 7.2 The impact of sample misalignment with the sample table of the Profile NMR-MOUSE (Nuclear Magnetic Resonance Mobile Universal Surface Explorer). The spatial resolution is investigated in terms of the full width at half maximum (FWHM) across a skin-cream layer between glass slides and a silicone rubber film covering a wooden board for different misalignment angles [21]. The calculated slice thickness agrees with the extrapolated slice thickness of 80 μm. The misalignment angle of the sample table with the sensitive slice is 0.5°.

[23]. The latter approach has been realized with the Profile NMR-MOUSE (Figure 7.1b), by arranging four magnet blocks on an iron yoke with well-defined gaps in between to produce a flat sensitive slice [18]. Yokes are favorable elements in stray-field magnet assemblies, because they confine the stray field at the passive sides of the sensor where it poses an avoidable hazard to the user and the environment.

The direction of the magnetic field emanating from the active side of the sensor is another important feature of stray-field sensors. In anisotropic materials, the relaxation rates depend on the angle between the magnetic field and the order axis. This has been demonstrated by measurements of the relaxation anisotropy of tendon [24] and of strained natural rubber [25] employing the NMR-MOUSE for which the magnetic field direction is parallel to the sensor surface (Figure 7.1b). If a sensor like the barrel magnet (Figure 7.1a) or a bar magnet [26], having a field orientation along the depth direction, had been employed in these measurements, only the isotropic part of the relaxation rate would have been measured, and the anisotropic part with the magic angle effect would have been inaccessible in the measurement.

The orientation of the stray field $\boldsymbol{B}_0$ with respect to the sensor surface (Figure 7.1) also imposes constraints on the type of radiofrequency coil that can be used [20]. Gradiometer coils like the double-D coil (Figure 7.1a), a variant of the figure-8 coil [12,26], are the basic elements in coil arrays that aim at canceling far-field noise. Their disadvantage is that at distances close to the coil, the signal is collected from different regions in the object. If the sensitive volume needs to be well localized, the simple wire-loop coil (Figure 7.1b) performs best, albeit it lacks noise-canceling properties. There is a simple way to map the shape of the sensitive volume by shifting a signal-bearing object with a straight edge across the sensor from different directions and acquiring the signal amplitude for each position. A two-dimensional map of the sensitive volume can be reconstructed from the radial signal traces by differentiation of each signal trace and subsequent back projection of all derivative traces [27].

While the default acquisition scheme in stray-field NMR is a multi-echo sequence like the CPMG (Carr–Purcell–Meiboom–Gill) sequence, one should bear in mind that many other measurement schemes are possible. For example, multipolar spin states can be excited [28], diffusion can be measured [29], and two-dimensional Laplace experiments can be performed [1,4,6]. Relaxation and diffusion NMR signals used to be analyzed in the time domain by fitting model functions [6]. Today they are preferably analyzed in terms of distributions of relaxation times and diffusion coefficients, following procedures effectively applied in the oil industry [1,3]. Such distributions are derived from multiexponential signal decays by algorithms referred to as inverse Laplace transformation (cf. Figure 7.4b). Standard depth profiles acquired with the Profile NMR-MOUSE resemble depth scans from logging tools (Figure 7.3b). Both show distributions of transverse relaxation times T_2 versus depth, but the depth decrement in well logging is on the meter scale, while for the NMR-MOUSE it is on the sub-millimeter scale, typically 0.1 mm.

7.3 Car Tires

One of the earliest applications of the NMR-MOUSE was to characterize car tires in an effort to reduce racetrack testing during tire development [11,15]. In this application the sensitivity of transverse relaxation to the cross-link density of vulcanized rubber is central [4,6,11,15]. The transverse magnetization decay of rubber is inherently nonexponential due to the presence of residual dipolar couplings among the protons resulting from anisotropic motion of the

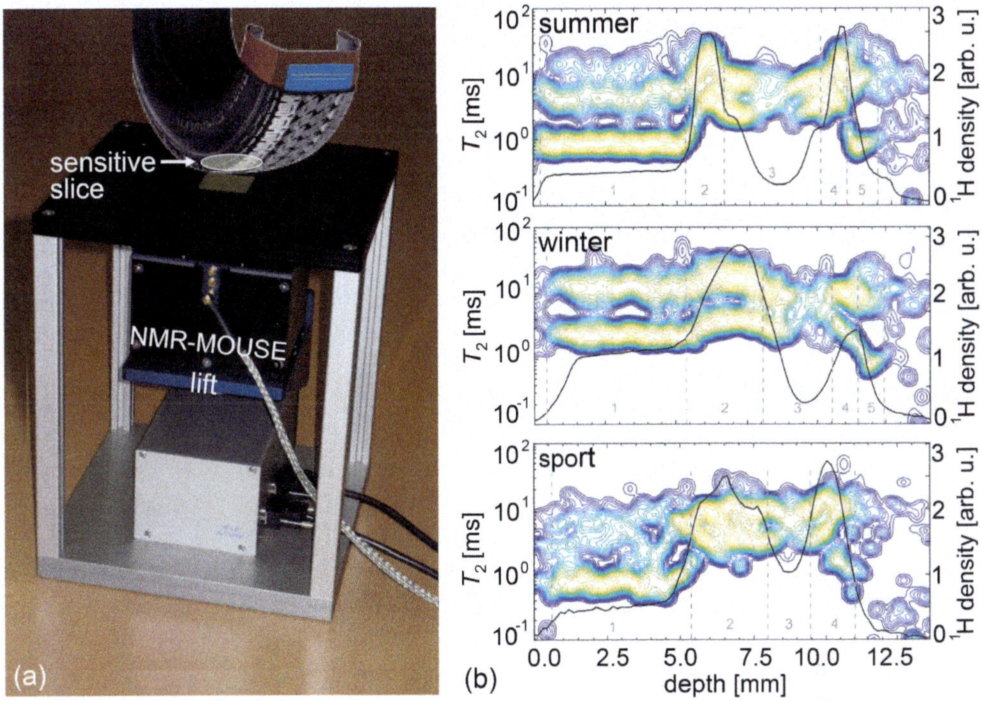

Figure 7.3 Tire analysis with the NMR-MOUSE (Nuclear Magnetic Resonance - Mobile Universal Surface Explorer). (a) To scan a depth profile of the tread, the tire is mounted on a positioning table under which the NMR-MOUSE rests on a precision lift, so that the sensitive slice can be stepped through the different layers of the tire [4]. (b) The depth-resolved relaxation-time distributions (colored maps) and the spin-density profiles (black lines) of a summer, a winter, and a sports tire differ significantly in the thickness and properties of the layers [30]. The gray numbers mark different material layers up to the first steel belt. Bernhard Blümich (author).

macromolecular chains between physical and chemical cross-links. Nevertheless, a phenomenological decomposition of the magnetization decay into nonorthogonal exponential basis functions by inverse Laplace transformation provides a useful fingerprint of the rubbery material that can be interpreted in detail with the help of a reference library of relaxation-time distributions. Depth-resolved distributions of relaxation times at selected sections of the tire provide far richer information on the stacked rubber layers from car tires than single-point measurements. To illustrate the issue, depth profiles through the tread of three different tires, a summer, a winter, and a sports tire, have been measured for comparison (Figure 7.3) [30].

Profiling the tire tread is straightforward. The tire is placed on the sample table of the NMR-MOUSE lift (Figure 7.3a), whereby it is attracted by the magnetic force between the magnet and the steel belt in the tire. This interaction decreases the gradient of the stray field and moves the position of the sensitive slice somewhat further away from the sensor surface, so that the position axis in the depth profiles reports apparent depth and not actual depth. To scan a depth profile of the tread, the NMR-MOUSE mounted to a precision lift is moved up and down, so that the sensitive slice can be stepped through the different layers of the tire. The depth-resolved relaxation-time distributions and spin-density profiles of a summer, a winter, and a sports tire differ significantly in the thickness and properties of the layers (Figure 7.3b) [30]. For example, the tread of a winter tire is softer and that of a sports tire is harder than that of a summer tire. This is evidenced in the first 5 mm of the depth profiles, where the apparent spin density of the winter tire is highest, and the relaxation-time distribution shows intense peaks near 2 and 10 ms. The distributions for the summer and the sports tire differ in that for the summer tire the lowest peak of the tread (at small depths) is below 1 ms and the higher peak is less intense and appears broadened or even split in two. Moreover, the thicknesses and properties of the layers underneath the outer tread layer vary. The material properties are determined by the formulation, the processing, and the vulcanization conditions. Given the performance levels of the tires from driving tests, these properties can be optimized with reference to NMR relaxation-time distributions replacing many of the expensive driving tests by more economical NMR measurements.

7.4 Violins

The most treasured objects in museums are rarely allowed to leave the site. Even when moved on the premises, they are tightly guarded. Examples of such valuable objects are the unique violins of the famous masters from the Amati, Guarneri, and Stradivari workshops at the Ashmolean Museum in Oxford, UK, and the Museo del Violino in Cremona, Italy. Violin soundboards are made from fir and violin backs from maple wood. The violin backs of a total of 15 master violins were investigated at both locations with the Profile NMR-MOUSE in terms of depth-resolved relaxation-time distributions [21,31]. To this end an NMR-MOUSE with 5 mm depth of access was mounted on a translation stage and operated by a Magritek Kea spectrometer. It was shipped to the museums a few days prior to the measurement campaigns in September 2016 and October 2018 for studies within the mobile laboratory MOLAB of the Horizon 2020 project IPERION CH of the European Union. To analyze the violins, they were unlocked from the safety-display cases and mounted upright in a custom-made frame (Figure 7.4a) in a nearby examination room by conservation scientists following tight security measures.

The transverse relaxation-time distribution of the hydrogen nuclei in maple wood acquired with the NMR-MOUSE shows two peaks (Figure 7.4b). The one at short T_2 near 0.09 ms is attributed to protons associated with amorphous cellulose and hemicellulose in the wood matrix, and the one at long T_2 near 1 ms to bound water [21]. Wood also contains glassy lignin and crystalline cellulose, but the relaxation times of their protons are so short that the signal disappears in the dead time of the sensor, here the echo time of the CPMG pulse sequence [6]. Depending on the state of the wood, the positions and peak intensities in the relaxation-time distribution vary.

If the peak integrals are denoted by I_s and I_l for the peak at short and at long relaxation time, respectively, then two useful parameters can be extracted from the distribution. One is the integral sum $I_s + I_l$ corresponding to the amplitude of the CPMG decay or the *hydrogen density*. The other is the ratio I_s/I_l. It defines a *relaxation contrast* that reports material stiffness or brittleness, which varies with the bound water content and possibly shifts in the wood relaxation peak to shorter times. For simplicity the peak integrals were approximated by integrating the relaxation-time distributions from 1 μs to 210 μs and 230 μs to 2 ms. The measurement parameters were $t_p = 7$ μs pulse length, $t_E = 40$ μs echo time, 512 CPMG scans with 64 echoes, and a depth decrement of 100 μm resulting in a total experiment time of two to three hours for one depth profile depending on the thickness of the wood.

The depth-resolved relaxation-time distribution of a well-preserved violin back is constant across the entire depth range apart from measurement noise. This is the case for the Amati

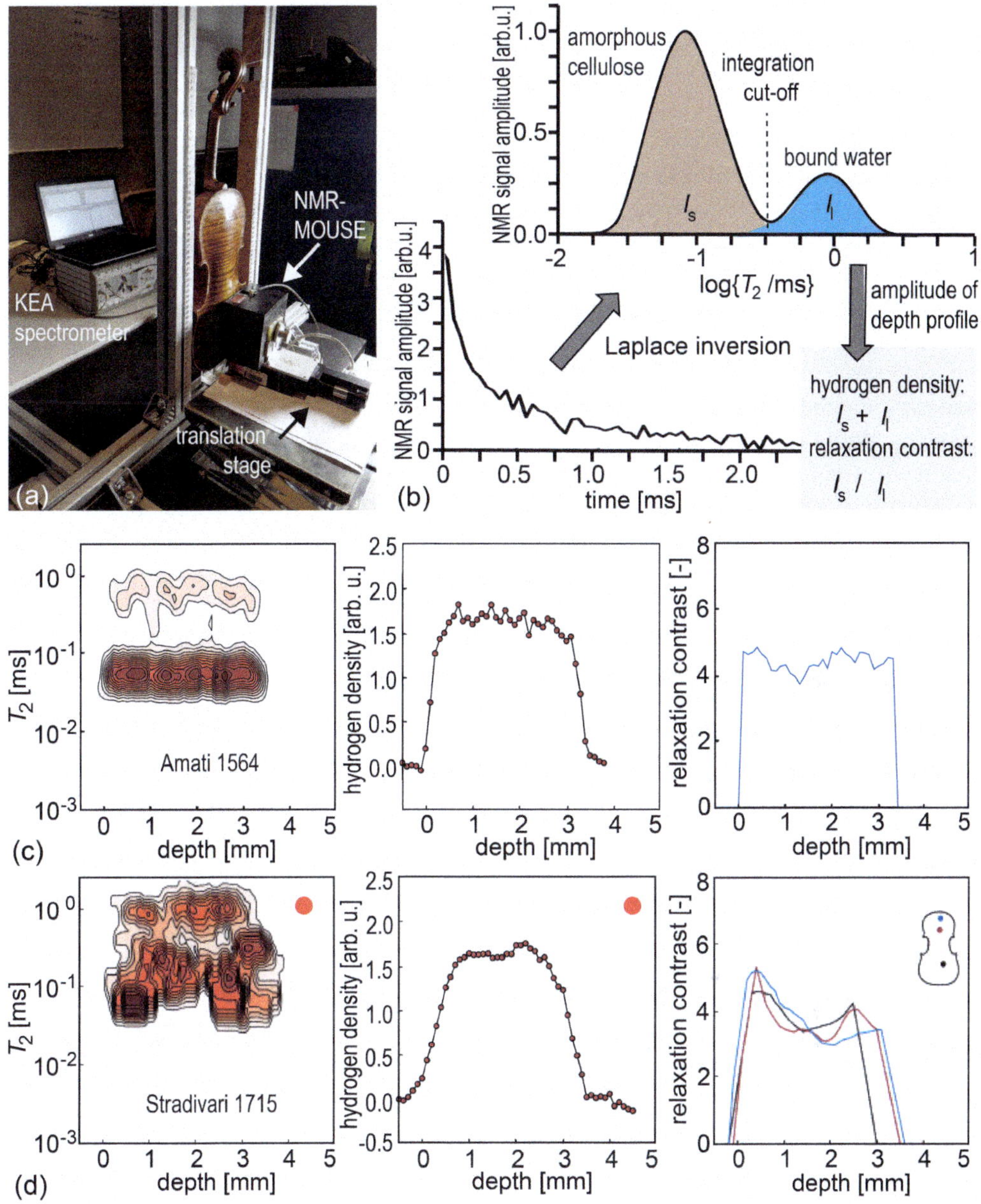

Figure 7.4 Profiling violin backs [21,31]. (a) Setup in the Museo del Violino in Cremona, Italy. (b) The CPMG (Carr–Purcell–Meiboom–Gill) decay signal is decomposed into a distribution of relaxation times by inverse Laplace transformation to extract depth-profile amplitudes. (c) Example of a well-preserved violin back without indications of surface treatments. (d) Example of a violin back bearing the signature of age and surface treatments. The left plots in (c) and (d) show 12 linearly spaced and individually assigned contour lines between zero and the maximum in each graph. The color is used to guide the eye. Bernhard Blümich (author).

1564 violin at the Ashmolean Museum (Figure 7.4c, left). Accordingly, also the spin-density (Figure 7.4c, middle) and the relaxation-contrast (Figure 7.4c, right) profiles are inconspicuous. There are no indications of wood properties altered by surface treatments, neither on the varnished outside nor on the inside of the back plate. This is in stark contrast to the profiles of the Stradivari 1715 violin from the Museo del Violino in Cremona (Figure 7.4d). The trace of the bound water peak in the depth-resolved relaxation-time distributions (Figure 7.4d, left) is more intense but fades at the inside wall. Moreover, the peak from wood is shifted to longer relaxation times in the middle of the back plate and on the inside wall. The spin-density profile (Figure 7.4d, middle) provides little more information than the width of the back plate at the point measured, but the relaxation contrast varies strongly (Figure 7.4d, right). It peaks on both the outer and the inner surfaces, indicating stiffer material than in the center. These variations can originate from several sources. While it is unlikely that the wood itself was originally flawed, the more intense bound water signal can, in principle, result from storage at higher ambient humidity or a different kind of maple wood, or, more likely from some surface treatment either at the time of manufacturing or later at some restoration event. The striking wrinkles in the relaxation peak from wood across depth are hypothesized to arise from softening of the cellulose matrix due to biological degradation or mechanical fatigue from vibrations during extended playing along with stiffening as a result of surface treatments. The details of how the different physical and chemical loads impact the NMR signature of the wood have not been investigated yet, but the overall variations are in line with other mechanical, spectroscopic, and tomographic studies [21].

When comparing the NMR depth profiles of 10 violins from Oxford and 5 violins from Cremona, those from Cremona appear more wrinkled than those from Oxford. Most of the Oxford violins have been donated to the Ashmolean Museum by members of the Hill family of violin makers and restorers who collected them from the mid-1700s onwards being aware of the fact that constant playing and restoration degrade the instruments [32]. The Cremona violins, on the other hand, have a more heterogeneous past in view of the collection housed in the Museo del Violino having been started only in 1961 with the acquisition of the Stradivari 1715 from Desmond Hill in London [33]. This difference in the history of use and restoration is uncovered by the NMR depth profiles of the violin back plates [21].

7.5 Heritage Buildings

Buildings define the pride of a nation. They are attacked in wars, reconstructed, and preserved as symbols of national identity. Buildings from the past are appreciated as important heritage objects that link our present culture to that of our ancestors. Stray-field NMR sensors are important tools to study building materials like stone, brick, cement, and concrete nondestructively to gain insights into the transformation and the state of the material properties during hydration, aging, preservation, and restoration [6,34–37]. Typically, hydrogen signal of free and bound water is detected, and a fraction of the nuclear magnetization with transverse relaxation times shorter than the dead time of sensor remains hidden. This dark magnetization relates to protons close to paramagnetic centers and protons bound in hydrates, dry salts, stone strengthener, as well as conservation agents and macromolecular additives. Recent adventures with the NMR-MOUSE applied to heritage buildings are summarized in the following.

7.5.1 The Mackintosh Library

The Mackintosh Building in Glasgow, Scotland, housed the library of the Glasgow School of Art until recently. It was designed by C.R. Mackintosh (1868–1928), a Scottish architect, designer, and artist known for his art nouveau and secessionism. The building was completed in 1909 and defines a new style of architecture in the twentieth century. On 23 May 2014 it was destroyed by a fire, then rebuilt and destroyed again by another fire on 16 June 2018. As part of another MOLAB measurement campaign, the low-density sandstone structure of the large window wall from the devastated reading room was investigated with the NMR-MOUSE in 2015, when it could be accessed via a full-scale scaffold that reached from the floor to the top (Figure 7.5b, inset).

The setup of the equipment and the measurements were challenged by the damaged floor and the >10 cm distance of the installed scaffold from the wall close to which the dedicated NMR-MOUSE scaffold (Figure 7.5a) carrying the Kea spectrometer and the sliding table with the sensor had to be mounted. Moreover, the equipment had to be shielded from electrical far-field noise by covering it with silver-coated and electrically grounded parachute silk (Figure 7.5a). Once set up, the spot of the wall in question was repeatedly sprayed with demineralized water for about 20 min. This time is a rough value based on experience. The impact of spraying time and amount of water on the signal properties has not been investigated in detail so far but is expected to depend on the pore size distribution, pore connectivity, and pore-surface properties. The water signal was measured right afterward at 20 mm depth with 2048 scans employing a CPMG sequence on a PM25 Profile NMR-MOUSE tuned to 13.7 MHz and fitted with a 5-mm spacer. The dwell time was set to 0.5 μs, the pulse length to 16 μs, and the echo time to 130 μs. The resultant signal decay was fitted with a mono-exponential decay function to extract T_2.

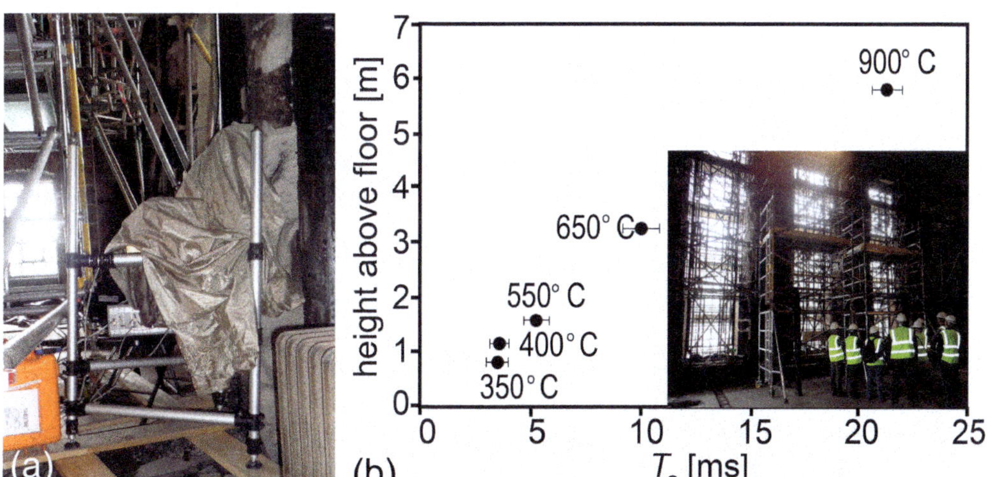

Figure 7.5 Study of sandstone heat damage in the Mackintosh Library in Glasgow. (a) Measurement setup. The scaffold carrying the NMR-MOUSE is balanced on the damaged floor. The sensor is hidden underneath grounded parachute silk to provide an electrical shield. (b) The NMR relaxation time T_2 reports the temperature of the fire at different height. Bernhard Blümich (author).

Measurements were carried out at different heights of the wall and on sandstone samples prepared in the laboratory at the University of West Scotland by heating them in an oven to temperatures ranging from 300° to 900°C [38]. In view of the fact that the spots on the wall and the reference samples were only partially water saturated with minor control over the saturation time and the amount of water taken up, the surprising result is that the relaxation times extracted from mono-exponential fits vary strongly between 1 and 21 ms, whereby longer relaxation times were measured at higher height. On the other hand, the laboratory study produced a correlation between relaxation time and exposure temperature of the sandstone. Given the laboratory calibration, the relaxation times from the wall measurements provide an estimate of the temperature to which the sandstone from the wall has been exposed during the fire (Figure 7.5b). According to the NMR measurements, the temperature at low height of 1–2 m above the floor was about 350° to 400° C, while it was about 900° C at close to 7 m height. The underlying dependence of the average relaxation time on the pore chemistry, the pore size and connectivity, and the water saturation as a function of the exposure temperature warrants further studies.

7.5.2 In Search of a Hidden Wall Painting

Giotto di Bondone (1267–1337) was a great painter of the early Italian Renaissance. Some of his most famous works are fresco paintings in Padua. During a restoration of the Sala del Capitolo within the complex of the Basilica di Sant'Antonio in Padua, wall paintings were found and uncovered by partially removing a later painting showing a portal. These early paintings are postulated to be remnants of Giotto's first work in Padua. MOLAB was asked to assist with nondestructive measurements to guide rescue and restoration efforts. To this end, the wall was investigated for structural integrity with holographic laser-speckle interferometry and NMR depth profiling among others.

The NMR depth profiles were measured with a PM25 NMR-MOUSE operating at 13.61 MHz with a 90° pulse width of 22 μs. The roughly square-shaped sensitive slice was about 40 mm in extent and 0.2 mm thick. The NMR signal was acquired with CPMG sequences at an echo time of 160 μs with 200 scans per position. CPMG decays were also recorded with 2250 scans with improved signal-to-noise ratio at selected depths identified from relaxation-weighted spin-density profiles to analyze them in terms of relaxation-time distributions. A relaxation weight was calculated as the ratio of two partial integrals of a CPMG signal decay by dividing the integral from a later part of the decay by that of the initial part [6]. Profiles of this parameter often provide better contrast than simple amplitude or spin-density profiles.

Some measurements had to be performed high up with the height-adjustable scaffold secured on a construction scaffold in front of the painted wall (Figure 7.6a) like the investigation of the fire damage in the Mackintosh Library. To measure a depth profile, the NMR-MOUSE was first placed a few millimeters close to the wall and then retracted in steps of 0.25 mm between CPMG acquisitions during a depth scan. Each depth scan lasted about 2.5 h. During this time even small vibrations of the scaffold had to be avoided, some of which were caused by air draft when opening the access door opposite to the wall under investigation. Moreover, temperature variations causing thermal expansion and contraction of the scaffold had to be taken into account in scheduling the measurements, in particular, when the room was being warmed up for the evening prayer of the monks. Therefore, long measurements were carried out overnight and remotely controlled via Teamviewer software (Teamviewer GmbH, Göppingen, Germany).

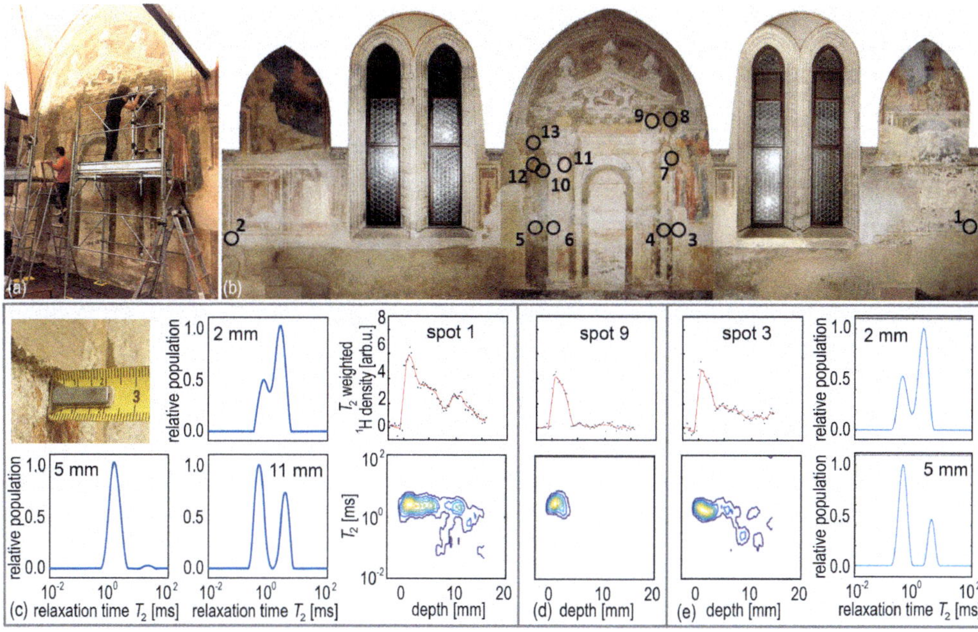

Figure 7.6 Wall paintings in the Sala del Capitolo of the Basilica di Sant'Antonio in Padua, Italy. (a) Setup to scan an NMR depth profile (right) and for laser-speckle interferometry (left). (b) Locations of spots measured with the NMR-MOUSE. (c) Spot 1 disclosed a covered-up older painting underneath a newer one. The depth-resolved relaxation-time distribution and the relaxation-weighted spin density revealed signal to depths beyond 15 mm. At 2 and at 11 mm depth, the relaxation-time distribution exhibited two peaks, whereas it showed only one at 5 mm depth. (d) The depth profiles from most spots gave signal only in the upper few millimeters like that of spot 9 with an uninformative relaxation signature. (e) Spot 3 was the only other spot that produced signal from lower depth. It also exhibited two relaxation peaks deeper down albeit a depth smaller than that of spot 1. Bernhard Blümich (author).

Depth profiles were acquired at the end of November 2017 at 12 different spots on the painted wall (Figure 7.6b) [38]. Before each depth scan, the spot was iteratively moistened with a fine mist of demineralized water, ensuring that no water drops could form, run down the fragile painting, and leave a trace on the dusty surface. This process did not follow strict rules but was guided by the observations at the spot in question. Some spots absorbed the water mist readily; others appeared to repel the water. As a rule of thumb, the water spray should be allowed to permeate the wall for about 20 min to get observable signal at 18 mm depth. In the right corner of the wall, a small section of a covered-up older wall painting could be seen where the top plaster layer was crumbling down (spot 1, Figure 7.6c). The depth-resolved relaxation-time distribution and the relaxation-weighted spin density measured close to that position revealed signal up to depths beyond 15 mm. At 2 and at 11 mm depth, the relaxation-time distribution showed two peaks, whereas it showed only one at 5 mm depth, reporting different consistencies of the mortar layers. This along with the depth-resolved relaxation-time distribution provided the reference NMR signature to expect of a painting hidden at other spots. Most other spots produced depth profiles like that from spot 9 (Figure 7.6d), essentially not providing any signal from below the first few millimeters. Moreover, at most spots the wall was difficult to moisten with water, and in many instances, it sounded hollow, indicating that there was a gap between the top mortar layer and the wall underneath. These areas of mortar ablation could be mapped with laser-speckle holography conducted in parallel to the NMR-MOUSE study. Spot 3 was the only other spot that produced signal from lower depth (Figure 7.6e). Interestingly, two relaxation peaks were detected also at larger depth albeit not as deep as where the second peak splitting was observed at spot 1. This suggests that another mortar layer is underneath the top one with a composition like that of the wall painting covered at spot 1. Whether or not that layer is painted cannot be derived from the NMR measurements. Moreover, the nature of the wall at the other blind spots remains elusive, because surface-spray water did not migrate to deeper depths due to some barrier posed either by a gap between the outer mortar layer and the wall or a water-blocking interface like wax that could have been applied to treat an earlier wall painting at some time before the current outer mortar layer with the new painting was applied.

7.5.3 Covered Roman Frescoes

The former harbor city of Ostia was founded in the seventh century BCE as a military settlement at the mouth (Latin: ostium) of the river Tiber to protect the city of Rome and control trade across the sea. Already in the third century BCE it had become one of the most important ports of the capital city of the Roman Empire. Most buildings preserved until today date from the second century BCE. In the second and third centuries CE the population of Ostia peaked at 100 000, and with the decline of Rome and the Roman Empire, it dwindled until the city was abandoned in the ninth century. From 1939 to 1942 the ancient city was excavated under Mussolini. Today the site is an archeological park known as Ostia Antica. The city was predominantly inhabited by merchants and workers as opposed to the antique town of Herculaneum at the foot of Mount Vesuvius, which was a resort town to the affluent Romans.

Several frescoes and fresco fragments were investigated in Ostia Antica by NMR depth profiling in the context of a long-term effort with Eleonora Del Federico from the Pratt Institute in Brooklyn to analyze wall paintings and Roman frescoes in an art historical context [39–42]. From all the measurements in Ostia Antica [38,42], only the ones relating to covered frescoes are summarized in view of the previous chapter reporting the search for a hidden wall painting in the Sala del Capitolo of the Basilica di Sant'Antonio. The measurements were conducted with a PM25 Profile NMR-MOUSE mounted on a stage to translate it horizontally (Figure 7.7b, left; Figure 7.7c, left) employing CPMG sequences with an echo time t_E of 120 μs. Before acquisition of a depth profile, the spot under investigation was moistened several times with a mist of demineralized water from a spray bottle. It took at least 20 min for the water to permeate to depths of 18 mm or more. In order to accelerate the measurement process, the NMR-MOUSE's movement was reversed to step forward between scans in the direction of the water migration into the object. In this way, three to four depth profiles could be acquired in a day's work.

A fresco fragment from the new depository at the site showed two painted layer sets, an older one from the first century BCE covered with a newer one from the end of the first century CE (Figure 7.7a, left). Covering older wall paintings with newer ones was customary as the walls accumulated the traces of time and the style of decoration changed along with the taste of the inhabitants. The thickness of the newer layer set was just a bit less than the scan range of the NMR-MOUSE so that the beginning of the older layer set was within reach. The amplitude profile reporting the relative water saturation and the depth-resolved distribution of relaxation times indeed show a discontinuity at the visually measured depth of 17 mm (Figure 7.7a, right). While the fragment could be saturated with water from both sides, frescoes on walls can only be moistened from one side. In consequence, the water saturation and the signal intensity decrease with depth following a half

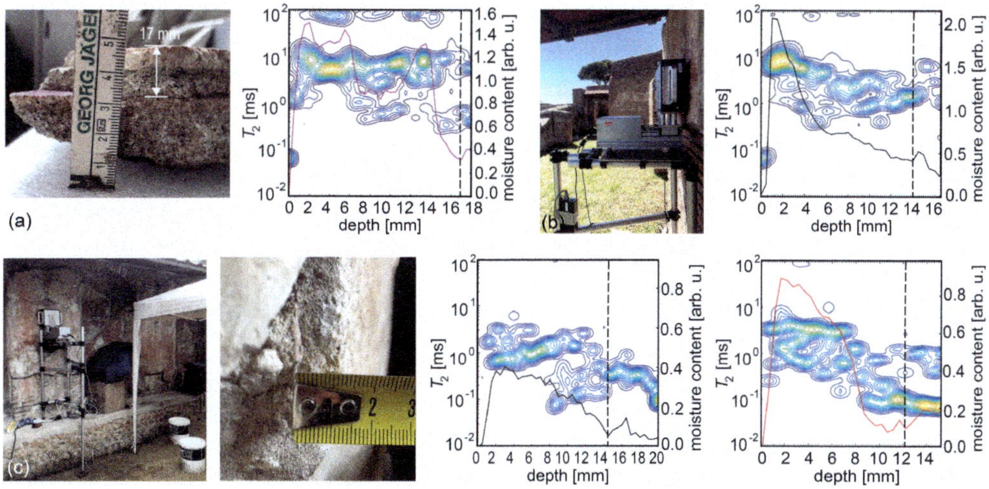

Figure 7.7 Frescoes and their depth profiles. (a) A fragment from the Nuovi Depositi showing two layers of wall paintings. (b) The outside wall from the Case a Giardino, which appeared to have a doubling of the mortar-layer stratigraphy. (c) The Terme del Bagnino Buticosus was built at the turn of the first century CE and refurbished at the end of the second century CE. Bernhard Blümich (author).

Gaussian-like overall profile. This was the case for an outside wall of the garden-house complex (Figure 7.7b) and the thermal bath of the bath superintendent Buticosus (Figure 7.7c). The garden houses are an integrated complex of living spaces and shops erected between 123 and 125 CE. A painted outside wall was easily accessible with the fresco still well attached to the original brick wall. A crumbled edge gave way to the unexpected insight that the outer painted layer set appeared to have been doubled. While the mortar stratigraphy of this wall is simpler than that of the fragment in the depository, a second peak, like that of the painted outside layer set starts to appear at 14 mm depth. It has lower amplitude than the outer one, because there is less water.

A similar signature is found for the west wall of room 1 of the thermal baths of the bath superintendent Buticosus. The wall with niches had been refurbished with a newer, uneven and painted mortar layer covering an older painted layer that surfaced at a damaged edge on the left side of the wall (Figure 7.7c, left). Heavy rain challenged the measurement, but the high humidity assisted in moistening the spots of interest. Depth profiles were acquired from two spots spaced a niche apart (Figure 7.7c, right). Both show similar moisture-content profiles with a second peak beginning to build up at 15 mm and 12 mm depth like the one observed for the garden-house wall. This peak is the first evidence so far in NMR depth profiles for a hidden fresco. Along with the hydrogen-density profiles the relaxation-time distributions provide a further clue to the presence of a new set of layers, in particular, that the peak at long relaxation time is discontinuous in the depth profile at the interface with the covered painted wall (Figure 7.6, Figure 7.7).

7.6 Biofilms in Yellowstone National Park

Yellowstone National Park is situated on top of a giant dormant volcano in the northwest of the United States embracing parts of Idaho, Montana, and Wyoming. It is known for its numerous geysers, hot springs and water holes, and other hydrothermal features. The hot waters host an abundance of algae and bacteria producing biofilms with colors characteristic of the water temperature. These thermophilic bacteria represent some of the most primitive forms of life. Among others, the bacterium *Thermus aquaticus* has been discovered there, which is the source of the key enzyme Taq polymerase needed to replicate DNA in the laboratory. The search for new bacteria capable of living under extreme conditions is pursued with the aim of developing new biotechnological processes, for example, to clean up certain water contaminants. While many bacterial biofilms can be recognized in the water by their beautiful colors, others reside under a cover of sediments, looking like rock. It was the aim of a depth-profiling mission with the NMR-MOUSE in Yellowstone National Park in September 2018 to investigate whether or not biofilms in the wilderness bear a clear NMR signature [43,44] and can be detected even when covered by sediment [45].

With permission of the park authorities and a guide, all equipment, including three 12 V car batteries, was carried from a nearby parking lot a few kilometers into the wilderness near Five Sisters Spring and Octopus Spring (Figure 7.8a). Safety regulations required carrying strong pepper spray against bears and wearing rubber boots to protect against hot water when crossing creeks and wetlands. For the measurements, a PM10 NMR-MOUSE was modified to operate at temperatures around 70 °C, wrapped in pond-liner foil, and attached to the Magritek lift of the NMR-MOUSE with a camera tripod extension rod passing through the radiofrequency coil hole in the sample table. The whole assembly was turned upside down and attached to the

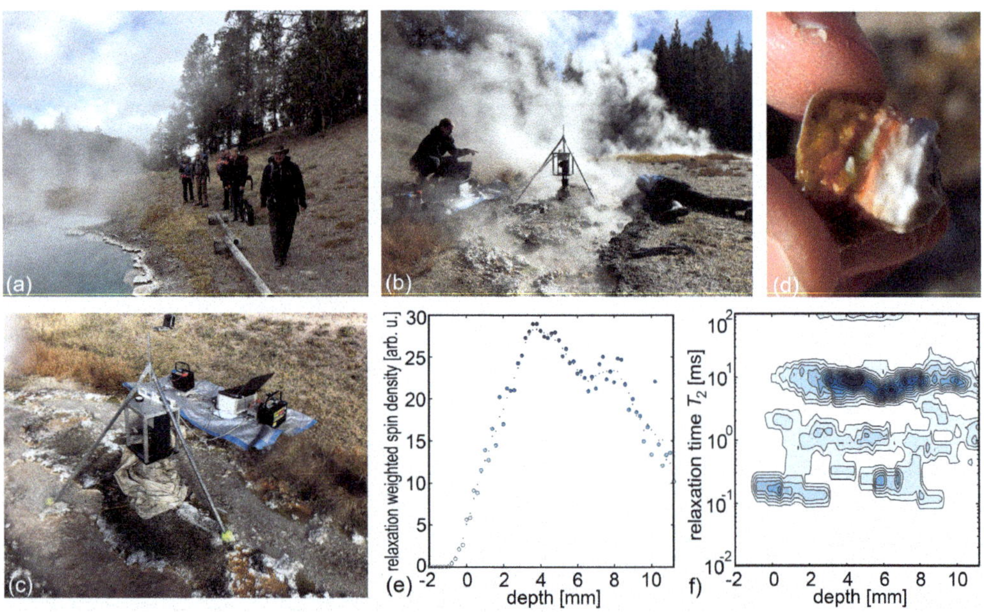

Figure 7.8 NMR depth profiling of biofilms in Yellowstone National Park. (a) Carrying the equipment past Octopus Spring. (b) Setting up the NMR-MOUSE at a runoff from the Octopus Spring. (c) The setup in operation while measuring a depth profile. (d) A core plug drawn from the measurement spot showing colorful biofilm layers covered by a thin sediment crust. (e) Relaxation-weighted spin-density profile across a 10-mm depth range. (f) Depth-resolved relaxation-time distribution bearing the signature of the biofilm layers. Bernhard Blümich (author).

tripod of an inexpensive swinging barbeque grill with a threaded rod. The legs of the tripod were covered with pinched tennis balls to minimize markings in the fragile ground (Figure 7.8b,Figure 7.8c). Before each depth scan, the sensor had to be adjusted parallel to the biofilm stratigraphy, a cumbersome procedure at times on account of the difficulty in adjusting a narrow gap between the ground and the sensor with the steam from the hot water clouding the brisk autumn air (Figure 7.8b). Despite being remote from civilization it was necessary to electrically shield the equipment with conducting parachute silk, a procedure that was most successful when the silk was in contact with the ground water (Figure 7.8c). Another issue was condensation water from the steam of the runoff water collecting inside the pond liner, which had to be minimized.

Seven depth profiles were acquired along with biofilm samples collected from the different measured spots to analyze them later. For example, the depth profile from the runoff of Octopus Spring reveals a rich stratigraphy already in the relaxation-weighted spin-density profile (Figure 7.8e) and even more so in the depth-resolved relaxation-time distribution (Figure 7.8f). A sample plug drawn from that exact spot confirms a complex structure of biofilm layers (Figure 7.8d). This came as a surprise to the untrained eye because the sediment-like appearance from the outside suggested it to be simply dead rock or sediment and not a form of life. It is the subject of ongoing research at Montana State University to correlate the NMR depth profiles with the biofilm properties, a challenging endeavor facing the difficulty of cutting the deep-frozen biofilm plugs into thin slices to analyze their DNA biochemically [45].

7.7 Outlook

Mobile NMR refers to NMR measurements at the site of the object. It is enabled by compact instruments that can be carried along over large distances. So far, mobile NMR instruments are essentially stray-field relaxometers used to log oil wells and test materials nondestructively. The acceptance of stray-field sensors like the NMR-MOUSE to analyze material properties has grown over the years and it has also become part of the portfolio of instruments to diagnose cultural heritage [46,47]. Shipping the equipment across borders and between continents is possible, albeit at the expense of import/export-tax bureaucracy, so long as the magnet is well shielded with steel plates. Important assets for measuring in museums and at excavation sites are the horizontal translation stages (Figure 7.4a, Figure 7.7b) and the height-adjustable scaffold (Figure 7.5a, Figure 7.7c), which can be disassembled for transport into a set of aluminum pipes and connectors. Moreover, the equipment can be operated from a car battery or a fuel-powered electrical generator. A major disadvantage of the current state of the art is that the equipment has been designed for laboratory use. The sensor, the console, and the translation stage come as separate modules that are connected by several cables. These can form ground loops and attract electrical noise, and the connectors are fragile and can break during transport or setup. To transport the equipment and measure at the site and outdoors it would be desirable to have the sensor, the console, and the translation stage integrated into one portable device. Moreover, investigations of material properties would greatly benefit from shorter dead time, for example, an echo time of 16 μs, where the protons of crystalline and glassy polymers deliver detectable signal [48], and from distance and alignment sensors to facilitate the instrument setup up when scanning high-resolution depth profiles (Figure 7.7b).

Acknowledgments

Three of the reported measurement campaigns were executed as part of the MOLAB activity within the IPERION CH grant no. 654028 of the H2020 Program of the European Union. They are (i) the study of the violins in Oxford and Cremona in collaboration with Colin Harrison, Marco Malagodi, Claudia Invernizzi, Maria Baias, and Christian Rehorn; (ii) the study of the Mackintosh Library in Glasgow organized by John Hughes, who also prepared the heat-treated sandstone samples in the laboratory together with Thorsten Howind; and (iii) the search for a hidden Giotto fresco in Padua in collaboration with Gilberto Artioli. Markus Küppers coordinated all MOLAB activities at RWTH Aachen University and participated in the measurements in Padua. Valeria Gabrielli investigated the accuracy of the sensor alignment. The fresco study in Ostia Antica would not have been possible without the support and initiative of Eleonora Del Federico, Paolo Tomassini, and Denis Jaschtschuk. The excursion to Yellowstone National Park was a collaborative endeavor together with Sarah Codd, Catherine Kirkland, Brent Peyton, Dana Skorupa, and Joseph Seymour.

References

1 Hürlimann, M.D. and Heaton, N.J. (2016). NMR well logging. In: *Mobile NMR and MRI* (eds. M. Johns, E.J. Fridjonssen, S. Vogt, and A. Haber), 11–85. Cambridge, UK: Royal Society of Chemistry.
2 Blümich, B. (2019). Low-field and benchtop NMR. *Journal of Magnetic Resonance* 306: 27–35.
3 Coates, G.R., Xiao, L., and Prammer, M.G. (1999). *NMR Logging: Principles and Applications.* Houston, TX: Halliburton Energy Services.
4 Blümich, B. (2019). *Essential NMR*, 2e. Cham, Switzerland: Springer Nature.
5 Fukushima, E. (2016). Introduction. In: *Mobile NMR and MRI* (eds. M. Johns, E.J. Fridjonssen, S. Vogt, and A. Haber), 1–10. Cambridge, UK: Royal Society of Chemistry.
6 Blümich, B., Haber-Pohlmeier, S., and Zia, W. (2014). *Compact NMR*. Berlin, Germany: de Gruyter.
7 Blümich, B. and Singh, K. (2018). Desktop NMR and its applications from materials science to organic chemistry. *Angewandte Chemie International Edition* 17: 6996–7010.
8 Hürlimann, M.D. and Griffin, D.D. (2000). Spin dynamics of Carr–Purcell–Meiboom–Gill-like sequences in grossly inhomogeneous B_0 and B_1 fields and application to NMR well logging. *Journal of Magnetic Resonance* 143: 120–135.
9 Balibanu, F., Hailu, K., Eymael, et al. (2000). NMR in inhomogeneous magnetic fields. *Journal of Magnetic Resonance* 145: 246–258.
10 Callaghan, P.T. (1991). *Magnetic Resonance Microscopy*. Oxford, UK: Clarendon Press.
11 Blümich, B. (2000). *NMR Imaging of Materials*. Oxford, UK: Clarendon Press.
12 Utsuzawa, S. and Fukushima, E. (2017). Unilateral NMR with a barrel magnet. *Journal of Magnetic Resonance* 282: 104–113.
13 Manz, B., Coy, A., Dykstra, R. et al. (2006). A mobile one-sided NMR sensor with a homogeneous magnetic field: The NMR-MOLE. *Journal of Magnetic Resonance* 183: 25–31.
14 Blümich, B. and Bruder, M. (2003). Mobile NMR zur Qualitätskontrolle von Elastomerprodukten. *KGK Kautschuk Gummi Kunststoffe* 56: 90–94.

15 Zimmer, G., Guthausen, A., and Blümich, B. (1998). Characterization of cross-link density in technical elastomers by the NMR-MOUSE. *Solid State Nuclear Magnetic Resonance* 12: 183–190.
16 Blümich, B., Teymouri, Y., and Clark, R. (2019). NMR on the road: Non-destructive characterization of the crumb-rubber fraction in asphalt. *Applied Magnetic Resonance* 50: 497–509.
17 Eidmann, G., Savelsberg, R., Blümler, P. et al. (1996). The NMR MOUSE: A mobile universal surface explorer. *Journal of Magnetic Resonance A* 122: 104–109.
18 Perlo, J., Casanova, F., and Blümich, B. (2005). Profiles with microscopic resolution by single-sided NMR. *Journal of Magnetic Resonance* 176: 64–70.
19 Casanova, F. and Perlo, J. (2011). NMR in inhomogeneous fields. In: *Single-Sided NMR* (eds. F. Casanova, J. Perlo, and B. Blümich), 11–56. Berlin, Germany: Springer.
20 Blümler, P. and Casanova, F. (2016). Hardware developments: Single-sided magnets. In: *Mobile NMR and MRI* (eds. M. Johns, E.J. Fridjonssen, S. Vogt, and A. Haber), 110–132. Cambridge, UK: Royal Society of Chemistry.
21 Blümich, B., Baias, M., Rehorn, C. et al. (2020). Comparison of historical violins by non-destructive MRI depth profiling. *Microchemical Journal* 158: 105219.
22 Van Landeghem, M., Danieli, E., Perlo, J. et al. (2012). Low-gradient single-sided NMR sensor for one-shot profiling of human skin. *Journal of Magnetic Resonance* 215: 74–84.
23 Blümich, B., Rehorn, C., and Zia, W. (2018). Magnets for small-scale and portable NMR. In: *Micro and Nano Scale NMR* (eds. J. Anders and J.G. Korvink), 1–20. Weinheim, Germany: Wiley-VCH.
24 Haken, R. and Blümich, B. (2000). Anisotropy in tendon investigated in vivo by a portable NMR scanner, the NMR-MOUSE. *Journal of Magnetic Resonance* 144: 195–199.
25 Hailu, K., Fechete, R., Demco, D.E., and Blümich, B. (2002). Segmental anisotropy in strained elastomers detected with a portable NMR scanner. *Solid State Nuclear Magnetic Resonance* 22: 327–343.
26 Blümich, B., Anferov, V., Anferova, S. et al. (2002). Simple NMR-MOUSE with a bar magnet. *Concepts in Magnetic Resonance: Magnetic Resonance Engineering* 15: 255–261.
27 Überrück, T., Rehorn, C., Höhner, R. et al. (2018). Visualizing the detection area of a unilateral NMR sensor using deconvolution and back-projection. *Journal of Magnetic Resonance* 296: 169–175.
28 Wiesmath, A., Filip, C., Demco, D.E. et al. (2002). NMR of multipolar spin states excited in strongly inhomogeneous magnetic fields. *Journal of Magnetic Resonance* 154: 60–72.
29 Rata, D.G., Casanova, F., Perlo, J. et al. (2006). Self-diffusion measurement by a mobile single-sided NMR sensor with improved magnetic field gradient. *Journal of Magnetic Resonance* 180: 229–235.
30 Jaschtschuk, D. (2019). Advances in Low-field NMR relaxometry. PhD dissertation. RWTH Aachen University.
31 Invernizzi, C., Fiocco, G., Iwanicka, M. et al. (2020). Non-invasive mobile technology to study the stratigraphy of ancient Cremonese violins: OCT, NMR-MOUSE, XRF and reflection FT-IR spectroscopy. *Microchemical Journal* 155: 104754.
32 Whiteley, J. (2008). *Stringed Instruments: Viols, Violins, Citterns, and Guitars in the Ashmolean Museum*. Oxford, UK: Ashmolean Museum.
33 Cacciatori, F. (2014). *Lo Scrigno Dei Tesori*. Cremona, Italy: Edizione Museo del Violino.
34 McDonald, P.J., Aptaker, P.S., Mitchell, J. et al. (2007). A unilateral NMR magnet for sub-structure analysis in the built environment: The surface GARField. *Journal of Magnetic Resonance* 185: 1–11.

35 Blümich, B., Casanova, F., Dabrowski, M. et al. (2011). Small-scale instrumentation for nuclear magnetic resonance of porous media. *New Journal of Physics* 13: 015003.
36 Schulte Holthausen, R. and Weichold, O. (2017). Nondestructive evaluation of thermal damage in concrete by single-sided nuclear magnetic resonance. *Journal of Infrastructure Systems* 23: B4016006.
37 Brizi, L., Camaiti, M., Bortolotti, V. et al. (2018). One and two-dimensional NMR to evaluate the performance of consolidants in porous media with a wide range of pore sizes: Applications to cultural heritage. *Microporous and Mesoporous Materials* 269: 186–190.
38 Rehorn, C.W.G. (2019). Studies of tangible cultural heritage with portable stray-field NMR. PhD dissertation. RWTH Aachen University.
39 Blümich, B., Haber, A., Casanova, F. et al. (2010). Noninvasive depth profiling of walls by portable nuclear magnetic resonance. *Analytical and Bioanalytical Chemistry* 397: 3117–3125.
40 Haber, A., Blümich, B., Souvorova, D. et al. (2011). Ancient Roman wall paintings mapped non-destructively by portable NMR. *Analytical and Bioanalytical Chemistry* 401: 4011441–4011452.
41 Rehorn, C., Kehlet, C., Del Federico, E. et al. (2018). Automatizing the comparison of NMR depth profiles. *Strain* 54: e12254.
42 Blümich, B., Del Federico, E., Jaschtschuk, D., et al. (2021). Nondestructive analysis of wall paintings at Ostia Antica. *Heritage* 4: 4421–4438.
43 Herrling, M.P., Lackner, S., Nirschl, H. et al. (2019). Recent NMR/MRI studies of biofilm structures and dynamics. *Annual Reports on NMR Spectroscopy* 97: 163–213.
44 Herrling, M.P., Weisbrodt, J., Kirkland, C.M. et al. (2017). NMR investigation of water diffusion in different biofilm structures. *Biotechnology and Bioengineering* 114: 2857–2867.
45 Blümich, B., Codd, S., Jaschtschuk, D., et al. (2022). Manuscript in preparation.
46 Capitani, D., Di Tullio, V., and Proietti, N. (2012). Nuclear magnetic resonance to characterize and monitor cultural heritage. *Progress in Nuclear Magnetic Resonance Spectroscopy* 64: 29–69.
47 Rehorn, C. and Blümich, B. (2018). Cultural heritage studies with mobile NMR. *Angewandte Chemie International Edition* 57: 7304–7312.
48 Oligschläger, D., Glöggler, S., Watzlaw, J. et al. (2015). A miniaturized NMR-MOUSE with a high magnetic field gradient (Mini-MOUSE). *Applied Magnetic Resonance* 46: 181–202.

Part II

Applications in Chemical Engineering

8

Ultrafast MR Techniques to Image Multi-phase Flows in Pipes and Reactors

Bubble Burst Hydrodynamics

Andrew J. Sederman, Andi Reci, and Lynn F. Gladden

Department of Chemical Engineering and Biotechnology, University of Cambridge, Cambridge, UK

8.1 Introduction

Historically, the development of an understanding of a process in chemical engineering has often had to be accomplished through measurements made external to or on the surface of the process of interest, such as taking samples at the entrance and exit of unit operations or taking temperature and pressure measurements at a single location on the surface of a reactor. Therefore, there is significant interest in the use of noninvasive imaging methodologies that can provide local measurements of the state of a system. A number of imaging techniques are now available to researchers including X-ray tomography, electrical tomography methods (resistance, capacitance, and impedance), and particle imaging velocimetry (PIV), but none of these provides the range of possible contrast, and therefore depth of information, that magnetic resonance imaging (MRI) methods can provide. The text by Paul Callaghan [1] provides an outstanding and comprehensive introduction to MRI in engineering and the physical sciences, though there are many other excellent texts [2–4]. The sensitivity of MRI to chemical species, surface interactions, diffusion, and coherent motion make it an extremely powerful measurement method for investigation in chemical processes such as reactors. In such systems local species concentration and mass transport to and from surface reaction sites are of fundamental importance to the efficiency of the process. After discussing the motivation for and different approaches to ultrafast MR imaging, this chapter proceeds to describe the first use of MR velocimetry to investigate bubble bursting dynamics at the liquid–gas interface in a pipe – an application of particular relevance to bioreactors and the large-scale production of bio-pharmaceutics. Velocity fields are acquired with a time resolution of 4 ms by using spiral velocity imaging and a new reconstruction technique appropriate for velocity imaging. The images are acquired in two orthogonal planes (axial and transverse) and show the time evolution of flow structures after the bubble burst event.

8.1.1 The Motivation for Ultrafast MR Imaging in Chemical Engineering

One disadvantage of MRI compared with several other imaging techniques is that image acquisition times using standard MR imaging sequences are relatively long. When we consider that many processes of interest in flowing systems will have parameters that vary with time and space, the timescales of the image acquisition become an important consideration. One MR approach is to make the timescales of the acquisition much longer than the temporal fluctuations in the

Magnetic Resonance Microscopy: Instrumentation and Applications in Engineering, Life Science and Energy Research, First Edition. Edited by Sabina Haber-Pohlmeier, Bernhard Blümich, and Luisa Ciobanu.

system, such that a time-averaged value can be accurately measured. Even in these cases, quantitative information about the underlying fluctuations of the system may also be measured as is described in the next section. Another approach, utilized in imaging the bubble burst hydrodynamics in this chapter, is to reduce the image acquisition times such that they can be considered short compared with the timescales of the fluctuations occurring in the system. This will often be on the order of seconds or less and so the imaging duration is likely to be less than ~100 ms. Such imaging methods are usually classed as ultrafast imaging techniques in MR. These ultrafast imaging methods are therefore key to measuring many transient engineering applications that vary on such timescales.

8.1.2 Time-averaged Approach

A common approach to MR imaging of processes that fluctuate over a short timescale is to image over a timescale much longer than the temporal fluctuations themselves and to generate time-averaged measurements. Depending on the timescale of the fluctuations and the timescales associated with the MR pulse sequence, it may be possible to extract useful quantitative information about the nature of these fluctuations as well. This has, perhaps, most usefully and extensively been used to measure turbulence and the transition to turbulence in single-phase Newtonian fluid flow, either in a pipe, or more complex geometries. Kuethe [5] measured the local turbulent diffusivities of fluid at a sudden expansion from a nozzle using a measurement similar to that used to measure diffusion – where signal is attenuated due to the level of diffusion. Newling et al. [6] showed how the SPRITE (Single-Point Ramped Imaging with T_1-Enhancement) imaging method could be adapted to image velocities and turbulent diffusivities for even higher velocities and Reynolds numbers. Further information about the nature of the turbulence can be extracted by altering the pulse sequence timings, as shown by Kuethe and Gao [7] where measurements of the correlation time associated with turbulence were obtained by altering the time (observation time) between velocity-encoding gradients. At observation times much shorter than the temporal fluctuations, displacements will increase linearly with time while the local dispersion will increase. As the observation time becomes longer then the correlation time of the fluctuations, the dispersion reaches an asymptotic value. A similar approach has been applied to granular flow to measure the average time between collisions of particles [8,9]. Further characterization of turbulent flow can be obtained by displacement measurements in multiple directions such as by using the ICOSA6 (six-directional icosahedral flow encoding) [10,11] encoding method to obtain the Reynolds stress tensor components. In multi-phase (gas–liquid, solid–fluid) systems, where local concentration is usually an important parameter to measure, it is necessary to take into account the effect that motion can have on the MR signal intensity in order to produce quantitative data [12,13]. Another possible source of error comes from velocity fluctuations on similar timescales to the timescales associated the velocity-encoding and imaging sequence. Where these may be similar it is necessary to sample the data for a sufficiently long time to ensure that an accurate average value is obtained as investigated in turbulent pipe flow [14] and in a pulsatile laminar pipe flow [15].

8.1.3 Ultrafast Imaging for Engineering Applications

As discussed in Section 8.1.2, time-averaged measurements can provide a range of data about the nature of the temporal fluctuations as well as the average velocity, but repeated measurements over timescales much shorter than the fluctuations can provide greater insight. Therefore, there is a clear motivation to acquire data on these short timescales, which may require ultrafast imaging. There are two broad approaches to increasing the speed of image acquisition in MRI, which can be combined and used to reduce the imaging timescales further. These are, first, to find acquisition protocols, or pulse sequences, that acquire the required data more quickly, and second, to reduce the number of data points that need to be acquired to be able to produce an image of the required quality. It should be noted that this second approach can lead to simple forms of acquisition that are designed purely to extract the important information rather than first reconstructing an image and then extracting data from the image [16]. This chapter focuses on ultrafast spiral velocity imaging with compressed sensing reconstruction of phase angles in particular but the next two sections briefly outline the general approaches that can be used to speed up image acquisition.

8.1.3.1 Ultrafast Image Acquisition Protocols

There are myriad different acquisition techniques with each spectrometer manufacturer producing slightly different versions with different names, which can be confusing to the newcomer. Fortunately, there are many excellent review articles and texts, along with web resources, in which the differences and similarities between these are detailed [3,17,18]. An excellent text for readers interested in the further details of MRI pulse sequence design is the book by Brown et al. [3]. Ultrafast imaging approaches can usually be divided into those that aim to acquire data for an entire image, from a single excitation followed by a duration to allow for T_1 magnetization recovery, and those that allow only minimal time for T_1 recovery with a fast recycle time, TR, but only use a fraction of the magnetization per excitation, often depleting the magnetization over the timescale of the image. When designing sequences for engineering applications, consideration must be given to the growth and decay of magnetization through T_1, T_2, and T_2^* relaxation as well as possible in-flow and out-flow effects as these will strongly influence the choice of imaging methodology. These parameters will influence the nuclear magnetic resonance (NMR) signal and may introduce errors into the final images unless their effect is accounted for. This is of particular importance in engineering applications where obtaining quantitative images with known errors is critical for their correct interpretation.

Where the lifetime of the signal is long enough, a whole image may be acquired from a single excitation. Echo planar imaging, or EPI [19], and similar variants are the fastest of these and can be chosen when all the required data points in $\boldsymbol{k}$-space can be acquired in a time approximately equal to T_2^*, where T_2^* is the time constant of the decay of the signal in the transverse plane due to both T_2 and field inhomogeneities. The early implementations of EPI were done using a rectilinear $\boldsymbol{k}$-space acquisition, though using a spiral trajectory [20] can produce a further increase in speed and is used in the velocity imaging of bubble burst events investigated later in this chapter. In engineering applications, it is more common for the T_2^* to be relatively short due to the magnetic field inhomogeneities that are an unavoidable consequence of samples consisting of multiple phases with different magnetic susceptibilities. In these cases it is most likely that 180° refocusing pulses will also be needed to continually refocus spin dephasing due to the presence of the magnetic field inhomogeneities. A technique such as rapid acquisition with relaxation enhancement (RARE) [21] is commonly used for this in which a refocusing pulse is applied between the acquisition of each single line of $\boldsymbol{k}$-space. However, there exist a number of other techniques, such as GRASE (gradient and spin echo) [22], which fit somewhere between these two methods. In addition, EPI acquisition can be, and is, often done during the formation of a spin echo after a 180° pulse.

When the lifetime of the signal in the transverse plane is too short for whole-image acquisition, it may be possible to acquire images fast enough with multiple excitation methods. The FLASH (Fast Low Angle SHot) [23] method was introduced in 1986 whereby a simple gradient echo imaging sequence is used with a small excitation tip angle. A small tip angle is used for each excitation so that longitudinal magnetization remains in the spin system, which can be used in subsequent excitations. The need to leave longitudinal magnetization for subsequent excitations usually precludes the use of multiple radiofrequency (RF) pulses and so the sequences tend to be relatively simple and robust, but lack the signal-to-noise ratio (SNR) that can be achieved in the best whole-image acquisition methods. The ultrashort echo time (UTE) [24–26] sequence fits into this approach and enables the imaging of materials with very short T_2 and T_2^* as may be found in some engineering applications, though it is relatively complex to implement. Zero echo time (ZTE) [27] is a similar method that is simpler to implement, though it does not allow easy slice selection for 2D imaging.

8.1.3.2 Ultrafast Imaging with Reduced Data Acquisition and Compressed Sensing Reconstruction

In addition to using fast image acquisition protocols as discussed in the previous sections, there are a number of techniques related to reducing the number of data points that need to be acquired. Compressed sensing has been a rapidly developing field in mathematics that uses the general underlying structures in images to reconstruct them with less data than would be required for a "fully sampled" image. Compressed sensing has been introduced to MRI as a method of reducing acquisition time without decreasing spatial resolution [28–31]. Depending on the nature of the image, accurate reconstruction can be achieved with far fewer data points than indicated by the Nyquist sampling theory without introducing artifacts [31]. The compressed sensing reconstruction method aims to maximize the sparsity of the image transformed into a domain in which the image is known to be sparse, i.e. a domain in which there are many "zero" data points. This must be done while requiring that the reconstructed image accurately represents the acquired data, i.e. there must be fidelity with the acquired data. In practice, minimizing of the L_1-norm of the sparse representation of the image is performed where the L_1-norm acts as a proxy for sparsity. In a simple case, the reconstruction is obtained by solving the following constrained optimization problem:

$$\min \|\Psi \boldsymbol{P}\|_1 \quad s.t. \|\mathcal{F}\boldsymbol{P} - \boldsymbol{Y}\|_2 < \varepsilon, \tag{8.1}$$

where $\boldsymbol{P}$ is the reconstructed magnitude image, Ψ is the operator that transforms from image representation to a sparse representation, $\mathcal{F}$ is the undersampled Fourier transform to $\boldsymbol{k}$-space, $\boldsymbol{Y}$ are the $\boldsymbol{k}$-space measurements, and ε is a threshold set depending on the required fidelity of the reconstruction to the original data and may be set to the expected noise level. The key challenges to implementing compressed sensing are the correct choice of a domain in which the image can be represented sparsely and the correct image reconstruction algorithm [32,33]. Compressed sensing reconstruction is usually done on the magnitude of the image, which is usually the desired end product; however the velocity information is encoded in the phase of our image and therefore a slightly different approach is needed as will be discussed later in application to bubble burst hydrodynamics.

8.2 Bubble Burst Hydrodynamics Captured with MR Velocimetry

In this section, undersampling of $\boldsymbol{k}$-space data, a recently proposed MR velocimetry processing method, and a carefully designed experimental system are used to highly accelerate the acquisition of MR velocimetry data. This has enabled the first experimental observation of the spatially resolved hydrodynamics of a bubble burst event, the hydrodynamics of which is important in areas as diverse as bioreactor design and geophysical processes. Three-component velocity fields are acquired every 4 ms using a spiral velocity imaging technique. The experimental data, acquired in two orthogonal planes, show the evolution of the flow fields after the bubble burst event.

8.2.1 Introduction to Bubble Burst Hydrodynamics

The burst of a bubble is an important event in many areas of science and engineering. A major transport mechanism of microorganisms [34,35], salt [36–38], and charged aerosols [39] from the ocean to the atmosphere is through bubbles bursting at the water–air interface. In sparged bioreactors, the largest rate of cell death has been experimentally observed to occur at the free surface of the reactor [40–42]. This has been attributed to the entrapment of cells in the liquid film formed between bubbles and the free air surface, and their subsequent high acceleration during the film rupture and bubble burst [43–46]. Bubbles bursting at an oil–water interface lead to the formation of submicrometer emulsions of oil in water [47], which is being used to produce nanomaterials with tunable rheological properties [48,49]. Bubble bursting is also important for the release and reach of flavors and aroma from carbonated drinks, such as champagne [50,51].

The processes occurring leading up to, during, and after a bubble burst event have been considered widely in the literature and are only briefly discussed here. When a bubble reaches a liquid–air interface, after an initial potential bouncing [52,53], the bubble rests on the surface, separated by a liquid film from the free air. The liquid film subsequently drains – a process that is explained by thinning film theories [54–56]. When the film reaches a critical thickness, a hole is punched in the film. If the hole is of a size above approximately twice the thickness of the film [57], it expands. Most of the retracting film is subsequently concentrated in a toroidal rim of liquid that bounds the unbroken film [58]. The rim motion follows to a good approximation the path defined by the original shape of the bubble [59], moving at a large retracting velocity on the order of 10 m/s [60,61]. The process is characterized by film drop shedding [62–65], although this may not always be present [66]. The cavity collapses under the influence of converging capillary waves. The wave ripples formed in the cavity surface have been observed experimentally [67,68] and captured by numerical methods [69–71]. At the convergence point of the capillary waves, a high capillary pressure point is formed, which leads to the formation of an upward and downward jet. The upward jet moves in the empty space left by the bubble cavity and may lead to jet drops being ejected [72–74]. The angle of the downward jet to a vertical axis depends on the position in which the bubble gets punctured. Once the upward jet achieves the maximum height, it falls back into the liquid pool and, after some oscillations, liquid motion dies out.

The main experimental observation technique used to validate the predictions of numerical methods about the shape of the cavity, evolution of the upward jet, and the phenomenon of jet drop shedding during bubble burst has been fast photography [38,65–68,71,73–75], although X-ray imaging [76] has also been used. Fast photography has also been used to validate the predictions of numerical methods about the ejection speed of the upward jet, by tracking the change in shape between subsequent frames. However, apart from Doppler-meter measurements on a small volume element [68], observations of the spatially resolved velocity field in the vicinity of the bursting bubble have been limited to flow visualization using a dye [67,68] or seeded particles [71]. Further, it is noted that measurement of the spatially resolved velocity field is important in validating the predictions of numerical methods, particularly in cases when different numerical methods disagree.

In this work, MR velocimetry is used to provide the first experimental spatially resolved velocity field of a bubble burst event.

Section 8.2.2 describes the experimental system and the signal acquisition scheme. Section 8.2.3 describes a recently proposed compressed sensing processing technique for the reconstruction of phase information in the MR image and is therefore suitable for velocimetry data. The experimental results of the bubble burst study and the discussion are presented in Section 8.2.4.

8.2.2 Materials and Methods

A polytetrafluoroethylene (PTFE) cylindrical pipe of inner diameter $D = 15.4$ mm was mounted vertically and filled up to a certain level with water, to which 17 mM $DyCl_3.6H_2O$ was added ($T_1 \approx T_2 = 55$ ms). This concentration of paramagnetic ions ensures fast relaxation of the NMR signal, magnetic susceptibility matching of the solution and air, and negligible effect on the surface tension of water [77]. The contact angle of water with the PTFE surface is >90°, resulting in an upward curving meniscus being formed between water and air, as illustrated in Figure 8.1a. Air was slowly injected into the pipe using a Harvard Apparatus Model 22 syringe pump. The air passes through a nozzle of 1 mm inner diameter, pinching off to consistently form bubbles of radius 2.4 mm. A bubble rises through the water and rests on the free surface for some time, before bursting. The upward curving meniscus forces the bubble to stay in the center of the pipe, being the point of highest liquid level [78]. The consistent size of the bubbles and their position at the center of the pipe allow reproducible bubble burst experiments, which is utilized in accelerating the acquisition of the experimental data, as discussed later. The subsequent bubble burst hydrodynamics was studied using MR velocimetry.

All experiments were conducted on an AV400 Bruker magnet, operating at a resonant frequency of 400.25 MHz for ^{1}H observation with an RF coil of 25 mm diameter. The maximum magnetic field gradient amplitude available in each spatial direction was 146 G/cm.

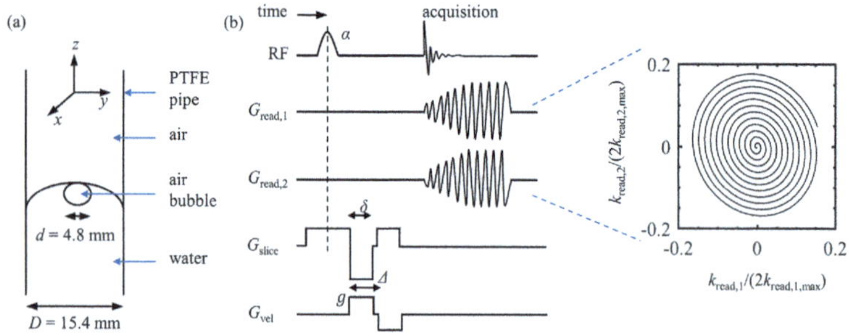

Figure 8.1 (a) Schematic of the experimental setup. (b) Pulse sequence used for MR velocimetry acquisitions and the corresponding ***k***-space traversal.

The 2D MR spiral imaging technique developed by Tayler et al. [77], shown in Figure 8.1b, was employed for acquiring velocity images. The spiral was designed according to the method of Hargreaves et al. [79] for an image comprising 64 × 64 pixels and of resolution 265 × 265 μm. The 2D images were acquired with a slice thickness of 150 μm. A total of 1024 complex points were acquired along the spiral with a dwell time of 2 μs, corresponding to 25% sampling percentage and imaging time of 2.05 ms. The optimal spiral density was close to being constant and the spiral stretches to ~ 40% of $\boldsymbol{k}_{\max}$, as illustrated in Figure 8.1b. The velocity-encoding gradients were designed with $\delta = 0.26$ ms, $\Delta = 0.49$ ms and $g = 29$ G/cm. The RF excitation pulse flip angle was 11° and the total image repetition time, TR, was 4.0 ms.

Three-component velocity maps on a 2D image were obtained for a transverse slice (perpendicular to the axis of the pipe) and a longitudinal slice (parallel to the axis of the pipe). Both slices cut through approximately the center of the bubble. For a given slice direction (transverse or longitudinal) and a given direction of the velocity to be measured (x, y, or z), a one-component velocity map (v_x, v_y, or v_z) was acquired every 4 ms using the pulse sequence in Figure 8.1b with the velocity-encoding gradient in the respective direction (x, y, or z) and with alternating polarity between consecutive pulse sequences (from $\pm g$ to $\mp g$). By taking the difference between the phase of the MRI images reconstructed from the acquired $\boldsymbol{k}$-space data of consecutive pulse sequences, and with reference to a zero-flow experiment, the one-component velocity maps at 4 ms temporal resolution are reconstructed, as discussed in the next section. The maps of the other components of velocity are obtained in a similar way at temporal resolutions of 4 ms but with the pulse sequences applied to different bursting bubbles. The size, position, and the hydrodynamics of different bursting bubbles at the free surface are shown to be consistent in the results section. As a result, the one-component velocity maps in different directions, v_x, v_y, or v_z, are superimposed, by aligning the point in time at which the bubble burst starts, to effectively give three-component velocity maps at a temporal resolution of 4 ms.

8.2.3 Reconstruction Technique

The acquired $\boldsymbol{k}$-space data using the spiral imaging technique is not sampled on a regular Cartesian grid. As a result, a nonuniform fast Fourier transform (NUFFT) operator [80], $\mathcal{F}$, constructed with the aid of the spiral path measurement technique of Duyn et al. [81] and Voronoi compensation diagrams [82], is used to relate the $\boldsymbol{k}$-space data, $\boldsymbol{Y}_{\mathrm{v}}$, to the complex image, $\boldsymbol{U}_{\mathrm{v}}$, and the noise, $\boldsymbol{E}_{\mathrm{v}}$, according to:

$$\boldsymbol{Y}_{\mathrm{v}} = \mathcal{F}\boldsymbol{U}_{\mathrm{v}} + \boldsymbol{E}_{\mathrm{v}}. \tag{8.2}$$

For MR velocimetry data, the complex image, $\boldsymbol{U}_{\mathrm{v}}$, is composed of the magnitude, $\boldsymbol{P}_{\mathrm{v}}$, and the phase, $\boldsymbol{\Phi}_{\mathrm{v}}$, such that $\boldsymbol{U}_{\mathrm{v}} = \boldsymbol{P}_{\mathrm{v}} \cdot \exp(\mathrm{i}\boldsymbol{\Phi}_{\mathrm{v}})$, where the dot notation refers to element-wise multiplication. The magnitude of the image is encoded with the ^{1}H density map, while the phase of the image is encoded with the velocity of the fluid and therefore the focus is on the reconstruction of $\boldsymbol{\Phi}_{\mathrm{v}}$ in this application.

A naive reconstruction technique for $\boldsymbol{\Phi}_{\mathrm{v}}$ is to first obtain $\boldsymbol{U}_{\mathrm{v}}$ from Zero-Filling:

$$\boldsymbol{U}_{\mathrm{v}} = \mathcal{F}^{-1}\boldsymbol{Y}_{\mathrm{v}} \tag{8.3}$$

and then extract $\boldsymbol{\Phi}_\mathrm{v}$ from the phase of $\boldsymbol{U}_\mathrm{v}$. This is typically suboptimal for undersampled $\boldsymbol{k}$-space data because the reconstructed $\boldsymbol{\Phi}_\mathrm{v}$ suffers from undersampling artifacts, which, for spiral imaging, consist of spiral ripples throughout the image.

More commonly, $\boldsymbol{\Phi}_\mathrm{v}$ is obtained by first reconstructing the real and the imaginary part of $\boldsymbol{U}_\mathrm{v}$ separately through common regularization techniques, such as total variation (TV) regularization [83], and subsequently combining them to give $\boldsymbol{\Phi}_\mathrm{v}$ [30,84,85]. However, it is not clear that the reconstructed phase should be accurate when the real and imaginary parts are reconstructed separately.

Reconstruction methods that regularize the phase of the image, based on some prior knowledge of the phase, have recently been developed [86–91]. A complication arises from the fact that the velocity images are generally acquired from the phase difference of different MR images. While there might be some prior knowledge about the velocity from fluid mechanics, there is generally no prior knowledge about the phase of individual MR images, since they are corrupted by, for example, field inhomogeneity artifacts. As a result, the regularization needs to be imposed on the phase difference of different MR images, rather than separately on the phase of each MR image. The reconstruction method of Benning et al. [86], outlined in the text that follows and used here, was found to be suitable for addressing this complication.

To acquire a single-component velocity image, four MR images are used, corresponding to: flow and ±g polarity for the velocity-encoding gradient $\left(\boldsymbol{P}_{\mathrm{v},1}\cdot\exp\left(\mathrm{i}\boldsymbol{\Phi}_{\mathrm{v},1}\right)\right)$; flow and ∓g polarity the velocity-encoding gradient $\left(\boldsymbol{P}_{\mathrm{v},2}\cdot\exp\left(\mathrm{i}\boldsymbol{\Phi}_{\mathrm{v},2}\right)\right)$; no flow and ±g polarity for the velocity-encoding gradient $\left(\boldsymbol{P}_{\mathrm{v},3}\cdot\exp\left(\mathrm{i}\boldsymbol{\Phi}_{\mathrm{v},3}\right)\right)$; and no flow and ∓g polarity the velocity-encoding gradient $\left(\boldsymbol{P}_{\mathrm{v},4}\cdot\exp\left(\mathrm{i}\boldsymbol{\Phi}_{\mathrm{v},4}\right)\right)$. The individual magnitudes, $\boldsymbol{P}_{\mathrm{v},j}$, are obtained through TV regularization, using the primal–dual hybrid gradient method (PDHGM) [92]:

$$\boldsymbol{P}_{\mathrm{v},j}=\arg\min_{\boldsymbol{P}_{\mathrm{v},j}}\left(\frac{1}{2}\|\mathcal{F}\boldsymbol{P}_{\mathrm{v},j}-\boldsymbol{Y}_{\mathrm{v}.j}\|_2^2+\alpha\|\left[\boldsymbol{R}\boldsymbol{P}_{\mathrm{v},j}\right]_2\|_{2,1}\right) \tag{8.4}$$

where $\boldsymbol{R}$ represents the TV transform operator.

Once the magnitude images are computed, the individual phase images, $\boldsymbol{\Phi}_{\mathrm{v},j}$, and the phase image corresponding to the velocity, $\boldsymbol{\Phi}_\mathrm{v}=\frac{1}{2}\left[\left(\boldsymbol{\Phi}_{\mathrm{v},1}-\boldsymbol{\Phi}_{\mathrm{v},2}\right)-\left(\boldsymbol{\Phi}_{\mathrm{v},3}-\boldsymbol{\Phi}_{\mathrm{v},4}\right)\right]$, are reconstructed using the prior knowledge that the velocity field has smooth spatial features. This is numerically achieved using the linearized Bregman iteration reconstruction technique [86]. The iteration steps are as follows:

$$\begin{cases}\boldsymbol{\Phi}_\mathrm{v}^{(k+1)}=\arg\min_{\boldsymbol{\Phi}_\mathrm{v}}\left\{\tau\langle\boldsymbol{\Phi}_\mathrm{v}-\boldsymbol{\Phi}_\mathrm{v}^{(k)},\nabla H\left(\boldsymbol{\Phi}_\mathrm{v}^{(k)}\right)\rangle+D_J^{p^{(k)}}\left(\boldsymbol{\Phi}_\mathrm{v},\ \boldsymbol{\Phi}_\mathrm{v}^{(k)}\right)\right\}\\ \boldsymbol{p}^{(k+1)}=\boldsymbol{p}^{(k)}-\tau\nabla H\left(\boldsymbol{\Phi}_\mathrm{v}^{(k)}\right)\end{cases} \tag{8.5}$$

where:

$$H\left(\boldsymbol{\Phi}_\mathrm{v}^{(k)}\right)=\frac{1}{2}\sum_j\|\mathcal{F}(\boldsymbol{P}_{\mathrm{v},j}\cdot\exp\left(\mathrm{i}\,\boldsymbol{\Phi}_{\mathrm{v},j}^{(k)}\right)-\boldsymbol{Y}_{\mathrm{v}.j}\|_2^2 \tag{8.6a}$$

$$J\left(\boldsymbol{\Phi}_\mathrm{v}^{(k)}\right)=\frac{1}{2}\|\boldsymbol{\Phi}_\mathrm{v}^{(k)}\|_2^2+\alpha\|[\boldsymbol{Q}\,\boldsymbol{\Phi}_\mathrm{v}^{(k)}]_2\|_{2,2} \tag{8.6b}$$

The Bregman distance for a function J is defined as:

$$D_J^p(\boldsymbol{x}_1,\boldsymbol{x}_2) = J(\boldsymbol{x}_1) - J(\boldsymbol{x}_2) - \langle \boldsymbol{p}, \boldsymbol{x}_1 - \boldsymbol{x}_2 \rangle, \tag{8.7}$$

with $\boldsymbol{p} \in \partial J(\boldsymbol{x}_2)$, and $\partial J(\boldsymbol{x}_2)$ being the so-called sub-differential, which is a generalization of the classical differential for convex functions. The second-order derivative matrix, $\boldsymbol{Q}$, enforces smoothness in the phase image corresponding to the velocity, Φ_v. The algorithmic parameter, τ, controls the convergence and stability of the numerical solution. In this work, $\tau = 1.9$ was found to give a good compromise between stability and convergence speed following the convergence criteria for τ found in the work of Benning et al. [32,86]. It is clearly difficult to validate this reconstruction technique against MR velocimetry data of the bursting bubble because the ground truth of the velocity map is not known; however, the reconstruction technique has been validated against experimental data of fully developed Poiseuille flow in a cylindrical pipe, for which the ground truth for the axial velocity is well known [93]. The regularization parameter, α, was chosen to be equal to the optimal regularization parameter used for method validation.

8.2.4 Results and Discussion

In this section, single-component transverse velocity maps of the bubble burst event are initially presented. It is then shown how these can be combined to produce velocity vector maps where all three components of the velocity vector are included. Velocity maps in the transverse plane are also acquired.

8.2.4.1 Bubble Burst Velocity Maps

The evolution of the map of the velocity in the x-direction, v_x, for a transverse plane that cuts through the middle of the bubble is shown in Figure 8.2. The evolution is given at times $t = 0$, 4, 8, and 12 ms, where $t = 4$ ms corresponds to the time when the bubble burst event starts. At $t = 0$ and 4 ms, the bubble or remains of the bubble are still within the imaging plane at the center of the tube and are shown as blank regions as no signal is obtained from the gas. A strong inward flow toward the bursting bubble is observed, which is consistent with the boundary-layer surface flow theory of MacIntyre [68]. The inward flow weakens with time and gets further away from

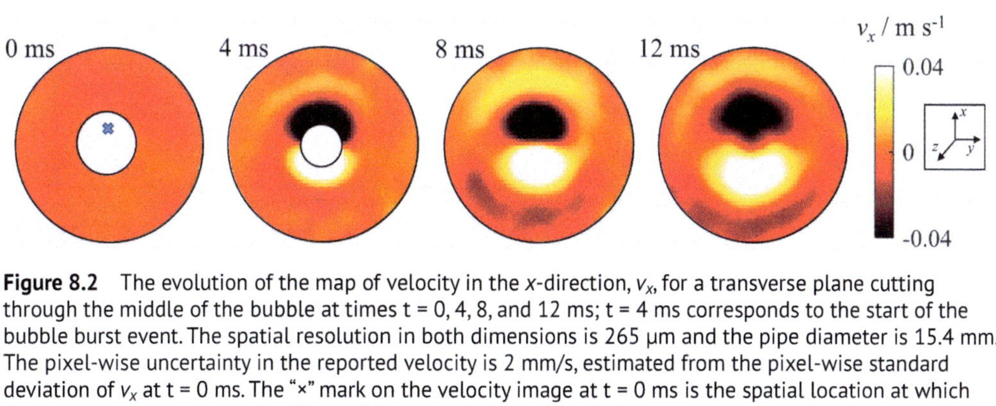

Figure 8.2 The evolution of the map of velocity in the x-direction, v_x, for a transverse plane cutting through the middle of the bubble at times t = 0, 4, 8, and 12 ms; t = 4 ms corresponds to the start of the bubble burst event. The spatial resolution in both dimensions is 265 μm and the pipe diameter is 15.4 mm. The pixel-wise uncertainty in the reported velocity is 2 mm/s, estimated from the pixel-wise standard deviation of v_x at t = 0 ms. The "×" mark on the velocity image at t = 0 ms is the spatial location at which the measurements reported in Figure 8.3 are made. The white circles seen in the middle of the images at 0 and 4 ms correspond to regions of the image filled with air, which does not give an MRI signal.

the center of the bubble burst, eventually reaching the pipe surface. The upward jet is not observed in the v_x map, but it is observed in the longitudinal plane images.

For the 2D images, measurements of v_x were recorded every 4 ms for a time interval 0–280 ms. The imaging of the bubble burst process was repeated for three other bubbles, and measurements of v_x at a particular spatial location, indicated by a cross in the $t = 0$ ms image in Figure 8.2, are shown in Figure 8.3. It is observed that the velocity evolutions for the different bubble burst events are very similar. Though not shown here, the same conclusion is obtained for all other spatial locations. The implication of this observation is that the bubble burst hydrodynamics are highly reproducible. As a result, if the map of the velocities in the x-direction, v_x, y-direction, v_y, and z-direction, v_z, are measured individually for three different bubbles, then, to a good approximation, these three components of the velocity can be combined to construct a three-component velocity map. This is the method employed for constructing the three-component velocity maps reported in this work.

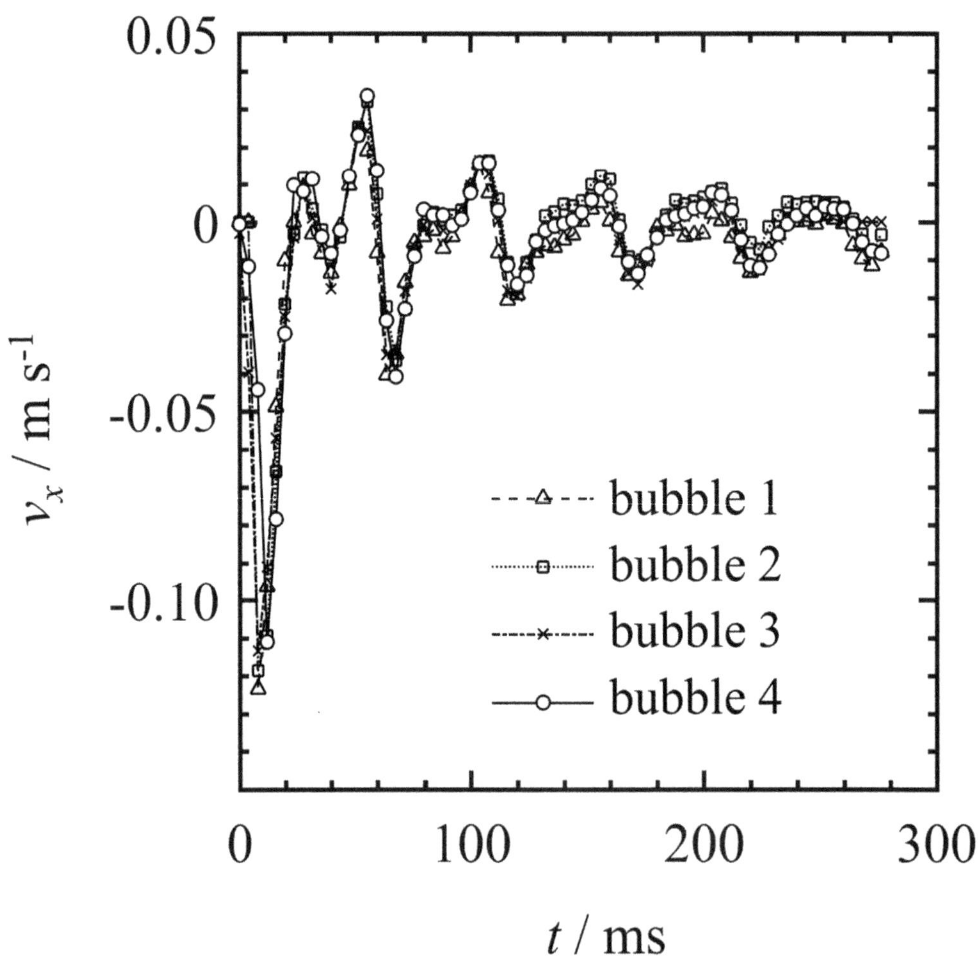

Figure 8.3 The evolution of the velocity in the x-direction, v_x, at the spatial location marked by "×" in Figure 8.2 for a time interval 0–280 ms; t = 4 ms corresponds to the start of the bubble burst event. The uncertainty in the reported velocity is 2 mm/s, estimated from the method described in Figure 8.2.

The evolution of the three-component velocity map at times $t = 0, 4, 8$, and 12 ms for a transverse plane that cuts through the middle of the bubble is shown in Figure 8.4. In addition to the inward flow observed in Figure 8.2, Figure 8.4 shows a central region with an initial large velocity in the z-direction, which corresponds to an upward jet being ejected; the maximum ejection speed detected is 0.15 m/s. The jet achieves the maximum height at $t = 12$ ms (the jet is more clearly seen in Figures 8.5 and 8.6) after which, the jet falls back; there is some periodicity to this movement as is also evident from the data in Figure 8.3. The measured maximum ejection speed is considerably slower than the predictions of numerical methods [69], which is attributed to the 4 ms time averaging and the 265 μm spatial averaging of the MR measurements.

The evolution of the map of the velocity in the z-direction, v_z, for a longitudinal plane that cuts through the middle of the bubble is shown in Figure 8.5. The evolution is given at times $t = 0$, 4, 8, and 12 ms. Apart from the features of bubble burst already observed in Figure 8.4, the longitudinal velocity image clearly shows a downward jet that is created in conjunction with the upward jet ($t = 12$ ms). This agrees with numerical works on bubble burst [69,70]. The longitudinal velocity images also give information about the change in the cavity shape during the bubble collapse, although highly spatially averaged as compared with what fast photography can achieve.

Applying the same method as for the transverse velocity images, the three-component velocity map on a longitudinal plane is reconstructed and is shown in Figure 8.6 at evolution times of $t =$ 0, 4, 8, and 12 ms. The main features of the three-component velocity maps were already observed in Figures 8.2, 8.4, and 8.5. The out-of-plane velocity, v_x, is small compared with the other components of velocity as is to be expected from the symmetry of the flow. The small, but nonzero, values of v_x arise from the imperfect experimental symmetry.

8.3 Conclusions

The principal general approaches to ultrafast MR imaging have been introduced in this chapter with particular consideration of their application to unsteady-state chemical engineering systems. In many cases, time-averaged MR imaging techniques will be able to provide some useful information, but ultrafast imaging techniques are often able to provide further details of the process being studied. It is important to understand the timescales of fluctuations in the process before using ultrafast imaging techniques and the intention is usually to acquire images on a

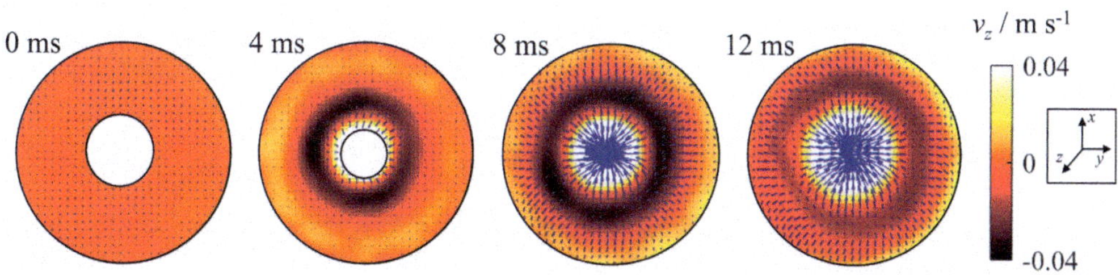

Figure 8.4 The map of the out-of-plane velocity, v_z (color coded), and the in-plane velocities, v_x and v_y (arrow coded), for a transverse plane cutting through the middle of the bubble at times t = 0, 4, 8, and 12 ms; t = 4 ms corresponds to the start of the bubble burst event. The largest arrow corresponds to a velocity of 0.15 m/s. The largest reported ejection speed is v_z = 0.15 m/s. The spatial resolution in both dimensions is 265 μm and the pipe diameter is 15.4 mm. The uncertainties in the estimate of the v_x, v_y, and v_z are 2, 2, and 1 mm/s, respectively.

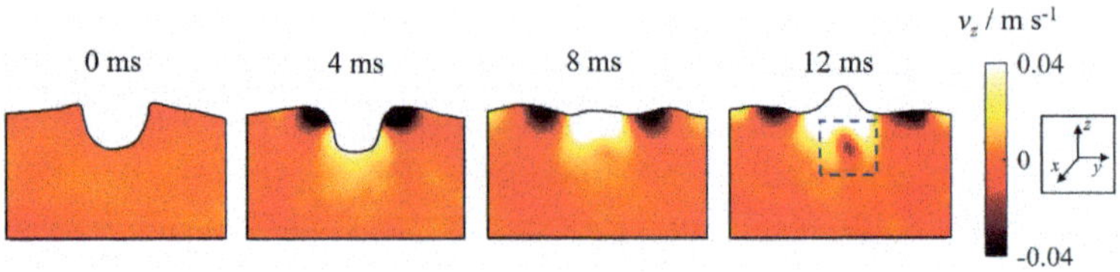

Figure 8.5 The evolution of the map of velocity in the z-direction, v_z, for a longitudinal plane cutting through the middle of the bubble at times t = 0, 4, 8, and 12 ms; t = 4 ms corresponds to the start of the bubble burst event. The spatial resolution in both dimensions is 265 μm and the pipe diameter is 15.4 mm. The pixel-wise uncertainty in the stated velocity is 3 mm/s. The highlighted feature at t = 12 ms corresponds to the downward jet.

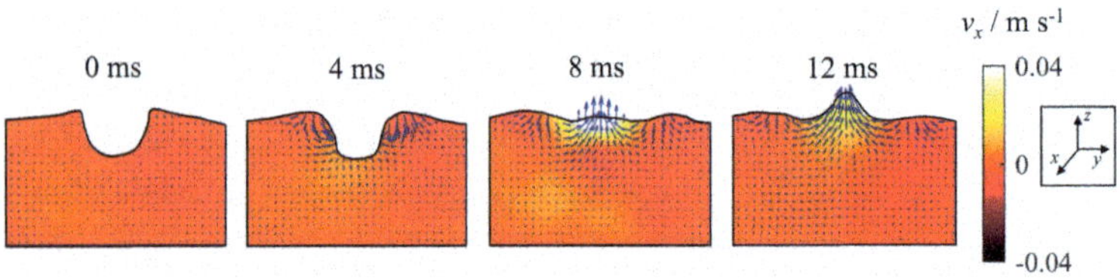

Figure 8.6 The map of the out-of-plane velocity, v_x (color coded), and the in-plane velocities, v_y and v_z (arrow coded), for a longitudinal plane cutting through the middle of the bubble at times t = 0, 4, 8, and 12 ms; t = 4 ms corresponds to the start of the bubble burst event. The largest arrow corresponds to a velocity of 0.13 m/s. The largest reported ejection speed is v_z = 0.13 m/s. The spatial resolution in both dimensions is 265 μm and the pipe diameter is 15.4 mm. The uncertainties in the estimate of the v_x, v_y, and v_z are 2, 2, and 3 mm/s, respectively.

timescale that is short compared with these fluctuations. Imaging times vary, but range from as long as several hundred milliseconds down to just a few milliseconds as shown in the results in this chapter. There are many different methods of ultrafast imaging and each method will have a particular sensitivity to parameters such as NMR relaxation times, diffusion, and flow. The sensitivity of the method to these parameters is a further important consideration when choosing which method to utilize for a particular application.

In this chapter, ultrafast MR velocimetry has been applied to study the hydrodynamics of bubble burst events, which is investigated in more detail. The first experimental measurements of spatially resolved velocity field generated in the liquid when a bubble bursts at a liquid–air interface have been reported. Bubble burst dynamics have been shown to be highly reproducible when velocity measurements from four bubble burst events were compared. Using this assumption, three-component velocity maps obtained at 4 ms temporal resolution have revealed features that were up to now only predicted by numerical methods. Measurements have been shown in two orthogonal planes that show the formation of both an upward and downward jet and show the approximate radial symmetry of the system.

This ultrafast imaging approach is based on spiral-EPI and is therefore not suited to highly magnetically heterogeneous systems, but the readout duration of ~2 ms is short enough to be applicable to some multi-phase applications such as the bubble burst shown here. It could therefore be used in a wide range of applications where $T_2^* >$ ~2 ms and image speed is important. These are likely to include gas or liquid unsteady single-phase flows as well as liquid–gas or liquid–solid systems.

Acknowledgments

The authors would like to acknowledge Martin Benning for providing the original optimization code. A.R. acknowledges Gates Trust Cambridge for financial support.

References

1 Callaghan, P.T. (1993). *Principles of Nuclear Magnetic Resonance Microscopy*. Oxford, UK: Oxford University Press.

2 Callaghan, P.T. (2011). *Translational Dynamics and Magnetic Resonance: Principles of Pulsed Gradient Spin Echo NMR*. Oxford, UK: Oxford University Press.

3 Brown, R.W., Cheng, Y.-C.N., Haacke, E.M. et al. (2014). *Magnetic Resonance Imaging: Physical Principles and Sequence Design*. New York: John Wiley & Sons, Inc.

4 Blumich, B. (2000). *NMR Imaging of Materials*. Oxford, UK: Oxford University Press.

5 Kuethe, D.O. (1989). Measuring distributions of diffusivity in turbulent fluids with magnetic-resonance imaging. *Physical Review A* 40 (8): 4542–4551.

6 Newling, B., Poirier, C.C., Zhi, Y. et al. (2004). Velocity imaging of highly turbulent gas flow. *Physical Review Letters* 93 (15): 154503.

7 Kuethe, D.O. and Gao, J.-H. (1995). NMR signal loss from turbulence: Models of time dependence compared with data. *Physical Review E* 51 (4): 3252–3262.

8 Holland, D.J., Müller, C.R., Dennis, J.S. et al. (2008). Spatially resolved measurement of anisotropic granular temperature in gas-fluidized beds. *Powder Technology* 182 (2): 171–181.

9 Seymour, J.D., Caprihan, A., Altobelli, S.A. et al. (2000). Pulsed gradient spin echo nuclear magnetic resonance imaging of diffusion in granular flow. *Physical Review Letters* 84 (2): 266–269.

10 Haraldsson, H., Kefayati, S., Ahn, S. et al. (2017). Assessment of Reynolds stress components and turbulent pressure loss using 4D flow MRI with extended motion encoding. *Magnetic Resonance in Medicine* 79 (4): 1962–1971.

11 Zwart, N.R. and Pipe, J.G. (2013). Multidirectional high-moment encoding in phase contrast MRI. *Magnetic Resonance in Medicine* 69 (6): 1553–1563.

12 Borup, D.D., Elkins, C.J., and Eaton, J.K. (2019). Effects of motion on MRI signal decay from micron-scale particles. *Journal of Magnetic Resonance* 305: 152–161.

13 Tayler, A.B., Holland, D.J., Sederman, A.J. et al. (2012). Applications of ultra-fast MRI to high voidage bubbly flow: Measurement of bubble size distributions, interfacial area and hydrodynamics. *Chemical Engineering Science* 71: 468–483.

14 Li, T.-Q., Seymour, J.D., Powell, R.L. et al. (1994). Turbulent pipe flow studied by time-averaged NMR imaging: Measurements of velocity profile and turbulent intensity. *Magnetic Resonance Imaging* 12 (6): 923–934.

15 Uludag, Y., Powell, R.L., and McCarthy, M.J. (2004). Effects of periodic flow fluctuations on magnetic resonance flow images. *AIChE Journal* 50 (8): 1662–1671.

16 Holland, D.J., Blake, A., Tayler, A.B. et al. (2011). A Bayesian approach to characterising multi-phase flows using magnetic resonance: Application to bubble flows. *Journal of Magnetic Resonance* 209 (1): 83–87.

17 IMAIOS. (n.d.). eMRI: MRI step-by-step interactive course on magnetic resonance imaging. https://www.imaios.com/en/e-Courses/e-MRI (accessed 10 November 2021).

18 Bitar, R., Leung, G., Perng, R. et al. (2006). MR pulse sequences: What every radiologist wants to know but is afraid to ask. *RadioGraphics* 26 (2): 513–537.

19 Mansfield, P. (1977). Multi-planar image formation using NMR spin echoes. *Journal of Physics C: Solid State Physics* 10 (3): L55–L58.

20 Ahn, C.B., Kim, J.H., and Cho, Z.H. (1986). High-speed spiral-scan echo planar NMR imaging-I. *IEEE Transactions on Medical Imaging* 5 (1): 2–7.

21 Hennig, J., Nauerth, A., and Friedburg, H. (1986). RARE imaging: A fast imaging method for clinical MR. *Magnetic Resonance in Medicine* 3 (6): 823–833.

22 Oshio, K. and Feinberg, D.A. (1991). GRASE (gradient-and spin-echo) imaging: A novel fast MRI technique. *Magnetic Resonance in Medicine* 20 (2): 344–349.

23 Haase, A., Frahm, J., Matthaei, D. et al. (1986). FLASH imaging. Rapid NMR imaging using low flip-angle pulses. *Journal of Magnetic Resonance (1969).* 67 (2): 258–266.

24 Bergin, C.J., Pauly, J.M., and Macovski, A. (1991). Lung parenchyma: Projection reconstruction MR imaging. *Radiology* 179 (3): 777–781.

25 Robson, M.D., Gatehouse, P.D., Bydder, M. et al. (2003). Magnetic resonance: An introduction to ultrashort TE (UTE) imaging. *Journal of Computer Assisted Tomography* 27 (6): 825–846.

26 Fabich, H.T., Benning, M., Sederman, A.J. et al. (2014). Ultrashort echo time (UTE) imaging using gradient pre-equalization and compressed sensing. *Journal of Magnetic Resonance* 245: 116–124.

27 Madio, D.P. and Lowe, I.J. (1995). Ultra-fast imaging using low flip angles and FIDs. *Magnetic Resonance in Medicine* 34 (4): 525–529.

28 Candes, E.J., Romberg, J., and Tao, T. (2006). Robust uncertainty principles: Exact signal reconstruction from highly incomplete frequency information. *IEEE Transactions on Information Theory* 52 (2): 489–509.

29 Gamper, U., Boesiger, P., and Kozerke, S. (2008). Compressed sensing in dynamic MRI. *Magnetic Resonance in Medicine* 59 (2): 365–373.

30 Holland, D.J., Malioutov, D.M., Blake, A. et al. (2010). Reducing data acquisition times in phase-encoded velocity imaging using compressed sensing. *Journal of Magnetic Resonance* 203 (2): 236–246.

31 Lustig, M., Donoho, D., Pauly, J.M., and Sparse, M.R.I. (2007). The application of compressed sensing for rapid MR imaging. *Magnetic Resonance in Medicine* 58 (6): 1182–1195.

32 Benning, M., Gladden, L., Holland, D., et al. (2014). Phase reconstruction from velocity-encoded MRI measurements – A survey of sparsity-promoting variational approaches. *Journal of Magnetic Resonance* 238: 26–43.

33 Tropp, J.A., Laska, J.N., Duarte, M.F. et al. (2010). Beyond Nyquist: Efficient sampling of sparse bandlimited signals. *IEEE Transactions on Information Theory* 56 (1): 520–544.

34 Baylor, E., Peters, V., and Baylor, M. (1977). Water-to-air transfer of virus. *Science* 197 (4305): 763–764.

35 Blanchard, D.C. and Syzdek, L. (1970). Mechanism for the water-to-air transfer and concentration of bacteria. *Science* 170 (3958): 626–628.

36 Boyce, S.G. (1951). Source of atmospheric salts. *Science* 113 (2944): 620–621.

37 Mason, B.J. (1954). Bursting of air bubbles at the surface of sea water. *Nature* 174 (4427): 470–471.

38 Woodcock, A.H., Kientzler, C.F., Arons, A.B. et al. (1953). Giant condensation nuclei from bursting bubbles. *Nature* 172 (4390): 1144–1145.

39 Blanchard, D.C. (1963). The electrification of the atmosphere by particles from bubbles in the sea. *Progress in Oceanography* 1: 73–202.

40 Chisti, Y. (2000). Animal-cell damage in sparged bioreactors. *Trends in Biotechnology* 18 (10): 420–432.

41 Handa-Corrigan, A., Emery, A.N., and Spier, R.E. (1989). Effect of gas–liquid interfaces on the growth of suspended mammalian cells: Mechanisms of cell damage by bubbles. *Enzyme and Microbial Technology* 11 (4): 230–235.

42 Kunas, K.T. and Papoutsakis, E.T. (1990). Damage mechanisms of suspended animal cells in agitated bioreactors with and without bubble entrainment. *Biotechnology and Bioengineering* 36 (5): 476–483.

43 Chalmers, J.J. and Bavarian, F. (1991). Microscopic visualization of insect cell-bubble interactions. II: The bubble film and bubble rupture. *Biotechnology Progress* 7 (2): 151–158.

44 Cherry, R.S. and Hulle, C.T. (1992). Cell death in the thin films of bursting bubbles. *Biotechnology Progress* 8 (1): 11–18.

45 Walls, P.L.L., Bird, J.C., and Bourouiba, L. (2014). Moving with bubbles: A review of the interactions between bubbles and the microorganisms that surround them. *Integrative and Comparative Biology* 54 (6): 1014–1025.

46 Wu, J. (1995). Mechanisms of animal cell damage associated with gas bubbles and cell protection by medium additives. *Journal of Biotechnology* 43 (2): 81–94.

47 Feng, J., Roché, M., Vigolo, D. et al. (2014). Nanoemulsions obtained via bubble-bursting at a compound interface. *Nature Physics* 10 (8): 606–612.

48 Feng, J., Nunes, J.K., Shin, S. et al. (2016). A scalable platform for functional nanomaterials via bubble-bursting. *Advanced Materials* 28 (21): 4047–4052.

49 Gupta, A., Eral, H.B., Hatton, T.A. et al. (2016). Nanoemulsions: Formation, properties and applications. *Soft Matter* 12 (11): 2826–2841.

50 Liger-Belair, G., Cilindre, C., Gougeon, R.D. et al. (2009). Unraveling different chemical fingerprints between a champagne wine and its aerosols. *Proceedings of the National Academy of Sciences* 106 (39): 16545–16549.

51 Liger-Belair, G., Polidori, G., and Jeandet, P. (2008). Recent advances in the science of champagne bubbles. *Chemical Society Reviews* 37 (11): 2490–2511.

52 Sanada, T., Watanabe, M., and Fukano, T. (2005). Effects of viscosity on coalescence of a bubble upon impact with a free surface. *Chemical Engineering Science* 60 (19): 5372–5384.

53 Zawala, J., Dorbolo, S., Vandewalle, N. et al. (2013). Bubble bouncing at a clean water surface. *Physical Chemistry Chemical Physics* 15 (40): 17324–17332.

54 Chesters, A.K. and Hofman, G. (1982). Bubble coalescence in pure liquids. *Applied Scientific Research* 38: 353–361.

55 Kočárková, H., Rouyer, F., and Pigeonneau, F. (2013). Film drainage of viscous liquid on top of bare bubble: Influence of the Bond number. *Physics of Fluids* 25 (2): 022105.

56 Marrucci, G. (1969). A theory of coalescence. *Chemical Engineering Science* 24 (6): 975–985.

57 Taylor, G.I. and Michael, D.H. (1973). On making holes in a sheet of fluid. *Journal of Fluid Mechanics* 58 (4): 625–639.

58 Rayleigh, L. (1891). Some applications of photography. *Nature* 44 (1133): 249–254.

59 Pandit, A.B. and Davidson, J.F. (1990). Hydrodynamics of the rupture of thin liquid films. *Journal of Fluid Mechanics* 212 (1): 11–24.

60 Culick, F.E.C. (1960). Comments on a ruptured soap film. *Journal of Applied Physics* 31 (6): 1128–1129.

61 Ranz, W.E. (1959). Some experiments on the dynamics of liquid films. *Journal of Applied Physics* 30 (12): 1950–1955.

62 Blanchard, D.C. and Syzdek, L.D. (1988). Film drop production as a function of bubble size. *Journal of Geophysical Research* 93 (C4): 3649–3654.

63 Knelman, F., Dombrowski, N., and Newitt, D.M. (1954). Mechanism of the bursting of bubbles. *Nature* 173 (4397): 261.

64 Lhuissier, H. and Villermaux, E. (2009). Bursting bubbles. *Physics of Fluids* 21 (9): 091111.

65 Spiel, D.E. (1998). On the births of film drops from bubbles bursting on seawater surfaces. *Journal of Geophysical Research: Oceans* 103 (C11): 24907–24918.

66 Ghabache, E., Liger-Belair, G., Antkowiak, A. et al. (2016). Evaporation of droplets in a champagne wine aerosol. *Scientific Reports* 6 (1): 25148.

67 Macintyre, F. (1968). Bubbles. Boundary-layer "microtome" for micron-thick samples of a liquid surface. *The Journal of Physical Chemistry* 72 (2): 589–592.

68 MacIntyre, F. (1972). Flow patterns in breaking bubbles. *Journal of Geophysical Research* 77 (27): 5211–5228.

69 Deike, L., Ghabache, E., Liger-Belair, G. et al. (2018). Dynamics of jets produced by bursting bubbles. *Physical Review Fluids* 3 (1): 013603.

70 Duchemin, L., Popinet, S., Josserand, C. et al. (2002). Jet formation in bubbles bursting at a free surface. *Physics of Fluids* 14 (9): 3000–3008.

71 Walls, P.L.L., McRae, O., Natarajan, V. et al. (2017). Quantifying the potential for bursting bubbles to damage suspended cells. *Scientific Reports* 7 (1): 15102.

72 Blanchard, D.C. (1989). The size and height to which jet drops are ejected from bursting bubbles in seawater. *Journal of Geophysical Research* 94 (C8): 10999–11002.

73 Ghabache, E. and Séon, T. (2016). Size of the top jet drop produced by bubble bursting. *Physical Review Fluids* 1 (5): 051901(R).

74 Walls, P.L.L., Henaux, L., and Bird, J.C. (2015). Jet drops from bursting bubbles: How gravity and viscosity couple to inhibit droplet production. *Physical Review E* 92 (2): 021002(R).

75 Kientzler, C.F., Arons, A.B., Blanchard, D.C. et al. (2016). Photographic investigation of the projection of droplets by bubbles bursting at a water surface. *Tellus* 6 (1): 1–7.

76 Lee, J.S., Weon, B.M., Park, S.J. et al. (2011). Size limits the formation of liquid jets during bubble bursting. *Nature Communications* 2 (1): 367.

77 Tayler AB, Holland DJ, Sederman AJ, et al. (2012). Exploring the origins of turbulence in multiphase flow using compressed sensing MRI. *Physical Review Letters* 108 (26): 264505.

78 Vella, D. and Mahadevan, L. (2005). The "Cheerios effect". *American Journal of Physics* 73 (9): 817–825.

79 Hargreaves, B.A., Nishimura, D.G., and Conolly, S.M. (2004). Time-optimal multidimensional gradient waveform design for rapid imaging. *Magnetic Resonance in Medicine* 51 (1): 81–92.

80 Fessler, J.A. and Sutton, B.P. (2003). Nonuniform fast Fourier transforms using min-max interpolation. *IEEE Transactions on Signal Processing* 51 (2): 560–574.

81 Duyn, J.H., Yang, Y., Frank, J.A. et al. (1998). Simple correction method for ***k***-space trajectory deviations in MRI. *Journal of Magnetic Resonance* 132 (1): 150–153.

82 Rasche, V., Proksa, R., Sinkus, R. et al. (1999). Resampling of data between arbitrary grids using convolution interpolation. *IEEE Transactions on Medical Imaging* 18 (5): 385–392.

83 Rudin, L.I., Osher, S., and Fatemi, E. (1992). Nonlinear total variation based noise removal algorithms. *Physica D: Nonlinear Phenomena* 60 (1–4): 259–268.

84 Roberts, T., Kingsbury, N., and Holland, D.J. (2013). Sparse recovery of complex phase-encoded velocity images using iterative thresholding. *2013 IEEE International Conference on Image Processing*, Melbourne, Australia (15–18 September). IEEE, 350–354.

85 Trzasko, J. and Manduca, A. (2009). Highly undersampled magnetic resonance image reconstruction via homotopic ℓ_0-minimization. *IEEE Transactions on Medical Imaging* 28 (1): 106–121.

86 Benning, M., Betcke, M.M., Ehrhardt, M.J., et al. (2016). Gradient descent in a generalised Bregman distance framework. *arXiv*: 161202506v2.

87 Feng, Z., Noll, D.C., Nielsen, J.F. et al. (2012). Separate magnitude and phase regularization via compressed sensing. *IEEE Transactions on Medical Imaging* 31 (9): 1713–1723.

88 Fessler, J.A. and Noll, D.C. (2002). Iterative image reconstruction in MRI with separate magnitude and phase regularization. 2nd IEEE International Symposium on Biomedical Imaging: Nano to Macro (IEEE Cat No. 04EX821) 1: 209–212. doi:10.1109/ISBI.2004.1398511

89 Ong, F., Uecker, M., Tariq, U. et al. (2015). Robust 4D flow denoising using divergence-free wavelet transform. *Magnetic Resonance in Medicine* 73 (2): 828–842.

90 Zibetti, M.V.W. and De Pierro, A.R. (2010). Separate magnitude and phase regularization in MRI with incomplete data: Preliminary results. 2010 IEEE International Symposium on Biomedical Imaging: From Nano to Macro: 736–739. doi:10.1109/ISBI.2010.5490069.

91 Zibetti, M.V.W. and De Pierro, A.R. (2016). Improving compressive sensing in MRI with separate magnitude and phase priors. *Multidimensional Systems and Signal Processing* 28 (4): 1109–1131.

92 Chambolle, A. and Pock, T. (2010). A first-order primal-dual algorithm for convex problems with applications to imaging. *Journal of Mathematical Imaging and Vision* 40 (1): 120–145.

93 Reci, A. (2019). *Signal Sampling and Processing in Magnetic Resonance Applications*. Cambridge, UK: University of Cambridge.

9

Magnetic Resonance Imaging of Membrane Filtration Processes

Denis Wypysek[1,2] *and Matthias Wessling*[1,2]

[1]*RWTH Aachen University, Chemical Process Engineering, Aachen, Germany*
[2]*DWI – Leibniz-Institute for Interactive Materials, Aachen, Germany*

9.1 Introduction to Membranes and Membrane Processes

9.1.1 Membranes

Synthetic membranes have emerged as a key technology for molecular separations. They act as a semipermeable or permeable barrier between two phases and are impermeable for at least one component [1]. Membranes can be classified based on their fabrication and material, such as synthetic (liquid, solid, organic, inorganic, etc.) or biological membranes. Depending on the application, some kinds of membranes are more advantageous than others. In the case of water treatment, for example, solid synthetic membranes dominate the market as they are easier to fabricate in a vast amount. These membranes may differ in their inner structure and can be porous, dense, or a denseporous composite depending on the substances that are to be separated [1].

Another possibility to classify membranes is their geometry in modules for filtration processes. Two main membrane geometries are distinguished: flatsheet membranes and hollow fiber or tubular membranes. Flatsheet membranes are comparable to a flat and thin permeable, semipermeable, or dense surface. In contrast to flat-sheet membranes, hollow fiber membranes have a tubular shape. Further development of hollow fiber membranes has resulted in multichannel membranes. Here, one support structure contains several lumen channels [1].

9.1.2 Membrane Module and Membrane Processes

While membrane materials are improving, the housing and thus the hydrodynamic conditions inside membrane modules are emerging as a bottleneck. Figure 9.1a depicts a schematic drawing of a membrane separation unit. In membrane processes, the membrane module is the sum of the membrane and housing with all connections for inlet and outlet streams. The inlet or feed stream is the educt that shall be treated by the process. The membrane, which acts as a barrier, separates the feed into two fractions: the permeate passes the membrane and the retentate is held back. The product of a membrane process can either be the retentate or the permeate. Hence, processes where only the permeate is required can be conducted without a retentate volume flow. This flow

Magnetic Resonance Microscopy: Instrumentation and Applications in Engineering, Life Science and Energy Research, First Edition. Edited by Sabina Haber-Pohlmeier, Bernhard Blümich, and Luisa Ciobanu.

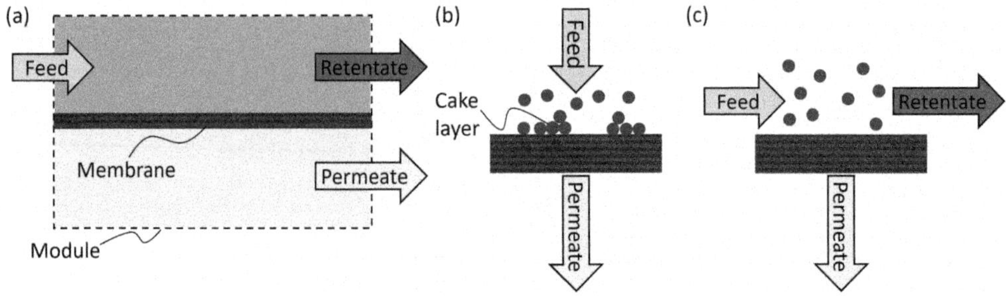

Figure 9.1 (a) Scheme of a basic membrane module, including membrane, feed inlet, permeate outlet, and retentate outlet. (b) Dead-end arrangement. (c) Cross-flow arrangement.

arrangement is called dead-end and is visualized in Figure 9.1b. If permeate and retentate outlets are open, the flow arrangement is called cross-flow (Figure 9.1c) [1].

Inside membrane modules, the necessary driving force for the process can be applied in different ways. In general, these driving forces are either a gradient of pressure (e.g. reverse osmosis, micro/ultrafiltration), concentration (e.g. dialysis), temperature (e.g. membrane distillation), or electrical potential (e.g. electrodialysis) [1]. The performance of membrane processes is determined by the transport phenomena inside and around the membrane. One important parameter is the flux. The flux is defined as the permeate flow rate per membrane unit area. It can be calculated for each feed component individually or the entire mixture. The flux decreases with increasing transport resistances and decreasing driving forces. Equation 9.1 depicts this general relationship [1].

$$\text{Flux} = \frac{\text{Driving Force}}{\text{Transport Resistance}} \tag{9.1}$$

The reciprocal value of the membrane transport resistance is defined as the membrane's permeability. For water and wastewater treatment applications, pressure-driven filtration processes dominate the market. The driving force for pressure-driven processes is the pressure difference between the feed/retentate side and the permeate side, the transmembrane pressure (TMP). It can be calculated from the pressures p of feed, retentate, and permeate using Equation 9.2.

$$\Delta p = \text{TMP} = \frac{p_{\text{Feed}} + p_{\text{Retentate}}}{2} - p_{\text{Permeate}}. \tag{9.2}$$

Pressure-driven filtration processes can be further classified depending on the separated particle sizes and applied pressure gradients.

9.1.3 Performance Decreasing Phenomena in Membrane Processes and Countermeasures

All membrane processes suffer from a decrease in performance (namely the flux) during operation. Besides the unavoidable resistance of the membrane itself and pressure losses on the feed and the permeate side, as well as in the porous structure, several resistances lower the performance of a membrane filtration over time, namely (i) concentration polarization on the feed and permeate side, (ii) fouling, scaling, and pore blocking, and (iii) temperature polarization on the feed and permeate side. According to Song [2] the total resistance can be calculated using the resistance-in-series model, Equation 9.3:

$$R_{\mathrm{Total}} = R_{\mathrm{Membrane}} + R_{\mathrm{Adsorption}} + R_{\mathrm{Pore}} + R_{\mathrm{Layer}}, \tag{9.3}$$

where R_{Pore} is the resistance of blocked pores and R_{Layer} the resistance of fouling and/or scaling layers (e.g. Figure 9.1b). In pressure-driven membrane processes, concentration polarization and layer formation are the main reasons for a decline in performance [3].
Figure 9.2a illustrates the phenomenon of concentration polarization. Due to the semipermeable character of the membrane, one substance passes through the membrane preferentially. Therefore, the concentration of the separated substance increases above the bulk value at the membrane, causing a concentration gradient. The concentration gradient results in a diffusive backflow from the membrane toward the bulk phase. Since the diffusive flow directed away from the membrane is significantly slower than the opposing convective flow, the separated substance accumulates at the membrane surface. The increased chemical potential counteracts the pressure difference. The result is a lower permeate flux and a decrease in purity [4]. Concentration polarization is the main reason for a buildup of fouling or scaling layers. While fouling describes a layer of undesirably accumulated retained particles in and at the membrane, scaling describes the formation of a crystalline layer on the membrane surface [1].

To increase the process performance, concentration polarization has to be reduced [1]. The reduction of concentration polarization and fouling can be achieved by changing the hydrodynamic conditions inside the membrane module. One measure is the use of cross-flow filtration compared with dead-end filtration, as shown in Figure 9.1b and c [1]. Other measures focus on increasing the vorticity within a membrane module, such as the generation of secondary flows, vibrating or rotating membranes that induce higher shear rates, inserts like spacers, or geometrical changes inside the membrane to distract the flow [5]. These measures are often accompanied by a higher energy demand, which shows that a tradeoff between higher energy consumption and better process performance has to be found. Another aspect of this tradeoff is the need for cleaning or replacing the membrane if the cake layer becomes too thick and a critical TMP is reached. In industrial applications, membranes are cleaned by permeate backwashing [1]. The

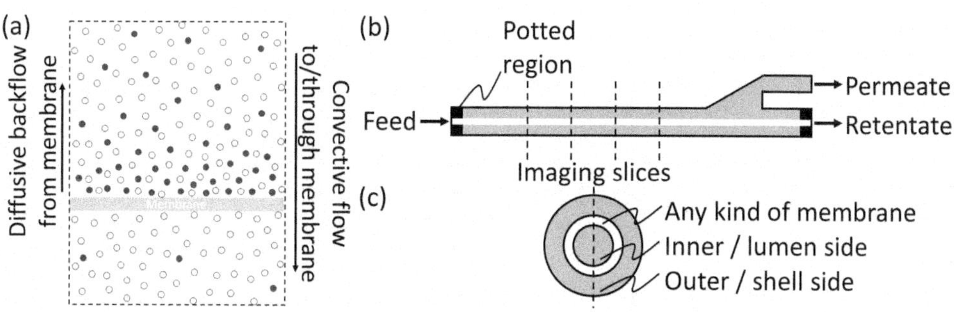

Figure 9.2 (a) Illustration of concentration polarization. (b) Membrane module side view. (c) Cross-sectional view of an arbitrary membrane module in cross-flow, inside-out mode.

flow over the membrane is reversed by increasing the pressure on the permeate side, and the cake layer is released from the membrane. Further washing in cross-flow mode removes the loosened particles from the membrane module [1].

9.1.4 Magnetic Resonance Imaging in Membrane Processes

The overall performance of membrane processes is easily monitored by measuring volume flows or pressures of feed, retentate, and permeate streams. These parameters do not leave much room for improvement. However, to understand the phenomena that occur during filtration, it is necessary to reveal the hydrodynamics inside the membrane module. A powerful technique for the visualization of internal structures and flow dynamics noninvasively and in situ during operation is magnetic resonance imaging (MRI). It enables, for example, the observation of flow profiles and polarization layers [6], the observation of layer buildup and particle deposition [7], and the analysis of systems that prevent concentration polarization and fouling [8]. Figure 9.2b and c schematically show how most of the membrane modules were built in the studies reported in this chapter. When all these phenomena mentioned above are entirely understood, membrane processes and especially membrane modules can be improved to counteract fluid resistances more efficiently. These parameters obtained from experimental data also help to develop, validate, and improve numerical models. The following sections present selected applications of MRI in filtration processes.

9.2 Oil–Water Emulsions and Silica Suspensions in Polymeric Hollow Fiber Modules

To characterize concentration polarization and deposit layers and, thus, draw conclusions about inhibiting factors during filtration, MRI can be used. Prangle et al. [9] measured and analyzed laminar flow in porous tube and shell systems. Later, in the 1990s, the working group around Yao, Pope, and Fane [6,10–13] used ^{1}H nuclear magnetic resonance (NMR) micro-imaging to study these effects. For all studies, similar filtration modules and feed concentrations were used.

Yao et al. [6] observed flow profiles and polarization layers in hollow fiber membrane modules. Their module consisted of five membranes sealed into a glass tube. For flow velocity imaging, distilled water was used, and the concentration polarization experiments utilized an oil–water emulsion. The feed was fed to the shell side, and the retentate was recirculated through the outer envelope so that a constant flow with outside-in filtration in dead-end mode was achieved. Images were obtained by dynamic NMR microscopy with modified chemical shift selective, spin echo Fourier imaging pulse sequences.

Figure 9.3 depicts magnitude (Figure 9.3a.i) and velocity (Figure 9.3b.i to b.iii) images of pure water filtration experiments in a membrane module with five membranes. As expected, parabolic flow profiles develop in each lumen channel. Yao et al. [6] showed good agreement of set volume flow rates and those obtained from MRI measurements, in positive and negative directions. These findings were also valid for a membrane module with higher packing density (containing 15 fibers) and confirmed the robustness of the used pulse sequences. The denser-packed membrane module showed channeling on the shell side and, thus, an uneven flow distribution. These uneven flow distributions can lead to performance and efficiency losses.

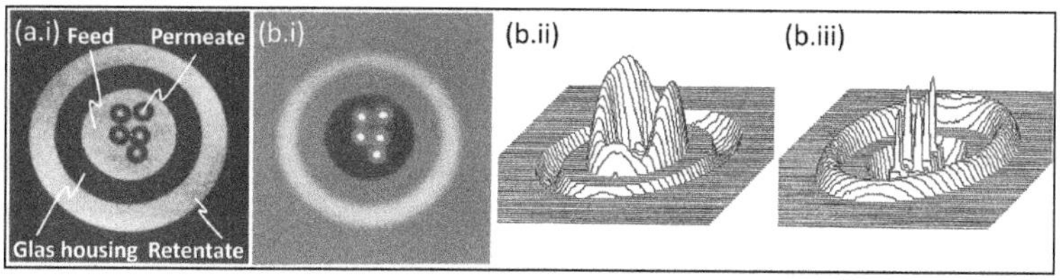

Figure 9.3 Flow through a hollow fiber membrane module containing five fibers. (a.i) Magnitude image showing the disposition of the hollow fiber membranes. (b.i) Contour plot of flow velocity distribution (bright shades, positive flow; dark shades, negative flow). Stackplot showing (b.ii) negative (max. 17.5 mm/s) and (b.iii) positive (max. 21.3 mm/s) velocity components. Slice thickness 1 mm. Adapted from Yao (1995) [6] with permission from Elsevier.

Figure 9.4 shows chemical shift selective images of the filtration of the oil–water emulsion in a module containing five fibers. Yao et al. [6] showed that it is possible to visualize only the oil phase (Figure 9.4a to c) or the water phase (Figure 9.4d), respectively. A layer of oil surrounded each membrane from the outside. The thickness of these oil layers is not constant. However, in regions with low feed flows and, thus, low shear velocities, the oil polarization layers were the thickest. Figure 9.4 also visualizes the dependence of the oil layer thickness on different pressure regimes. The higher the TMP (increasing from Figure 9.4a toward Figure 9.4c), the thicker is the oil polarization layer in areas with low velocities. In regions with high velocities, however, the thickness remains largely unchanged.

Additionally, the authors recognized high swelling of the membranes during filtration experiments with the oil–water emulsion. Using the resistance-in-series model (Equation 9.3), these oil polarization layers are recognized as the dominant resistances and flux-limiting factors. Yao et al. [6] also combined chemical shift imaging and flow mapping and showed that the oil polarization layer is rather stationary and does not move within the measurement error. Another argument for a stagnant layer is the oil layer acting as a boundary for water flow resulting in a velocity close to zero at the oil–water interface.

Pope et al. [12] and Airey et al. [13] continued the work of Yao et al. [6]. While Pope et al. investigated fouling layer thicknesses of oil–water emulsions, Airey et al. studied colloidal silica (Ludox, amorphous) suspensions. Both used chemical shift selective micro-imaging (described in detail by Yao et al. [6]). Both groups utilized only one hydrophobic, tubular polypropylene membrane rather than a bundle of membranes. Its bigger diameter facilitated recognition of polarization layers.

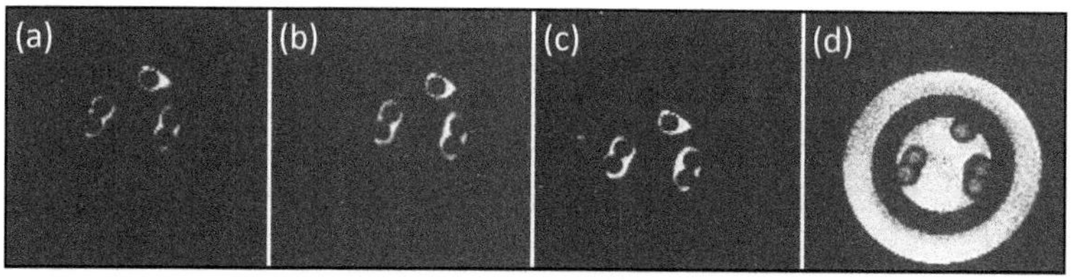

Figure 9.4 Chemical shift images of the oil polarization layer dependence on feed pressure and flow rate. Oil images obtained for (a) 2.9 kPa or 1.38 ml/min, (b) 4.5 kPa or 2.10 ml/min, (c) 7.6 kPa or 3.75 ml/min, respectively. (d) Water image corresponding to the intermediate pressure head value. Slice thickness 1 mm. Adapted from Yao (1995) [6] with permission from Elsevier.

Pope et al. [12] used a tubular membrane module under "well-fouled" conditions, which means that the flux decline during operation does not change significantly anymore (quasi-stationary condition). This was achieved by filtering an oil–water emulsion dispersed in water. A cross-flow, inside-out filtration setup with a recirculating system to maintain a constant TMP was used in this study. Figure 9.5 illustrates time-dependent oil-selective MR images, taken 25 cm behind the feed inlet during operation at 65 kPa and a cross-flow Reynolds number of 110.

The authors saw an increase in oil concentration polarization layer thickness during the filtration process. However, due to buoyancy effects, the oil layer thickness was not uniform over the membrane's whole circumference. Especially in Figure 9.5f it shows a higher thickness on the top part of the membrane's lumen channel. Moreover, the signal intensity within the membrane increases, which means that oil droplets enter the membrane and cause internal fouling. This internal fouling can lead to a (long-term) flux decrease due to higher resistances. The effect of the increasing resistances can be seen in Figure 9.6a, which shows the permeate flux J over filtration time under three different operating conditions. With increasing filtration time, the flux in all three cases follows an approximately exponential decay. Furthermore, the higher the cross-flow Reynolds number and the lower the applied pressures, the higher was the achieved steady-state flux and, thus, the smaller the polarization layer. In this study, a (maximum) critical Reynolds number of 1350 was found for a TMP of 65 kPa, after which no oil polarization layer was observed anymore.

Based on a pixel-by-pixel analysis of the obtained MR images, the authors calculated the oil polarization layer thickness over time in Figure 9.6b and the averaged intensity over time in Figure 9.6c. The averaged thickness increased exponentially during filtration and reached steady state after three hours, whereas the intensity reached a plateau region earlier. The authors were able to calculate a specific resistance of the oil polarization layer that was 145 times higher than that of the clean membrane. Additionally, fitting the measured polarization layer thicknesses with a mass transfer–polarization model led to good agreement between observed flux and flux predicted by Brownian diffusion. Thus, Brownian diffusion was found to be the prevailing back-transport mechanism in this filtration setup.

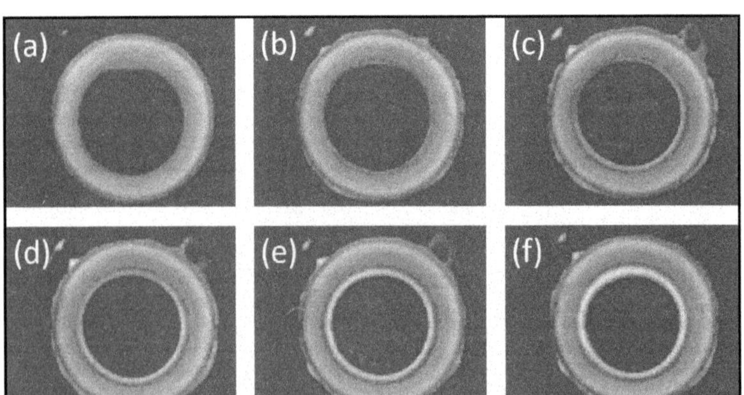

Figure 9.5 Two-dimensional oil-selective cross-sectional magnetic resonance images of a tubular membrane module, 25 cm behind the inlet. (a) "Well-fouled" membrane before the feed was turned on, after (b) 10 min, (c) 40 min, (d) 70 min, (e) 115 min, and (f) 205 min of inside-out filtration. Adapted from Pope (1996) [12] with permission from Elsevier.

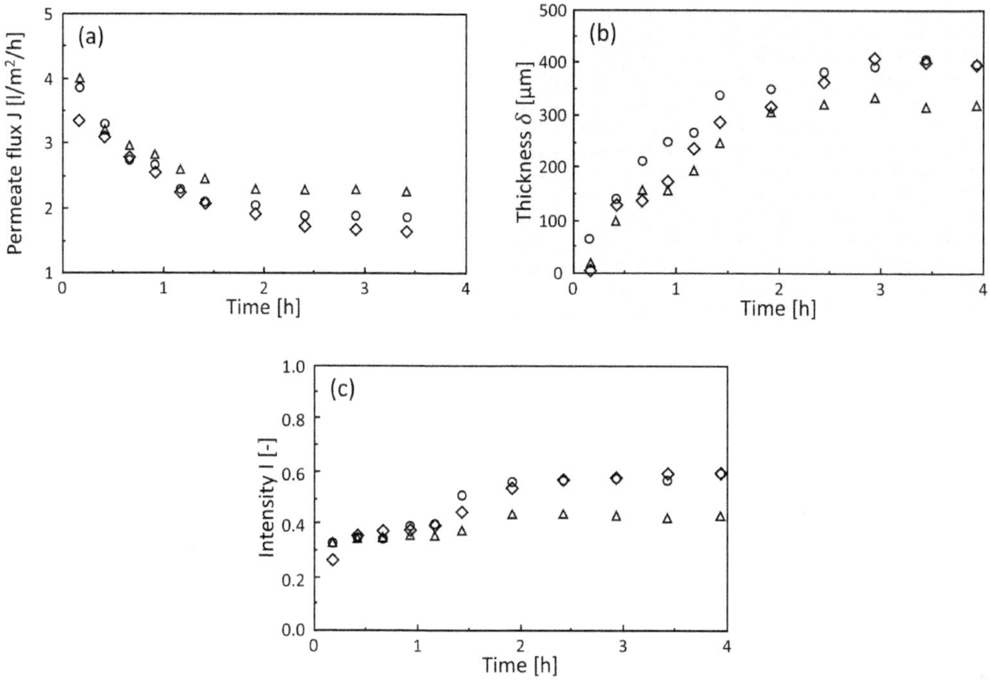

Figure 9.6 (a) Length averaged permeate flux J versus time for three different operating conditions. Plots of (b) oil layer thickness δ, and (c) intensity I of the oil layer versus time calculated from magnetic resonance images for three different operating conditions. (o) Δp = 75 kPa, Re = 110, (◊) Δp = 65 kPa, Re = 110, and (Δ) Δp = 65 kPa, Re = 365. Adapted from Pope (1996) [12] with permission from Elsevier.

Airey et al. [13] investigated the formation and dissipation of concentration polarization phenomena of colloidal silica suspensions. Their setup was similar to that of Pope et al. [12]. Cross-flow inside-out filtration was used in a pump-driven recirculation system. In contrast to the previous studies, colloidal silica particles (Ludox, amorphous) were used as model particles in a suspension in distilled water. The authors found linearly decreasing T_1 and T_2 relaxation times with an increase in silica suspension concentration. Thus, it is possible to measure silica concentrations online and in situ in MR images using a suitable calibration curve. In this study, however, the MRI measurements indicated a maximum packing density for the polarization layers. Figure 9.7 shows cross-sectional intensity images of the membrane module before (Figure 9.7a) and after (Figure 9.7b) flow of the silica suspension was turned on.

A highlighted circle on the membrane's inner side can clearly be seen in Figure 9.7b. Additionally, the peaks in the signal intensity profile close to the membrane wall in Figure 9.7d confirm the possibility of visualizing silica polarization layers. These peaks are not visible in the intensity profile before the silica suspension filtration in Figure 9.7c. In this study, the thickness of silica polarization layers was highly nonuniform. A thick deposition layer of silica particles on the bottom side of the membrane (bright shade, Figure 9.7b) gives a hint that gravitational forces influenced the fouling process under these conditions. This phenomenon disappeared during vertical operation of the membrane module instead of horizontal operation. The silica layer growth during filtration and the corresponding permeate flux can be obtained from the graph in Figure 9.8a. The permeate flux decreased exponentially with filtration time, whereas the maximum concentration polarization layer thickness growth was reached between the first and

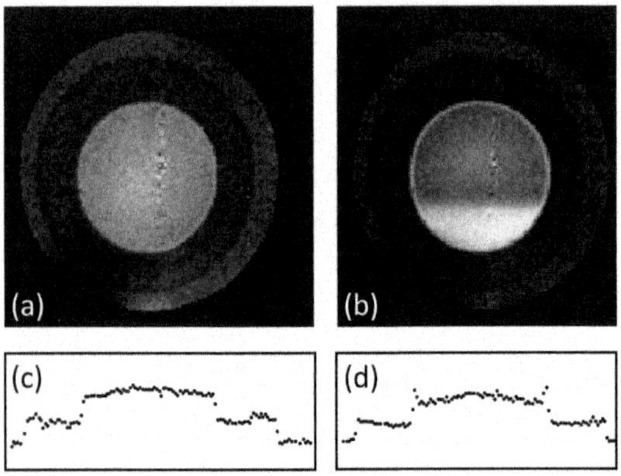

Figure 9.7 Two-dimensional cross-sectional T_1-weighted ^1H spin density images obtained from the tubular membrane module ~25 cm behind the inlet. Magnetic resonance images (a) before and (b) after flow of the 5 wt% colloidal silica feed suspension was turned on. Profiles of the signal intensity along a horizontal line through the centers (c) before and (d) after filtration of the silica suspension. Flow conditions: $\Delta p = 75$ kPa, cross-flow Re = 45. Adapted from Airey (1998) [13] with permission from Elsevier.

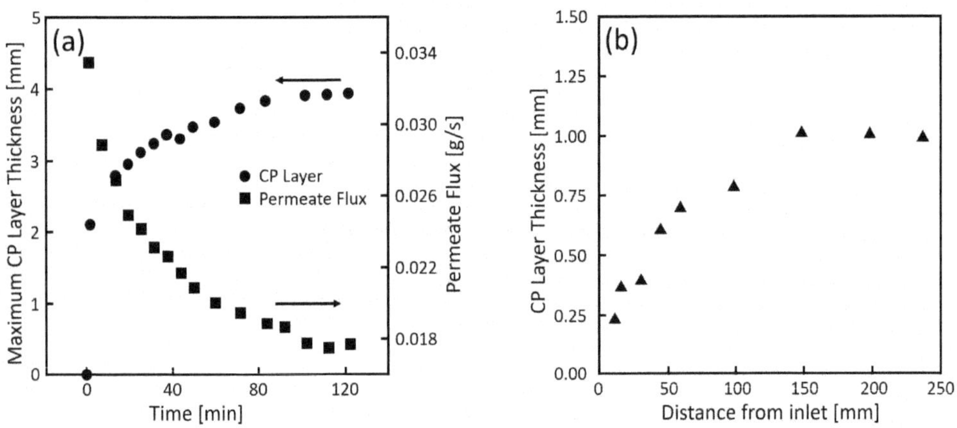

Figure 9.8 (a) Growth in the thickest part of the polarization layer with time and the corresponding decline in the flux rate. Images were taken 135 mm behind the feed inlet with a cross-flow Reynolds number of Re = 37, and a transmembrane pressure (TMP) $\Delta p = 75$ kPa. (b) Change in the steady-state thickness of the thick part of the polarization layer with distance from the inlet. Approximately half the total length of the filter (460 mm) was accessible for imaging. $\Delta p = 7$ kPa, Re = 74. Adapted from Airey (1998) [13] with permission from Elsevier.

second measurement, followed by limited growth. There is no clear relation between maximum silica layer thickness and permeate flux rate since most of the flux leaves through the less fouled parts. However, steady-state fouling layers could be observed using MRI. Figure 9.8b shows the variation of maximum layer thickness according to the distance from the feed inlet. In the first part of the membrane module, the polarization layer was not fully developed. After ~150 mm behind the feed inlet, the authors observed a constant layer thickness. The applied TMP also influenced these layers. With increasing TMP, the layer thickness increased initially.

Based on the previous studies, Airey et al. [13] and Yao et al. [10,11] investigated the fluidity of such polarization layers. While Airey et al. analyzed colloidal silica particles, Yao et al. looked further into oil–water emulsions. Both groups used the same membrane and module. Airey et al. [13] saw a discontinuity in the velocity profiles between feed and silica layer. There was a significant axial flow inside the particle layer up to half of the layer thickness. They observed a velocity in the boundary separating the polarization layer and the bulk of about 4 mm/s, falling to ~0 mm/s at half of the layer thickness. Contrary, Yao et al. [10,11] recognized a rather stationary polarization layer within the measurement error of ±7 μm/s. This shows that the fluidity of polarization layers is highly dependent on the composition of the foulant. However, both studies present a stagnant polarization layer close to the membrane.

In contrast to the studies mentioned above, Buetehorn et al. [14] applied MRI to visualize and measure the distribution of local permeate flux and local cake layer growth in a hollow fiber membrane module that was placed vertically (not horizontally) in an MRI system. Thus, the effect of gravitational forces influencing the filtration process was negligible in this study. The authors used a polyether sulphone membrane. The flow of deionized water doped with $CuSO_4$ was visualized. Cake growth experiments contain additional colloidal silica particles (Bindzil, amorphous) and an outside-in filtration mode with a stagnant feed on the shell side of the membrane was applied. The velocity was measured with a 3D phase-encoding pulse sequence and cake growth with a 3D gradient echo pulse sequence. Buetehorn et al. [14] found a position-dependent behavior of the permeate (lumen-sided) velocity: the further away from the module's outlet, the lower the measured velocities. Additionally, an increase in average permeate flux caused the maximum velocity to increase linearly. This linearity showed that the lumen-sided pressure loss is lower than the pressure loss through the porous structure. The lumen-sided pressure loss was neglected in the estimations, which is why their model slightly underestimated the measured velocities.

The effect of lumen-sided pressure losses can be seen in Figure 9.9a by analyzing the local cumulative permeate flux. The local cumulative permeate flux results from subtracting two consecutive measured lumen volume flows from each other. The graph shows a decreasing average permeate flux. With increasing distance from the permeate outlet, a decreasing cumulative permeate flux indicates lumen-sided pressure losses, which affect the flux. This negative slope increases with increasing average permeate flux. Figure 9.9b illustrates the evolution of the cake layer thickness (three identical experiments) over filtration time. The cake layer growth is faster at the beginning of the filtration and slows down with increasing filtration time, caused by varying TMPs over time. In this case, the cake layer increases logarithmically. The relationship between TMP and cake thickness is depicted in Figure 9.9c. The TMP grows exponentially with increasing cake layer thickness. The authors gave different possible explanations. As a result of the growing cake layer, it might have been compacted or compressed reducing the porosity of the cake layer. Moreover, uneven growth along the whole membrane or the effect of a steadily reducing growth rate on the membrane's circumference might have caused a higher increase in TMP later in the filtration process. Figure 9.9d shows the ratio of transmembrane pressure and cake layer thickness versus filtration time and accounts for the TMP that is required to overcome 1 μm of cake thickness. This linear characteristic explains the dynamics of cake formation. The possible hypotheses of cake compaction, uneven cake layer growth, or reduced cake growth rate while the cake layer continuously grows on the curved membrane surface would explain this.

Buetehorn et al. [14] also investigated cake growth for different positions in the membrane module. They found thicker cake layers at the outlet position during the entire filtration time than at positions further away from the outlet. It seems that the cake growth started next to the outlet position (active permeate extraction point) and continued growing away from that point. The authors also analyzed different average permeate fluxes in dependence on different solid

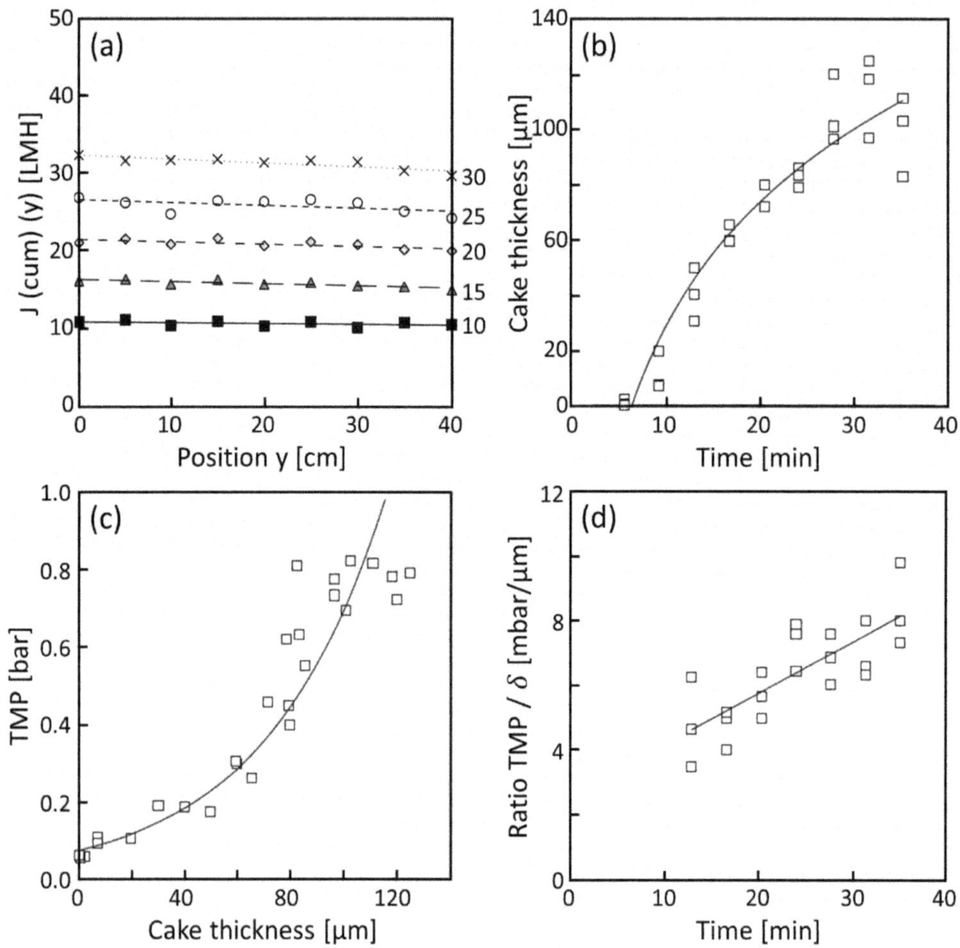

Figure 9.9 (a) Local cumulative permeate flux as a function of vertical position and average permeate flux with linear trend lines (J_{avg}: (×) 30, (○) 25, (◇) 20, (Δ) 15, and (■) 10 l/(m^{2}h), v_{max} = 2.15 · v_{avg}). (b) Evolution of cake layer thickness with a logarithmic trend line. (c) Transmembrane pressure (TMP) versus cake layer thickness with an exponential trend line. (d) Evolution of TMP/cake-height ratio with a linear trend line (data points of three different test runs, J = 20 l/(m^{2}h), c_{SiO2} = 0.4 vol.%, y = 0 cm). Adapted from Buetehorn (2011) [14] with permission from Elsevier.

concentrations and measured the time until a cake layer of 50 μm built up. With an increasing average permeate flux and an increase in solid concentration, the time to reach a cake thickness of 50 μm decreased from 325.0 min at 10 LMH and 0.2 vol.% to 5.4 min at 30 LMH and 0.6 vol.%. After fouling of the membrane, Buetehorn et al. [14] successfully applied a shell-sided air bubbling to remove the cake layer and restore the membrane's performance. With increasing aeration pressure and with increasing aeration time, the cake layer thickness decreased. The authors proposed a critical aeration pressure that has to be overcome to remove more than 30% of the formed cake layer in just one short aeration circle. Above this value, the duration of backwashing circles is less sensitive to aeration times.

Çulfaz et al. [3] used the same experimental setup and MRI pulse sequences as Buetehorn et al. [14] to investigate the influence of differently charged particles (Ludox, amorphous, nominal diameter ~23 nm and Bindzil, amorphous, nominal diameter ~52 nm) and sodium alginate on the deposition behavior at self-made hollow fiber membranes of two shapes, round and structured. Details on the membrane manufacturing process can be obtained in [15]. The structured fiber's surface was ~60% thicker than that of the round fiber. The membranes were operated in constant flux filtration and backwashing experiments. MR images of the fouling behavior over filtered permeate volume of round and structured hollow fiber membranes can be seen in Figure 9.10. The deposition behavior of the two membranes differed a lot. With increasing filtered permeate volume, the resistances for both round and structured fiber increased as well. However, the increase in resistance for the structured fiber was greater than that of the round fiber for all fluxes and particle concentrations measured. In the case of the Bindzil filtration, a continuous, homogeneously distributed cake layer with the same thickness for round and structured membrane grows on the outer surface. In the Ludox filtration case, the cake layer grows symmetrically on the round fiber, while on the structured fiber, the grooves are filled first, and the fins are hardly covered with particles. The authors explained the different

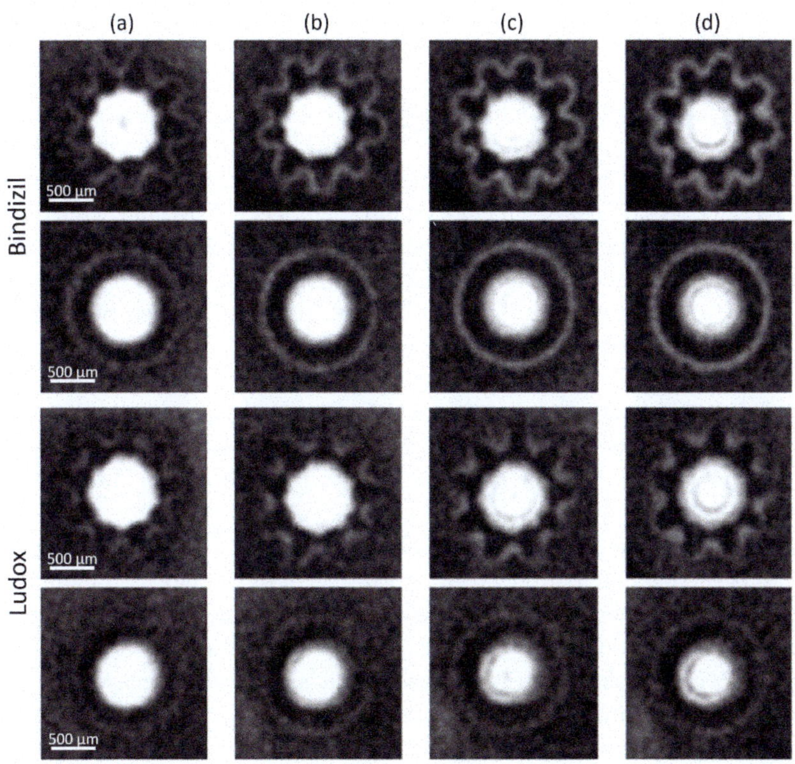

Figure 9.10 Particle deposition on structured and round fibers at 70 LMH with 0.1 wt% Ludox and Bindzil sols. Images are from filtered permeate feed V_f = 0.01 (a), 0.02 (b), 0.04 (c), and 0.06 m^3/m^2 (d). Adapted with permission from Çulfaz et al. [3] with permission from American Chemical Society.

deposition behavior by unequal particle concentrations in the cake layer resulting from differences in back diffusion. While Bindzil particles formed a smaller but highly concentrated cake layer, Ludox particles formed a larger and looser cake layer. Due to the high resistance in the Bindzil cake layer, the permeation started to self-regulate earlier and caused a homogeneous cake layer thickness. The Ludox particles, however, were trapped at the bottom of the grooves due to the low resistance of the cake layer (approximately the same resistance everywhere on the membrane's surface) and, thus, a delayed self-regulation permeation and a reduced back diffusion or more dominant convective flow took place. In another experiment, the authors showed better fouling reversibility of alginate fouling for structured fibers compared with round fibers due to size-dependent deposition (a looser structure of the formed cake layer). However, this process was not monitored using MRI. In situ monitoring of fouling and cleaning processes with sodium alginate by MRI will be discussed in the next section.

This section summarized how MRI measurements were implemented in the field of membrane filtration processes and revealed its possibilities. The information gained from these measurements, like cake and polarization layer thicknesses, flow distributions, swelling of membranes, and the behavior of different foulants, leads to a better understanding of simple membrane processes and underlying physical phenomena and helps to optimize them in the future in terms of process conditions and materials used.

9.3 Sodium Alginate Solutions in Ceramic Hollow Fiber Modules

This section condenses methods and possibilities for using MRI with regard to other types of membranes and foulants, which again leads to a better understanding of filtration processes and elucidates the behavior of fouling layers: dense, compressible gel layers and reversible, loose, and fluffy layers. The working group of Arndt et al. [16,17], Schork et al. [18], and Schuhmann et al. [19] investigated filtration processes of sodium alginate in ceramic hollow fiber modules. In all the studies, similar membranes made of Al_2O_3 were used. A sodium alginate solution was used as a model foulant. To investigate the influence of gel layer buildup, $CaCl_2$ was added to the fouling solution; the solution without $CaCl_2$ resembled a concentration polarization layer structure. Additionally, magnetic iron oxide nanocrystallites (MIONs) were added to the feed to increase the contrast in the MR images. In one of their latest studies, the groups around Schork et al. [20] and Schopf et al. [21] investigated deposition layers during skim milk filtration (for more details, see [20] and [21]). All membrane modules were placed vertically into the MRI tomograph.

Arndt et al. [16] performed inside-out dead-end and cross-flow experiments to analyze the effects of the contrast agent and different flow regimes. For this purpose, they used multi-slice multi-echo imaging (MSME) and flow measurements via phase-contrast imaging (FLOWPC). The images before and after filtration without the addition of MIONs resemble each other. In contrast, the addition of MIONs makes it possible to distinguish between alginate layer and bulk solution, as MION particles accumulated at the surface of the gel layer. Thus, it was possible to measure layer thicknesses and the composition of the layer. The authors claimed that in the presence of Ca^{2+}, the layer was dense enough and formed a gel layer that hindered particles from penetrating the layer. These findings are in agreement with filtration data linked to the cake-filtration model [16]. In situ flow measurements during cross-flow mode showed a significant difference in flow profiles. The flow velocities in a membrane fouled with calcium ions increased as a dense layer was built on the membrane surface (~0 mm/s gel layer velocity) in the gel layer leading to a decrease in the inner diameter of the membrane. This behavior was not visible for the membrane fouled without calcium ions, which performed like a new membrane.

The subsequent study of Arndt et al. [17] focused on time-dependent cake layer growth during inside-out, constant-pressure experiments. Flow during dead-end filtration was measured by an MSME sequence and cake layer thicknesses during cross-flow filtration by a rapid acquisition with relaxation enhancement sequence (RARE). To achieve industrial relevant cross-flow velocities, the authors utilized MRI outflow effects to increase the contrast and visualize cake formation. The results showed an increase in cake layer thickness for both solutions, in the presence and the absence of Ca^{2+} ions during filtration. Yet, the increase was significantly higher for the filtration without Ca^{2+}. However, in the absence of calcium ions, the fouling layer formed a loose and fluffy structure increasing the permeation flux. Investigating the fouling mechanisms revealed a dense gel layer formation for filtration in the presence of Ca^{2+}, whereas in the case without Ca^{2+}, a deviation from a linear trend was observed indicating other influencing factors such as pore blocking and adsorption phenomena. These fouling mechanisms also influenced the flow profiles inside the membrane, shown in Figure 9.11. With increasing filtration time, the maximal velocity decreases for both cases. In the case of filtration with Ca^{2+} (Figure 9.11a), the flow profiles develop as laminar (tubular) flow and meet the predicted profiles (lines). The channel's width seems to decrease over time, which accounts for a relatively stable layer with velocities close to zero. In the case of filtration without Ca^{2+} (Figure 9.11b), the velocity profiles deviate from the ones calculated for simple laminar flow, and a stagnant region next to the membrane wall is observable. These phenomena support the interpretation of a fluffy and looser structure. In cross-flow experiments, the high velocities (and thus shear rates) are the reason for thinner cake layers. The authors saw only little fouling in the absence of Ca^{2+}, which again indicates the development of a fluffy, unstable-concentration polarization layer rather than a dense gel layer.

In the next step, Schork et al. [18] looked at dead-end fouling processes and subsequent backwashing and flushing steps to regenerate the membrane. Figure 9.12 shows MRI images (taken

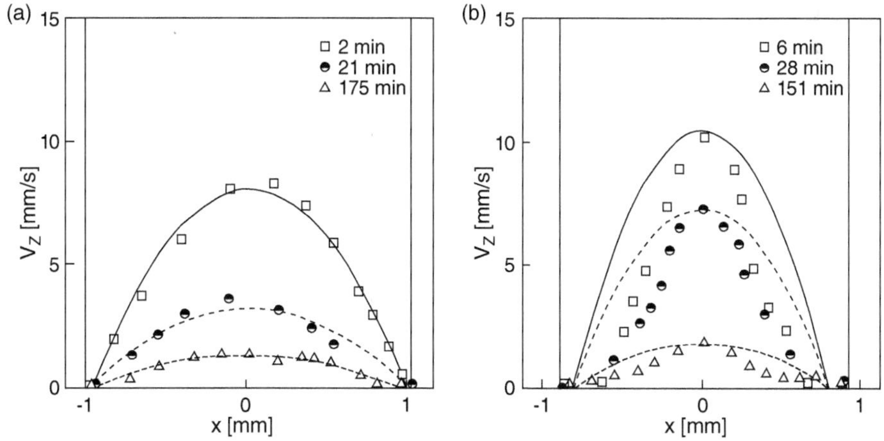

Figure 9.11 Dead-end filtration (a) with Ca^{2+} and (b) without Ca^{2+}. Data points were obtained from magnetic resonance imaging measurements. Lines visualize the velocity profiles according to the Hagen–Poiseuille equation. Adapted from Arndt (2017) [17] with permission from Elsevier.

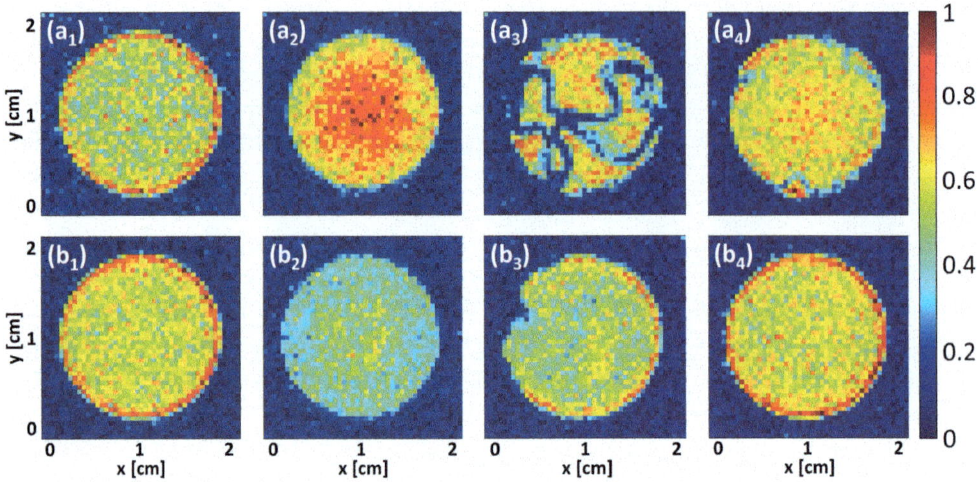

Figure 9.12 MRI images at different filtration steps of Feed A (with Ca^{2+}) and Feed B (without Ca^{2+}) for the first four process steps. (a_1/b_1) before filtration, (a_2/b_2) after 20 min dead-end filtration, (a_3/b_3) after backwashing, (a_4/b_4) after flushing. Adapted from Schork (2018) [18] with permission from Elsevier.

with a RARE pulse sequence) with (Feed A) and without (Feed B) Ca^{2+} for the different process steps. Figure 9.12a_1 and b_1 show the reference state before the filtration. In the presence of Ca^{2+}, a compact and round fouling layer built up, whereas, in the absence of calcium ions, a loose and fluffy structure developed (Figure 9.12a_2 and b_2). After applying the backwashing flux (Figure 9.12a_3 and b_3), the difference in cake layer composition became clear. Figure 9.12a_3 reveals a gel-like layer in the presence of calcium ions that still sticks partly to the membrane surface. This layer was removed during the flushing step. However, incomplete backwashing was visible over two subsequent filtration cycles. Contrarily, the loose concentration polarization layer that built up in the absence of Ca^{2+} ions could be removed (Figure 9.12b_3) and flushed away (Figure 9.12b_4) entirely after the first cycle so that the membrane was regenerated completely. However, after two subsequent filtration and backwashing cycles, the layer could not be removed anymore. In conclusion, for the chosen imaging slice and the applied conditions, the regeneration of the membrane filtered without calcium ions was more successful.

Schuhmann et al. [19] extended the earlier work on fouling behavior characterization of alginate solutions by compressed sensing RARE imaging increasing the time resolution drastically. During in situ dead-end inside-out filtration experiments, the same behavior for filtration experiments with and without calcium ions was observed. Cake layer growth in the lumen was analyzed quantitatively, exploring the membrane's symmetry. Modeling the intensity of these measurements led to similar observations of different fouling mechanisms as in the studies before: a gel-like, compressible layer was obtained for the filtration with Ca^{2+} and a reversible, fluffy, loose concentration polarization layer for the filtration without Ca^{2+}. The authors also observed fouling along the membrane's length. Gel layer formation experiments revealed a small increase in time and position toward the permeate outlet. This increase was also visible for concentration polarization measurements. However, the increase was much more pronounced, and the layer thickness at the position closest to the outlet is ~11 times higher than that of the gel layer.

9.4 Sodium Alginate Solutions and Silica Suspensions in Polymeric Multichannel Membrane Modules

The techniques mentioned earlier are adaptable to polymeric multichannel membranes. Multichannel membranes consist of several lumen channels embedded in one porous support structure, which reduces fiber breakage and increases the active surface-to-volume ratio. Due to their special geometry, these membranes behave differently compared with hollow fiber membranes. MRI makes it possible to analyze flow distributions inside such membranes and helps to understand flow phenomena and initial wetting behavior online and in situ. Schuhmann et al. [22] and Wypysek et al. [7] analyzed flow distribution and fouling behavior in such membrane systems, and Wypysek et al. [23] investigated initial wetting behavior using MRI.

Schumann et al. [22] performed inside-out dead-end measurements using pure water and an alginate solution as a model foulant doped with gadolinium-diethylene-triaminepenta-acetic acid (gadopentetic acid, Gd-DTPA) as a contrast agent. Figure 9.13 illustrates cross-sectional MR intensity images obtained during filtration without (top row) and with (bottom row) calcium ions in the feed solution. During filtration without calcium ions, a concentration polarization layer developed, which increased with filtration duration. At the end of filtration, the deposit was located in the whole lumen area. The inner channel seemed less fouled compared with the outer channels. One explanation for this finding is the higher mass transfer resistance due to the longer pathway of the inner lumen channel. During the filtration with calcium ions, a dense gel layer

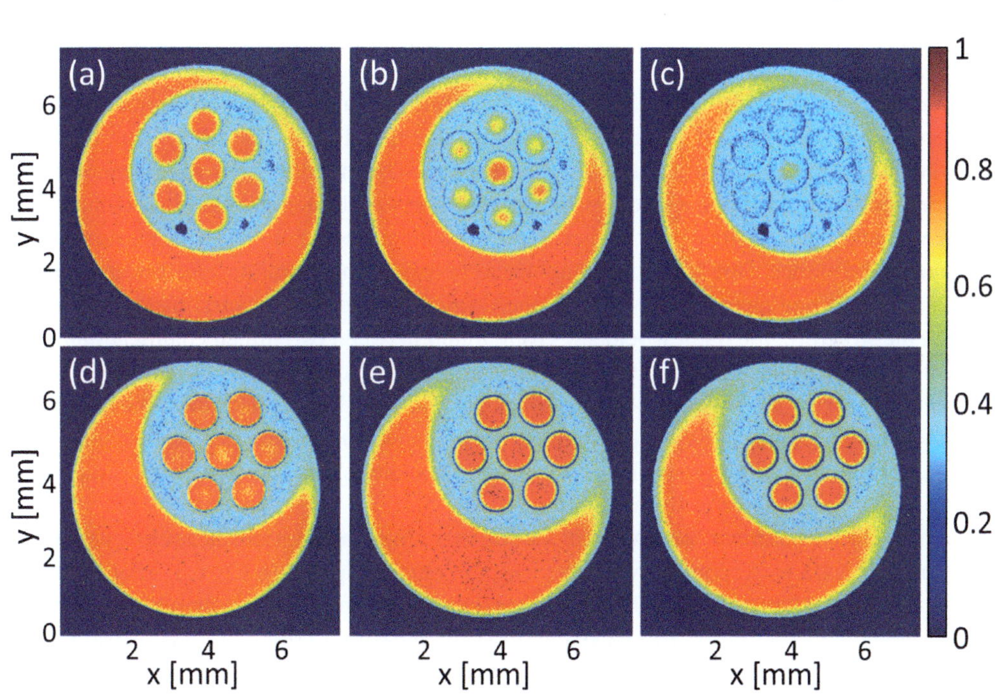

Figure 9.13 T_1- and T_2-weighted magnetic resonance images of an aqueous solution of sodium alginate (calginate = 200 mg/l) without Ca^{2+} (a) after 2 min 25 s, (b) after 29 min 26 s, (c) after 43 min 35 s and with Ca^{2+} ($c_{Ca^{2+}}$ = 2 mmol/l) (d) after 4 min 44 s, (e) after 24 min 45 s, and (f) after 51 min 39 s filtration time. Adapted from Schuhmann (2019) [22] with permission from Elsevier.

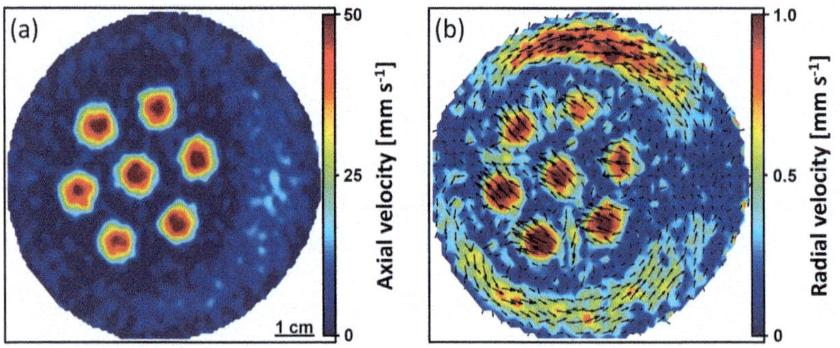

Figure 9.14 MRI of pure water dead-end filtration. (a) Axial and (b) radial flow patterns. Sagging of the membrane introduces secondary flows in the lumen channel and a directed flow on the shell side. Adapted from Wypysek (2019) [7] with permission from Elsevier.

was formed, increasing in thickness over filtration time. However, the layer stays stable and does not cover the whole lumen area. With this filtration mechanism, no significant difference in fouling between the single channels is observable. However, for both filtration solutions, the highest velocities and volume flows were measured for the center bore, showing that the velocity distribution between the different bores is uneven.

Wypysek et al. [7] also investigated multichannel membranes with a colloidal silica suspension as a model foulant. Figure 9.14a depicts axial and Figure 9.14b radial velocity distributions of inside-out, dead-end, pure water filtration experiments. The axial velocities on the shell side had the highest velocities at a maximal distance between membrane and shell wall due to small local pressure differences on the shell side. In the lumen channels, a parabolic flow profile developed in every channel. However, the total maximal velocities differed from each other. Thus, the channels did not contribute equally to the filtration process. This might arise from the membrane's eccentric position inside the module or slight differences in lumen geometry. Looking at the radial velocities, an eccentric position generates a secondary flow field on the shell side toward the maximal distance. Due to the bending of the polymeric membrane, velocities in all bores were directed in one direction, which influenced the deposition of particles. Inside the porous membrane structure, the highest velocities were found between two lumen channels, being the path of least resistance. Backwashing with pure water showed that the flow preferentially enters the membrane from the bottom side and reaches every lumen channel. However, the flow distribution was unequal inside the lumen channels, which can influence the backwashing efficiency. Moving further away from the backwashing inlet, the backwashing flux was lower, indicating a decrease in backwashing effectiveness.

Figure 9.15a.i to c.i shows MR images of fouling with silica particles and two subsequent backwashing steps with corresponding velocity measurements in Figure 15a.ii to c.ii. After fouling in dead-end mode, MR images show blocked bores only on the bottom of the membrane. Channel one, two, and six seem not to be blocked at all at the measured position. However, taking the velocity images into account, no fluid flowing through channel two could be detected. This means that the channel had to be blocked at a position closer to the inlet. In contrast, channel one shows the highest velocities of >100 mm/s in the measured cross-section, and channel six shows lower velocities of ~50 mm/s. This unequal behavior negatively influences filtration performance. After one backwashing step, the MR image shows a nearly fully recovered membrane. However,

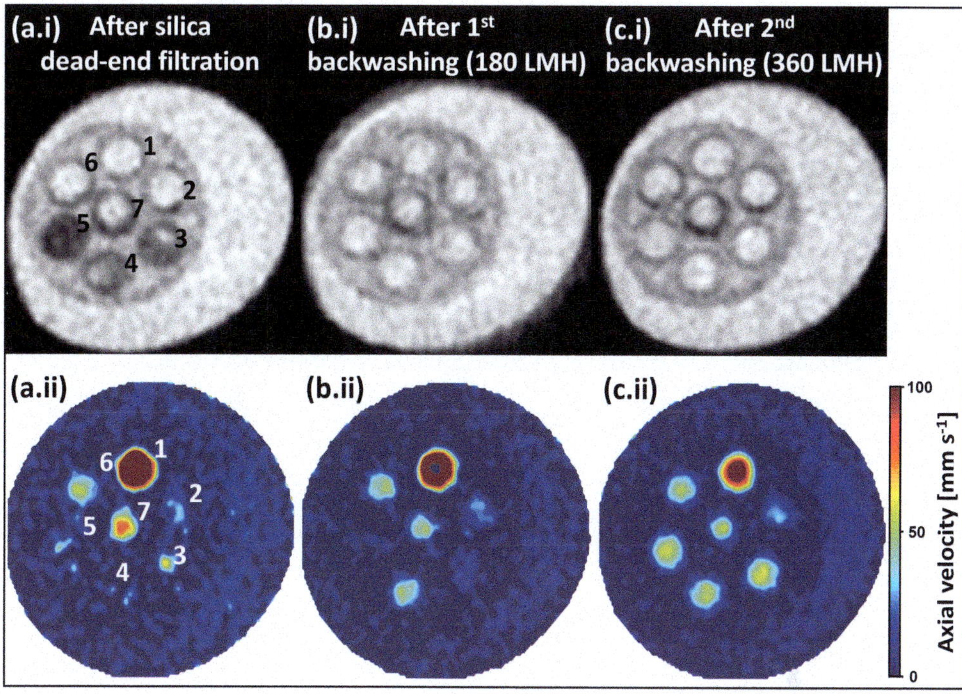

Figure 9.15 Top: images showing the deposition of silica particles after filtration and after two subsequent backwashing steps. Bottom: magnetic resonance imaging velocimetry. Not all lumen channels are sufficiently cleaned after first backwashing step. The channels must be recovered with a higher backwash flux. Adapted from Wypysek (2019) [7] with permission from Elsevier.

velocity images show again that some channels did not work correctly. Even the second backwashing step with a higher backwashing flux could not restore the same conditions as in a new membrane. On the other hand, fouling in cross-flow mode showed particle deposition on the top right corner in each channel with increasing thickness during filtration. The reasons for this are higher drag forces as the permeate outlet channel is located at this position. This study emphasized that only the combination of MR images and velocity maps fully reveals the hydrodynamic properties in such kinds of membranes.

Especially micro- and ultrafiltration membranes in the delivery state are impregnated with a pore-stabilizing agent located in the membrane's porous structure. This agent has to be removed before the filtration task to ensure a maximal performance of the membrane (see dark blue areas in the support structure in Figure 9.13a, b, and c). Wypysek et al. [23] investigated this early-stage removal of various prewetting agents in multichannel membrane modules in three different packing densities during dead-end permeation. Due to the high contrast in signal strength during MRI measurements, the prewetting fluids can be distinguished from the feed solution. Figure 9.16a.i to c.i shows MR images and Figure 9.16a.ii to c.ii shows wetting degrees (WD) over time for three different positions in a membrane module containing 16 delivery-state fibers during permeation at 200 LMH with deionized water. After three hours of permeation, only position A-A (Figure 9.16a.i and a.ii, closest to the feed inlet) is wetted completely (WD = 1). This can be seen as no blackish areas are visible in the porous structure anymore. From Figure

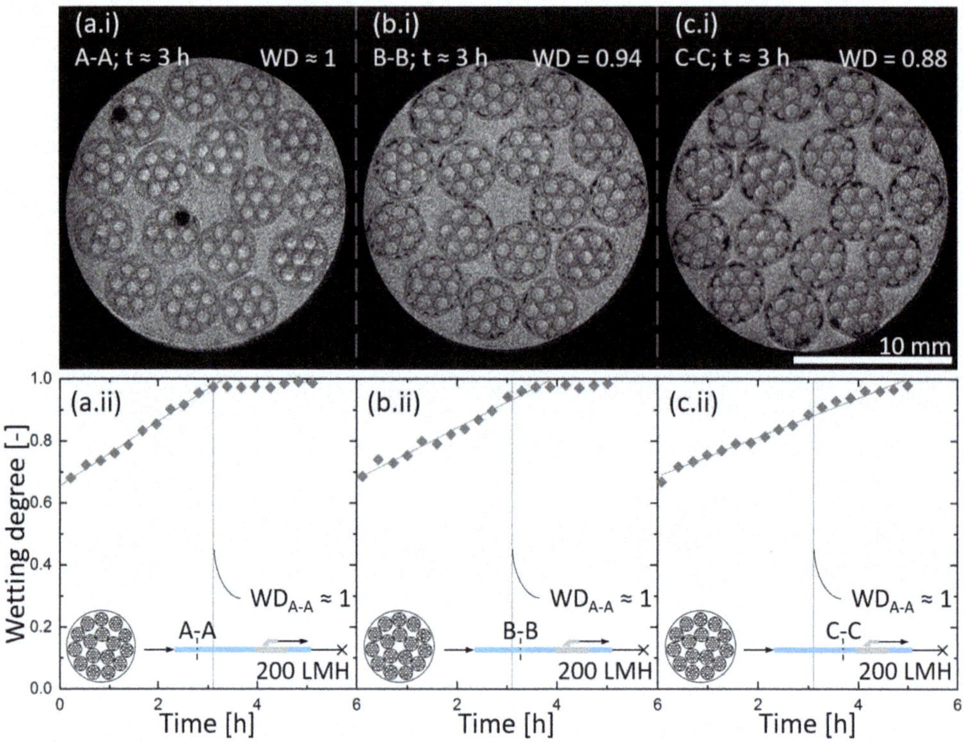

Figure 9.16 (a.i) to (c.i) Magnetic resonance images of a 16-fiber module at 3 different positions after 3 h of permeation at 200 LMH. (a.ii) to (c.ii) Corresponding evaluation of wetting degree over filtration time. Adapted from Wypysek (2021) [23] with permission from Elsevier.

9.16a to Figure 9.16c, the WD after three hours decreases, and the necessary time for complete wetting increases. The authors found that complete wetting of the membranes takes several hours when using water as a feed fluid. This time reduces drastically (to several minutes) when ethanol is used as an invading fluid. All in all, the authors showed that the knowledge of the wetting dynamics in the first filtration cycle is essential to run membrane filtration processes at maximum performance.

9.5 Measures Preventing Concentration Polarization and Fouling

The studies described in Section 9.4 explained and visualized problems in membrane filtration like fouling and concentration polarization in several material systems. This section deals with modules and geometries to tackle these problems. An enhancement in filtration performance is typically measured by an increase in flux. Often, the reasons for the enhanced flux cannot be clarified entirely. Online and in situ measurements via MRI can reveal these reasons.

Wiese et al. [24] studied flow at the submillimeter scale in a 3D-printed, meandering channel reactor with incorporated staggered herringbone structure (herringbone: height 2 mm, width 0.75 mm). For shear rate analysis, a 3D expansion of the FLIESSEN [25] pulse sequence was used. Figure 9.17 shows different MR imaging slices of the herringbone structure for velocity and shear rate distributions (due to low signal-to-noise ratio, some parts are not visible). These measurements are compared with computational fluid dynamics simulations (Figure 9.17b.ii and c.ii). Figure 17a.i reveals that a wave-like velocity profile developed above the herringbone structure that changed periodically due to the height differences in this structure. This variation of higher and lower velocity magnitudes enhances mixing effects. The velocity distribution within the grooves of the structure (Figure 17a.ii) was rather chaotic, as described by Stroock et al. [26]. Especially, velocities perpendicular to the main fluid flow emerged, which favor mixing in the fluid channel and thus reduce polarization effects. These transversal velocities, which were caused by small, local pressure differences, can be seen in Figure 9.17b.i. Here, upward and downward velocities were plotted from which a vortex buildup can be recognized. The center of the vortex is at the position where negative velocities meet positive velocities. Figure 9.17c.i shows measured shear stress patterns. These patterns occurred above the herringbone grooves and the channel surface and were dependent on the herringbone tip orientation. The alternating patterns enhanced mixing in the channel and, thus, increased the overall mass exchange. The obtained MRI results are in good agreement with an accompanying computational fluid dynamics study (Figure 9.17b.ii and Figure 9.17c.ii) [24].

Luelf et al. [27] fabricated twisted tri-bore fibers to introduce vorticity during filtration processes and with that decrease concentration polarization and fouling. Figure 9.18a depicts axial, and Figure 9.18b and c radial velocity components measured by MRI. The membrane itself is visible in the axial velocity image (Figure 9.18a). White spots indicate trapped air. Inside these macrovoids, the membrane was not wetted completely. This can reduce the performance of the membrane. Therefore, membranes should be prewetted with a wetting agent such as ethanol. Parabolic flow profiles were visible inside the lumen channels and the shell side. Transversal velocities can be seen in Figure 9.18b and c. Due to the twisted geometry, regions with a counter-movement of high radial flow in the shell and on the lumen side occurred. These phenomena cause mixing and, thus, enhance mass transfer during filtration in these regions.

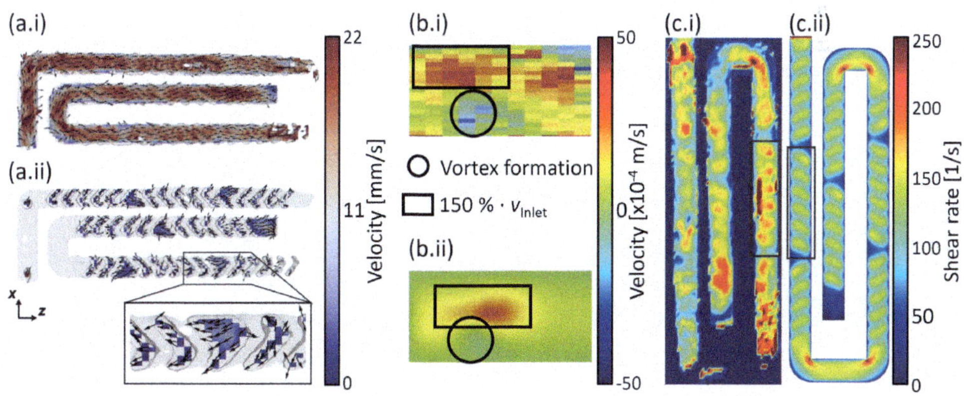

Figure 9.17 (a) Two-dimensional velocity maps above (a.i) and within (a.ii) the herringbone structure. (b.i) Measured and (b.ii) simulated two-dimensional velocity maps in height direction for the cross-sectional view. (c.i) Analytically derived and (c.ii) simulated shear stress patterns alter depending on tip orientation. Adapted from Wiese (2018) [24] with permission from Elsevier.

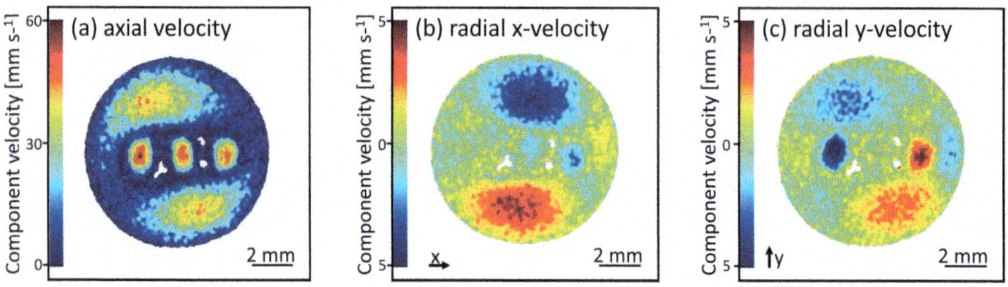

Figure 9.18 Cross-sectional view of the polymeric helical tri-bore hollow fiber resulting from flow-magnetic resonance imaging. (a) Axial velocities. (b) Radial velocities in x-direction. (c) Radial velocities in y-direction. Adapted from Luelf (2018) [27] with permission from Elsevier.

Wiese et al. [8] also investigated a membrane that introduces secondary flows in a low-field MRI tomograph using a spin echo pulse sequence with velocity encoding. In contrast to the previously mentioned study, this study focused on a straight, tubular membrane with a double-helical structure on the inside. This tubular membrane was compared with a membrane without

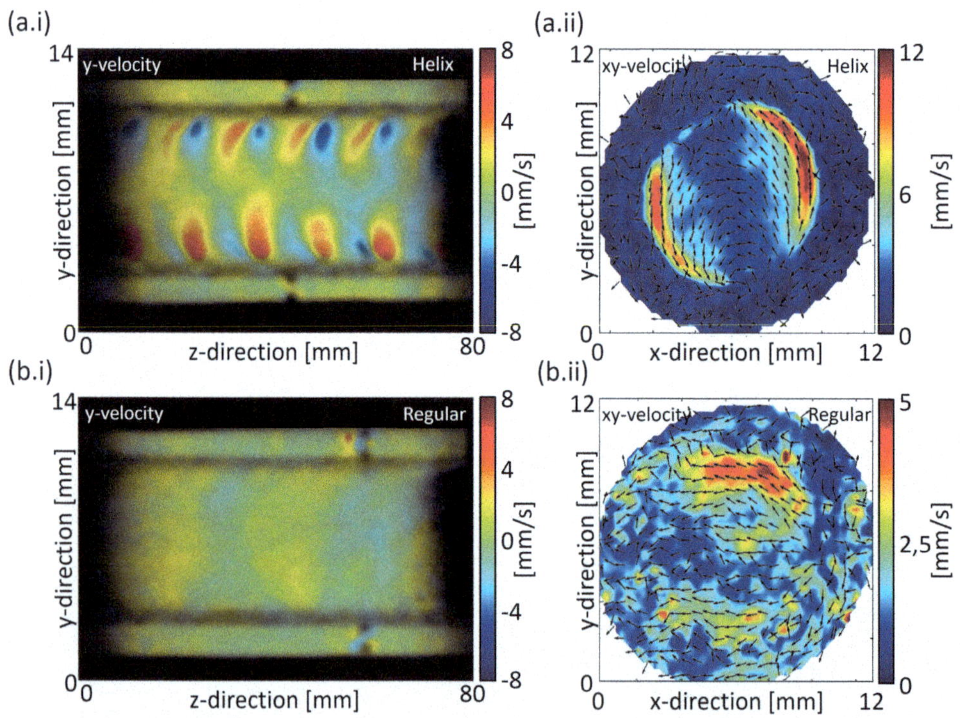

Figure 9.19 Flow-magnetic resonance imaging of the y-velocity in a helix (a.i) and regular (b.i) membrane module at Re = 1000. Enhanced back transport of fluid from the membrane toward the bulk phase is imposed by the corrugated spiraling double helix. (a.ii) and (b.ii): Vortex formation in dependence on a double helix showing the huge impact of a 0.5 mm thick helix surface patterning. Vortices turn along with the double helix favoring mixing in the bulk phase. Adapted from Wiese (2019) [8] with permission from Elsevier.

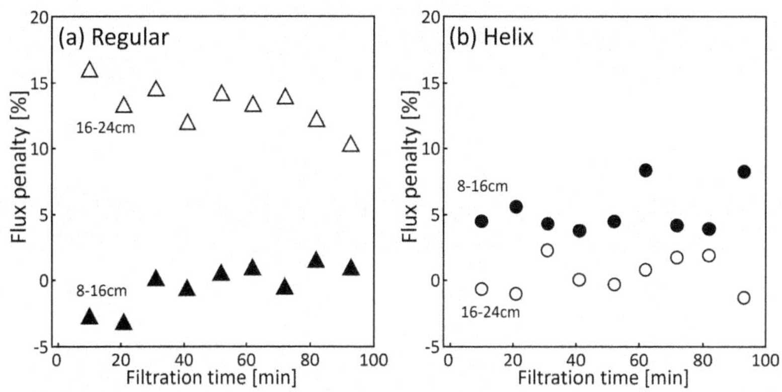

Figure 9.20 Lumen volume flow evolution measured by flow-magnetic resonance imaging. (a) Regular membrane. (b) Helix membrane. Higher increase in lumen volume flow indicates stronger particle rejection. The helix membrane module suffers less from fouling. Adapted from Wiese (2019) [8] with permission from Elsevier.

the internal helical structure. MRI data of cross-flow inside-out filtration experiments were a basis to explain obtained classical filtration data. A microgel solution was used as a model foulant for soft colloids. Fouling of these colloids followed the "gel model" and suggested a ~50% higher mass transfer coefficient and a steady-state flux for the membrane with helical structure twice as high as for the membrane with a regular surface due to the mixing effect caused by the helix. These mixing effects are shown in Figure 9.19a.i and a.ii. The left side of Figure 9.19 illustrates longitudinal slices of the helix and regular membranes. Velocities toward and away from the membrane can be seen that were up to eight times higher in the helix membrane than in the regular membrane. Particles that flow toward the membrane are redirected into the bulk phase again and, thus, fouling is reduced. An additional swirling flow can be seen in cross-section measurements (Figure 9.19a.ii for the helix and b.ii for the regular membranes), which changes in dependence on the helix position and enhancing mixing. This swirling flow diminishes along the axial position with increasing distance from the inlet. However, for the regular membrane, no mixing was visible.

Evaluated flow-MRI measurements during microgel fouling at Re = 1000 at three positions are shown in Figure 9.20a for the regular, and in Figure 9.20b for the helix membrane. In these graphs, a higher flux penalty (difference in volume flows in two measurements 8 cm away from each other) indicates more severe fouling. There was only little fouling in the first part and strong fouling in the second part of the regular membrane. In contrast, no significant flux penalty for the helix membrane was observed. It is worth mentioning that in classical filtration experiments, both membranes performed similarly at low Reynolds numbers (here Re = 1000). However, flow-MRI measurements showed a huge difference in fouling behavior, emphasizing that only the combination of classical filtration and MRI experiments completely unravels all phenomena.

Membrane modules in industrial processes are equipped with more than one membrane. Hence, another option to increase shear rates and vorticity is to alter the membrane's outer geometry in highly packed modules. For a membrane distillation process, Yang et al. [28] analyzed four kinds of membrane modules: a randomly packed module with straight hollow fibers, a module with straight hollow fiber membranes knitted into a spacer structure, a module with curly fiber membranes, and a module with half curly fibers and half straight fibers. The flow in the curly membrane modules and the spacer knitted resulted in higher transverse velocities, better mixing, and filtration performance than in the randomly packed one.

9.6 Analysis of Medical Products

While studies in the previous sections dealt with self-made modules and the analysis of fundamental fouling and hydrodynamic phenomena, this section summarizes studies with commercially available products. Even commercially available products sometimes suffer from bad flow distributions and consequently a reduced filtration performance. MRI can help to find weaknesses and thus help optimize membrane modules.

9.6.1 Hemodialysis – Flow Distribution in Hemodialyzers

MRI can be used in different medical applications. One big field of application is the analysis of hemodialyzers. Hemodialyzers are membrane modules consisting of up to several thousands of polymeric hollow fiber membranes for the removal of toxins from the human blood. Typically, they are operated in counter-current cross-flow mode with blood inside the membranes and a dialysate solution on the shell side. The higher the uniformity of fluid distribution on the lumen and shell side, the better the removal of toxins and the better the prediction of dialysis processes.

Pangrle et al. [29] analyzed a Fresenius F80 dialyzer module in "closed-shell" mode for different Reynolds numbers using spin echo, time-of-flight measurements to obtain velocity distributions for the inlet and outlet sections at which the membranes are potted into the housing. The authors found a maldistribution throughout the module, which hinders toxin removal. At the inlet, the fluid flowed through the center part, whereas at the outlet, the flow moved to the outer membranes with only little flow through the central membranes. Donoghue et al. [30] used the same dialyzer to successfully visualize cell growth inside such a module using a diffusion damping pulse sequence.

Zhang et al. [31] improved the previously mentioned measurements by magnetic resonance Fourier velocity imaging of flow distributions in the shell and lumen side simultaneously. They compared three dialyzers with different packing densities with each other (CA170, Baxter; CT190, Baxter; and F60A, Fresenius). The flow distribution in all three dialyzers was relatively uniform inside the lumen channel in the whole cross-section. In contrast, the flow distribution on the shell side was not uniform and skewed to higher flow rates. The authors reported no clear trend regarding the different packing densities of the dialyzers.

Laukemper-Ostendorf et al. [32] demonstrated a velocity-encoding and velocity-compensated spin echo MRI sequence to analyze two model dialyzers – one with and one without textile fibers as spacer yarns in between the hollow fiber membranes. Figure 9.21 depicts the flow distributions on the shell and lumen sides in these dialyzers. Due to the different packing densities, the shell-sided flow in the module with textile fibers (bottom) was much more homogeneously distributed (left) and narrowed (right) than in the module without spacers (top). Additionally, the authors found that due to swelling of the membranes in the potted area, the inner diameter of the membranes decreased. Thus, the lumen velocity increased, influencing the filtration behavior.

Poh et al. further analyzed these findings by investigating the influence of inserted spacer yarns [33] and installed flow baffles [34] on shell-sided flow using two-dimensional phase-contrast (2DPC) and two-dimensional Fourier transform (2DFT) velocity imaging. In agreement with

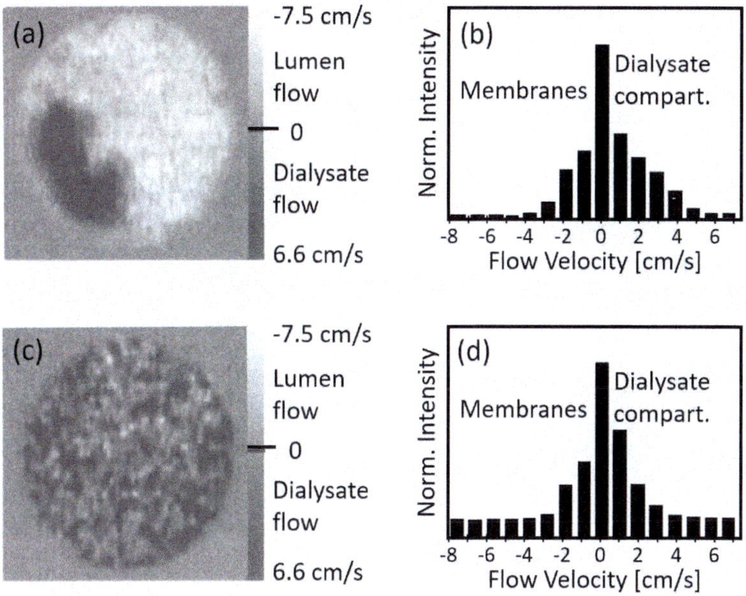

Figure 9.21 Velocity-weighted spin-density images measured on the model (a) without and (b) with textile fibers. The images were taken from the middle section of the model hemodialyzers. (b) and (d) are corresponding flow-velocity distributions. Adapted from Laukemper-Ostendorf (1998) [32] with permission from Elsevier.

previous findings, the flow distribution was not uniform on the shell side, and channeling occurred in the peripheral region. However, spacer yarns between the membranes increased the uniform distribution on the shell side of the studied dialyzers. Additionally, due to higher packing densities, the maximum velocities were higher, and, thus, the removal of toxins was better. All analyzed baffles failed to provide identical flow conditions throughout the cross-section.

Osuga et al. followed the approach of using contrast agents like Gd-DTPA [35] or heavy water [36] to investigate the flow distribution on the shell side of hemodialyzers. Figure 9.22 shows MR images of consecutive injection levels of a Gd-DTPA solution (black) into a dialyzer filled with a dialysate solution (white). The lumen channels were closed so that only flow through the shell side was possible. The fluid motion was faster at the peripheral area and decreased in velocity toward the center of the module, symmetrically in the cross-section with small nonuniformities and channeling between the fibers. Subsequent flushing with dialysate solution revealed that the Gd-DTPA tracer was still present inside the hollow fibers, although the inlet and outlet channels were closed. This back-filtration in real dialysis treatments can force toxins to enter the bloodstream again, which should be prevented. The authors obtained similar effects for experiments with heavy water [36], which acts as a contrast agent. However, they reported sedimentation of the heavy water as it flowed beneath the present dialysate solution due to higher density. Still, the authors concluded there was uniform flow distribution throughout the dialyzer.

Weerakoon et al. [37,38] also used Gd-DTPA as a contrast agent in contrast-enhanced spin echo T_1-weighted MR images to analyze the blood compartment of hemodialyzers. The authors stated a relatively uniform flow distribution over the cross-section of the measured dialyzers close to the blood inlet and a more heterogeneous one close to the blood outlet. Additionally, they found a

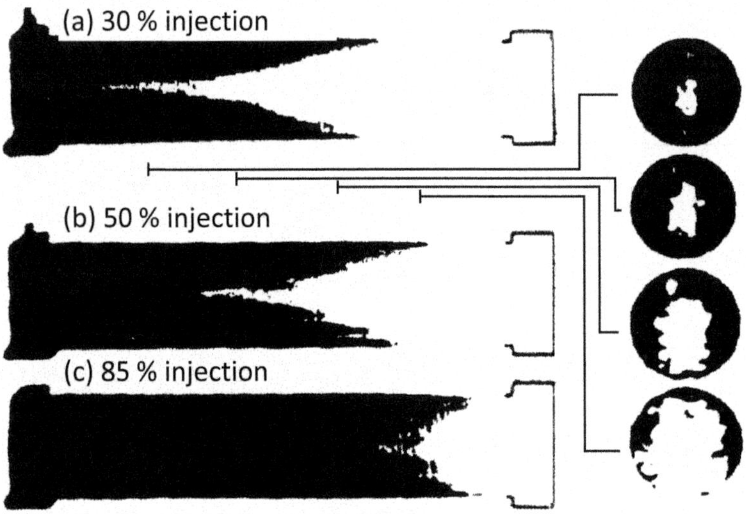

Figure 9.22 Magnetic resonance imaging scans of central vertical section of the dialyzer after commencing the injection of the gadopentetic acid (Gd-DTPA) solution. Injection levels of (a) 30%, (b) 50%, and (c) 85% were taken after 6.0, 10.0, and 16.0 s, respectively. Black region, Gd-DTPA solution; white region, present saline. Adapted from Osuga (2004) [35] with permission from Elsevier.

decrease in signal intensity toward the outlet, indicating the enrichment of substances from the inside toward the outside of the fibers.

Hardy et al. [39] investigated a whole hemodialyzer system (SYNTRA 160, Baxter) with counter-current flow in the lumen and shell sides to analyze the aforementioned back-filtration processes. They succeeded in measuring the local ultrafiltration rate in hemodialyzers using MR Fourier velocity imaging. The authors estimated a uniformly decreasing ultrafiltration rate in accordance with the TMP decreasing along the dialyzer. As the manufacturer only gives an average value for the whole module, this technique helps optimize dialyzers and dialysis processes.

9.6.2 Hemodialysis – Flow Distribution in Endotoxin Adsorbers

During dialysis treatment, the use of sterile filters with the ability to adsorb endotoxins has emerged in the last years. On the one hand, these filters pretreat the water used for dialysis, and, on the other hand, they reduce the amount of endotoxins, which increases the survival rate during dialysis treatment. Wiese et al. [40] investigated such a dead-end filter in an MRI study, performing pure water filtration experiments and fouling measurements with colloidal silica particles using a spin echo pulse sequence with velocity encoding. Figure 9.23 shows pure water flow measurements of the cross-sections close to the inlet (Figure 9.23a.i), next to the permeate outlet (Figure 23a.ii), and a slice in perpendicular orientation (Figure 9.23b). The three flow baffles in Figure 9.23a.i and a.ii (white area in the top part) that strengthen the housing of the membrane module impact the inflow negatively. The main fluid flow was located in the middle two sections, whereas a backflow develops in the two outer sections. This maldistribution affected the flow in the 18 permeate channels below the membrane seen by the different maximal flow velocities in the channels. Additionally, this maldistribution led to a dead zone forming in the first part of the

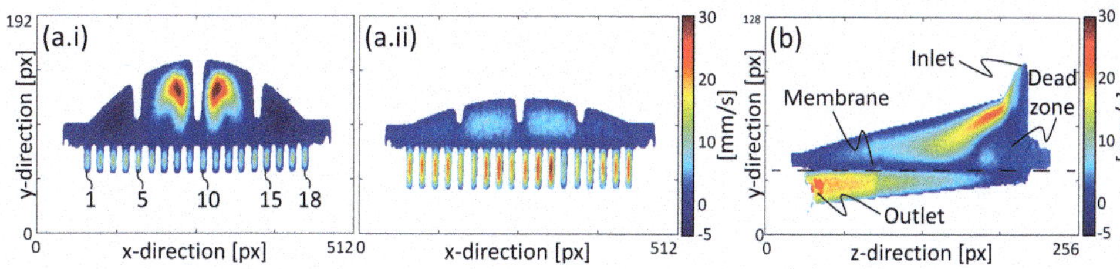

Figure 9.23 Cross-sectional view of the z-velocity in the membrane filter (a.i) at the inlet, and (a.ii) close to the outlet. (b) Contour plot showing the total velocity. At the position where the inlet stream touches the membrane, a stream forward and backflow emerge. In the backflow zone, no permeation occurs, showing the inactivity of the membrane. Adapted from Wiese (2018) [40] with permission from Elsevier.

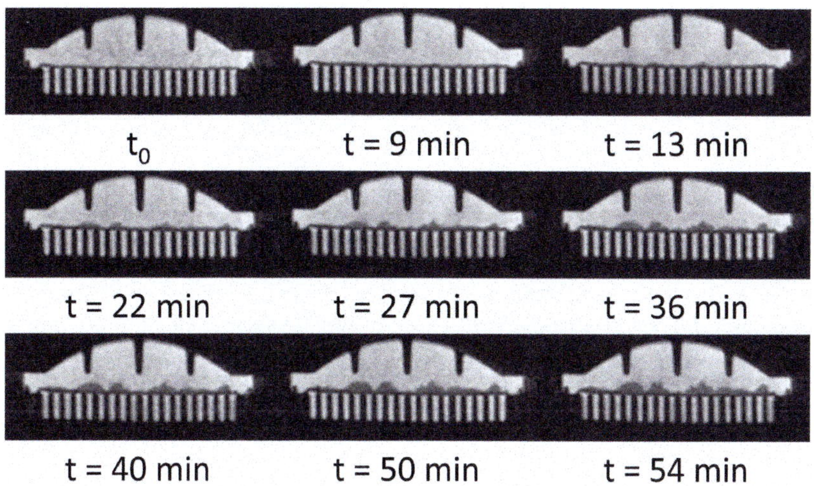

Figure 9.24 Temporal evolution of cake layer formation on the membrane surface. Magnetic resonance imaging shows the cross-sectional plane in the center of the membrane module. The largest silica particle depositions seem to develop below the left and right baffles. Adapted from Wiese (2018) [40] with permission from Elsevier.

module (Figure 9.23b), and thus to a loss of effective filtration area and a reduction in filtration performance.

Figure 9.24 shows the deposition of silica particles on the membrane surface during filtration. The amount of deposited particles on the membrane increased with increasing filtration time, with stronger accumulations in the membrane center and below the baffles. However, this deposition was not homogeneous over the whole membrane. A buildup of valleys and hills occurred caused by the maldistribution of flow. Subsequent flow measurements also showed decreased permeation and filtration performance. The only active permeate channels remaining were the two outer ones. The authors proved the independence from gravity of the deposition behavior by measurements of a module rotated by 180°. A similar distribution of silica particles was observed in both analyzed orientations.

9.6.3 Connectors – Flow Distribution in Medical Connectors

All the abovementioned medical devices have to be connected to a filtration setup. For this purpose, several medical needleless connectors are available. Nybo et al. [41] investigated flow velocity maps inside six different types of medical connectors. Every design has a unique flow distribution, and complex flow paths with secondary flows develop throughout each connector. This study demonstrates the strength of MRI that provides basic data for further bacteria or colloid transport studies. With this dataset, such connectors can be optimized, and computational fluid dynamics (CFD) studies can be validated.

9.7 Analysis of Forward and Reverse Osmosis Modules

Desalination in membrane filtration is an essential process for drinking water production. Two examples are the so-called forward and reverse osmosis. In these processes, spiral-wound modules are often used. A grid structure serves as a spacer between two membrane layers [1]. Besides other membrane processes, forward and reverse osmosis suffer from fouling and scaling, especially biofouling. Several research groups have focused on the investigation of biofouling using ^{1}H NMR [42–46]. Recently, ^{23}Na NMR was presented as a method for analyzing desalination processes [47].

Von der Schulenburg et al. [42] were the first analyzing spiral-wound modules regarding the spatial biofilm distribution (feed: tap water supplemented with nutrient solution) with RARE imaging, the velocity field inside the module with a standard phase-shift velocity-encoding pulse sequence, and the distributions of molecular displacement of tracers in a lab cell, as well as in an industrial membrane module (mass transport displacement statistics) with z-displacement propagators (pulsed-field gradient). They found a decreasing active membrane surface area after biofouling. Furthermore, biofouling preferentially occurred at spacer junctions (Figure 9.25a). Only a small amount of biofouling was enough to cause stagnant zones and a profoundly changing flow field in experiments in a flow cell with spacer structures (Figure 9.25b).

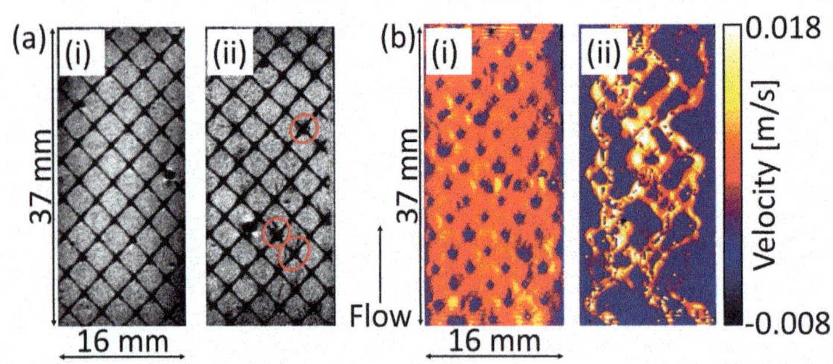

Figure 9.25 The grid structure serves as a spacer between two membrane layers. (a) Magnetic resonance imaging (MRI) structural images of (i) an unfouled and (ii) a fouled flow cell. (b) MRI velocity images of (i) an unfouled and (ii) a fouled flow cell. Adapted from Creber (2010) [43] with permission from Elsevier.

Creber et al. [43] used the same flow cell to investigate fouling and cleaning processes in spiral-wound membrane modules. Figure 9.25 shows similar results as seen by von der Schulenburg et al. [42]: even a little fouling had a large impact on the active surface area and the flow field. Cleaning the membrane with a combination of sodium dodecyl sulfate and sodium hydroxide showed a broadening effect of existing flow channels. It was not possible to remove the biofilm completely. However, cleaning at an early stage of biofouling seemed more efficient than cleaning a fouling layer that had aged longer.

Fridjonsson et al. [44] presented an approach to investigate this kind of early-stage biofouling and additionally evaluated cleaning efficiencies. The authors used Earth's field NMR methods to detect biofouling in commercial spiral-wound modules. With the NMR signal increasing under operational conditions, detecting early-stage biofouling at any position of the spiral-wound module was possible. This direct measurement technique can also be applied during cleaning processes to reduce the amount of chemicals used in industrial applications in which several spiral-wound modules are utilized in a glass fiber pressure vessel.

The setup of spiral-wound modules inside a pressure vessel was investigated by Ujihara et al. [45]. The authors succeeded in imaging the inside of a spiral-wound reverse osmosis module in a pressure vessel during operation. Moreover, they were able to enhance the contrast between flowing and stagnant water and could also locate the foulant using an Earth's field tomograph. This technique is much cheaper in operating and investment costs than high-field systems. Although high-field MRI resolution is higher, low-field imaging was sufficient to visualize the desired phenomena.

Bristow et al. [46] used a high-field system to look at fouling at different positions in a spiral-wound module inside a pressure vessel during filtration with a sodium alginate solution in two different concentrations. Directly at the module inlet, the highest heterogeneities and channeling could be observed. The authors found a highly variable velocity distribution after fouling experiments with the low concentrated solution at the inlet. In contrast to that, only small changes in the middle and at the end of the module occurred. Furthermore, after filtration with the highly concentrated solution, the flow was bypassing the membrane module, thus reducing the module's filtration performance significantly as the fouling was irreversible.

Zargar et al. [47] proposed another approach to analyze the performance of reverse osmosis processes. They did not investigate fouling in membrane modules but rather the saline feed solution. In their proof of concept study, the authors used ^{23}Na nuclear MRI to measure the sodium ion concentration inside the membrane and close to its surface. Additionally, they succeeded in measuring the concentration polarization during desalination that decreased the filtration performance. This is the next step to better predict desalination processes, carry out simulations, or optimize them.

9.8 Conclusion and Alternative Techniques

Due to its flexible methodologies (different pulse sequences), MRI can be used for many different purposes in the field of chemical engineering, such as characterizing material or rheological properties, investigating dynamic systems and reactions, analyzing electrochemical systems, and measuring porous media and flow in membranes and membrane modules [48].

This chapter shows the strengths of MRI for studying membrane processes. MRI is able to spatially resolve flow velocity distributions and, at least partially, porosities and permeabilities in any selected plane. Its noninvasiveness makes it possible to elucidate structures and different components in opaque systems in situ. The examples presented for the visualization of concentration polarization layers, fouling depositions, and fluid flow distributions in several membrane modules show that MRI provides valuable information on governing mechanisms and enables the testing of theoretical models. This information is of the highest interest as it uncovers how fouling can be controlled and minimized and how to design backwashing and cleaning processes.

The application of MRI for studying hydrodynamics within membrane processes is well established. However, as effects on the membrane surface can occur very fast and on a smaller scale, MRI suffers from its spatial resolution of some microns and its time resolution in the range of up to several minutes. In addition, the use of metallic/magnetic parts inside an MRI tomograph is challenging. Still, the development of faster and more robust pulse sequences is ongoing [48].

For investigations where MRI comes to its limit, other techniques may be deployed, including the experimental microfluidic observation of fouling layer behavior [49,50] or computational fluid dynamics – discrete element method (CFD-DEM) simulations of fouling and backwashing mechanisms [51,52]. These publications give insights into the spatio-temporal behavior of filter cake deposits in relevant hydrodynamic conditions. Additionally, computerized tomography for in situ measurements of concentration distributions [53], or light deflection, radioisotope labeling, electron diode array microscopy, and direct pressure measurement for in situ monitoring of concentration polarization layers can be used [54]. In situ monitoring of cake formation and membrane fouling can be realized via direct observation methods, laser triangulometry, optical laser measurements, ultrasonic time-domain reflectometry, and electrical impedance spectroscopy [54]. Small-angle neutron and X-ray scattering may reveal pore blockage and cake structures [54].

References

1 Baker, R.W. (2004). *Membrane Technology and Applications*. 3e. Newark, NJ: John Wiley & Sons.

2 Song, L. (1998). Flux decline in crossflow microfiltration and ultrafiltration: Mechanisms and modeling of membrane fouling. *Journal of Membrane Science* 139 (2): 183–200. https://doi.org/10.1016/S0376-7388(97)00263-9.

3 Çulfaz, P.Z., Buetehorn, S., Utiu, L. et al. (2011). Fouling Behavior of microstructured hollow fiber membranes in dead-end filtrations: Critical flux determination and NMR imaging of particle deposition. *Langmuir* 27 (5): 1643–1652. https://doi.org/10.1021/la1037734.

4 Scott, K. and Hughes, R. (2013). *Industrial Membrane Separation Technology*. 2e. Dordrecht, the Netherlands: Springer.

5 Jaffrin, M.Y. (2012). Hydrodynamic techniques to enhance membrane filtration. *Annual Review of Fluid Mechanics* 44 (1): 77–96. https://doi.org/10.1146/annurev-fluid-120710-101112.

6 Yao, S., Costello, M., Fane, A.G. et al. (1995). Non-invasive observation of flow profiles and polarisation layers in hollow fibre membrane filtration modules using NMR micro-imaging. *Journal of Membrane Science* 99 (3): 207–216. https://doi.org/10.1016/0376-7388(94)00219-O.

7 Wypysek, D., Rall, D., Wiese, M. et al. (2019). Shell and lumen side flow and pressure communication during permeation and filtration in a multibore polymer membrane module. *Journal of Membrane Science* 584: 254–267. https://doi.org/10.1016/j.memsci.2019.04.070.

8 Wiese, M., Nir, O., Wypysek, D. et al. (2019). Fouling minimization at membranes having a 3D surface topology with microgels as soft model colloids. *Journal of Membrane Science* 569: 7–16. https://doi.org/10.1016/j.memsci.2018.09.058.

9 Pangrle, B.J., Walsh, E.G., Moore, S.C. et al. (1992). Magnetic resonance imaging of laminar flow in porous tube and shell systems. *Chemical Engineering Science* 47 (3): 517–526. https://doi.org/10.1016/0009-2509(92)80001-S.

10 Yao, S., Pope, J.M., and Fane, A.G. (1995). Chemical shift selective flow imaging: Applications in membrane filtration of oil-water emulsions. *Bulletin of Magnetic Resonance* 17 (1-4): 220–221.

11 Yao, S., Fane, A.G., and Pope, J.M. (1997). An investigation of the fluidity of concentration polarisation layers in crossflow membrane filtration of an oil-water emulsion using chemical shift selective flow imaging. *Magnetic Resonance Imaging* 15 (2): 235–242. https://doi.org/10.1016/S0730-725X(96)00325-6.

12 Pope, J.M., Yao, S., and Fane, A.G. (1996). Quantitative measurements of the concentration polarisation layer thickness in membrane filtration of oil-water emulsions using NMR micro-imaging. *Journal of Membrane Science* 118 (2): 247–257. https://doi.org/10.1016/0376-7388(96)00107-X.

13 Airey, D., Yao, S., Wu, J. et al. (1998). An investigation of concentration polarization phenomena in membrane filtration of colloidal silica suspensions by NMR micro-imaging. *Journal of Membrane Science* 145 (2): 145–158. https://doi.org/10.1016/S0376-7388(98)00051-9.

14 Buetehorn, S., Utiu, L., Küppers, M. et al. (2011). NMR imaging of local cumulative permeate flux and local cake growth in submerged microfiltration processes. *Journal of Membrane Science* 371 (1): 52–64. https://doi.org/10.1016/j.memsci.2011.01.018.

15 Culfaz, P.Z., Rolevink, E., van Rijn, C. et al. (2010). Microstructured hollow fibers for ultrafiltration. *Journal of Membrane Science* 347 (1): 32–41. https://doi.org/10.1016/j.memsci.2009.

16 Arndt, F., Roth, U., Nirschl, H. et al. (2016). New insights into sodium alginate fouling of ceramic hollow fiber membranes by NMR imaging. *AIChE Journal* 62 (7): 2459–2467. https://doi.org/10.1002/aic.15226.

17 Arndt, F., Schuhmann, S., Guthausen, G. et al. (2017). In situ MRI of alginate fouling and flow in ceramic hollow fiber membranes. *Journal of Membrane Science* 524: 691–699. https://doi.org/10.1016/j.memsci.2016.11.079.

18 Schork, N., Schuhmann, S., Arndt, F. et al. (2018). MRI investigations of filtration: Fouling and cleaning processes. *Microporous and Mesoporous Materials* 269: 60–64. https://doi.org/10.1016/j.micromeso.2017.05.042.

19 Schuhmann, S., Schork, N., Beller, K. et al. (2018). In-situ characterization of deposits in ceramic hollow fiber membranes by compressed sensing RARE-MRI. *AIChE Journal* 64 (11): 4039–4046. https://doi.org/10.1002/aic.1620110.003.

20 Schork, N., Schuhmann, S., Nirschl, H. et al. (2019). In situ measurement of deposit layer formation during skim milk filtration by MRI. *Magnetic Resonance in Chemistry* 57 (9): 738–748. https://doi.org/10.1002/mrc.4826.

21 Schopf, R., Schork, N., Amling, E. et al. (2020). Structural characterisation of deposit layer during milk protein microfiltration by means of in-situ MRI and compositional analysis. *Membranes* 10 (4): 59. https://doi.org/10.3390/membranes10040059.

22 Schuhmann, S., Simkins, J.W., Schork, N. et al. (2019). Characterization and quantification of structure and flow in multichannel polymer membranes by MRI. *Journal of Membrane Science* 570–571: 472–480. https://doi.org/10.1016/j.memsci.2018.10.072.

23 Wypysek, D., Kalde, A.M., Pradellok, F. et al. (2021). In-situ investigation of wetting patterns in polymeric multibore membranes via magnetic resonance imaging. *Journal of Membrane Science* 622: 119026. https://doi.org/10.1016/j.memsci.2020.119026.

24 Wiese, M., Benders, S., Blümich, B. et al. (2018). 3D MRI velocimetry of non-transparent 3D-printed staggered herringbone mixers. *Chemical Engineering Journal* 343: 54–60. https://doi.org/10.1016/j.cej.2018.02.096.

25 Amar, A., Blümich, B., and Casanova, F. (2010). Rapid multiphase flow dynamics mapped by single-shot MRI velocimetry. *ChemPhysChem* 11 (12): 2630–2638. https://doi.org/10.1002/cphc.201000062.

26 Stroock, A.D., Dertinger, S.K.W., Ajdari, A. et al. (2002). Chaotic mixer for microchannels. *Science* 295 (5555): 647–651. https://doi.org/10.1126/science.1066238.

27 Luelf, T., Rall, D., Wypysek, D. et al. (2018). 3D-printed rotating spinnerets create membranes with a twist. *Journal of Membrane Science* 555: 7–19. https://doi.org/10.1016/j.memsci.2018.03.026.

28 Yang, X., Fridjonsson, E.O., Johns, M.L. et al. (2014). A non-invasive study of flow dynamics in membrane distillation hollow fiber modules using low-field nuclear magnetic resonance imaging (MRI). *Journal of Membrane Science* 451: 46–54. https://doi.org/10.1016/j.memsci.2013.09.015.

29 Pangrle, B.J., Walsh, E.G., Moore, S. et al. (1989). Investigation of fluid flow patterns in a hollow fiber module using magnetic resonance velocity imaging. *Biotechnology Techniques* 3: 67–72. https://doi.org/10.1007/BF01876224.

30 Donoghue, C., Brideau, M., Newcomer, P. et al. (1992). Use of magnetic resonance imaging to analyze the performance of hollow-fiber bioreactors. *Annals of the New York Academy of Sciences* 665: 285–300. https://doi.org/10.1111/j.1749-6632.1992.tb42592.x.

31 Zhang, J., Parker, D.L., and Leypoldt, J.K. (1995). Flow distributions in hollow fiber hemodialyzers using magnetic resonance Fourier velocity imaging. *Asaio Journal* 41 (3): M678–82. https://doi.org/10.1097/00002480-199507000-00097.

32 Laukemper-Ostendorf, S., Lemke, H.D., Blümler, P. et al. (1998). NMR imaging of flow in hollow fiber hemodialyzers. *Journal of Membrane Science* 138 (2): 287–295. https://doi.org/10.1016/S0376-7388(97)00241-X.

33 Poh, C.K., Hardy, P.A., Liao, Z. et al. (2003). Effect of spacer yarns on the dialysate flow distribution of hemodialyzers: A magnetic resonance imaging study. *ASAIO Journal* 49 (4): 440–448.

34 Poh, C.K., Hardy, P.A., Liao, Z. et al. (2003). Effect of flow baffles on the dialysate flow distribution of hollow-fiber hemodialyzers: A nonintrusive experimental study using MRI. *Journal of Biomechanical Engineering* 125 (4): 481–489. https://doi.org/10.1115/1.1590355.

35 Osuga, T., Obata, T., and Ikehira, H. (2004). Detection of small degree of nonuniformity in dialysate flow in hollow-fiber dialyzer using proton magnetic resonance imaging. *Magnetic Resonance Imaging* 22 (3): 417–420. https://doi.org/10.1016/j.mri.2003.08.008.

36 Osuga, T., Obata, T., and Ikehira, H. (2004). Proton magnetic resonance imaging of flow motion of heavy water injected into a hollow fiber dialyzer filled with saline. *Magnetic Resonance Imaging* 22 (3): 413–416. https://doi.org/10.1016/j.mri.2003.07.003.

37 Weerakoon, B.S. and Osuga, T. (2015). The applicability of Gd-DTPA magnetic resonance imaging contrast agent for the evaluation of blood compartment flow distribution in hollow fiber

hemodialyzers. *Journal of Biomedical Science and Engineering* 08 (11): 789–796. https://dx.doi.org/10.4236/jbise.2015.811075.
38 Weerakoon, B.S. and Osuga, T. (2016). Characterization of flow distribution in the blood compartment of hollow fiber hemodialyzers with contrast-enhanced spin echo magnetic resonance imaging. *Applied Magnetic Resonance* 47 (4): 453–469. https://doi.org/10.1007/s00723-016-0766-8.
39 Hardy, P.A., Poh, C.K., Liao, Z. et al. (2002). The use of magnetic resonance imaging to measure the local ultrafiltration rate in hemodialyzers. *Journal of Membrane Science* 204 (1): 195–205. https://doi.org/10.1016/S0376-7388(02)00038-8.
40 Wiese, M., Malkomes, C., Krause, B. et al. (2018). Flow and filtration imaging of single use sterile membrane filters. *Journal of Membrane Science* 552: 274–285. https://doi.org/10.1016/j.memsci.2018.02.002.
41 Nybo, E., Maneval, J.E., Codd, S.L. et al. (2018). Flow velocity maps measured by nuclear magnetic resonance in medical intravenous catheter needleless connectors. *Journal of Pharmaceutical and Biomedical Analysis* 152: 1–11. https://doi.org/10.1016/j.jpba.2018.01.037.
42 Graf von der Schulenburg, D.A., Vrouwenvelder, J.S., Creber, S.A. et al. (2008). Nuclear magnetic resonance microscopy studies of membrane biofouling. *Journal of Membrane Science* 323 (1): 37–44. https://doi.org/10.1016/j.memsci.2008.06.012.
43 Creber, S.A., Vrouwenvelder, J.S., van Loosdrecht, M.C.M. et al. (2010). Chemical cleaning of biofouling in reverse osmosis membranes evaluated using magnetic resonance imaging. *Journal of Membrane Science* 362 (1): 202–210. https://doi.org/10.1016/j.memsci.2010.06.052.
44 Fridjonsson, E.O., Vogt, S.J., Vrouwenvelder, J.S. et al. (2015). Early non-destructive biofouling detection in spiral wound RO membranes using a mobile earth's field NMR. *Journal of Membrane Science* 489: 227–236. https://doi.org/10.1016/j.memsci.2015.03.088.
45 Ujihara, R., Fridjonsson, E.O., Bristow, N.W. et al. (2018). Earth's field MRI for the non-invasive detection of fouling in spiral-wound membrane modules in pressure vessels during operation. *Desalination and Water Treatment* 12 (135): 16–24.
46 Bristow, N.W., Vogt, S.J., O'Neill, K.T. et al. (2020). Flow field in fouling spiral wound reverse osmosis membrane modules using MRI velocimetry. *Desalination* 491: 114508. https://doi.org/10.1016/j.desal.2020.114508.
47 Zargar, M., Ujihara, R., Vogt, S.J. et al. (2020). Imaging of membrane concentration polarization by NaCl using ^{23}Na nuclear magnetic resonance. *Journal of Membrane Science* 600: 117868. https://doi.org/10.1016/j.memsci.2020.117868.
48 Benders, S. and Blümich, B. (2019). Applications of magnetic resonance imaging in chemical engineering. *Physical Sciences Reviews* 4 (10): 20180177. https://doi.org/10.1515/psr-2018-0177.
49 Linkhorst, J., Beckmann, T., Go, D. et al. (2016). Microfluidic colloid filtration. *Scientific Reports* 6: 22376. https://doi.org/10.1038/srep22376.
50 Linkhorst, J., Rabe, J., Hirschwald, L.T. et al. (2019). Direct observation of deformation in micro-gel filtration. *Scientific Reports* 9: 18998. https://doi.org/10.1038/s41598-019-55516-w.
51 Lohaus, J., Perez, Y.M., and Wessling, M. (2018). What are the microscopic events of colloidal membrane fouling? *Journal of Membrane Science* 553: 90–98. https://doi.org/10.1016/j.memsci.2018.02.023.

52 Lohaus, J., Stockmeier, F., Surray, P. et al. (2020). What are the microscopic events during membrane backwashing? *Journal of Membrane Science* 602: 117886. https://doi.org/10.1016/j.memsci.2020.117886.

53 Frank, A., Lipscomb, G.G., and Dennis, M. (2000). Visualization of concentration fields in hemodialyzers by computed tomography. *Journal of Membrane Science* 175 (2): 239–251. https://doi.org/10.1016/S0376-7388(00)00421-X.

54 Chen, J.C., Li, Q., and Elimelech, M. (2004). In situ monitoring techniques for concentration polarization and fouling phenomena in membrane filtration. *Advances in Colloid and Interface Science* 107 (2): 83–108. https://doi.org/10.1016/j.cis.2003.10.018.

10

Whither NMR of Biofilms?

Joseph D. Seymour[1,2], Gisela Guthausen[3], and Catherine M. Kirkland[1,4]

[1] *Center for Biofilm Engineering, Montana State University, Bozeman, MT, USA.*
[2] *Department of Chemical and Biological Engineering, Montana State University, Bozeman, MT, USA.*
[3] *Institute for Mechanical Engineering and Mechanics and Engler-Bunte Institute Water Science and Technology, KIT, Karlsruhe, Germany*
[4] *Department of Civil Engineering, Montana State University, Bozeman, MT, USA.*

10.1 Introduction

Microbial activity in nature has been shown to be dominated by biofilms, communities of surface-attached bacteria [1,2]. Biofilms are composed of microbial cells surrounded by extracellular polymeric substances that form a hydrogel from various biomolecules [3,4]. One of the first applications of nuclear magnetic resonance (NMR) imaging (magnetic resonance imaging [MRI]) specifically to biofilms occurred some three decades ago [5]. Since then significant applications using NMR/MRI have been undertaken and these applications have been comprehensively reviewed recently [6]. In this chapter we contemplate future directions for NMR biofilm studies. In doing so, we do not provide a comprehensive review of the literature on the NMR of biofilms or the NMR methods. Rather, we consider a few select current areas of interest in biofilm research and indicate NMR methods, some of which have already been applied to these systems and some of which have not, that could have an impact on enhancing understanding of these complex fluid materials [7].

10.2 Pulsed Gradient Spin Echo NMR and Porous Media Design

In many systems such as filtration devices, the earth's subsurface, and water treatment facilities, biofilms grow in and on a porous substrate. The function of filtration devices and ecological hydrology is dependent on the transport of solvated molecules in a liquid stream that often contains suspended colloidal particulates or precipitated components. The fluid dynamics-based mass transport through the system is a function of the transport phenomena driving forces, the structure of the porous media substrate, and the impact of the biofilm growth. Biofilm growth has been demonstrated to generate a transition in the porous media structure–transport relation from that of homogeneous to heterogeneous porous media [8]. Homogeneous porous media, where local Fickian transport dominates, result in Gaussian advection diffusion equation (ADE) dynamics. In heterogeneous porous media, however, nonlocal effects induce long-range spatial and temporal correlations, which result in non-Gaussian ADE dynamics. This change in transport behavior due to biofouling can be controlled by slight variations of the underlying porous media structure, which has significant implications for design of filtration systems.

Magnetic Resonance Microscopy: Instrumentation and Applications in Engineering, Life Science and Energy Research, First Edition. Edited by Sabina Haber-Pohlmeier, Bernhard Blümich, and Luisa Ciobanu.

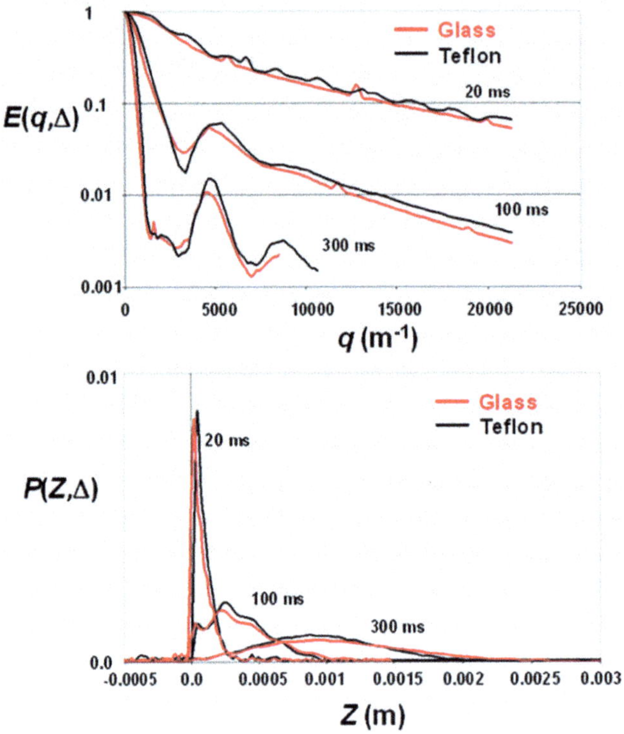

Figure 10.1 Echo attenuation and propagator data as a function of displacement time Δ for 241 mm monodisperse beads packed in a low radius tolerance Teflon cylinder and a high radius tolerance glass cylinder. Both packings exhibit flow diffraction effects in the echo attenuation data indicating the structural similarity of the clean bead packs. The propagators show the classical advection diffusion transition from the distribution of displacements within a pore at Δ = 20 ms to asymptotic Gaussian behavior at 300 ms.

The data in Figure 10.1 are for two packed bed systems composed of identical monodispersed 241 μm diameter polymer spheres – one packed in a 5 mm diameter glass cylinder of high diameter tolerance ($D = 5 \pm 0.1$ mm) and the other in a low-diameter tolerance 5 mm Teflon cylinder, which results in slightly altered ordering of the particle packing due to wall effects. The pulsed gradient spin echo (PGSE) measured signal attenuation $E(q, \Delta)$ and propagators $P(Z, \Delta)$ [9] for flow through the two packed beds as a function of displacement time Δ exhibit similar dynamics. The propagators show the classic advection dispersion Gaussian distributionof displacements at $\Delta = 300$ ms. Here, multiple pores are sampled and the $\boldsymbol{q}$-space data reveal flow diffraction effects due to the ordered structure, which generates phase coherence at the reciprocal particle diameter [10].

The two different packings respond differently to biofouling by a *Pseudomonas aeruginosa* biofilm as seen in Figure 10.2. The slightly less ordered Teflon cylinder system generates highly non-Gaussian anomalous transport behavior associated with a change to heterogeneous structure due to the formation of preferential flow paths or channeling after biofilm growth [8]. In the more highly ordered glass cylinder system, a no-flow peak arises, along with a Gaussian propagator of larger mean displacement and broader width due to increased hydrodynamic dispersion. The hydrodynamics within the more ordered bead structure generates a uniform biofilm distribution over the bead surfaces. The resulting smaller effective, but still homogeneous, pore

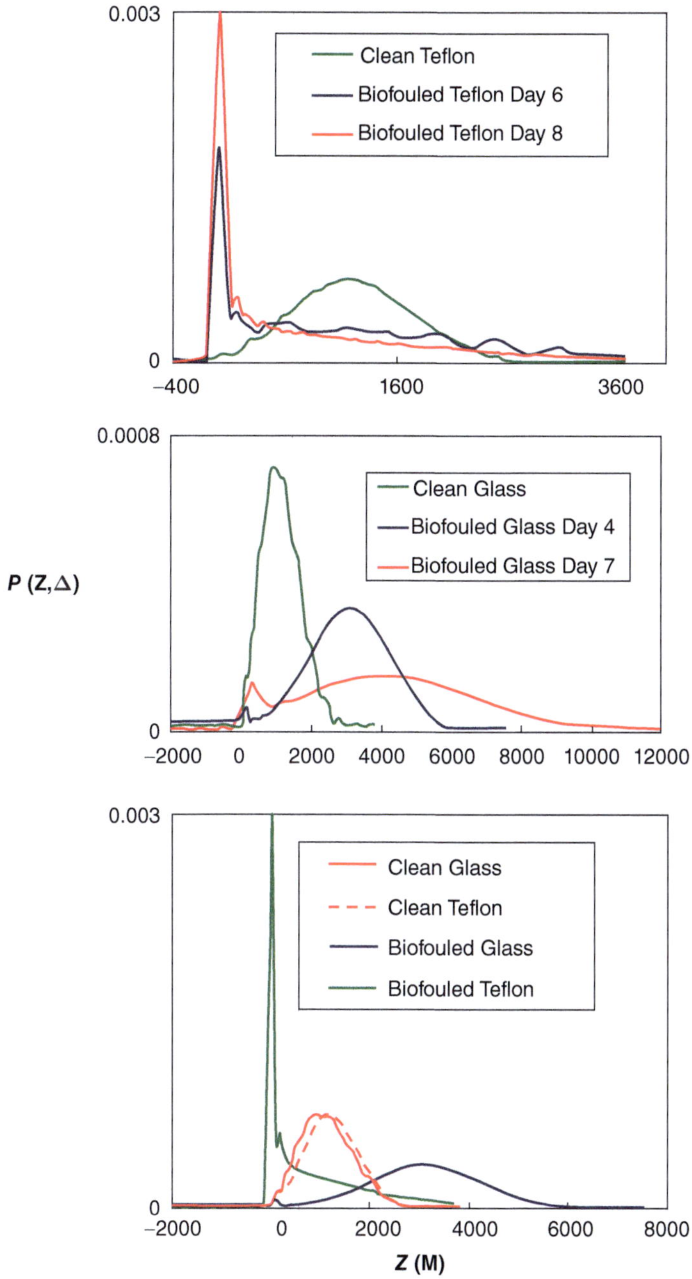

Figure 10.2 Propagators at Δ = 300 ms as a function of biofouling for the Teflon (top) and glass (center) columns, and comparison of the clean and fully biofouled propagators for the two columns (bottom).

structure increases the mean displacement $< v > \Delta$ for fixed volumetric flow rate and enhances Gaussian dispersion. This indicates the utility and underutilized potential of PGSE NMR in the design of porous media filtration devices whose underlying structure can be used to control post-biofouling dynamics to maintain filter function and mass transport.

10.3 ^{19}F NMR Oxymetry

The affinity of molecular oxygen for liquid fluorocarbon phases produces a linear dependence of the ^{19}F T_1 relaxation rate on the local oxygen concentration [11,12]. Simkins et al. have demonstrated the ability to use ^{19}F T_1 relaxation to probe oxygen distribution in microbial bacteria and biofilm-impacted model porous media systems [13–15]. The data in Figure 10.3 demonstrate the oxygen consumption from a single bead inoculated with *Escherichia coli* during flow of nutrients through a packed bed of alginate beads dispersed with perfluorooctylbromide (PFOB) [15]. The method has allowed for determination of the impact of antibiotic treatment on oxygen consumption [14] and analysis of engineering models of mass transfer [15]. Currently the method is limited to spherical alginate particles, and studies of microbes with varying motility and biofilm-forming capacity under a range of flow transport conditions in packed beds have great potential. Attempts in the Montana State University MR Laboratory to generate planar alginate gel substrates with dispersed perfluorinated hydrocarbons resulted in the strong migration of the fluorinated compounds away from the air–gel interface. For biofilm systems, the development of such planar sensors will extend the applications of this MR-based method to study spatial heterogeneity of oxygen consumption and corresponding metabolism in individual biofilms.

10.4 MR Elastography and Rheo-NMR

Recent research on biofilm morphology has developed chemomechanical models that couple the stresses in the biofilm matrix to stress and nutrient distributions in the soft agar substrate the biofilm is growing on in order to correlate spatiotemporally varying mechanical properties with biofilm morphology and composition [16]. This advance indicates the potential to apply MR elastography [17] methods to study biofilm growth, which to our knowledge has not yet been attempted. Figure 10.4 shows the complex structure of a biofilm granule, which is of significant interest in a novel wastewater treatment method [18]. Current NMR studies of this type of biofilm system have focused on characterization of the translational and rotational water dynamics within the structures evident in T_2 contrast images [18,19]. These biofilm granules represent just one system of high interest where MR elastography during growth could provide novel experimental data for testing of morphology stress models. Other potential applications where MR elastography might suggest biofilm mitigation strategies include membrane biofouling [20] and chronic wound biofilms [21]. While the impact of shear stress on biofilms has been measured for flow in circular conduits [22], studies of the interplay of shear and biofilm deformation under more controlled shear have not been conducted. Rheo-NMR [23] of biofilms in geometries such as a Couette concentric cylinder or cone and plate geometries with highly controlled shear and combining displacement phase encoding over a range of magnetic field gradient strengths will allow simultaneous velocimetry and elastography.

10.5 Singlet NMR Diffusion

Optical light sheet microscopy affords biofilm researchers a means to image fast cell processes within intact and thick biofilms while at the same time maintaining cell viability and decreasing image acquisition times [24,25]. The method has recently elucidated the dynamics of cells during

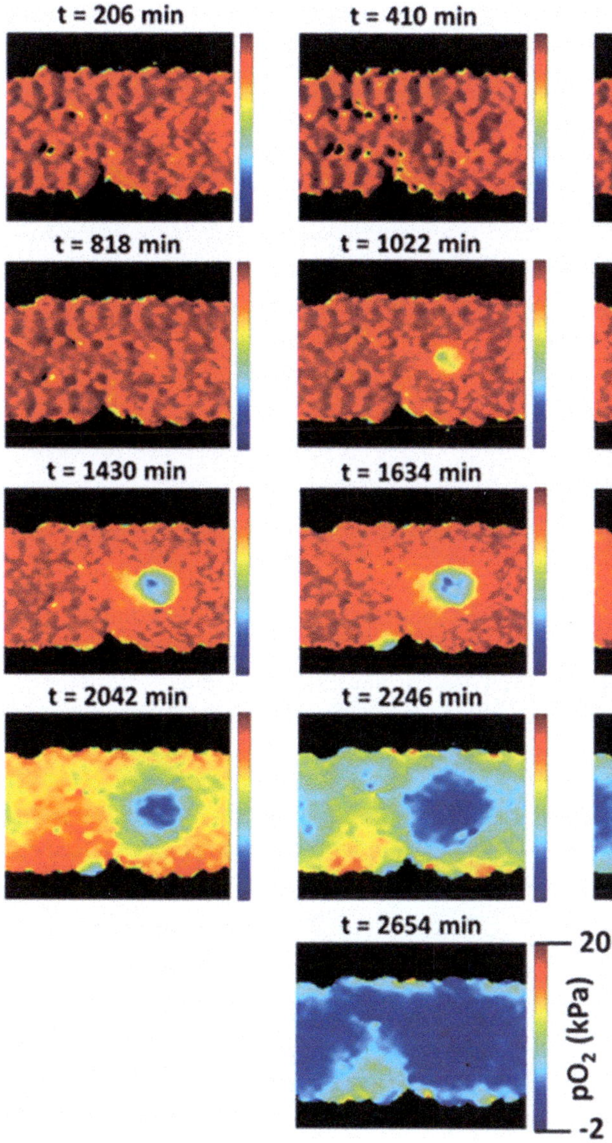

Figure 10.3 Magnetic resonance O_2 concentrations maps generated from calibration with T_1 relaxation of ^{19}F resonance in particles generated by emulsifying and entrapping perfluorooctylbromide (PFOB) into alginate gel. The resulting oxygen-sensing alginate beads are then used as the solid matrix of a packed bed. Time series of sagittal oxygen maps in kPa for a single bead inoculated with *Escherichia coli* HB101 and 25 ml/h nutrient solution flow. Fields of view for the oxygen maps are 15 × 12 mm^2 with a matrix size of 128 × 128 for a resolution of 117 × 94 μm^2/pixel [15] Simkins et al. (2020), figure 03[p.06]/John Wiley & Sons, Inc.

biofilm growth with temporally resolved cell trajectories indicating a transition from Brownian motion in the 2D plane during early-stage growth to ballistic cell motion during later-stage vertical expansion and 3D structural development [26]. The length and timescales of the dynamics generate velocities of the order of 10^{-2} µm/s and mean squared displacements (MSD) of the order

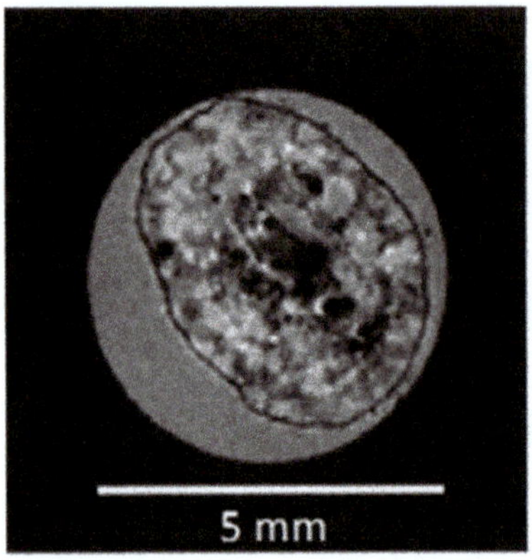

Figure 10.4 Magnetic resonance image of a biofilm wastewater granule acquired at 950 MHz with a spatial resolution 47 μm × 47 μm × 100 μm, t_E = 15.9 ms, T_r = 550 ms. [18] Kirkland et al. (2020), figure 02[p.632]/IWA Publishing/CC BY 4.0.

of 1 μm^2 on timescales of hundreds of seconds [26]. These scales mean standard PGSE methods are ineffective to study in situ cell colonization and growth dynamics as biofilms initiate due to the $T_1 \sim 1$ s limit of stimulated echo displacement times Δ. While light sheet microscopy can provide novel insights into fast processes like cell replication, developments in singlet NMR promise new means to examine slow biochemical processes and significantly larger MSD than is possible with traditional NMR methods.

Singlet NMR is based on long-lived singlet order, which is not relaxed by intramolecular dipole–dipole interactions, and thus allows for relaxation times $T_S \sim 100$ s [27]. Singlet-assisted diffusion (SAD) NMR has been demonstrated in model porous media using probe molecules to study dynamics in porous media on timescales of hundreds of seconds and length scales of hundreds of micrometers [28,29]. The longest displacement-time studies in biofilms by PGSE NMR to date (up to $\Delta = 30$ s) have used ^{13}C NMR [30], and SAD NMR would allow significant extensions for slow advective transport processes like nutrient or biocide transport in environmental or medical applications, respectively, where some biomolecules like aspartate, glutamine, and adenosinediphosphate (ADP) can support singlet states [31]. Furthermore, SAD NMR could be applied to support heteronuclear hyperpolarization, discussed in the next section, to advance the field of NMR metabolomics and molecular imaging via the use of artificial prosthetic singlet tags.

The primary challenge to application of singlet NMR to biofilms is in the development and selection of these tags for biofilm systems. Appropriate molecular tags must be stable in the presence of dissolved oxygen and nondeuterated solvents and under biologically compatible pH and temperature conditions. Biologically compatible tags such as a synthetic bromoacrylate (BrAc-STAG) were used to label several amino acids as well as a glucose derivative and produced T_S times approximately four times greater than the measured T_1 values [31]. In addition to the synthetic singlet tag, the metabolite phosphoenolpyruvate (PEP), whose acrylic moieties provide a path to singlet states, was also found to produce a T_S value 3.6 times longer than T_1 [31]. PEP is a glycolysis pathway metabolite that is up-regulated in some cancer cells and is also found naturally in biofilms [32]. These singlet states, whether synthetic like BrAc-STAG or naturally occurring like PEP, were demonstrated in a relatively strong radiofrequency field (11.7 T and 22.2 T) [31] and are thus currently limited to in vitro applications. Singlet NMR has, to our knowledge, not yet been applied to biofilms.

Ultimately, and with further development, there is potential for SAD NMR to be applied to elucidate fundamental mass transport behavior of biomacromolecules with spatial and temporal resolution in complex biofilm structures. Coupling this type of information with that gleaned from complementary methods like optical light sheet microscopy would represent a significant advancement in our understanding of the biofilm mode of growth and could suggest avenues to either promote or inhibit biofilm growth.

10.6 Heteronuclear NMR and MRI

Metabolomics in biofilms is an important developing research area that has been demonstrated to provide correlations between gene expression and metabolic pathways on one hand, and biofilm formation and structure on the other [33]. Because the biofilm phenotype enables enhanced cell survival in environments inhospitable to planktonic cells, and because these morphological shifts are reflected in the metabolites produced by biofilms, metabolomics has an important role in development of antimicrobial treatments [32,33]. Metabolomics by NMR of biofilms has primarily been done using bulk NMR spectroscopy of processed and concentrated samples, with a few exceptions [33,34].

Gjersing et al. applied traditional ^{1}H NMR spectroscopy to liquid supernatant from planktonic and biofilm cultures, as well as ^{1}H high-resolution magic angle spinning (HRMAS) NMR spectroscopy to concentrated cell pellets derived from each culture [34]. Principal component analysis (PCA) was used to evaluate the statistical significance of observed differences between the spectra of the different culture phenotypes and cultivation methods. While the liquid samples showed only quantitative spectral differences, the ^{1}H HRMAS data yielded both quantitative and qualitative differences of statistical significance between planktonic and biofilm intracellular metabolic profiles [34].

A major advance beyond the application of ^{1}H HRMAS NMR would combine MRI with NMR metabolite spectroscopy to measure spatial distribution of metabolic function. Since NMR deals with small energy quanta, which leads to intrinsic insensitivity, direct MRI of metabolic processes with natural abundance ^{15}N or ^{13}C is not possible without signal enhancement. Diverse approaches using various spin physics to increase signal are active areas of research. Hyperpolarization techniques lead to significant signal enhancements, which open up new possibilities to study novel aspects of materials and of processes [35–40].

A primary method of interest for biofilms is parahydrogen-induced polarization (PHIP) and its continuous exchange-based extension SABRE (signal amplification by reversible exchange) [41]. Heteronuclear NMR/MRI on native biofilm systems becomes more feasible due to the large factors of sensitivity gain. Hyperpolarization techniques applied to biofilms can overcome the restriction of low sensitivity due to low concentrations and low gyromagnetic ratios of ^{15}N [38,39] and ^{13}C [40] in biofilm systems. NMR and MRI applications beyond metabolomics, for example the study of the antibiofilm and antibiofouling behavior of 2-aminoimadazoles and their synergistic effects with antibiotics, should be readily achievable as well [39,42].

10.7 NMR and MRI on Electrochemical Active Biofilm Systems

Conventional battery research primarily has focused on Li-ion batteries due to good technical performance, but with disadvantages concerning environmental and safety issues. An important topic in current biofilm research, however, concerns the electrochemical activity of some biofilms. The idea to generate electric power during wastewater processing and cleaning is convincing [43,44]. The processes involved require detailed investigations to understand and describe the elementary biochemical activity responsible for generating electrical activity and NMR has the potential to contribute significantly to this inquiry. NMR/MRI has been applied for electrolyte characterization [45], and on complete batteries to investigate the electrochemical processes [46–48]. Apart from novel paper-based or polymer batteries [49], NMR and MRI applications to non-Li-ion systems are limited. MRI studies of biofilm fuel cells have been even more limited. One of the few applications used 3D MRI to answer questions about biofilm distribution, geometry, and morphology on the rough and irregularly shaped surface of activated carbon anodes in a microbial fuel cell [50]. It was demonstrated that T_1 contrast weighting provided better biofilm resolution on the activated carbon than T_2 contrast. They applied turbo spin echo rapid acquisition with relaxation enhancement (RARE) MRI with short echo time (2.92 ms) and optimized T_1 weighting repetition time (0.7 s) at 14.1 T to generate (28 μm)^3 images from which biomass volume surface coverage was calculated and correlated to total produced electric charge [50]. The study was conducted on different activated carbon anodes removed from an operating microbial fuel cell as a function of fuel cell operation time over 22 days and the authors indicate the potential and need for in situ measurements to further quantify the operational biomass electric charge correlation [50]. Electrophoretic NMR techniques provide a potential means for making such measurements [51]. Combining the experience of NMR and MRI in conventional battery research, recent developments in electrophoretic NMR [51] and the general knowledge about biofilm characterization provided by NMR and MRI techniques will open up new fields of development for and applications of microbial fuel cells.

10.8 Conclusions

The use of NMR and MRI techniques to study biofilms has contributed significantly to enhancing understanding of these important microbial systems over the past 30 years [6]. To extend these advances in understanding further it is important for NMR and MRI experimentalists to integrate with current biofilm research to provide complementary and novel data not available from other techniques. This chapter has indicated several fruitful directions that require modification of established methods to focus on the biophysical and biochemical aspects of biofilm formation and function.

References

1 Costerton, J.W., Stewart, P.S., and Greenberg, E.P. (1999). Bacterial biofilms: A common cause of persistent infections. *Science* 284 (5418): 1318–1322.

2 Costerton, J.W. and Stewart, P.S. (2001). Battling biofilms. *Scientific American* 285 (1): 1–9.

3 Sutherland, I.W. (2001). Biofilm exopolysaccharides: A strong and sticky framework. *Microbiology* 147: 3–9.
4 Flemming H-C, W.J. (2010). The biofilm matrix. *Nature Reviews Microbiology* 8 (9): 623–633.
5 Lewandowski, Z., Altobelli, S.A., Majors, P.D. et al. (1992). NMR imaging of hydrodynamics near microbially colonized surfaces. *Water Science and Technology* 26 (3-4): 577–584.
6 Herrling, M.P., Lackner, S., Nirschl, H. et al. (2019). Recent NMR/MRI studies of biofilm structures and dynamics. In: *Annual Reports on NMR Spectroscopy*, (ed. G.A. Webb), 163–213. London: Academic Press.
7 Wilking, J.N., Angelini, T.E., Seminara, A. et al. (2011). Biofilms as complex fluids. *MRS Bulletin* 36 (5): 385–391.
8 Seymour, J.D., Gage, J.P., Codd, S.L. et al. (2004). Anomalous fluid transport in porous media induced by biofilm growth. *Physical Review Letters* 93 (19): 198103.
9 Callaghan, P.T. (2011). *Translational Dynamics and Magnetic Resonance Principles of Pulsed Gradient Spin Echo NMR*. New York: Oxford University Press.
10 Seymour, J.D. and Callaghan, P.T. (1996). "Flow-diffraction" structural characterization and measurement of hydrodynamic dispersion in porous media by PGSE NMR. *Journal of Magnetic Resonance* 122: 90–93.
11 Eidelberg, D., Johnson, G., Barnes, D. et al. (1988). 19F NMR imaging of blood oxygenation in the brain. *Magnetic Resonance in Medicine* 6 (3): 344–352.
12 Mason, R.P., Rodbumrung, W., and Antich, P.P. (1996). Hexafluorobenzene: A sensitive 19F NMR indicator of tumor oxygenation. *NMR in Biomedicine* 9 (3): 125–134.
13 Simkins, J.W., Stewart, P.S., and Seymour, J.D. (2018). Spatiotemporal mapping of oxygen in a microbially-impacted packed bed using ^{19}F nuclear magnetic resonance oximetry. *Journal of Magnetic Resonance* 293: 123–133.
14 Simkins, J.W., Stewart, P.S., Codd, S.L. et al. (2019). Non-invasive imaging of oxygen concentration in a complex in vitro biofilm infection model using ^{19}F MRI: Persistence of an oxygen sink despite prolonged antibiotic therapy. *Magnetic Resonance in Medicine* 82 (6): 2248–2256.
15 Simkins, J.W., Stewart, P.S., Codd, S.L. et al. (2020). Microbial growth rates and local external mass transfer coefficients in a porous bed biofilm system measured by ^{19}F magnetic resonance imaging of structure, oxygen concentration, and flow velocity. *Biotechnology and Bioengineering* 117 (5): 1458–1469.
16 Fei, C., Mao, S., Yan, J., et al. (2020). Nonuniform growth and surface friction determine bacterial biofilm morphology on soft substrates. *Proceedings of the National Academy of Sciences.* 117(14):7622.
17 Muthupillai, R., Lomas, D.J., Rossman, P.J. et al. (1995). Magnetic resonance elastography by direct visualization of propagating acoustic strain waves. *Science* 269: 1854–1857.
18 Kirkland, C.M., Krug, J.R., Vergeldt, F.J. et al. (2020). Characterizing the structure of aerobic granular sludge using ultra-high field magnetic resonance. *Water Science and Technology* 82 (4): 627–639.
19 Van Den Berg, L., Kirkland, C.M., Seymour, J.D. et al. (2020). Heterogeneous diffusion in aerobic granular sludge. *Biotechnology and Bioengineering* 117: 3809–3819.
20 Bristow, N.W., Vogt, S.J., O'Neill, K.T., et al. (2020). Flow field in fouling spiral wound reverse osmosis membrane modules using MRI velocimetry. *Desalination* 491: 114508.
21 Lone, A.G., Atci, E., Renslow, R. et al. (2015). *Staphylococcus aureus* induces hypoxia and cellular damage in porcine dermal explants. *Infection and Immunity* 83 (6): 2531–2541.

22 Manz, B., Volke, F., Goll, D. et al. (2003). Measuring local flow velocities and biofilm structure in biofilm systems with magnetic resonance imaging (MRI). *Biotechnology and Bioengineering* 84 (4): 424–432.

23 Callaghan, P.T. (1999). Rheo-NMR: Nuclear magnetic resonance and the rheology of complex fluids. *Reports in Progress Physics* 62: 599–670.

24 Baumgart, E. and Kubitscheck, U. (2012). Scanned light sheet microscopy with confocal slit detection. *Optics Express* 20 (19): 21805–21814.

25 Chen, B.-C., Legant, W.R., Wang, K. et al. (2014). Lattice light-sheet microscopy: Imaging molecules to embryos at high spatiotemporal resolution. *Science* 346 (6208): 1257998.

26 Qin, B., Fei, C., Bridges, A.A. et al. (2020). Cell position fates and collective fountain flow in bacterial biofilms revealed by light-sheet microscopy. *Science* 369 (6499): 71.

27 Levitt, M.H. (2012). Singlet nuclear magnetic resonance. *Annual Review of Physical Chemistry* 63 (1): 89–105.

28 Torres, A.M., Ghadirian, B., and Price, W.S. (2012). Diffusion–diffraction using singlet spin states and various NMR coherences in a J-coupled AX spin system. *RSC Advances* 2 (8): 3352–3360.

29 Tourell, M.C., Pop, I.-A., Brown, L.J. et al. (2018). Singlet-assisted diffusion-NMR (SAD-NMR): Redefining the limits when measuring tortuosity in porous media. *Physical Chemistry Chemical Physics* 20 (20): 13705–13713.

30 Graf Von Der Schulenburg, D.A., Akpa, B.S., Gladden, L.F. et al. (2008). Non-invasive mass transfer measurements in complex biofilm-coated structures. *Biotechnology and Bioengineering* 101 (3): 602–608.

31 Glöggler, S., Elliott, S.J., Stevanato, G. et al. (2017). Versatile magnetic resonance singlet tags compatible with biological conditions. *RSC Advances* 7 (55): 34574–34578.

32 Zhang, B. and Powers, R. (2012). Analysis of bacterial biofilms using NMR-based metabolomics. *Future Medicinal Chemistry* 4 (10): 1273–1306.

33 Lu, H., Que, Y., Wu, X. et al. (2019). Metabolomics deciphered metabolic reprogramming required for biofilm formation. *Scientific Reports* 9 (1): 13160.

34 Gjersing, E.L., Herberg, J.L., Horn, J. et al. (2007). NMR metabolomics of planktonic and biofilm modes of growth in *Pseudomonas aeruginosa*. *Analytical Chemistry* 79 (21): 8037–8045.

35 Münnemann, K., Amor, N., Zanker, P.P. et al. (2009). Magnetic resonance imaging of dissolved hyperpolarized (129)Xe using a membrane-based continuous flow system. *Journal of Magnetic Resonance* 201 (1): 93–99.

36 Coffey, A.M., Kovtunov, K.V., Barskiy, D.A. et al. (2014). High-resolution low-field molecular magnetic resonance imaging of hyperpolarized liquids. *Analytical Chemistry* 86 (18): 9042–9049.

37 Bouchard, L.S., Burt, S.R., Anwar, M.S. et al. (2008). NMR imaging of catalytic hydrogenation in microreactors with the use of para-hydrogen. *Science* 319 (5862): 442–445.

38 Skovpin, I.V., Svyatova, A., Chukanov, N. et al. (2019). 15N hyperpolarization of dalfampridine at natural abundance for magnetic resonance imaging. *Chemistry – A European Journal* 25 (55): 12694–12697.

39 Shchepin, R.V., Barskiy, D.A., Coffey, A.M. et al. (2016). 15N Hyperpolarization of imidazole-15N2 for magnetic resonance pH sensing via SABRE-SHEATH. *ACS Sensors* 1 (6): 640–644.

40 Zhou, Z., Yu, J., Colell, J.F.P. et al. (2017). Long-lived 13C2 nuclear spin states hyperpolarized by parahydrogen in reversible exchange at microtesla fields. *Journal of Physical Chemistry Letters* 8 (13): 3008–3014.

41 Adams, R.W., Aguilar, J.A., Atkinson, K.D. et al. (2009). Reversible interactions with para-hydrogen enhance NMR sensitivity by polarization transfer. *Science* 323 (5922): 1708–1711.

42 Rogers, S.A., Huigens, R.W., Cavanagh, J. et al. (2010). Synergistic effects between conventional antibiotics and 2-aminoimidazole-derived antibiofilm agents. *Antimicrobial Agents and Chemotherapy* 54 (5): 2112–2118.

43 Rhoads, A., Beyenal, H., and Lewandowski, Z. (2005). Microbial fuel cell using anaerobic respiration as an anodic reaction and biomineralized manganese as a cathodic reactant. *Environmental Science & Technology* 39 (12): 4666–4671.

44 Carmona-Martínez, A.A., Harnisch, F., Kuhlicke, U. et al. (2013). Electron transfer and biofilm formation of *Shewanella putrefaciens* as function of anode potential. *Bioelectrochemistry* 93: 23–29.

45 Wiemers-Meyer, S., Winter, M., and Nowak, S. (2019). NMR as a powerful tool to study lithium ion battery electrolytes. In: *Annual Reports on NMR Spectroscopy*, (ed. G.A. Webb), 121–162. London: Academic Press.

46 Krachkovskiy, S.A., Foster, J.M., Bazak, J.D. et al. (2018). Operando mapping of Li concentration profiles and phase transformations in graphite electrodes by magnetic resonance imaging and nuclear magnetic resonance spectroscopy. *Journal of Physical Chemistry C* 122 (38): 21784–21791.

47 Freytag, A.I., Pauric, A.D., Krachkovskiy, S.A. et al. (2019). In situ magic-angle spinning 7Li NMR analysis of a full electrochemical lithium-ion battery using a jelly roll cell design. *Journal of the American Chemical Society* 141 (35): 13758–13761.

48 Morales Escalante, J., Ko, W., Foster, J.M. et al. (2020). Discerning models of phase transformations in porous graphite electrodes: Insights from inverse modelling based on MRI measurements. *Electrochimica Acta* 349: 136290.

49 Nguyen, T.H., Fraiwan, A., and Choi, S. (2014). Paper-based batteries: A review. *Biosensors and Bioelectronics* 54: 640–649.

50 Caizan-Juanarena, L., Krug, J.R., Vergeldt, F.J. et al. (2019). 3D biofilm visualization and quantification on granular bioanodes with magnetic resonance imaging. *Water Research* 167: 115059.

51 Fang, Y., Yushmanov, P.V., and Furó, I. (2020). Improved accuracy and precision in electrophoretic NMR experiments. Current control and sample cell design. *Journal of Magnetic Resonance* 318: 106796.

11

MRI of Transport and Flow in Plants and Foods

Maria Raquel Serial[1], *Camilla Terenzi*[1], *John van Duynhoven*[1,2], *and Henk Van As*[1]

[1] *Laboratory of Biophysics, Wageningen University, Wageningen, the Netherlands*
[2] *Unilever Global Foods Innovation Centre, Wageningen, the Netherlands*

11.1 Introduction

One of the most compelling challenges for humankind in the next decades is to feed the growing world population without exhausting our planet. Our current food supply chain produces food in an unsustainable manner and needs to be redesigned from farm to fork. The ecological burden of animal food production has reached its limits [1], and a move toward plant-based food production is called for. While a sustainable global food supply chain will also comprise animal-based food production [2, 3], most significant scientific challenges will be encountered within the drive toward more efficient agricultural crop production and an industrial plant-based food supply chain. In both fields, noninvasive assessment of transport and flow is critical for understanding dynamic processes, either in vivo in plants or in situ in foods. In this chapter, we will discuss how magnetic resonance imaging (MRI) has thus far been deployed in a wide range of applications in plants (Section 11.2) and foods (Sections 11.3 and 11.4), and we will also outline future perspectives (Section 11.5).

In a period of climate change, enhanced crop production must be achieved without additional land for farming and with reduced inputs of water, fertilizers, and pesticides. Plants with enhanced photosynthetic and water use efficiency are considered essential to meet future food needs. Thus, a significant increase in crop output is needed to produce higher biomass and yields on a per-plant basis [4]. Yield can be increased by an enhanced utilization of the photoassimilates by growing organs and by a higher carbohydrate production in leaves, in turn determined by light interception, number of leaves, total leaf area, breeding for stay-green traits, and photosynthesis efficiency. Despite the enormous importance of carbohydrates for growth and yield, we are just beginning to uncover and understand the mechanisms controlling the whole-plant distribution of carbon fixed by photosynthesis, particularly in crops [4]. There is a complex interplay between uptake and evaporation of water and transportation of photoassimilates, mostly sucrose, from source leaves to growing food tissues like fruits, seeds, cereal grains, nuts, and tubers. Long-distance water and mineral transport between roots and leaves occurs via the xylem, driven by evaporation, and sucrose transport between source leaves and sinks occurs via the phloem, driven by osmotically generated pressure differences. Because carbon transport is influenced by source and sink activity, it can integrate changes that occur

Magnetic Resonance Microscopy: Instrumentation and Applications in Engineering, Life Science and Energy Research, First Edition. Edited by Sabina Haber-Pohlmeier, Bernhard Blümich, and Luisa Ciobanu.

throughout the plant, potentially influencing growth and allocation, up to defense and reproduction [5]. Xylem and phloem are hydraulically connected. Xylem acts mostly as a water source for phloem, but also its water potential influences phloem transport. Measuring long-distance water or sucrose transport, as well as water and sucrose transport toward/inside the individual fruits, is very challenging. These transport processes should be studied under dynamic environmental conditions, controlling light intensity, temperature (-gradients), water availability, and developmental stage of the plant. Intact-plant nuclear magnetic resonance (NMR) and MRI methods have been developed over the past 40 years, to obtain key transport parameters and quantify the dynamics therein under variable environmental conditions [6]. MRI also has been used to measure fruit quality of fresh, postharvesting stored and processed fruits. These applications will not be covered here. We refer to a number of recent reviews and references cited therein [7–10].

Further in the farm-to-fork chain, the food industry is also facing significant challenges. The need to introduce sustainably produced plant-based lipids [11] and proteins [1] in food products that are appreciated by consumers drives the food industry to radically redesign food formulations and processes. Another industrial challenge is to produce, store, and distribute food products with less waste and reduced energy expenditure. This has led to a renewed interest in how water and oil, as constituents of virtually every food product, interact with other components in determining texture [12, 13]. In order to obtain textures that are appreciated by consumers, most food ingredients need to undergo transformation processes in the presence of water and/or oil, such as cooking and frying. These processed food products are typically thermodynamically unstable, leading to structural rearrangements that involve spatial redistribution of water and oil. For both such transformative food processing and shelf-life studies, MRI has proven itself as a valuable approach for noninvasive, real-time, and often quantitative assessment of water and oil migration [6]. In this chapter, we will discuss how challenges with respect to resolution in time, space, and molecular mobility have been overcome. Another critical factor that determines both efficient manufacturing and consumer appreciation of foods is flow behavior. This is the domain of food rheologists, who are constantly challenged by the heterogeneous and time-dependent flow behavior of food dispersions. In many cases, the conventional rheological methods fall short in describing the intricate non-Newtonian behavior of food materials and products. We will here outline the unique contributions of rheo-MRI in characterizing flow behavior of complex, optically opaque food systems under transient conditions and in confined geometries. We note that use of MRI for food quality control and assessment of microstructural features that underlie food texture is beyond the scope of this chapter.

11.2 Transport and Flow in Plants and into Fruits

There is a continuous effort in improving the quantity and quality of plant products like fruits, seeds, cereal grains and tubers, and vegetables. Despite the success of plant breeding programs, relatively little is known about the regulations of these plant products and the anatomical and physiological changes required to increase yield [14, 15]. Growth is a central determinant of plant yield. Available soil water, temperature, nutrients, light, and CO_2 are the key drivers of plant growth [16]. Water, temperature, and nutrients have fundamentally different effects as compared with light and CO_2, because the former can affect both sink and source activities, while light and CO_2 only affect the source activity by C-assimilation [16].

Leaves produce photoassimilates, e.g. sugars and amino acids, by photosynthesis. The photoassimilates are transported by the phloem to sinks such as roots, storage organs, or growing tissues like buds, new leaves, and fruits/seeds (Figure 11.1). This transport is driven by osmotically generated pressure differences due to the loading of sucrose from the sources and the unloading by the sinks. Plant photosynthesis depends on CO_2 uptake from the air by the leaves. This process is inevitably coupled to the evaporation of water, which drives the long-distance water transport in the xylem and uptake from the soil. Evaporation and CO_2 uptake depend on stomatal opening in leaves, which is under hormonal control and adjusts in response to an imbalance between water uptake in the roots and flux in the vascular system and the rate of evaporation. There is a complex coupling between photosynthetic activity, evaporation and CO_2 uptake, xylem/phloem transport (sap flow and hydraulic resistances), transport and allocation of photoassimilates, xylem transport and root water uptake, and growth/development/yield. This coupling includes complex transport and signaling processes between different organs and tissues at different locations in the plant [4, 17, 18].

The factors that determine how much carbon is allocated to an individual sink are largely unknown. Many researchers consider that the limiting factor in fruit growth is sink activity,

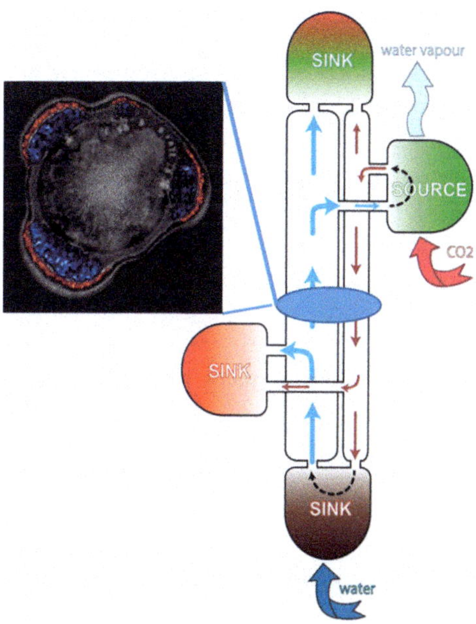

Figure 11.1 Schematic drawing of long-distance transport in a plant. Xylem (blue) connects uptake and transpiration, and phloem (red) connects source leaves to growing/storage tissues, the sinks. In the cross-sectional image of the stem of a tomato plant xylem and phloem flow is overlaid on a proton-density image. Details of MRI flowmetry (at 3 T): field of view: 3 × 3 cm^2, slice thickness 3 mm, 126 × 126 image matrix, phloem velocity scale: 0–0.6 mm/s (flow encoding time Δ = 150 ms), xylem velocity scale: 0–2.6 mm/s (flow encoding time Δ = 20 ms). Adapted from Prusova et al. [19].

which is determined by a variety of processes including phloem unloading, starch synthesis, sugar accumulation, cell cycle regulation, and cellular metabolism; others believe that carbon from the sources limits fruit growth [14]. Most probably both the source and sink play a role in carbon partitioning, and their relative importance depends on environmental parameters, developmental stage, and the aspect of plant productivity that is examined [14]. In addition, the questions of if and when the vascular transport capacity toward actively growing fruits can become the limiting factor are still open. In this debate it is important to discriminate between water and sucrose transport. Sucrose concentration depends on loading from the sources and uptake of sucrose along the transport pathway by lateral sinks. The concentration largely determines the viscosity of the phloem sap. What is needed is detailed information on flux (volume flow), velocity and flow-conducting area of water and sucrose, and sucrose concentration in the sap and the tissue surrounding the phloem.

11.2.1 Long-distance Water Flow

Since the introduction of NMR and MRI, many applications in intact plants have been reported, including imaging of plant development/growth: seed and bulb germination, seed development, fruit growth, root growth, and dynamic imaging of metabolites [20]. Intact-plant NMR and MRI methods have been developed to obtain key transport parameters and the dynamics therein, e.g. xylem and phloem flow and conductivity, under variable environmental conditions [6, 20–25].

Because of the dimensions of the flow-conducting elements in xylem and especially in phloem, and the velocities in phloem, MRI of water transport in xylem and phloem is extremely challenging [21, 22]. This challenge has been tackled by a MRI flowmetry method, based on propagator measurements by pulsed field gradient-stimulated echo rapid acquisition with relaxation enhancement (PFG-STE-RARE) imaging [26, 27]. The propagators can be analyzed in such a way that the flowing and nonflowing water, both contributing to the propagators within single pixels, can be separated and the flow characteristics calculated. Unique information such as the flow-conducting area, flow profile, velocity, volume flow rate, and the flow direction can be measured – all on a per-pixel basis. MRI flowmetry and relaxometry (T_2) imaging have been implemented on dedicated lab-based intact-plant MRI systems (with control/monitoring of microclimate parameters, light intensity, temperature, humidity, and plant watering). In this way an excellent tool to study xylem and phloem water transport in relation to the plant water balance and radial transport to storage pools has become available [19, 28]. Flow in the stem of a variety of herbaceous and woody plants, among others fully developed maize, tomato up to 10 m height, cucumber and grapevine plants, and young trees, e.g. oaks up to 4 m height, has been studied [29, 30].

Long-distance xylem and phloem flows have been studied in fruit-bearing tomato plants as a function of different manipulations. Source strength was manipulated by varying light intensity and photoperiod, and sink strength by fruit pruning; xylem–phloem interaction and leaf evaporation was altered by drought stress [19]. In addition, flow was studied in potato genotypes with different phloem area, developing differently sized tubers that differ in sink strength, in the root part [19], and in *Ricinus* plants during cold girdling of the stem, which affects water and sucrose transport [31] and flooding of the roots, resulting in anoxic root conditions that limit sucrose unloading from phloem [32]. In all these experiments, the source–sink relation was altered in an attempt to answer the question of whether phloem transport can be a limiting factor for photosynthesis efficiency and plant growth/development. The results revealed that xylem velocity and flux scale with transpiration, but phloem flux is regulated by varying the flow-conducting area, not the velocity [19, 32]. In a number of experiments, it was observed that at the same time the sucrose content of phloem sap, and thus the sap viscosity or the resistance in the phloem, varied. The changes in phloem sucrose concentration were observed on the basis of T_2 imaging at the higher B_0 field of the 3-T Plant MRI system [19]. Variations in sucrose content in the flow-conducting tracheids of phloem may stimulate exchange/radial transport to the nonflowing sap in companion cells and other lateral sinks [33]. If present, such exchange will have a strong effect on the observed flow characteristics measured using flow encoding time (Δ) values within 50–150 ms in the PFG part of the sequence [28]. Since propagators and flow results were independent of Δ, it could be concluded that the observed flow-conducting area and velocity values were not affected by exchange [19].

These results have some very intriguing implications. Intact-plant MRI under well-defined dynamic plant conditions can be used for in-depth plant phenotyping, for example based on dynamic transport traits. Maximum phloem flux seems to be limited by the phloem-conducting area present in plants, and can inhibit photosynthesis under optimal photosynthetic conditions. Breeding should include optimizing maximum phloem-conducting area in relation to optimizing photosynthesis efficiency. All this illustrates that MRI can be of help by providing information that is hardly available by other methods.

This MRI flowmetry method has also been applied to study the xylem and phloem water transport toward developing tomato fruits in the peduncle of a tomato truss, still attached to an intact tomato plant over an eight-week period [34]. The ratio between phloem and xylem fluxes toward fruits determines growth and quality. It is commonly estimated that about 90% of the water reaches the fruit via the phloem and the remaining 10% via the xylem. However, it was observed that under the conditions of the experiment over 75% of net water influx into the fruit occurred via the xylem. In contrast to accepted expectations, the xylem transport into the truss remained functional throughout the eight weeks of growth. Halfway through truss development, a xylem backflow appeared. As the truss matured, the percentage of xylem water that circulated into the truss and out again increased in comparison with the net uptake, but no net loss of water from the truss was observed. This application shows the strength and, at the same time, the weakness of this approach. Transport toward the fruits can be studied in detail by using the proper MRI hardware, but the experiments have to be repeated under different environmental conditions to really answer the questions about the ratio of xylem and phloem import. Light intensity and root water potential will most probably have a strong effect on that ratio. It is hard to repeat such long-lasting experiments in a dedicated plant MRI system for a couple of plants under different environmental conditions. For that we need cheaper and more dedicated MRI flowmetry sensors. Recently, the method has been implemented on an outdoor MRI system to measure flow in the stem of trees [35, 36], which shows the potential for in-field and greenhouse use. Even cheaper and less expensive systems have been developed [37, 38]. MRI flowmetry with these systems has still to be demonstrated.

11.2.2 Transport and Distribution of Metabolites

It may be clear that measurement of water flow in phloem is only part of the problem. To answer questions on C-allocation and the role of sucrose in the phloem in terms of concentration and regulations, it is necessary to be able to study sucrose transport and distribution over the different plant organs. Unraveling the factors determining the allocation of carbon to various plant organs is one of the great challenges of modern plant biology.

As already mentioned, at B_0 values above 2–3 T, the T_2 of the phloem sap becomes a function of sucrose content, because of proton exchange between water and sucrose [19]. This allows phloem flow characteristics and phloem sap sucrose content to be related to photosynthetic activity. Recently, a ^{1}H NMR method to measure both sucrose and water transport directly has been developed by combining localized STimulated Echo Acquisition Mode (STEAM) chemical shift resolved spectroscopy with the PFG propagator approach. The three slice-selective 90°-pulses in the STEAM sequence excite three orthogonal planes. The STEAM signal is a stimulated echo and results from protons at the (cuboid-shaped) intersection of the three planes that have experienced all three radiofrequency pulses. This signal is acquired in the absence of a readout gradient. For flow encoding two gradient pulses are applied. The first PFG is applied between the first and second 90°-pulses and the second PFG after the third 90°-pulse, respectively. This localized PFG-STEAM approach results in quantitative measurement of flow of metabolites, in addition to water, in complex systems, including plants (Figure 11.2). Measurements in an intact tomato plant in a 3-T dedicated intact-plant MRI and a 4-cm diameter radiofrequency coil resulted in a measurement time of a couple of minutes for water flow in a xylem and a phloem location (Figure 11.2b and c), and about 3.5 hours for sucrose in the same phloem location. The ^{1}H signal in the 3.5–3.9 ppm region in the phloem still contained a contribution of stagnant water in addition to the stagnant and flowing sucrose (Figure 11.2d), as was concluded on the basis of the two different diffusion coefficients needed to describe the propagator of the stagnant part of the propagator.

Allocation of metabolites in growing tissues like fruits, grains, tubers, etc. can also be followed by imaging metabolite composition and concentrations in certain tissues as a function of time. Such studies in general require MRI systems with relatively high magnetic fields, which however have limited bore space, and therefore limit the measurements to individual parts of a plant or to small, germinating or seedling plants. A number of approaches to study C-allocation by MRI has been reported and reviewed [6, 20, 39, 40]. These different approaches can be classified based on relaxation-time imaging (T_2 and/or T_1), chemical shift imaging (CSI), localized spectroscopy (e.g. point resolved spectroscopy [PRESS] or STEAM, see Figure 11.2), and chemical shift selective imaging (CSSI) [39]. Relaxation-time imaging can be used to image water and lipids in plant systems with high lipid content like seeds, avocado, olives, and oil palm fruits. This is even successful in low-field conditions [39, 40]. If T_2 of water and lipids overlap, differences in T_1 and/or diffusion can be used to separate the water and lipid signals. Spectroscopic imaging methods, like CSI, where the spins of the metabolites are directly detected on the basis of their chemical shift characteristics, are quite time consuming when low-concentrated compounds need to be detected with a high spatial resolution. Solutions for this problem are the use of high- or ultrahigh-field strengths. Indeed, the use of both CSSI, which reduces the information to a certain compound or all signal intensity within a narrow frequency range, and localized spectroscopy in combination with fast water imaging for tissue/location reference, has been demonstrated for studying different functional zones in root nodule of *Medicago truncatula* at 22.3 T (localized magnetic resonance spectroscopy voxels of $200 \times 350 \times 350\ \mu m^3$ at different positions, acquisition time 34 min per spectrum) [41]. An alternative proposed method is chemical exchange saturation transfer (CEST) [42]. CEST has been proven to be a promising approach for experiments where the signal-to-noise ratio is limited. CEST detects metabolites indirectly, by magnetization transfer via exchangeable protons of a metabolite with water. A substantial signal enhancement is achieved compared with the measurement of the protons of the metabolite directly. The gain in signal-to-noise ratio can be used to achieve higher temporal or spatial resolutions or the detection of lower

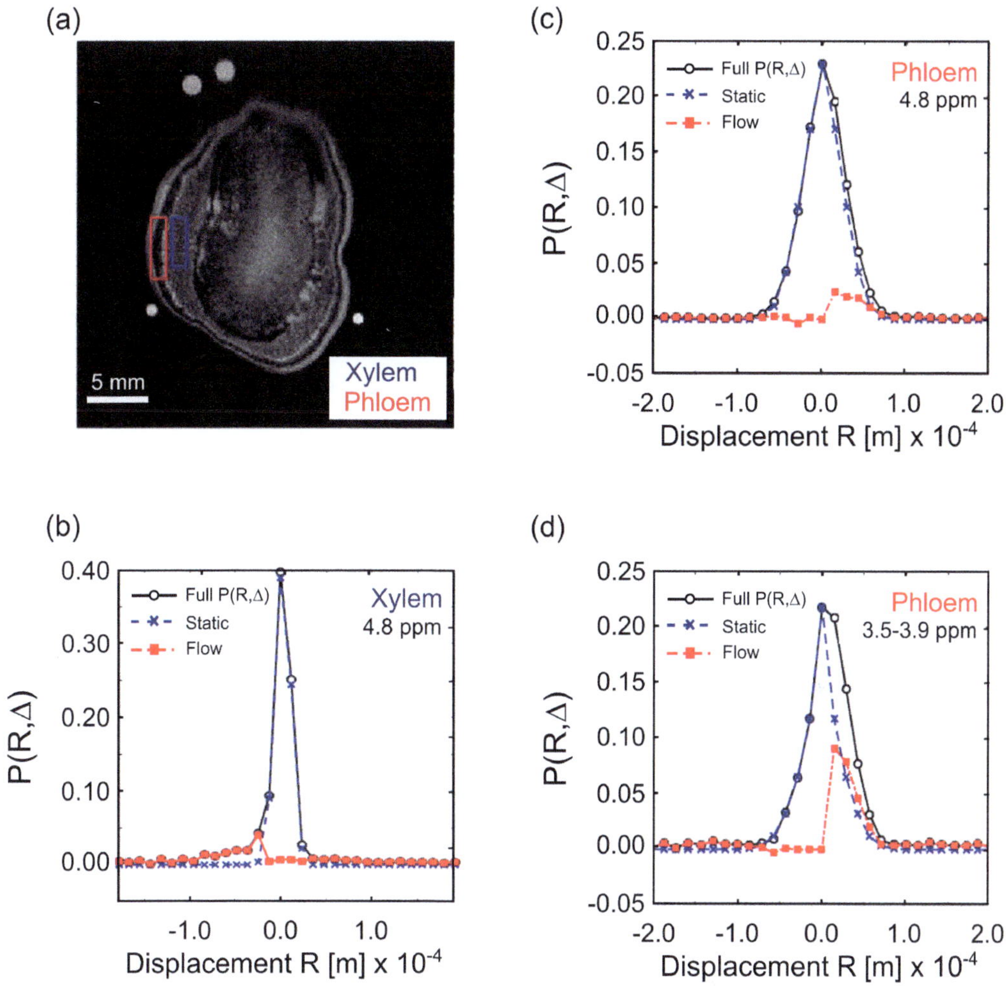

Figure 11.2 Localized flow measurements of water and sucrose in the stem of an intact tomato plant by use of a PFG-STEAM sequence and a dedicated 3 T Plant MRI system. The three selective 90°-pulses in the STEAM sequence excite three different planes. The echo signal results from the intersection of the three planes. The resulting localized spectra (in the phloem region, red rectangle in the image presented in (a) (voxel volume $5 \times 1.3 \times 6$ mm^3) clearly show water (~4.8 ppm) and sucrose (~3.5–3.9 ppm) signals. The sequence can be combined with a water suppression step to reduce the strong, partly sucrose overlapping, water peak. Propagator results from the xylem water (blue box in the image presented in (a), voxel volume $4 \times 1.2 \times 6$ mm^3) and the phloem water (4.8 ppm) and the signal from the 3.5–3.9 ppm region in the phloem are displayed in (b), (c), and (d), respectively. The static part of the propagators in (a) and (b) can be described by a diffusion coefficient (D) of 1.9×10^{-9} m^2/s and 2.1×10^{-9} m^2/s, respectively, typical for water. The 3.5–3.9 ppm signal in phloem (d) consists of two contributions with distinct D values, 2.4×10^{-9} m^2/s and 2.0×10^{-10} m^2/s, representing water and sucrose. The flowing part of the propagators represents velocities of 1.3 mm/s, 0.19 mm/s, and 0.15 mm/s in (b), (c), and (d), respectively. Results from van Kesteren et al. (work in preparation).

concentrated compounds. CEST can also be combined with imaging, e.g. CEST-RARE, as has been demonstrated for a cherry tomato to image glucose and glutamate [43]. However, the selectivity to a certain metabolite of interest is a challenge using CEST techniques, even at ultrahigh field of 22.3 T [43].

One of the difficulties in metabolite detection in plants using ^{1}H MR spectroscopic imaging techniques is the dominating presence of water and a large overlap in metabolite signals with water. Water and metabolites usually have a significantly broader line-width in plant tissues as compared with animal tissues, because of the prominent presence of local magnetic susceptibility differences due to intercellular air spaces. The water signal needs to be suppressed to enable detection of other ^{1}H-containing compounds, which generally have a concentration that is about a factor 1000 smaller than that of water. In addition, due to the normally broad lines, metabolite selectivity and water suppression can be improved by use of correlation spectroscopy and double quantum filtering (DQF) or coherence selection methods [20]. Both by direct and indirect detected ^{13}C gradient enhanced heteronuclear multiple quantum coherence (HMQC) imaging, sucrose allocation and conversion into and within seeds has been demonstrated in a 500-MHz system [44]. The sucrose distribution in the stem of sugar cane (*Saccharum officinarum*) has been imaged by DQF-CSI [45].

11.3 Transport in Foods: Water and Oil Migration

11.3.1 Water and Oil Migration During Food Preparation

Food processing is one of the oldest technologies known to man for preserving foods and making them suitable for consumption. During many food processes, the presence of water mediates critical phase transitions, in turn affected by the redistribution and status of water within the food product. Crystallization and glass transitions occur during drying and freezing, while reverse transitions and gelatinization occur during baking, frying, and cooking, with concurrent effects of oil migration. Although the thermodynamic and molecular basis [46] of the aforementioned phase transitions are well known, describing and predicting the kinetics of these processes in terms of foods' meso- and microstructures remains a big challenge for food engineers [47, 48]. Some of the major complications lie in the joint occurrence of mass/heat transfer with endothermic and/or exothermic phase transitions, and in the structural heterogeneity of foods over a large span of length scales.

In the past decade, ^{1}H MRI has been widely used, mostly in low-field (<2 T) conditions, for studying water migration during various processing conditions in a range of food materials: those published works, alongside a very few studies on oil migration, are classified in according to the food preparation method used. Remarkably, the MRI methods used are limited to RARE or spin echo sequences, with few or no method developments carried out to achieve faster or more informative measurements.

A very small selection of studies have focused on methodology developments, mainly with the aim to (i) make MRI more quantitative, to enable online calibration of water content during drying [87]; (ii) separate water–oil signals [88]; (iii) increase spatial and time resolution in kinetics studies, e.g. by the use of compressed sensing [7]; (iv) jointly deploy low-field time-domain NMR and MRI measurements [75, 77, 89]; and (v) include variable temperature conditions in MRI [90]. Notably, the majority of the works listed in Table 11.1 involve MRI in combination with, or alongside, low-field time-domain NMR relaxation of self-diffusion measurements. The joint deployment of MRI and low-field time-domain NMR has likely been adopted because it enables monitoring and discriminating spatially resolved changes in water and oil phases in terms of their content and mobility, while using low-cost, low-maintenance, and user-friendly NMR equipment also typically used in food industries.

Table 11.1 Overview of the past decade (2010–2020) of low- and high-field MRI studies of water or oil migration during different food preparation steps.

Liquid migrating	Preparation step	Purpose and food material	MRI method (magnetic field)	References
Water	Cooking	Ingress in rice kernels	GE (1 T), RARE (0.7 T)	[49–51]
		Structural rearrangements in meat	SSFP, FLASH (4.7 T)	[52]
		Water redistribution during cooking of pasta	SE (7 T); RARE (11.7 T); MSME (14 T)	[53–55]
		Water ingress during cooking of navy beans	MESE (1 T)	[56]
	Frying, boiling, stewing	Water redistribution in turbot	SE (0.5 T)	[57]
		Impact of salt on water ingress in noodles	MSME (9.4 T)	[58]
	Soaking, rehydration	Water ingress in rice kernels	GE (7.1 T; 11.7 T)	[59–61]
		Impact of drying routes on rehydration of carrot	RARE (7 T)	[62]
		Rehydration of dried sea cucumber	SE (0.5T)	[63]
	Brining, salting	Brining of feta cheese	MSSE (1 T)	[64]
		Salting of fish	RARE, MSME, FLASH, ADC (4.7 T)	[65]
	Curing	Impact on moisture distribution in ham	MSME (0.2 T)	[66]
	Drying	Broccoli	SE (0.5 T) RARE, MSME (3 T)	[67, 68]
		Apple, mushroom, oyster, eggplant, shrimp, corn, wheat, carrot	SE (0.5 T)	[69–77]
	De/rehydration	Abalone, clam	SE (0.5 T)	[78, 79]
	Freezethawing	Beef	SE (0.5 T)	[80]
	Ripening	Gran Padano cheese	SE (7 T)	[81]
	Freezing	Tofu	SE (0.5 T)	[82]
	Baking	Development of porous structure in bread	SE (2.35 T)	[83]
		Effect of emulsifier on porous structure bread	RARE (1.5 T)	[84]
Oil	Frying	Oil uptake in beef	SE (9.4 T)	[85]
		Oil uptake in potatoes	SE (0.5 T)	[86]

Acronyms for MRI methods used are: spin echo (SE), multi-slice multi echo (MSME), multi-echo spin echo (MESE), multi-slice single echo (MSSE), gradient echo (GE), steady-state free-precession (SSFP), Fast Low Angle SHot (FLASH), rapid acquisition with relaxation enhancement (RARE), and apparent diffusion coefficient (ADC).

11.3.2 Water and Oil Migration During Shelf Life

The shelf life of many food products is compromised by their inherent thermodynamic instability. For food products where water is immobilized by the matrix, this is mostly due to gradients in the chemical potential of water, which is described by food scientists as a thermodynamic mismatch of the so-called water activity [91]. Such thermodynamic instabilities are typically encountered for multi-textured foods [92], where components with different water activities are combined [93]. The thermodynamic (non-) equilibrium of such systems can be adequately described, but the kinetics of moisture migration can only be described in macroscopic terms, mostly without understanding the role of the involved mesostructures [47]. Migration of oil from porous foods due to network shrinkage or collapse is also often due to structural rearrangements toward a thermodynamically more stable situation. An example is fat blooming in chocolate, which is the result of coarsening of oil-containing percolating fat crystal networks induced by recrystallization [94, 95].

In the past decade, ^{1}H MRI has been used, in both low- and high-field conditions, for studying water and/or oil migration during shelf-life conditions. Relevant publications are reported in Table 11.2: it can be seen that they mostly involve use of spin echo-based sequences, with the exception of one paper on single-point imaging (SPI) [96].

Table 11.2 Overview of the past decade (2010–2020) of low- and high-field MRI studies of water or oil migration during shelf life.

Liquid migrating	Purpose and food material	MRI method (magnetic field)	References
Water	Verification of models describing moisture transport in porous cereal matrix	SPI (3 T)	[96]
	Moisture redistribution in Amaretti cookies	MSME (7 T)	[97]
	Verification of model for moisture migration in sandwich	SE (0.3 T)	[98]
	Effect of storage temperature on moisture migration in beef	SE (0.5 T)	[89]
	Effect of hydrocolloid on water distribution in apple jelly		[99]
	Effect of coating of fish on water migration		[100]
	Moisture migration in model lipids	SE (7 T)	[101]
Oil	Migration of oil from filling to chocolate coating	MSME (7 T); MSSE (1T); SE (3 T)	[102–105]
	Modelling of oil migration in chocolate confectionary	MSME (3 T); MSSE (1 T)	[106, 107]
	Oil migration in a cocoa butter/cream model system	MSSE (7 T)	[108]
	Effect of cocoa processing on oil migration	MSSE (7 T)	[109]
	Oil migration in chocolate confectionery	SE (7 T)	[110]
Water and oil	Frying of potatoes	SE, MSME (3 T)	[111]

Acronyms of MRI methods are as explained in the notes for Table 11.1.

11.4 MRI of Flow in Foods

The flow characterization of granular or fluid food products, often of non-Newtonian nature, is crucial for designing and optimizing the quality and sensory properties of foods, as well as their efficient processability. The intrinsic complexity of these materials demands the implementation of spatially resolved techniques that allow for the assessment of time-dependent and heterogeneous flow. The vast range of physical and spatial information that can be encoded during experiments has increased MRI applications to studying complex rheological properties [112]. In this section, we will outline how information regarding structural and density variations, as well as interparticle interaction effects of food materials under flow, can be spatially encoded during MRI experiments.

11.4.1 Shear-induced Spatial Heterogeneities Under Transient Flow Conditions

Under steady-state conditions, the flow of food dispersions can be described by classical rheological methods, using the empirical flow curve equation $\sigma(\dot{\gamma})$, which relates the applied shear stress σ to the shear rate $\dot{\gamma}$ [113] for a given material. Different so-called constitutive laws have been developed to describe flow behaviors, such as shear-thinning and shear-thickening, but they all require the assumption that both shear stress and particle concentration are homogeneous across the cell geometry. Food materials that undergo colloidal network formation or degradation under shear, such as fat dispersions or emulsions, exhibit time-dependent and/or localized structural changes under flow, which can result in thixotropy and shear-banding effects [114–116]. Under such transient conditions, the classical approach of applying a sweep in shear rate or stress does not enable the use of a steady-state constitutive law. This limitation can be overcome by obtaining local flow curves, $\sigma(\dot{\gamma}(r))$, that can be considered as a transient constitutive law, and where local shear stress $\sigma(r)$ and shear rate $\dot{\gamma}(r)$ are determined from separate measurements of torque and rheo-MRI velocity profiles, respectively [117]. This approach has been successfully applied to yield-stress food materials, such as emulsions and gels, generally employing wide-gap (centimeter) Couette cells at low-field and under steady-state conditions [118–120]. More recently, local flow curves were recorded at high magnetic field (7 T) to improve sensitivity and thus also temporal resolution of 1D rheo-MRI measurements (PFG spin echo sequence, 48.8 μm and 5 minutes spatial and temporal resolutions, respectively) [121]. This has enabled studying in detail the flow of complex foods, such as micronized fat crystal (MFC) dispersions, which are known to undergo rapid recrystallization and network formation under shear [122, 123]. A proper understanding of the interplay between these mechanisms during flow is of great importance for the industrial processing and formulation of fat products. Spatially heterogeneous local flow properties, induced by the formation and subsequent disruption of MFCs networks, were recorded every 5 minutes over 13 hours at a constant applied shear rate and unraveled (i) the presence of three flowing bands, and (ii) that the flow is described not by a single Herschel–Bulkley model, but by several constitutive laws (see Figure 11.3a) [121]. Local stress σ, local yield stress σ_y, and effective viscosity μ_{eff} also proved to be spatially heterogeneous, unraveling the complex interplay between stress heterogeneity and MFC network formation (Figure 11.3b).

The presence of structural inhomogeneities under shear can lead to incorrect interpretation of macroscopic flow curves. A well-known example is the phenomenon of shear-induced migration, which generally causes an underestimation of the apparent viscosity of the material [117]. In wide-gap geometries, particle migration results from shear heterogeneities that induce an accumulation of particles in zones of low shear. Optimizing and controlling the transport of concentrated food dispersions and emulsions remains a challenge, especially in the case of novel plant-based products for which often almost no experience in industrial processing exists. Because shear-induced migration can only be poorly described and predicted by existing models, it is important to detect and quantify this phenomenon in order to evaluate its impact on the material's flow behavior. MRI offers the possibility to spatially resolve particle concentration by recording signal intensity profiles under flow. Recently, rheo-MRI was employed to study the shear-induced migration of oil droplets in a dense vegan mayonnaise [124]. By suppressing the water signal coming from the continuous phase of the emulsion, rheo-MRI enabled spatially resolving the kinetics of oil droplet migration (Figure 11.3d, 48.8 μm spatial resolution and 2 minutes temporal resolution), and correlating the largest decrease of oil concentration to the interface between high- and low-shear bands. These measurements also confirmed the existence of a plug region, where no further oil droplets can migrate (position C in Figure 11.3c). Similar findings were obtained in the shear-induced migration of noncolloidal particles dispersed in a yield-stress fluid [125].

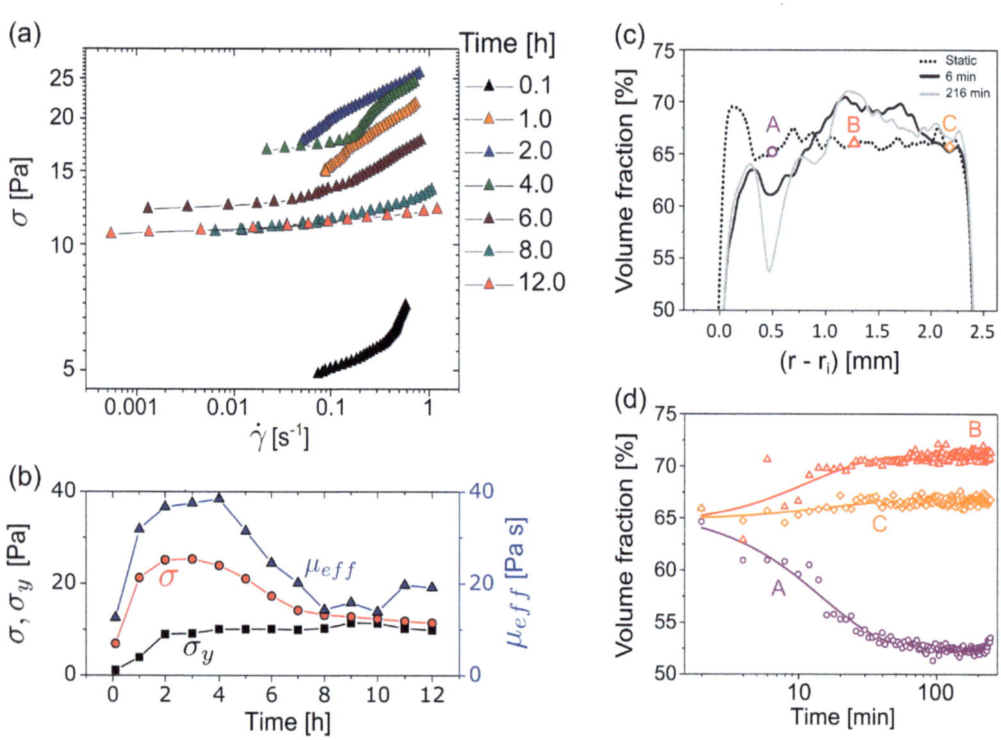

Figure 11.3 [1]H Rheo-MRI measurements carried out at 7 T in a 2.5-mm gap Couette cell of (a)–(b) MFC dispersion and (c–d) vegan mayonnaise. Time dependence of (a) local flow curves at a constant rotational velocity of $\Omega = 0.96$ rpm, and of (b) local stress σ, local yield stress σ_y, and effective viscosity μ_{eff} for a position close to the inner rotating cylinder. Adapted from Nikolaeva et al. [121] Copyright (2020) American Chemical Society. (c) Oil droplet volume fraction profiles as a function of the distance to the inner cylinder radius r_i and under static conditions (dotted line), after 6 minutes (dark gray line) and after 226 minutes (light gray line) of shear at $\dot{\gamma} = 4.56\,s^{-1}$; (d) at selected positions A, B, and C along the cell gap as a function of shearing time. Adapted with permission from Serial et al. [124] Copyright (2018) John Wiley & Sons Ltd.

11.4.2 Flow in Granular and Soft Food Materials

Granular media are widely present in food industry, whether as raw materials, intermediate, or final products. Common examples are sugar and cereals, as well as many food powders, grains, and seeds. Under flow, these materials tend to form static regions and cause segregation of blended ingredients, flow obstruction, and even spoiling of material. These phenomena are known industrial bottlenecks during discharge of storage silos and dispensing of ingredients by hoppers during food processing. The flow behavior inside these geometries depends strongly on the dimensions of the silo or hopper, the roughness of the wall, and on interparticle interactions. Even though many discrete and continuum models have been developed to investigate silo flow, validation measurements are still required. Contrary to optical microscopy methods, velocimetry-MRI allows for the measurement of bulk velocities of even optically opaque granular materials away from the silo walls, which are especially difficult to assess. Recently, Fullard et al. studied the flow behavior of poppy seeds flowing inside a discharging silo [126]. Two-dimensional vertical and horizontal velocity maps were measured changing the angle of the silo (23.5 mm wide and 6.5 mm inlet diameter) and testing the effect of rough walls on the flow dynamics (Figure 11.4a). Comparison with numerical simulations showed that the so-called friction law [127], commonly used to define granular viscosity, qualitatively predicts the flow pattern inside the silo for the mass and funnel flow regimes, but cannot quantify flow rates. Possible causes of this behavior are discussed by the authors, pointing toward compressibility and finite size effects, as well as cooperativity flow behavior commonly found in granular materials. The latter effect becomes significant when the dimensions of the geometry are comparable with the average particle diameter [128]. Recently, the nonlocal flow behavior of mustard seeds (~1.2 mm) was assessed by de Cagny et al. [129] employing rheo-MRI techniques in a 2-cm Couette cell geometry. Velocity profiles were shown to be in good agreement with the nonlocal fluidity model proposed by Bocquet et al. [130–132], and to yield cooperativity or correlation lengths ξ on the same order of particle size (right plot in Figure 11.4b). Proton-density profiles recorded simultaneously during shearing experiments allowed the investigation of the effect of concentration heterogeneities on nonlocal flow behavior (left plot in Figure 11.4b).

Nonlocality has also been observed in soft-structured materials, such as several emulsions, foams, and colloidal gels. Even though the nonlocal flow of these systems has been successfully demonstrated, few papers focus on the study of food-related materials due to their high complexity and optical opacity. During processing, these systems are known to flow partially through narrow-gap geometries, like pipes, gaps, and extruders, where flow is affected by interparticle cooperativity effects. Understanding how soft food materials flow under confinement is therefore important for food production, as well as for new emerging technologies, such as food 3D printing. Recently, de Kort et al. employed conventional rheology and rheo-MRI methods to investigate the flow of microfibrillated cellulose, employed in food industry as a food thickener [133]. Rheology results for these dispersions unraveled strong thixotropic behavior, supported also by rheo-MRI experiments in a cone-and-plate geometry, where banding behavior is expected for thixotropic fluids under homogeneous stress field conditions [134]. The evidence of smoothened banding effects, but of no discrete bands (Figure 11.5a), pointed toward the presence of nonlocal flow. Velocity curves were fitted employing the nonlocal fluidity model [133], yielding cooperativity correlation lengths of the order of 80 μm, in line with the expected dimensions of flocs of microfibrillated cellulose.

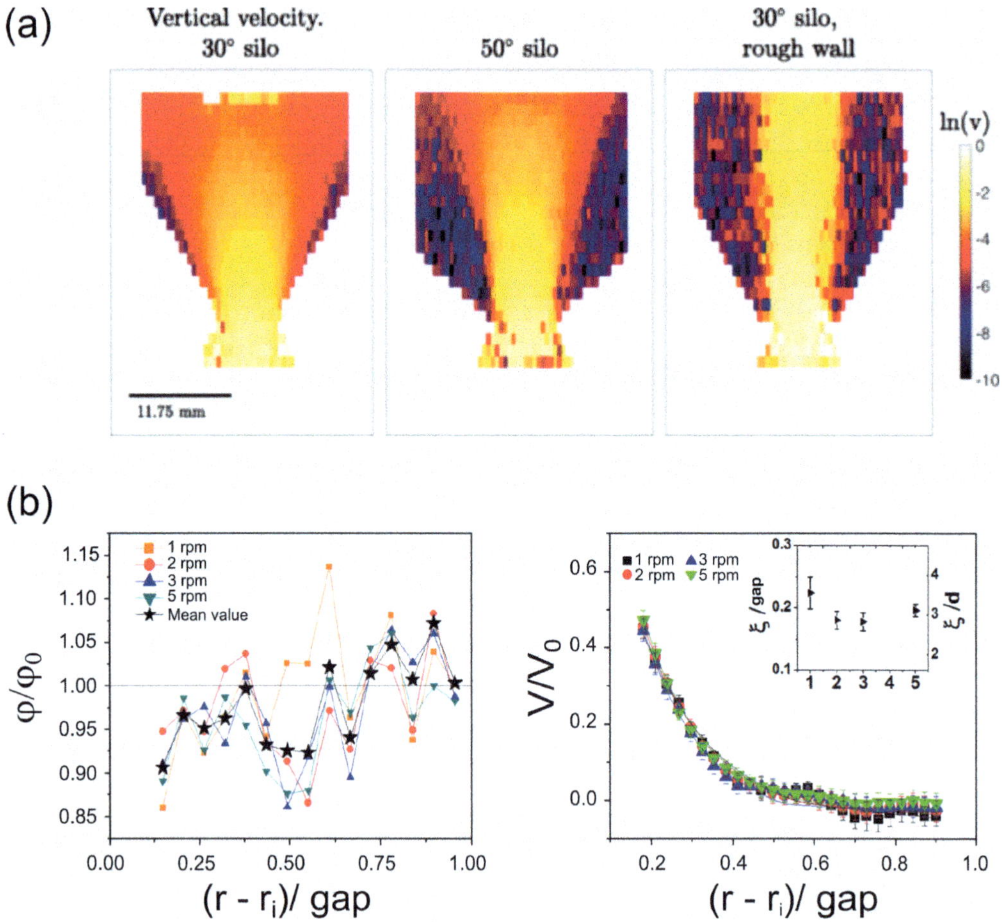

Figure 11.4 ^{1}H MRI velocimetry measurements of poppy seeds (a) and mustard seeds (b). (a) Vertical velocities at 9.4 T of poppy seeds flowing through 3D-printed silos of cone half-angles of 30° (left) and 50° (middle); the 30° silo experiment was repeated using a rough-wall condition (right). From left to right: mass, funnel, and rat-holing flow regimes are observed, respectively. Adapted from Figure 3 with permission from Fullard et al. [126]. Copyright (2019) by the American Physical Society. (b) Spatially resolved rheo-MRI measurements at 0.5 T of local relative concentration φ/φ_0 (left) and normalized velocity V/V_0 (right) in a 2-cm gap, inner radius r_i = 3 cm Couette cell at rotational speeds within 1–5 rpm. Inset in the right plot shows the fitted cooperativity lengths ξ normalized to the gap size and the particle diameter d obtained by the nonlocal fluidity model [131]. Adapted with permission from de Cagny et al. [129] Copyright (2015), The Society of Rheology.

Similar flow behavior can also be found in complex food systems such as stirred milk yogurt (Figure 11.5b). In this material nonlocal flow originates interactions between protein aggregates, typically with sizes in the range of 20–200 μm [135]. The size and shape of these aggregates strongly depends on processing conditions such as temperature, pH, and mechanical disruption. In order to observe and address nonlocal effects in such material, tailor-made cell geometries with gap sizes in the submillimeter range need to be employed. Moreover, such materials that exhibit time-dependent flow properties are suitable systems to characterize cooperative flow in the presence of thixotropy, particle density variations, and flow heterogeneities.

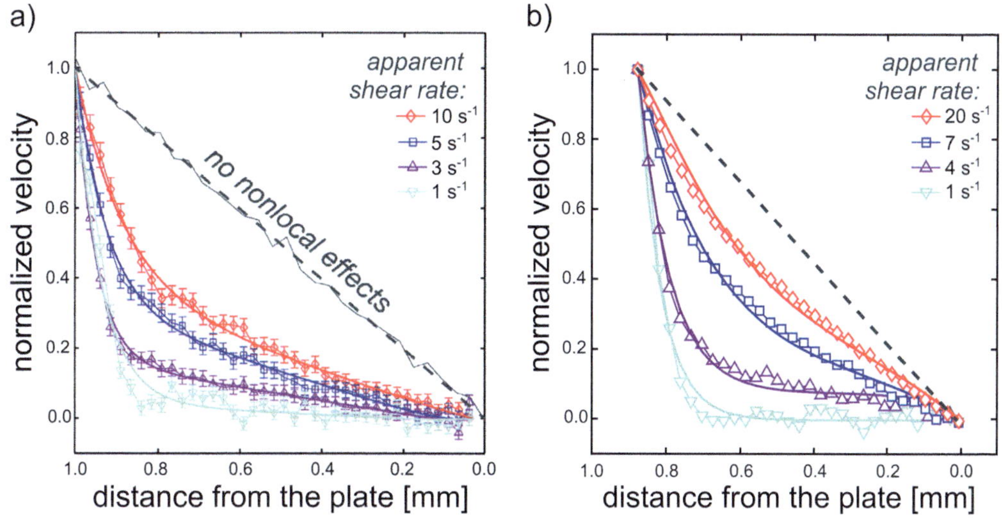

Figure 11.5 [1]H Rheo-MRI velocity profiles acquired at 7 T in a cone-and-plate geometry (velocity normalized to the cone velocity) for (a) a microfibrillated cellulose dispersion and (b) stirred milk yogurt. Solid lines are fits to the nonlocal model [132] without (a) or with (b) taking into account shear-banding effects. The dashed line represents the flow profile of nonthixotropic yield-stress fluids in a homogeneous stress field in a gap size much larger than the particle size [136]. (a) Republished with permission of the Royal Society of Chemistry, from de Kort et al. [133] permission conveyed through Copyright Clearance Center, Inc. (b) Results are from Serial et al.

11.5 Perspectives

From our review of recent application of MRI in plant science we have identified several methodological challenges ahead. Long-distance xylem and phloem transport in plants can be well studied by MRI flowmetry based on PFG-S(T)E-RARE. PFG-STE-RARE propagator measurements are time consuming, typically on the order of 1 hour for a 128 × 128 image matrix and 32 or 64 $\boldsymbol{q}$-steps. Recently it has been demonstrated that it can be speeded up by use of undersampling and compressed sensing [137]. This can be combined by noise reduction by proper filtering of the propagators before analysis [36]. In this way an unprecedented reduction of the measurement time by almost an order of magnitude can be achieved.

In addition to lab-bound equipment, outdoor portable MRI systems have already been demonstrated to be able to measure xylem and phloem transport in intact plants. Further hardware developments for cheaper and dedicated mobile MRI systems are foreseen. For flowmetry applications magnet homogeneity and stability is an important factor.

MRI of metabolites is very promising, but at the same time limited by hardware constraints. For the study of C-allocation lab-bound MRI systems with relatively high magnetic field are necessary, with the exception of imaging of oil in organs with relatively high oil content. In the latter case, low-field T_2 and/or T_1 MRI can be used, both using lab-bound and outdoor systems. Only with B_0 fields above around 2–3 T has it been demonstrated that T_2 can be used to monitor changes in sucrose content of the phloem sap, with local concentrations around 400 mm and higher. At lower field strength, the potential of CEST-PFG-S(T)E-RARE can be explored to study sucrose phloem transport. In all other cases, (ultra)high-field MRI systems are necessary. Only a very limited number of such expensive systems are in use with large diameter bores that allow the insertion of bigger plants in the magnet. In wide-bore systems plants can be studied that fit in the bore of the magnet, e.g. seedlings, small plants, or only part of the larger intact plant can be imaged (e.g. ears, small fruits, grains). Control of the climatic plant conditions in such magnets is a challenge, has hardly been reported, but is a prerequisite for useful results.

In our review of the literature on MRI applications in food science, it has emerged that most works rely on a limited number of spin echo-based sequences, deployed in a narrow window of environmental conditions such as temperature, humidity, and pressure. The introduction of wide-bore low-field MRI instruments [99, 100] will enable wider choices of environmental conditions. The use of spin echo sequences enables solely a view on phases with long transversal relaxation rates, and hence high molecular mobility. We envisage that by introducing pulse sequences, such as ultrashort echo time (UTE) MRI, with effective echo time limited only by the hardware's receiver dead time, phase transitions in the solid and semi-solid regime can also be characterized. This would be relevant for studying phenomena including phase transitions wherein crystalline or glassy states are involved. Due to the short T_2 NMR relaxation time of these phases these cannot be observed by spin echo-based MRI pulse sequences. SPI allows a view on these phases [96], but its broad application is hampered by long acquisition times. The use of UTE sequences would enable faster measurements and assessment of rapid structural changes, such as jamming transitions or crystallization effects. Faster measurements can also be achieved by combining single-shot MRI methodology and hardware modifications. An example is the recent rapid velocimetric assessment of granular systems by the deployment of 16 radiofrequency coils around the region of interest in a 3 T whole-body human scanner [138] resulting in a temporal resolution of 22.5 ms over a pixel size of 3 mm × 5 mm × 10 mm.

Narrow-gap geometries are of particular interest for rheological characterization of food dispersions, where particle clusters or flocs can be disrupted by the stresses induced during strongly confined flow. Phenomena like wall-slip, flow heterogeneities, and steady-state shear-banded flows result from the competition between shear flow and attractive forces, which in turn control both cluster size and shape. Such systems exhibit shear-induced structuring driven by hydrodynamic coupling between the flocs and the confining boundaries [139], but the underlying mechanisms are poorly understood and plague industrial process design. For such applications, the development of rheo-MRI setups for submillimeter, narrow-gap geometries with inner-to-outer cylinder ratio $\frac{R_i}{R_o} > 0.97$ [140] would enable molecular level-understanding of confined flow, including shear banding, nonlocal, and thixotropic effects, in food particulate suspensions. We foresee that for the assessment of flow in narrow-gap geometries, higher magnetic fields (>14 T), and even the use of sensitive cryogenic MRI probe heads, will be required.

Wide-gap geometries, with inner-to-outer cylinder ratio $\frac{R_i}{R_o} < 0.97$ [140], are often dismissed in conventional rheology due to the large stress variation therein, which in turn complicates the interpretation of flow. Yet, for many new applications of rheology in food industry, e.g. to large-scale food productions and texturization processes, access to wide-gap and upscaled flow cells has proven to be essential. A compelling example is the manufacturing of highly anisotropic fibrous meat alternatives via high-temperature shear processing of plant proteins in wide-gap Couette cells [141] or extruders [142]. Key to optimizing fibrousness in such products is believed to be shear- and heat-induced phase separation, but the underlying physico-chemical mechanisms are under debate [142, 143]. The development of lab-scale and upscaled wide-gap geometries for rheo-MRI would open new avenues to the verification of models describing shear-induced anisotropic structure formation. Critical will be operating in the range 30–100°C, the feasibility of which has been demonstrated by Laryea et al. [144], who developed the first temperature-controlled rheo-NMR Couette cell with a narrow gap (1 mm).

Recently, interest has also emerged in studying food materials by means of large amplitude oscillatory shear (LAOS) flow, which can offer better understanding of large deformations and structural changes occurring during different kinds of processing and consumption [145]. Tailoring the LAOS Rheo-MRI setup for both submillimeter and wide-gap geometries, in combination with the implementation of temperature control, would open new possibilities for characterizing the local rheology of complex food materials under varying flow conditions. The feasibility of implementing such a LAOS Rheo-MRI setup was recently demonstrated in spectroscopy mode in a 9-T wide-bore magnet for a 1-mm-gap Couette cell, where oscillatory shear was driven by a servo-stepper [146, 147]. Recently, the first integration of a low-field (0.7 T) pulsed NMR spectrometer into a commercially available high-end shear rheometer has been reported [148, 149]. The new in situ combination enabled full use of the rheometer to apply nonlinear deformation under both steady-shear and LAOS conditions. The next step of implementing affordable and routine rheo-MRI at low field would open up this capability for a wider range of users, beyond the current NMR expert domain.

Several of the scientific challenges that have been encountered during the radical move toward plant-based food production, both within agricultural crop production and in the industrial food supply chain, have been presented and discussed. We conclude that further developments of MRI methods and hardware to dynamically study transport and flow in vivo in plants, in operando in food processing, and in situ in food storage studies can contribute to the success of this transition.

References

1 Aiking, H. and de Boer, J. (2020). The next protein transition. *Trends in Food Science & Technology* 105: 515–522.

2 Van Zanten, H.H.E., Herrero, M., Van Hal, O. et al. (2018). Defining a land boundary for sustainable livestock consumption. *Global Change Biology* 24 (9): 4185–4194.

3 Leroy, F., Hite, A.H., and Gregorini, P. (2020). Livestock in evolving foodscapes and thoughtscapes. *Frontiers in Sustainable Food Systems* 4: 105.

4 Braun, D.M., Wang, L., and Ruan, Y.-L. (2014). Understanding and manipulating sucrose phloem loading, unloading, metabolism, and signalling to enhance crop yield and food security. *Journal of Experimental Botany* 65 (7): 1713–1735.

5 Savage, J.A., Clearwater, M.J., Haines, D.F. et al. (2016). Allocation, stress tolerance and carbon transport in plants: How does phloem physiology affect plant ecology? *Plant, Cell & Environment* 39 (4): 709–725.

6 Van As, H. and van Duynhoven, J. (2013). MRI of plants and foods. *Journal of Magnetic Resonance* 229: 25–34.

7 Nicolaï, B.M., Defraeye, T., De Ketelaere, B. et al. Nondestructive measurement of fruit and vegetable quality. *Annual Review of Food Science and Technology* 5: 285–312.

8 Pathmanaban, P., Gnanavel, B., and Anandan, S.S. (2019). Recent application of imaging techniques for fruit quality assessment. *Trends in Food Science & Technology* 94: 32–42.

9 Kirtil, E., Cikrikci, S., McCarthy, M.J. et al. Recent advances in time domain NMR & MRI sensors and their food applications. *Current Opinion in Food Science* 17: 9–15.

10 Srivastava, R., Talluri, S., Beebi, S.K. et al. (2018). Magnetic resonance imaging for quality evaluation of fruits: A review. *Food Analytical Methods* 11 (10): 2943–2960.

11 Khor, Y.L. (2011). The oil palm industry bows to NGO campaigns. *Lipid Technology* 23 (5): 102–104.
12 Schmidt, S.J. (2004). Water and solids mobility in foods. *Advances in Food and Nutrition Research* 48: 1–103.
13 Lewicki, P.P. (2004). Water as the determinant of food engineering properties. A review. *Journal of Food Engineering*. 61 (4): 483–495.
14 Savage, J.A., Haines, D.F., and Holbrook, N.M. (2015). The making of giant pumpkins: How selective breeding changed the phloem of *Cucurbita maxima* from source to sink. *Plant, Cell & Environment* 38 (8): 1543–1554.
15 White, A.C., Rogers, A., Rees, M. et al. (2016). How can we make plants grow faster? A source–sink perspective on growth rate. *Journal of Experimental Botany* 67 (1): 31–45.
16 Fatichi, S., Leuzinger, S., and Körner, C. (2014). Moving beyond photosynthesis: From carbon source to sink-driven vegetation modeling. *New Phytologist* 201 (4): 1086–1095.
17 Yu, S.-M., Lo, S.-F., and Ho, T.-H.D. (2015). Source–sink communication: Regulated by hormone, nutrient, and stress cross-signaling. *Trends in Plant Science* 20 (12): 844–857.
18 De Schepper, V. and Steppe, K. (2010). Development and verification of a water and sugar transport model using measured stem diameter variations. *Journal of Experimental Botany* 61 (8): 2083–2099.
19 Prusova, A. (2016). Light on phloem transport (an MRI approach). PhD thesis. Wageningen University.
20 Borisjuk, L., Rolletschek, H., and Neuberger, T. (2012). Surveying the plant's world by magnetic resonance imaging. *The Plant Journal* 70 (1): 129–146.
21 Van As, H. (2007). Intact plant MRI for the study of cell water relations, membrane permeability, cell-to-cell and long distance water transport. *Journal of Experimental Botany* 58 (4): 743–756.
22 Van As, H., Scheenen, T., and Vergeldt, F.J. (2009). MRI of intact plants. *Photosynthesis Research* 102 (2-3): 213.
23 Köckenberger, W. (2001). Functional imaging of plants by magnetic resonance experiments. *Trends in Plant Science* 6 (7): 286–292.
24 Reinders, J.E.A., Van As, H., Schaafsma, T.J. et al. (1988). Water balance in *Cucumis* plants, measured by nuclear magnetic resonance, II. *Journal of Experimental Botany* 39 (9): 1211–1220.
25 Van As, H., Reinders, J.E.A., de Jager, P.A. et al. (1994). In situ plant water balance studies using a portable NMR spectrometer. *Journal of Experimental Botany* 45 (1): 61–67.
26 Scheenen, T., Vergeldt, F.J., Windt, C.W. et al. (2001). Microscopic imaging of slow flow and diffusion: A pulsed field gradient stimulated echo sequence combined with turbo spin echo imaging. *Journal of Magnetic Resonance* 151 (1): 94–100.
27 Scheenen, T., Heemskerk, A., de Jager, P.A. et al. (2002). Functional imaging of plants: A nuclear magnetic resonance study of a cucumber plant. *Biophysical Journal* 82 (1): 481–492.
28 Homan, N., Windt, C.W., Vergeldt, F.J. et al. (2007). 0.7 and 3 T MRI and sap flow in intact trees: Xylem and phloem in action. *Applied Magnetic Resonance* 32 (1-2): 157–170.
29 Windt, C.W., Vergeldt, F.J., de Jager, P.A. et al. (2006). MRI of long-distance water transport: A comparison of the phloem and xylem flow characteristics and dynamics in poplar, castor bean, tomato and tobacco. *Plant, Cell & Environment* 29 (9): 1715–1729.
30 Copini, P., Vergeldt, F.J., Fonti, P. et al. (2019). Magnetic resonance imaging suggests functional role of previous year vessels and fibres in ring-porous sap flow resumption. *Tree Physiology* 39 (6): 1009–1018.

31 Peuke, A.D., Windt, C.W., and Van As, H. (2006). Effects of cold-girdling on flows in the transport phloem in *Ricinus communis*: Is mass flow inhibited? *Plant, Cell & Environment* 29 (1): 15–25.
32 Peuke, A.D., Gessler, A., Trumbore, S. et al. (2015). Phloem flow and sugar transport in *Ricinus communis* L. is inhibited under anoxic conditions of shoot or roots. *Plant, Cell & Environment* 38 (3): 433–447.
33 De Schepper, V., De Swaef, T., Bauweraerts, I. et al. (2013). Phloem transport: A review of mechanisms and controls. *Journal of Experimental Botany* 64 (16): 4839–4850.
34 Windt, C.W., Gerkema, E., and Van As, H. (2009). Most water in the tomato truss is imported through the xylem, not the phloem: A nuclear magnetic resonance flow imaging study. *Plant Physiology* 151 (2): 830–842.
35 Nagata, A., Kose, K., and Terada, Y. (2016). Development of an outdoor MRI system for measuring flow in a living tree. *Journal of Magnetic Resonance* 265: 129–138.
36 Terada, Y., Horikawa, Y., Nagata, A. et al. (2020). Dynamics of xylem and phloem sap flow in an outdoor zelkova tree visualized by magnetic resonance imaging. *Tree Physiology* 40 (3): 290–304.
37 Windt, C.W. and Blümler, P. (2015). A portable NMR sensor to measure dynamic changes in the amount of water in living stems or fruit and its potential to measure sap flow. *Tree Physiology* 35 (4): 366–375.
38 Meixner, M., Tomasella, M., Foerst, P. et al. (2020). A small-scale MRI scanner and complementary imaging method to visualize and quantify xylem embolism formation. *New Phytologist* 226 (5): 1517–1529.
39 Musse, M. and Van As, H. (2018). NMR imaging of air spaces and metabolites in fruit and vegetables. In: *Modern Magnetic Resonance*, 1765–1779. Springer International Publishing.
40 Munz, E., Jakob, P.M., and Borisjuk, L. (2016). The potential of nuclear magnetic resonance to track lipids in planta. *Biochimie* 130: 97–108.
41 Van Schadewijk, R., Krug, J.R., Shen, D. et al. (2020). Magnetic resonance microscopy at cellular resolution and localised spectroscopy of *Medicago truncatula* at 22.3 Tesla. *Scientific Reports* 10 (1): 1–11.
42 Cai, K., Haris, M., Singh, A. et al. (2012). Magnetic resonance imaging of glutamate. *Nature Medicine* 18 (2): 302–306.
43 Krug, J.R. (2020). Higher, better, faster!? Ultra-high field magnetic resonance imaging and spatially resolved spectroscopy. PhD thesis. Wageningen University.
44 Melkus, G., Rolletschek, H., Fuchs, J. et al. (2011). Dynamic $^{13}C/^{1}H$ NMR imaging uncovers sugar allocation in the living seed. *Plant Biotechnology Journal* 9 (9): 1022–1037.
45 Wolf, K., van der Toorn, A., Hartmann, K. et al. (2000). Metabolite monitoring in plants with double-quantum filtered chemical shift imaging. *Journal of Experimental Botany* 51 (353): 2109–2117.
46 Roos, Y.H. (1999). *Water Management in the Design and Distribution Quality of Foods*. CRC Press.
47 Guillard, V., Bourlieu, C., and Gontard, N. (2013). *Food Structure and Moisture Transfer: A Modeling Approach*. Springer Science & Business Media.
48 Guillard, V., Bourlieu, C., and Gontard, N. (2013). Relationship between multiscale food structure and moisture transfer properties. In: *Food Structure and Moisture Transfer*, 35–44. Springer.
49 Koizumi, M. and Kano, H. (2016). Water status in differently watered gelatinized rice (α-rice), dried rice grains after boiling and steaming, observed by micro-magnetic

resonance imaging. *American Journal of Food Science and Nutrition Research* 3 (6): 154.
50 Mohorič, A., Vergeldt, F.J., Gerkema, E. et al. (2004). Magnetic resonance imaging of single rice kernels during cooking. *Journal of Magnetic Resonance* 171 (1): 157–162.
51 Mohorič, A., Vergeldt, F.J., Gerkema, E. et al. (2009). The effect of rice kernel microstructure on cooking behaviour: A combined μ-CT and MRI study. *Food Chemistry* 115 (4): 1491–1499.
52 Bouhrara, M., Clerjon, S., Damez, J.-L. et al. (2011). Dynamic MRI and thermal simulation to interpret deformation and water transfer in meat during heating. *Journal of Agricultural and Food Chemistry* 59 (4): 1229–1235.
53 Sekiyama, Y., Horigane, A.K., Ono, H. et al. (2012). T2 distribution of boiled dry spaghetti measured by MRI and its internal structure observed by fluorescence microscopy. *Food Research International* 48 (2): 374–379.
54 Bernin, D., Steglich, T., Röding, M. et al. Multi-scale characterization of pasta during cooking using microscopy and real-time magnetic resonance imaging. *Food Research International* 66: 132–139.
55 Marti, A., Ragg, E.M., and Pagani, M.A. (2018). Effect of processing conditions on water mobility and cooking quality of gluten-free pasta. A magnetic resonance imaging study. *Food Chemistry* 266: 17–23.
56 Zhang, L. and McCarthy, M.J. (2013). NMR study of hydration of navy bean during cooking. *LWT – Food Science and Technology* 53 (2): 402–408.
57 Xia, K., Xu, W., Huang, L. et al. (2018). Water dynamics of turbot flesh during frying, boiling, and stewing processes and its relationship with color and texture properties: Low-field NMR and MRI studies. *Journal of Food Processing and Preservation* 42 (1): e13338.
58 Sangpring, Y., Fukuoka, M., and Ratanasumawong, S. (2015). The effect of sodium chloride on microstructure, water migration, and texture of rice noodle. *LWT – Food Science and Technology* 64 (2): 1107–1113.
59 Horigane, A.K., Suzuki, K., and Yoshida, M. (2013). Moisture distribution of soaked rice grains observed by magnetic resonance imaging and physicochemical properties of cooked rice grains. *Journal of Cereal Science* 57 (1): 47–55.
60 Horigane, A.K., Suzuki, K., and Yoshida, M. (2014). Moisture distribution in rice grains used for sake brewing analyzed by magnetic resonance imaging. *Journal of Cereal Science* 60 (1): 193–201.
61 Tong, L.-T., Zhu, R., Zhou, X. et al. Soaking time of rice in semidry flour milling was shortened by increasing the grains cracks. *Journal of Cereal Science* 74: 121–126.
62 Voda, A., Homan, N., Witek, M. et al. (2012). The impact of freeze-drying on microstructure and rehydration properties of carrot. *Food Research International* 49 (2): 687–693.
63 Geng, S., Wang, H., Wang, X. et al. (2015). A non-invasive NMR and MRI method to analyze the rehydration of dried sea cucumber. *Analytical Methods* 7 (6): 2413–2419.
64 Altan, A., Oztop, M.H., McCarthy, K.L. et al. (2011). Monitoring changes in feta cheese during brining by magnetic resonance imaging and NMR relaxometry. *Journal of Food Engineering* 107 (2): 200–207.
65 Gudjónsdóttir, M., Traoré, A., Jónsson, Á. et al. The effects of pre-salting methods on salt and water distribution of heavily salted cod, as analyzed by ^{1}H and ^{23}Na MRI, ^{23}Na NMR, low-field NMR and physicochemical analysis. *Food Chemistry* 188: 664–672.
66 Manzocco, L., Anese, M., Marzona, S. et al. (2013). Monitoring dry-curing of S. Daniele ham by magnetic resonance imaging. *Food Chemistry*. 141 (3): 2246–2252.

67 Xu, J.X., Zhang, L. et al. Investigation on water status and distribution in broccoli and the effects of drying on water status using NMR and MRI methods. *Food Research International* 96: 191–197.

68 Jin, X., van Boxtel,A.J.B., Gerkema, E. et al. (2012). Anomalies in moisture transport during broccoli drying monitored by MRI? *Faraday Discussions* 158 (0): 65–75.

69 Kamal, T., Song, Y., Tan, Z. et al. (2019). Effect of hot-air oven dehydration process on water dynamics and microstructure of apple (Fuji) cultivar slices assessed by LF-NMR and MRI. *Drying Technology* 37 (15): 1974–1987.

70 Su, D., Lv, W., Wang, Y. et al. (2020). Drying characteristics and water dynamics during microwave hot-air flow rolling drying of *Pleurotus eryngii*. *Drying Technology* 38 (11): 1493–1504.

71 Cheng, S., Li, R., Yang, H. et al. (2020). Water status and distribution in shiitake mushroom and the effects of drying on water dynamics assessed by LF-NMR and MRI. *Drying Technology* 38 (8): 1001–1010.

72 Cheng, S., Zhang, T., Yao, L. et al. (2018). Use of low-field-NMR and MRI to characterize water mobility and distribution in pacific oyster (*Crassostrea gigas*) during drying process. *Drying Technology* 36 (5): 630–636.

73 Adiletta, G., Iannone, G., Russo, P. et al. (2014). Moisture migration by magnetic resonance imaging during eggplant drying: A preliminary study. *International Journal of Food Science and Technology* 49 (12): 2602–2609.

74 Cheng, S., Tang, Y., Zhang, T. et al. (2018). Approach for monitoring the dynamic states of water in shrimp during drying process with LF-NMR and MRI. *Drying Technology* 36 (7): 841–848.

75 Lv, W., Zhang, M., Wang, Y. et al. (2018). Online measurement of moisture content, moisture distribution, and state of water in corn kernels during microwave vacuum drying using novel smart NMR/MRI detection system. *Drying Technology* 36 (13): 1592–1602.

76 Hu, X., Wu, P., Zhang, S. et al. (2019). Moisture conversion and migration in single-wheat kernel during isothermal drying process by LF-NMR. *Drying Technology* 37 (7): 803–812.

77 Wang, L., Xu, B., Wei, B. et al. Low frequency ultrasound pretreatment of carrot slices: Effect on the moisture migration and quality attributes by intermediate-wave infrared radiation drying. *Ultrasonics Sonochemistry* 40: 619–628.

78 Song, Y., Zang, X., Kamal, T. et al. (2018). Real-time detection of water dynamics in abalone (*Haliotis discus hannai* Ino) during drying and rehydration processes assessed by LF-NMR and MRI. *Drying Technology* 36 (1): 72–83.

79 Wang, S., Lin, Z., Xia, K. et al. (2017). Dynamics of water mobility and distribution in Sur clam (*Mactra chinensis*) during dehydration and rehydration processes assessed by low-field NMR and MRI. *Journal of Food Measurement and Characterization* 11 (3): 1342–1354.

80 Cheng, S., Wang, X., Li, R. et al. (2019). Influence of multiple freeze-thaw cycles on quality characteristics of beef semimembranous muscle: With emphasis on water status and distribution by LF-NMR and MRI. *Meat Science* 147: 44–52.

81 Mulas, G., Anedda, R., Longo, D.L. et al. (2016). An MRI method for monitoring the ripening of Grana Padano cheese. *International Dairy Journal* 52: 19–25.

82 Li, Y., Shi, W., Cheng, S. et al. (2017). Freezing-induced proton dynamics in tofu evaluated by low-field nuclear magnetic resonance. *Journal of Food Measurement and Characterization* 11 (3): 1003–1010.

83 Bajd, F. and Serša, I. (2011). Continuous monitoring of dough fermentation and bread baking by magnetic resonance microscopy. *Magnetic Resonance Imaging* 29 (3): 434–442.

84 Pojić, M., Musse, M., Rondeau, C. et al. (2016). Overall and local bread expansion, mechanical properties, and molecular structure during bread baking: Effect of emulsifying starches. *Food and Bioprocess Technology* 9 (8): 1287–1305.
85 Clerjon, S., Kondjoyan, A., Bonny, J.M. et al. (2012). Oil uptake by beef during pan frying: Impact on fatty acid composition. *Meat Science* 91 (1): 79–87.
86 Yang, D., Wu, G., Li, P. et al. (2019). Comparative analysis of the oil absorption behavior and microstructural changes of fresh and pre-frozen potato strips during frying via MRI, SEM, and XRD. *Food Research International* 122: 295–302.
87 Jiang, Y., Jin, X., Xu, F. et al. (2019). Quantifying food drying rates from NMR/MRI experiments: Development of an online calibration system. *Drying Technology* 37 (16): 2047–2058.
88 Hong, Y.-S., Cho, J.-H., Kim, N.-R. et al. (2009). Artifacts in the measurement of water distribution in soybeans using MR imaging. *Food Chemistry* 112 (1): 267–272.
89 Li, X., Wang, H., Mehmood, W. et al. (2018). Effect of storage temperatures on the moisture migration and microstructure of beef. *Journal of Food Quality* 2018: Article ID 3873179.
90 Song, Y., Cheng, S., Wang, H. et al. (2018). Variable temperature nuclear magnetic resonance and magnetic resonance imaging system as a novel technique for in situ monitoring of food phase transition. *Journal of Agricultural and Food Chemistry* 66 (3): 740–747.
91 Labuza, T.P. and Hyman, C. (1998). Moisture migration and control in multi-domain foods. *Trends in Food Science & Technology* 9 (2): 47–55.
92 Kilara, A. and Sengupta, T. (2014). Multi-textured foods. In: *Food Texture Design and Optimization* (ed. Y. Lal Da and J.M. Light).
93 Roudaut, G. and Debeaufort, F. (2011). 3 - Moisture loss, gain and migration in foods. In: *Food and Beverage Stability and Shelf Life* (ed. D. Kilcast and P. Subramaniam). Woodhead Publishing.
94 Delbaere, C., van de Walle, D., Depypere, F. et al. (2016). Relationship between chocolate microstructure, oil migration, and fat bloom in filled chocolates. *European Journal of Lipid Science and Technology* 118 (12): 1800–1826.
95 Maleky, F. (2018). Oil migration through fats – Quantification and its relationship to structure. In: *Structure–Function Analysis of Edible Fats*, 241–266. Elsevier.
96 Esveld, D.C., van der Sman, R.G.M., Witek, M.M. et al. (2012). Effect of morphology on water sorption in cellular solid foods. *Part II: Sorption in Cereal Crackers. Journal of Food Engineering.* 109 (2): 311–320.
97 Secchi, N., Stara, G., Anedda, R. et al. (2011). Effectiveness of sweet ovine whey powder in increasing the shelf life of Amaretti cookies. *LWT – Food Science and Technology* 44 (4): 1073–1078.
98 Gulati, T., Datta, A.K., Doona, C.J. et al. Modeling moisture migration in a multi-domain food system: Application to storage of a sandwich system. *Food Research International* 76: 427–438.
99 Kamal, T., Song, Y., Zhang, T. et al. Effect of hydrocolloid and processing potentiality on water migration in apple jellies of Yinduqing cultivar. *LWT – Food Science and Technology* 98: 381–389.
100 Li, N., Shen, Y., Liu, W. et al. (2018). Low-field NMR and MRI to analyze the effect of edible coating incorporated with map on qualities of half-smooth tongue sole (*Cynoglossus semilaevis* Günther) fillets during refrigerated storage. *Applied Sciences* 8 (8): 1391.
101 Paluri, S., Shavezipur, M., Heldman, D.R. et al. (2015). Analysis of moisture diffusion mechanism in structured lipids using magnetic resonance imaging. *RSC Advances* 5 (94): 76904–76911.

102 Lee, W.L., McCarthy, M.J., and McCarthy, K.L. (2010). Oil migration in 2-component confectionery systems. *Journal of Food Science* 75 (1): E83–E89.

103 Maleky, F., McCarthy, K.L., McCarthy, M.J. et al. (2012). Effect of cocoa butter structure on oil migration. *Journal of Food Science* 77 (3): E74–E9.

104 Altan, A., Lavenson, D.M., McCarthy, M.J. et al. (2011). Oil migration in chocolate and almond product confectionery systems. *Journal of Food Science* 76 (6): E489–E94.

105 Cikrikci, S. and Oztop, M.H. (2018). Oil migration in hazelnut paste/chocolate systems using magnetic resonance imaging. *Journal of Food Measurement and Characterization* 12 (3): 1460–1472.

106 Cikrikci, S. and Oztop, M.H. (2017). Mathematical modeling and use of magnetic resonance imaging (MRI) for oil migration in chocolate confectionery systems. *Food Engineering Reviews* 9 (1): 50–70.

107 Rumsey, T.R. and McCarthy, K.L. (2012). Modeling oil migration in two-layer chocolate–almond confectionery products. *Journal of Food Engineering* 111 (1): 149–155.

108 Wang, H. and Maleky, F. (2018). Effects of cocoa butter triacylglycerides and minor compounds on oil migration. *Food Research International* 106: 213–224.

109 Wang, H., Shi, X., Paluri, S. et al. (2016). Effects of processing and added ingredients on oil diffusion through cocoa butter using magnetic resonance imaging. *RSC Advances* 6 (91): 88498–88507.

110 Choi, Y.J., McCarthy, K.L., and McCarthy, M.J. (2005). Oil migration in a chocolate confectionery system evaluated by magnetic resonance imaging. *Journal of Food Science* 70 (5): E312–E317.

111 Isik, B., Sahin, S., and Oztop, M.H. (2018). Determination of oil and moisture distribution in fried potatoes using magnetic resonance imaging. *Journal of Food Process Engineering* 41 (6): e12814.

112 Coussot, P. (2020). Progress in rheology and hydrodynamics allowed by NMR or MRI techniques. *Experiments in Fluids* 61 (9): 1–20.

113 de Kort, D.W., Nikolaeva, T., and Dijksman, J.A. (2018). Rheo-NMR: Applications to food. In: *Modern Magnetic Resonance* (ed. G.A. Webb), 1589–1608. Cham, Switzerland: Springer International Publishing.

114 Marangoni, A.G., van Duynhoven, J.P., Acevedo, N.C. et al. (2020). Advances in our understanding of the structure and functionality of edible fats and fat mimetics. *Soft Matter* 16 (2): 289–306.

115 Singla, N., Verma, P., Ghoshal, G. et al. (2013). Steady state and time dependent rheological behaviour of mayonnaise (egg and eggless). *International Food Research Journal* 20 (4): 2009.

116 Raudsepp, A., Feindel, K.W., and Hemar, Y. (2010). Shear localisation in stirred yoghurt. *Rheologica Acta* 49 (4): 371–379.

117 Ovarlez, G., Bertrand, F., and Rodts, S. (2006). Local determination of the constitutive law of a dense suspension of noncolloidal particles through magnetic resonance imaging. *Journal of Rheology* 50 (3): 259–292.

118 Coussot, P., Tocquer, L., Lanos, C. et al. (2009). Macroscopic vs local rheology of yield stress fluids. *Journal of Non-Newtonian Fluid Mechanics*. 158 (1–3): 85–90.

119 Ovarlez, G., Rodts, S., Ragouilliaux, A. et al. (2008). Wide-gap Couette flows of dense emulsions: Local concentration measurements, and comparison between macroscopic and local constitutive law measurements through magnetic resonance imaging. *Physical Review E* 78 (3): 036307.

120 Mendes, R., Vinay, G., Ovarlez, G. et al. (2015). Reversible and irreversible destructuring flow in waxy oils: An MRI study. *Journal of Non-Newtonian Fluid Mechanics* 220: 77–86.

121 Nikolaeva, T., Vergeldt, F.J., Serial, R. et al. (2020). High field microMRI velocimetric measurement of quantitative local flow curves. *Analytical Chemistry* 92 (6): 4193–4200.

122 Nikolaeva, T., den Adel, R., van der Sman, R. et al. (2019). Manipulation of recrystallization and network formation of oil-dispersed micronized fat crystals. *Langmuir* 35 (6): 2221–2229.

123 Nikolaeva, T., den Adel, R., Velichko, E. et al. (2018). Networks of micronized fat crystals grown under static conditions. *Food & Function* 9 (4): 2102–2111.

124 Serial, M.R., Nikolaeva, T., Vergeldt, F.J. et al. (2019). Selective oil-phase rheo-MRI velocity profiles to monitor heterogeneous flow behavior of oil/water food emulsions. *Magnetic Resonance in Chemistry* 57 (9): 766–770.

126 Serial, M., Bonn, D., Huppertz, T., Dijksman, J., van der Gucht, J., van Duynhoven, J., Terenzi, C. (2021). Non-local effects in the shear banding of a thixotropic yield stress fluid. *Physical Review Fluids* 6: 113–301.

125 Gholami, M., Rashedi, A., Lenoir, N. et al. (2018). Time-resolved 2D concentration maps in flowing suspensions using X-ray. *Journal of Rheology* 62 (4): 955–974.

127 Fullard, L., Holland, D.J., Galvosas, P. et al. (2019). Quantifying silo flow using MRI velocimetry for testing granular flow models. *Physical Review Fluids* 4 (7): 074302.

128 Jop, P., Forterre, Y., and Pouliquen, O. (2006). A constitutive law for dense granular flows. *Nature* 441 (7094): 727–730.

129 Eringen, A.C. (2002). *Nonlocal Continuum Field Theories*. Springer Science & Business Media.

130 de Cagny, H., Fall, A., Denn, M.M. et al. (2015). Local rheology of suspensions and dry granular materials. *Journal of Rheology* 59 (4): 957–969.

131 Bocquet, L., Colin, A., and Ajdari, A. (2009). Kinetic theory of plastic flow in soft glassy materials. *Physical Review Letters* 103 (3): 036001.

132 Goyon, J., Colin, A., Ovarlez, G. et al. (2008). Spatial cooperativity in soft glassy flows. *Nature* 454 (7200): 84–87.

133 Goyon, J., Colin, A., and Bocquet, L. (2010). How does a soft glassy material flow: Finite size effects, non local rheology, and flow cooperativity. *Soft Matter* 6 (12): 2668–2678.

134 de Kort, D.W., Veen, S.J., Van As, H. et al. (2016). Yielding and flow of cellulose microfibril dispersions in the presence of a charged polymer. *Soft Matter* 12 (21): 4739–4744.

135 Moller, P., Fall, A., Chikkadi, V. et al. (2009). An attempt to categorize yield stress fluid behaviour. *Philosophical Transactions of the Royal Society A: Mathematical Physical and Engineering Sciences* 367 (1909): 5139–5155.

136 Rohart, A., Sieffermann, J.-M., and Michon, C. (2015). Effect of micro-gel shape and concentration on sensory perception of micro-gels-enriched stirred yoghurts. *Colloids and Surfaces A: Physicochemical and Engineering Aspects* 475: 94–102.

137 Ovarlez, G., Cohen-Addad, S., Krishan, K. et al. (2013). On the existence of a simple yield stress fluid behavior. *Journal of Non-Newtonian Fluid Mechanics* 193: 68–79.

138 de Kort, D.W., Hertel, S.A., Appel, M. et al. (2019). Under-sampling and compressed sensing of 3D spatially-resolved displacement propagators in porous media using APGSTE-RARE MRI. *Magnetic Resonance Imaging* 56: 24–31.

139 Penn, A., Tsui, T., Brunner, D.O. et al. (2017). Real-time probing of granular dynamics with magnetic resonance imaging. *Science Advances* 3: e1701879.

140 Varga, Z., Grenard, V., Pecorario, S. et al. (2019). Hydrodynamics control shear-induced pattern formation in attractive suspensions. *Proceedings of the National Academy of Sciences* 116 (25): 12193–12198.

141 Krintiras, G.A., Diaz, J.G., Van der Goot, A.J. et al. (2016). On the use of the Couette cell technology for large scale production of textured soy-based meat replacers. *Journal of Food Engineering* 169: 205–213.

142 Krintiras, G.A., Göbel, J., Van der Goot, A.J. et al. (2015). Production of structured soy-based meat analogues using simple shear and heat in a Couette cell. *Journal of Food Engineering* 160: 34–41.

143 Sandoval-Castilla, O., Lobato-Calleros, C., García-Galindo, H. et al. (2010). Textural properties of alginate–pectin beads and survivability of entrapped *Lb. casei* in simulated gastrointestinal conditions and in yoghurt. *Food Research International* 43 (1): 111–117.

144 Dekkers, B.L., Boom, R.M., and van der Goot, A.J. (2018). Structuring processes for meat analogues. *Trends in Food Science & Technology* 81: 25–36.

145 Laryea, E., Schuhardt, N., Guthausen, G. et al. (2018). Construction of a temperature controlled Rheo-NMR measuring cell – Influence of fluid dynamics on PMMA-polymerization kinetics. *Microporous and Mesoporous Materials* 269: 65–70.

146 Duvarci, O.C., Yazar, G., and Kokini, J.L. (2017). The comparison of LAOS behavior of structured food materials (suspensions, emulsions and elastic networks). *Trends in Food Science & Technology* 60: 2–11.

147 Kuczera, S., Gentile, L., Brox, T.I. et al. (2018). Multilamellar vesicle formation probed by Rheo-NMR and Rheo-SALS under large amplitude oscillatory shear. *Langmuir* 34 (28): 8314–8325.

148 Galvosas, P., Brox, T.I., and Kuczera, S. (2019). Rheo-NMR in food science – Recent opportunities. *Magnetic Resonance in Chemistry* 57 (9): 757–765.

149 Ratzsch, K.-F., Friedrich, C., and Wilhelm, M. (2017). Low-field rheo-NMR: A novel combination of NMR relaxometry with high end shear rheology. *Journal of Rheology* 61 (5): 905–917.

150 Colbourne, A., Blythe, T., Barua, R. et al. (2018). Validation of a low field Rheo-NMR instrument and application to shear-induced migration of suspended non-colloidal particles in Couette flow. *Journal of Magnetic Resonance* 286: 30–35.

Part III

Applications in Life Sciences

12

MRI of Single Cells Labeled with Superparamagnetic Iron Oxide Nanoparticles

Cornelius Faber

Translational Research Imaging Center, Clinic of Radiology, University of Münster, Münster, Germany

12.1 Introduction

The immense progress in biomedical research during recent decades has opened the door toward personalized medicine and treating disease by applying engineered cells that have been adapted to the patient's genetic and immunological profile. Widespread application of such cellular therapies is currently hampered by several technical issues, including the lack of powerful methods to monitor effects and fate of administered cells. A method that allows for noninvasively detecting and tracking single cells through a patient's body may be considered as the holy grail of diagnostics in personalized medicine. Monitoring therapeutic cells, detecting metastasizing tumor cells, identifying small bacterial vegetation, and following migrating immune cells in the body hold great promise for earliest diagnosis and dramatically improved therapeutic options. Among the variety of noninvasive imaging modalities, magnetic resonance imaging (MRI) stands out, due to its capability of resolving individual cells at reasonable temporal resolution, while at the same time providing image information of the surrounding tissue [1,2]. Alternative methods such as positron emission tomography (PET) or single-photon emission computed tomography (SPECT) afford better detection sensitivity than MRI, but suffer from poor spatial and temporal resolution. Radioisotope-based methods are further not suitable for long-term cell tracking, due to decaying radioactivity of the label. X-ray-based methods such as computed tomography (CT) require exorbitantly high doses of ionizing radiation to achieve the required microscopic resolution in longitudinal studies. Fluorescence and bioluminescence imaging are subject to pronounced sensitivity losses in deep tissue, and therefore only applicable for cell tracking in organisms up to the size of a mouse. Bioluminescence, in particular, requires genetic methods, which imposes a major obstacle for use in patients. Ultrasound imaging using microbubbles for cellular labeling provides high sensitivity, but is limited in penetration depth at the frequencies required for microscopic resolution in the images. Yet, current progress in the methodology of so-called functional ultrasound (fUS) may render this method a future tool for cell tracking [3]. In parallel to the developments of fUS, photoacoustic tomography (PAT, also called multispectral optoacoustic tomography – MSOT) has recently emerged as powerful option for cellular imaging [4]. The temporal resolution of this modality is outstanding. However, intrinsic soft-tissue contrast, penetration depth, and whole-body coverage cannot yet compete with MR microscopy.

Magnetic Resonance Microscopy: Instrumentation and Applications in Engineering, Life Science and Energy Research, First Edition. Edited by Sabina Haber-Pohlmeier, Bernhard Blümich, and Luisa Ciobanu.

12.2 Strategies for Detecting Cells by MRI

MR microscopy has previously been shown to be capable of resolving individual cells and even intracellular compartments in large cells [5]. In vivo, spatial resolutions of only a few tens of microns have been reported. Yet the direct imaging of individual cells in biomedical applications remains challenging for two reasons. First, it is obvious that in vivo MRI does not reach sufficiently high resolution to directly resolve individual cells. Mammalian cells have a diameter $<20\ \mu m$, most are actually substantially smaller, and therefore just below the feasible resolution in vivo. Second, exploiting differences in physical parameters between different cell types, or cell and interstitium, requires elaborate MR sequences that do not allow the ultimate resolution in feasible scan times. It is therefore mandatory to employ efficient labeling and signal amplification strategies for signal acquisition.

For this purpose, two fundamentally different cell labeling and detection approaches are possible [2]. First, similar to optical imaging or radioisotope imaging methods, a cellular marker is used for labeling, which can be directly detected against an, ideally, vanishing background signal. Such cellular imaging strategies are often referred to as positive contrast methods and directly detect the signal originating from the cellular label. The most prominent example for this type of approach is fluorine MRI using ^{19}F as the signal-generating moiety, which is not present in the mammalian body (at least not in an MR detectable form), and can be applied in a variety of molecular formulations [6,7]. Most efficient for cell labeling are nano-emulsions of perfluorinated carbon compounds. If administered in combination with suitable paramagnetic components, paramagnetic relaxation enhancement (PRE) can be exploited to amplify fluorine signal and reach optimized detection sensitivity [8,9]. Yet, the detection limit is not sufficient to observe single labeled cells.

The use of hyperpolarized compounds to label cells promises dramatic improvements in detection sensitivity, which in principle renders single-cell detection feasible. However, relaxation times of hyperpolarized substances are short on the cell tracking timescale, and therefore appear unsuitable for tracking single cells in vivo. Alternatively, PRE has been demonstrated to enable positive contrast imaging methods. Highly shifted proton MRI, also called heteronuclear proton MRI, has been shown to allow the detection of labeled cells specifically over several days. Directly detecting the strongly shifted resonance lines of methyl groups of thulium-1,4,7,10-tetra-azacyclododecane-α,α′,α″,α‴-tetramethyl-1,4,7,10-tetraacetic acid (Tm-DOTMA) resulted in a stable and highly specific cellular signal. However, despite low spatial resolution and long scan times, the detection limit for immune cells was estimated to be around several hundred cells in vivo [10].

The second and more popular approach is to label cells with (relaxation-enhancing) contrast agents. Labeled cells are not detected directly, but the effect of relaxation enhancement on the surrounding environment is observed. For this purpose, in principle, either enhancement of longitudinal or transverse magnetization can be employed. However, most often contrast agents acting predominantly on transverse relaxation are used. More specifically, agents are used that exploit enhanced T_2^* relaxation due to strong dipolar fields, originating from the agents and impacting on the surrounding bulk magnetization. Since the resulting magnetic field distortion, which can be converted into image contrast in T_2^* weighted images, greatly exceeds the physical size of the labeled cell, single cells become detectable in the images, even if spatial resolution is far coarser than cellular dimensions. Pivotal for achieving single-cell detection is that sufficient image contrast can be generated from a single cell. This requires that the contrast agent used gives rise to a sufficiently strong magnetic field distortion, which in turn requires a large magnetic moment. How these preconditions can be fulfilled is described in the following section.

12.3 Theoretical Considerations on Image Contrast

As has been mentioned, detection of labeled cells can be based on either longitudinal or transverse relaxation effects. Yet, since the label must be attached to the cellular surface or stored inside the cell, reduced water accessibility and high label density may reduce the efficiency of longitudinal relaxation enhancement. While also transverse relaxation may be attenuated inside the cell (as will be discussed at the end of this section), it is in most cases less affected than longitudinal relaxation, and thus more suitable for detection of single cells. The following considerations will therefore be limited to contrast agents that generate a strong dipolar field around the labeled cell, such as superparamagnetic nanoparticles.

12.3.1 Dipolar Fields and Image Contrast

To demonstrate how contrast is generated by superparamagnetic nanoparticles, we consider an isotropic and homogeneous medium with a labeled spherical cell with volume V at the origin. The magnetic susceptibility differs from the surroundings by $\Delta\chi$. Strictly, this description is only valid for a single homogeneous superparamagnetic particle, but we neglect intracellular distribution of the contrast agent for the moment (for discussion, see the end of this section). According to classical magnetostatics this cell gives rise to a static local dipolar field:

$$B_{\text{dip}}(\boldsymbol{r}) = \frac{VB_0\Delta\chi}{4\pi}\frac{3\cos^2\theta - 1}{r^3}, \tag{12.1}$$

where θ and r are the spherical coordinates of position, B_0 is the main magnetic field of the scanner oriented along $\theta = 0$ (i.e. in z direction). B_{dip} results in a frequency distribution $\Delta\omega(\boldsymbol{r}) = \gamma B_{\text{dip}}(\boldsymbol{r})$ as depicted in Figure 12.1. Along the longitudinal direction, the typical dipolar pattern is observed. Positive frequency shifts are observed in a cone above and below the cell, limited by the magic angle of $\theta = \pm\ 54.7°$, negative frequency shifts are observed laterally (Figure 12.1, upper left). In the transverse plane, only negative frequency shifts are observed (Figure 12.1, upper right). Frequency shifts of more than 10 Hz are limited to the close vicinity of a few hundred micrometers around the labeled cell, for this scenario of a 5-μm cell radius and a susceptibility difference of 5000 ppm, which may be achieved with efficient labeling.

To produce strong image contrast this frequency distribution is very effectively converted into T_2* contrast by gradient echo imaging. In the rotating frame, T_2* contrast can be calculated if any further interaction (diffusion, other relaxation mechanisms, imaging gradients) is neglected, by:

$$\frac{\text{d}\boldsymbol{M}(\boldsymbol{r},t)}{\text{d}t} = \gamma\boldsymbol{M}(\boldsymbol{r},t) \times \frac{\Delta\omega(\boldsymbol{r})}{\gamma}. \tag{12.2}$$

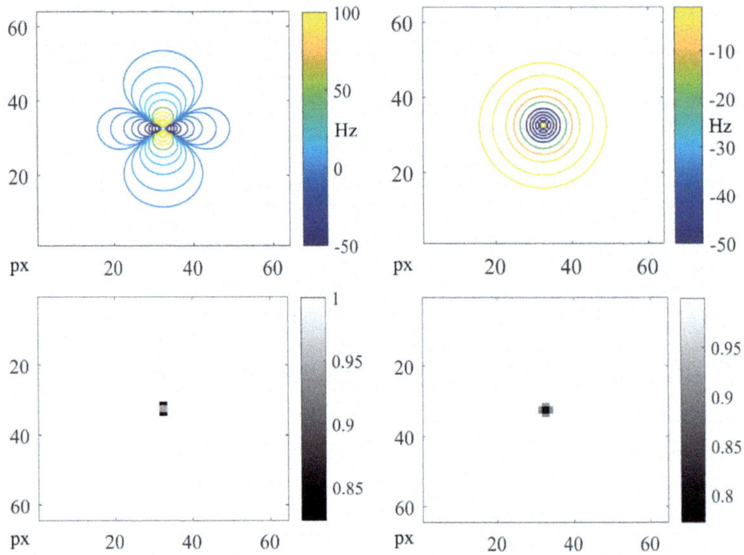

Figure 12.1 Frequency distortion and resulting signal reduction originating from a single labeled cell according to Equations 12.1 and 12.2. For these calculations a spherical cell shape with a radius of 5 μm, a susceptibility difference between cell and surrounding medium of $\Delta\chi$ = 5000 ppm, and a magnetic field B_0 = 9.4 T (along z) was assumed. Top row shows contour plots of B_{dip} originating from the single cell, for the x–z-plane (left) and the x–y-plane (right) of a 64^3 matrix representing a $(3.2\ \text{mm})^3$-volume, i.e. a spatial resolution of 50 μm per pixel. Bottom row shows signal reduction, relative to theoretically maximum available signal, after 1 ms dephasing time for the x–z-plane (left) and the x–y-plane (right), each averaged over three central slices, resulting in a slice thickness of 150 μm. Calculations and plots were performed using MATLAB (Mathworks Inc., Natick, USA). The cell, which is fivefold smaller than the voxel size, was placed in the origin, i.e. in the center between the eight central voxels of the matrix.

In a pictorial description as well as for the purpose of simulations, the sample is divided into small voxels, over which the field variations are assumed to be negligible. These isochromats precess at different frequencies according to Equation 12.2 and, thus, accumulate locally different phases of transverse magnetization. When integrating over a region of the sample, such as for example one pixel in an image, the different phases eventually lead to signal attenuation, equally for positive and negative differences. For our example, assuming a magnetic field of 9.4 T and a dephasing time (echo time) of 1 ms, the results of this simplified approach are illustrated in Figure 12.1 (lower left and lower right), depicting T_2* contrast from a single labeled cell. In the x–z-plane the typical dipole pattern is observed with stronger signal attenuation above and below the cell and less pronounced signal reduction laterally. In the transverse plane signal reduction is symmetrical around the origin. In a real measurement, most importantly diffusion has an additional effect, which further pronounces dephasing and blurs out phase oscillations. Therefore, labeled cells detected by gradient echo MRI sequences appear as signal voids or regions of reduced signal around the position of the cell, and may largely exceed the physical dimensions of the cells. Single labeled cells can thus be visualized, although the spatial resolution of the MR image may be more than an order of magnitude above the diameter of the cell.

This "enlargement" of cells in the images, which is sometimes called the "blooming effect," is exemplified in Figure 12.2, for a cell with 10 μm diameter and image resolutions of 50 μm. According to Equations 12.1 and 12.2, image contrast depends on a number of experimental parameters. First of all, B_0 has a pronounced effect as illustrated in Figure 12.2, which compares contrast from a single labeled cell at 1.5 T (left) and 9.4 T (right). Generally it can be stated that the higher the main magnetic field, the stronger the dipole field will be and the more pronounced the image contrast that can be obtained. While the cell in our simulated example is not detected for an echo time of 1 ms at 1.5 T (Figure 12.2, upper left), a weak but clearly discernible signal reduction is observed at 9.4 T (Figure 12.2, upper right). A second and experimentally easily accessible parameter is the echo time, TE. Longer TE dramatically raises contrast from labeled cells. Increasing TE from 1 ms to 10 ms, the cell becomes detectable at 1.5 T as a small region of reduced signal (Figure 12.2, central left). At 9.4 T, for TE of 10 ms a large signal void is observed over several hundred micrometers (Figure 12.2, central right). In vivo, however, the length of TE is often limited by short T_2* of surrounding tissue, and by increasing image artifacts. Particularly at high B_0, motion or large-scale susceptibility variations will result in signal loss if TE is too long.

A third and often neglected parameter is spatial resolution, both in-plane or in terms of slice thickness. To generate detectable contrast, a labeled cell must dephase a sufficiently large portion of the voxel in the image. This is illustrated in Figure 12.2 (lower row) for increasing slice thickness. While contrast from the cell is pronounced in a 250-μm slice (Figure 12.2, central row), it is substantially reduced in a 1-mm slice (Figure 12.2, lower row). This phenomenon is also observed for in-plane resolution (not shown). If too many spins inside one voxel are not subject to dephasing, signal reduction from the cell is compensated and may eventually vanish. Finally, the achievable contrast is determined by the effective magnetic moment of the labeled cells, which is defined by V and $\Delta\chi$, the product of volume and susceptibility difference. Since cell size cannot be varied experimentally, it is important to maximize $\Delta\chi$. For this purpose, contrast agents with higher magnetization are favorable, and loading of the cell has to be as efficient as possible. Here it has to be noted that the above treatment assumed a homogeneously magnetized cell, which is not given in a real experiment. Superparamagnetic nanoparticles, which are often used for cell labeling, are either attached to the cellular surface or stored in intracellular vesicles. Their distribution is therefore highly inhomogeneous on a microscopic scale. A correct treatment requires a more complex approach including the exact distribution of magnetization and effects of diffusion. Previous studies have shown that local concentration of nanoparticles into a few vesicles may substantially reduce relaxation effects as compared with a homogeneous distribution of the particles [11–14]. Independent of the exact spatial distribution of the particles inside the cell, it can be generally stated that a large magnetic moment of the cell will lead to pronounced image contrast. It is therefore beneficial to use particles with large magnetization and to try to achieve high intracellular loads [15].

12.4 Cell Labeling

One of the milestones in detecting single cells by MRI is the efficient labeling of the cells with an appropriate contrast agent. Superparamagnetic nanoparticles provide sufficiently large magnetic moments and further favorable properties that can be exploited during the labeling procedure. The vast majority of examples have used superparamagnetic iron oxide

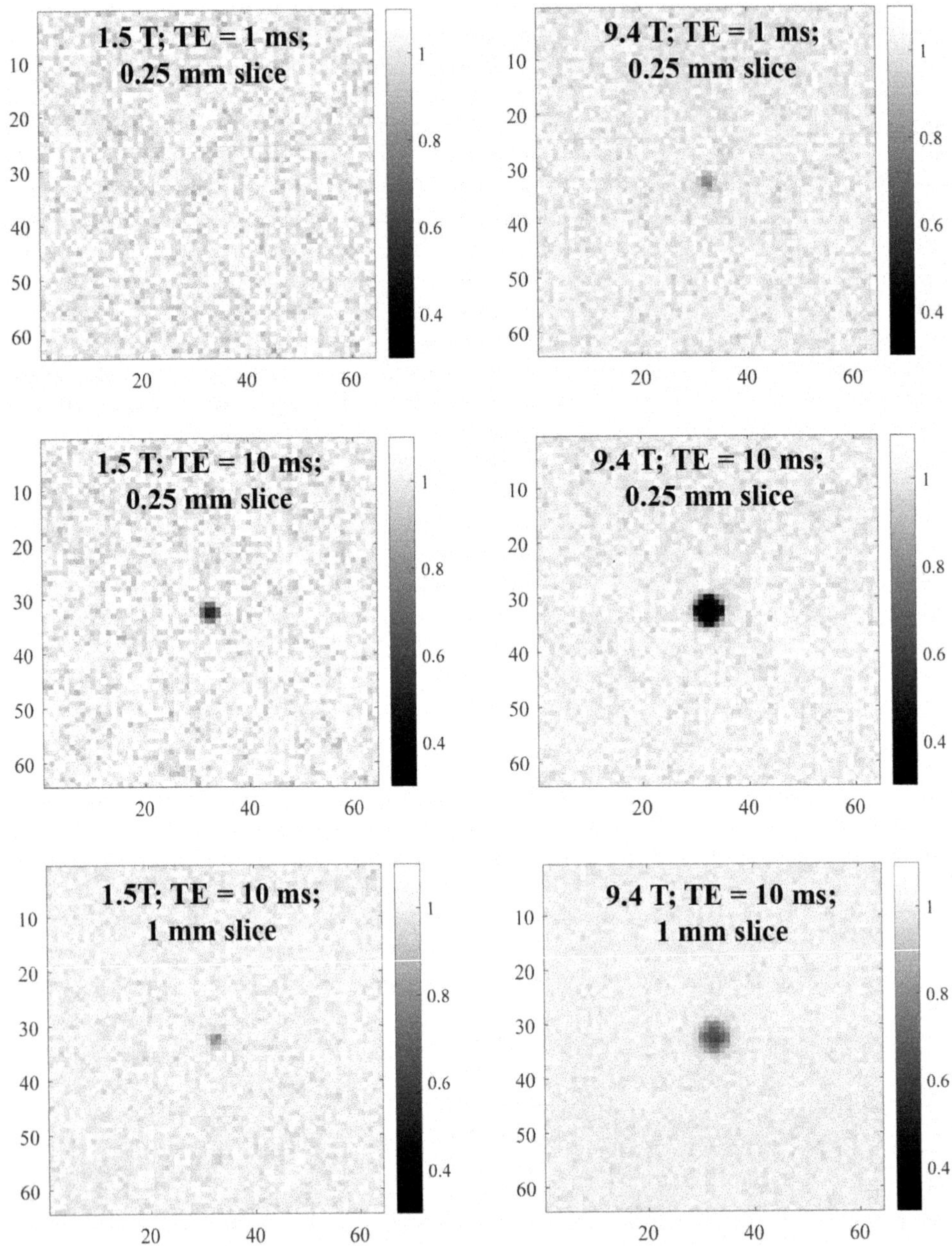

Figure 12.2 Simulation of image contrast of a labeled cell for different magnetic field strengths B_0, echo times TE, and slice thickness. For these calculations a spherical shape with a radius of 5 μm, a susceptibility difference between cell and surrounding medium of $\Delta\chi$ = 5000 ppm, and an in-plane resolution of 50 μm in a 64^3 matrix, representing a $(3.2\ \text{mm})^3$-volume, was assumed. Calculations for B_0 = 1.5 T are shown in the left column, and for B_0 = 9.4 T in the right column. Top row used TE = 1 ms and 250 μm slice thickness, middle row TE = 10 ms and 250 μm slice thickness; bottom row used TE = 10 ms and 1 mm slice thickness. Calculations and plots were performed using MATLAB (Mathworks Inc., Natick, USA). The cell was placed in the origin, i.e. in the center between the eight central voxels of the matrix. Images of 0.25-mm slices and 1-mm slices were obtained by averaging the complex magnetization over 5 and 20 slices, respectively. Images are scaled identically to their average intensity of 1. Gaussian white noise with zero mean and 1% variance for 9.4 T and 3% for 1.5 T was added to the image intensity.

nanoparticles (IONP). These are typically composed of a core consisting of small magnetite/maghemite crystals and a coating layer, such as for example dextrane, that grants stability of the particles and defines surface properties and hydrodynamic radius, which are crucial for interactions and cellular uptake. A large number of IONPs have been developed and were used for cell tracking by MRI. According to their diameter IONPs are often classified (see Table 12.1) as very small iron oxide particles (VSOPs, <10 nm) and ultrasmall iron oxide particles (USPIOs, up to 50 nm); both types have monocrystalline cores and are therefore also referred to as microcrystalline iron oxide nanoparticles (MIONs), superparamagnetic iron oxide particles (SPIOs, >50 nm), and micron-sized iron oxide particles (MPIOs, ~1 μm) [16]. The differently sized particles have been produced with a wide range of surface properties, which defines their suitability for particular applications. Specific physico-chemical properties render particles prone to be taken up by distinct cell types. If the particle surface is decorated with ligand molecules, such as for example antibody fragments, well-defined cellular populations can be labeled with high specificity. Generally, IONPs are taken up in the organism by cells via phagocytosis and pinocytosis, depending on the size and composition of the particles. While larger particles are phagocytosed, mostly by macrophages, smaller particles are taken up by different cell types. Overall, macrophages, monocytes, and neutrophils display a tendency toward the most efficient particle uptake; however, the actual uptake is strongly dependent on physico-chemical properties, on cell type, and also on the physiological status of the cells. In vivo uptake is therefore not easy to predict and may vary strongly with slightly altered experimental conditions. Previously, it has been observed that either very small IONPs (<10 nm) or larger IONPs (>50 nm) lead to more efficient uptake, while IONPs with diameters between roughly 10 and 50 nm appeared less well suited for cell labeling in vivo (see Table 12.1). However, specific surface coating or decoration by molecular patterns may lead to highly efficient uptake of particles of any size via a specific pathway [16].

It must be noted here that not only iron may be used to produce particles with high magnetic moments suitable for cell tracking. Balla et al. have designed ferromagnetic carbon-coated cobalt nanoparticles decorated with antibody fragments and successfully targeted pancreatic islets for single-cell in vivo visualization by high-field MRI [17].

Generally, there are three different approaches to efficient cell labeling:

Table 12.1 Types of iron oxide nanoparticle by size.

	Size (hydrodynamic diameter)	**Core**	**Suitability for cell labeling (cellular uptake)**[e]
VSOP[a]	<10 nm	Single crystal (4 nm)	High (dependent on coating)
USPIO[b]	10–50 nm	Single crystal (4–7 nm)	Low (dependent on coating)
SPIO[c]	>50 nm	Multiple crystals (4–7 nm)	High
MPIO[d]	>0.5 μm	Multiple crystals (varying size)	High

[a] Very small iron oxide particle.
[b] ultrasmall superparamagnetic iron oxide particle.
[c] superparamagnetic iron oxide particle.
[d] micron-sized superparamagnetic iron oxide particle.
[e] cellular uptake is generally strongly dependent on surface coating or decoration.

1. *Unspecific labeling*: IONPs are administered (usually into the bloodstream) in the living animal and subsequently taken up by different cells. This approach is most efficient for labeling (myeloid) immune cells, since phagocytosis is often most efficient.
2. *Targeted approach*: IONPs decorated with specific ligands are administered (usually into the bloodstream) in the living animal and subsequently preferentially taken up by a specific cell type.
3. *Prelabeling*: Cells are incubated in vitro with IONPs (either with unspecific coating or decorated with molecular patterns) and subsequently grafted into the animal. Application can either be performed in the bloodstream or directly into the tissue of interest.

12.5 History of Single-cell MRI

Each of these three labeling approaches has been employed successfully to resolve single cells by MRI. The first systematic studies, resolving single cells, were performed in vitro and used fluorescence microscopy to validate that it was actually single cells that were giving rise to the observed contrast [18–20]. The blooming effect of the IONP label was efficiently exploited to resolve single cells even at a clinical field strength of 1.5 T [21]. Based on theoretical work [22] and on experimental parameter optimization [23] the detection limit in vivo was gradually lowered from 500 cells [24], to well below 100 cells [25], to finally single-cell detection in the mouse brain [26,27]. Despite the substantial progress in MR instrumentation and methodology, in vivo single-cell detection is still a demanding task, which requires solid experience and experimental efforts [28–30]. With efficiently labeled cells, which for immune cells can be achieved both in vivo and in vitro, substantial signal reductions in the voxel containing the cell can be reached in reasonable scan times using high-resolution imaging protocols. Figure 12.3 (upper row) shows an example of monocytes that have been labeled in vitro with Resovist® IONPs, and were subsequently embedded in agar gel. A multi-slice 2D gradient echo sequence with roughly 60 μm resolution in a 300-μm slice enabled detection of single cells in only 8 min scan time. As illustrated by the line profile, signal in the central voxel dropped to about 35% [30]. Among several parameters, the attainable contrast in single-cell MRI depends on the amount of iron that can be loaded into the cell of interest. Most of the examples above have visualized myeloid cells, stem cells, or tumor cells, which all have diameters of ~10 μm or above. Efforts to detect very small cells such as bacteria with a diameter of <1 μm have not been successful, as of today. Hoerr et al. showed that *Staphylococcus aureus* can be labeled with 0.2 pg iron per cell using VSOPs. This iron load enabled detection of about 10^4 bacteria, which was sufficient to track migration of bacteria in living mice [31]. For monocytes, on the other hand, single-cell detection can also be achieved in vivo. As shown in Figure 12.3 (lower row), single monocytes become visible after in vivo labeling with Resovist® in a gradient echo image acquired in 8 min scan time. Also in vivo, the observed signal drop was 60–70%, and allowed for ready identification of single monocytes in the mouse brain [30].

12.6 Moving Cells

12.6.1 Impact of MR Data Acquisition

For static labeled cells that do not move during acquisition of the MR image, contrast can be estimated according to the above-mentioned treatment. In vivo, however, labeled cells may be

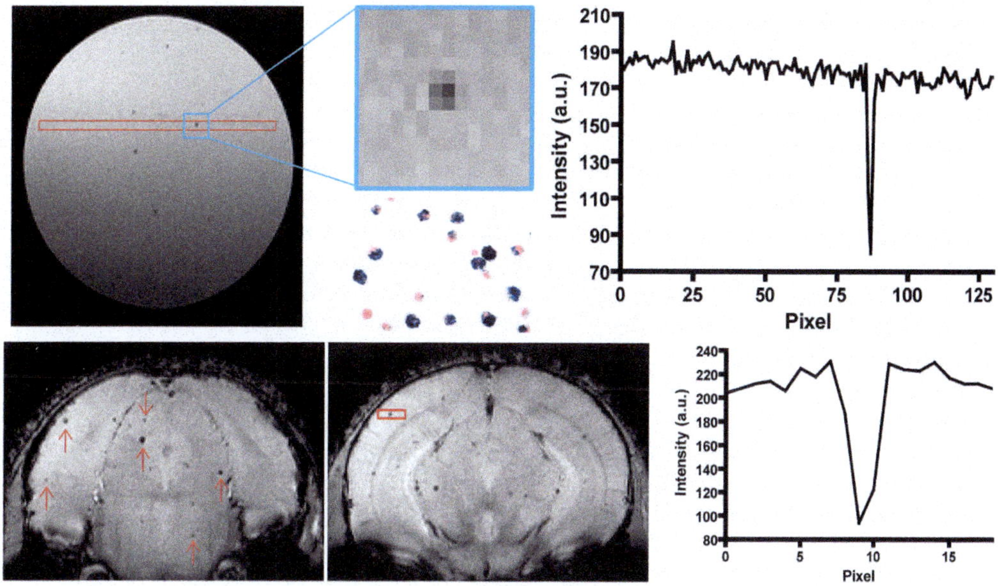

Figure 12.3 Magnetic resonance imaging (MRI) detection of single cells in vitro and in vivo. Top row: HoxB8 monocytes were labeled with Resovist® and embedded in an agarose gel phantom in a 1.5-ml tube. Labeling efficiency of the protocol using 100 μg Fe/ml can be appreciated in the central inset (40× magnification) by blue color indicating iron in the Prussian blue stain. The zoomed image shows details of the observed contrast pattern in the MR image. The intensity profile on the right was taken from the box indicated in the image on the left, and shows that the signal drops by about 60–70% at the position of the labeled cell. Lower row: in vivo MRI of the mouse brain after in vivo labeling of immune cells. Several hypointense spots (exemplarily marked with red arrows) identify labeled cells. Note that appearance of the central hypointensity in the left image closely resembles simulations in Figure 12.2. The intensity profile on the right was taken from the box indicated in the central image, and shows that also in vivo a signal drop by about 60–70% at the position of the labeled cell is observed. Both in vitro and in vivo MRI were acquired at 9.4 T using a T_2*-weighted gradient echo sequence with the following scan parameters: TR: 649 ms, TE: 8.0 ms, FA: 60°, averages: 4, matrix: 180 × 256, in-plane resolution: 61 × 55 μm^2, 38 contiguous slices, slice thickness: 300 μm, scan time: 8 min 12 s (single timeframe). Figure modified from Masthoff et al. [30], figure 01 & 03[p.03 & 04]/Springer Nature/CC BY 4.0).

subject to motion, for example migrating immune cells, homing stem cells, spreading tumor cells, or moving bacteria. Early reports on migrating cells observed motion that was slow in comparison with MR acquisition time. Cells were considered static during image acquisition and motion was deduced from a series of recordings taken over several days [24,32].

Real-time capture of cells in motion was for the first time realized by Mori et al. in 2014 [28]. In their seminal paper they implemented the concept of time-lapse imaging for MRI cell tracking. While the individual cells were still considered as static during the image acquisition, scan time was reduced to 18 min for whole-brain coverage in the mouse, and scans were continued for many hours. Subsequently images were concatenated to yield movies of migrating cells in the brain. Masthoff et al. have further refined image acquisition and achieved whole-brain time-lapse imaging of the brain with 8-min frames [30], resolving slowly migrating cells inside the brain vasculature (Figure 12.4).

Besides showing that time-lapse MRI can be performed either with prelabeled and grafted monocytes, or with in vivo labeled cells, Masthoff et al. also showed that time-lapse MRI is a

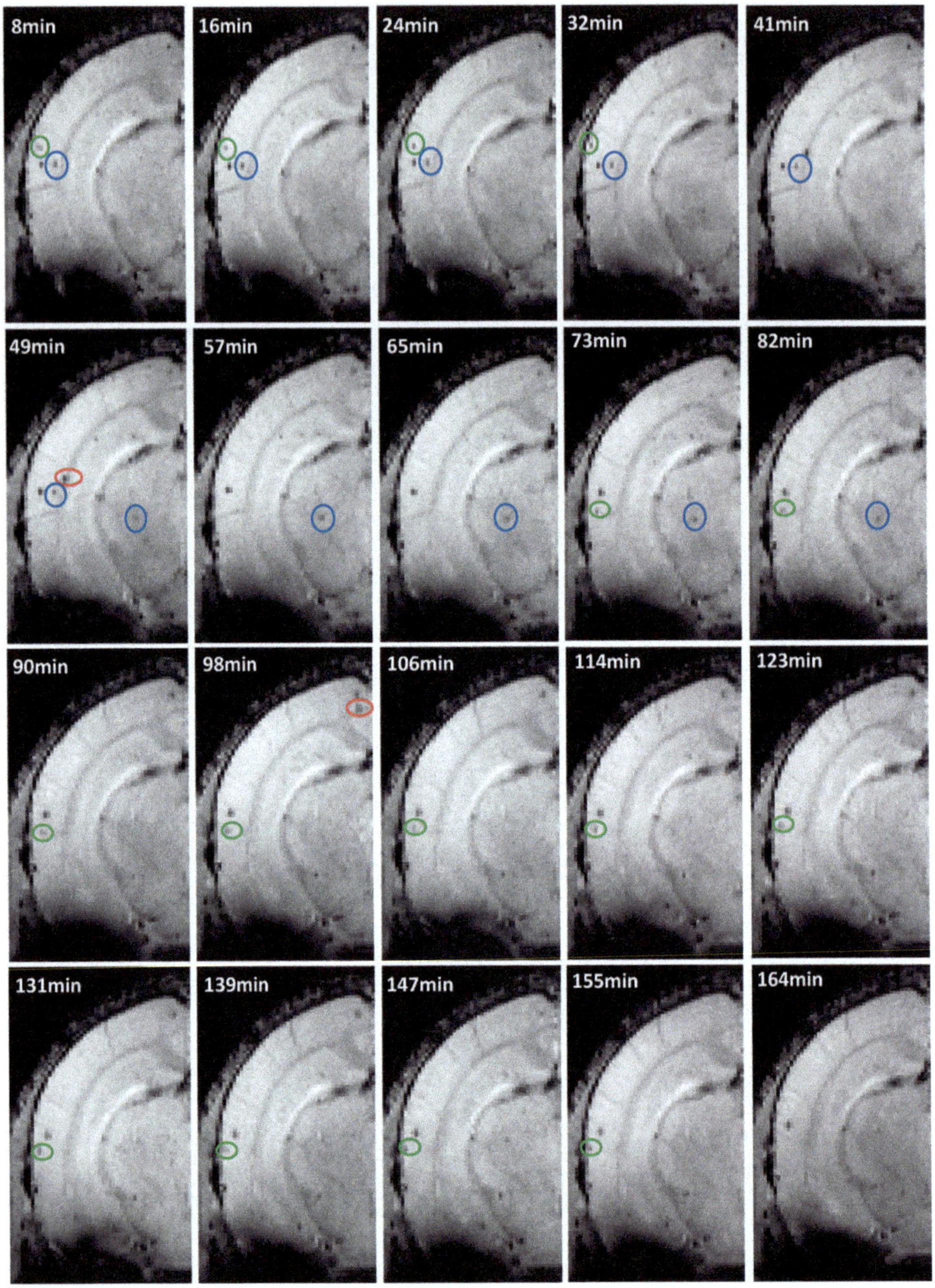

Figure 12.4 In vivo detected cells can be classified according to their velocities. Example of a time series of frames acquired during a time-lapse magnetic resonance imaging (MRI) measurement, showing one hemisphere of a mouse brain. Twenty 8 min timeframes of one slice of a naive mouse brain resolve single labeled cells as hypointense spots with different motion behavior. Time-lapse MRI is able to detect cells that appear and disappear during the measurement and are therefore only visible in one timeframe (red elongated ellipsoids), cells that remain static over many timeframes (blue circles), and cells that move from pixel to pixel on consecutive frames (green ellipsoids). Figure modified from Masthoff et al. [30], figure 04[p.05]/Springer Nature/CC BY 4.0.

novel tool for detecting activation of an immune response. In a model of experimental autoimmune encephalomyelitis substantially fewer cells were detected as compared to healthy controls. Even the transition from symptom-free to clinical state in immunized mice could be detected by time-lapse MRI. To explain this seemingly paradoxical observation, it was estimated at which velocities cells are detectable with time-lapse MRI. Evidently, the assumption of static cells during the time of image acquisition, which jump to their subsequent position in between the sequential images, is not a good model for quickly moving cells. Only for slow-motion cells may the static condition be assumed and image contrast can be estimated according to Equations 12.1 and 12.2. For higher-motion velocities, some kind of blurring in the images must be expected, similar to moving objects in photographic pictures. However, since MR images are generated by Fourier transformation of the acquired $\boldsymbol{k}$-space data, blurring effects differ from simple photography. As done by Masthoff et al., blurring can be simulated using synthetic $\boldsymbol{k}$-space data. As illustrated in Figure 12.5 (upper row), $\boldsymbol{k}$-space data can be composed of synthetic images of moving cells. Depending on the velocity of cell motion, different fractions of $\boldsymbol{k}$-space are filled from images with cells at the same position, before cells are assumed to have progressed to the next voxel, and $\boldsymbol{k}$-space data are generated from these synthetic images. For a static cell, the complete $\boldsymbol{k}$-space data are generated from one static image. For slow motion, assuming that cells progress only to the neighboring voxel while the full image is acquired, $\boldsymbol{k}$-space is generated from two consecutive synthetic images. For faster motion, $\boldsymbol{k}$-space data is generated from three, four, five, etc. consecutive images, depending on the number of voxels the cell is assumed to have moved during image acquisition.

As can be seen in Figure 12.5, contrast blurs out increasingly with higher cell velocities. Depending on the initial signal decrease observed for the static cell, contrast may still be observable, when cells are moving across up to around 20 voxels during the time 1 image is acquired. Yet, such observation requires efficient cellular labeling with an effective contrast agent. In vivo, the detection limit may be reduced to slower motion, since image artifacts and additional noise may reduce contrast. Tracking of cells with high velocities becomes extremely difficult, since location of signal reduction in the image strongly depends on timing, when $\boldsymbol{k}$-space center was acquired. It is therefore not generally possible to predict which motion velocities are detectable in the experiment.

12.7 Validation Methods for Labeled Cell Detection

As discussed in Section 12.2, positive contrast methods such as fluorine MRI, hyperpolarization, or highly shifted proton MRI, which unambiguously identify labeled cells, do not provide sufficient sensitivity for single-cell tracking. Labeling with superparamagnetic IONPs is currently the only feasible method that enables reliable single-cell tracking. Using this approach, one critical issue in single-cell MRI detection is validating that the observed contrast actually originates from the cell of interest. Two different aspects, a biological and a physical one, have to be considered. Due to the nature of the origin of the observed contrast, which is a dipolar field emanating from a local difference in magnetic susceptibility, the observed signal reduction or void does not necessarily result from the presence of a labeled cell. Any local susceptibility variation may give rise to a dipolar field, which becomes observable in the images as signal loss. Most importantly, blood clots from e.g. microbleedings following injections or surgical interventions, but also iron accumulations from the ample sources of iron in the body (ferritin, hemosiderin,

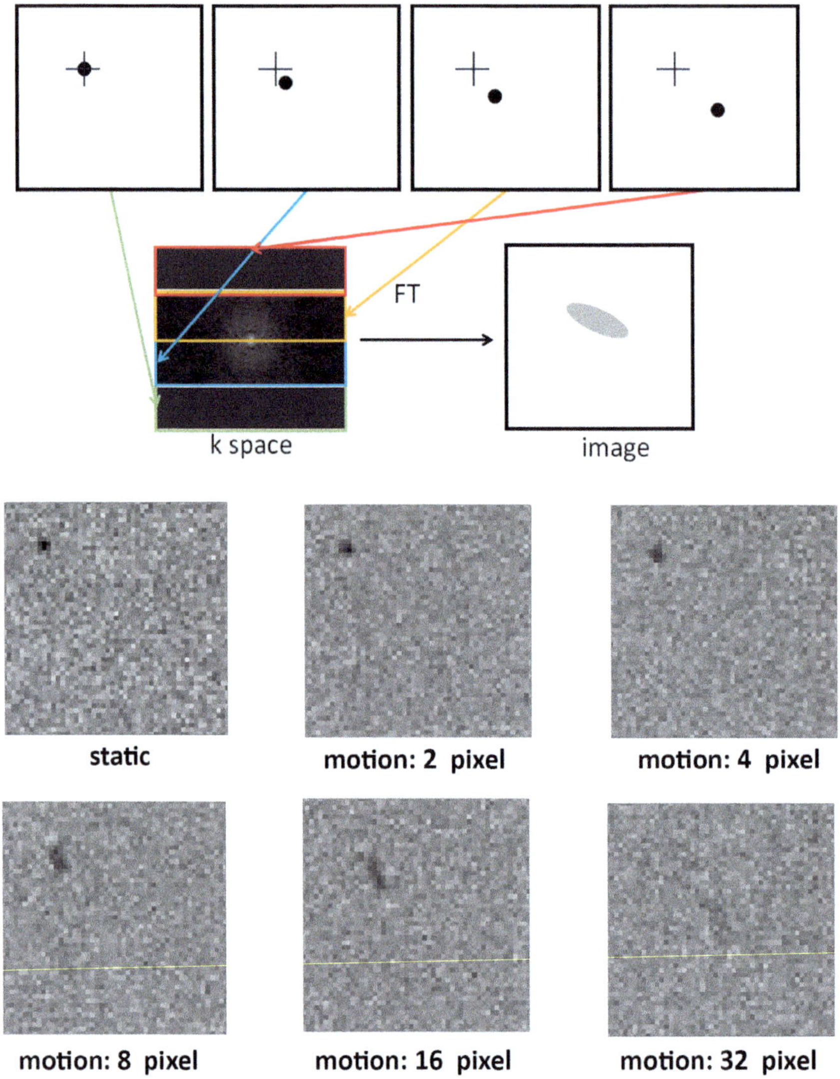

Figure 12.5 Simulation of image contrast of a single moving cell. Upper part of the figure illustrates the principle of how contrast of a moving cell is simulated. Four consecutive timeframes show different positions of the cell (black dot) during the time acquisition of one image required. Motion is considered stepwise, from pixel to pixel. Here, motion takes place to the diagonally adjacent pixel. The initial position is marked by a crosshair. In this example motion occurs across four pixels during image acquisition. Consequently, each pixel position provides data for one-fourth of ***k***-space. The completely filled ***k***-space is then Fourier transformed and yields the blurred image of the moving cell (gray). The lower part of the figure shows the results of an actual simulation, reproducing imaging parameters of a time-lapse magnetic resonance imaging (MRI) experiment. With a pixel dimension of 50 μm, the cell is observed as a 70% signal drop in the central pixel and 30% in the directly adjacent pixels, in the static image (top left). Images show a 50 × 50 pixel zoom of the 256 × 256 pixel image. From top left to bottom right distance traveled by the cell is increased stepwise. In the lower right panel the cell has traveled 32 pixels (diagonally), which means that only 3.25% (i.e. 1/32) of ***k***-space is filled with lines from each position. Note that the contrast decreases steadily. The center of contrast moves along the trajectory of the cell, and reflects the position of the cell at the time when the ***k***-space center is filled. In these simulations the cell is still observable in the second last image, which corresponds to a condition where the cell resides in one pixel for only 30 s during the 8 min of acquisition of one full image.

hemoglobin, transferrin, and myoglobin), or foreign objects, air bubbles, or simply tissue scars may be confounded with labeled cells in the MR images. Due to this fundamental physical ambiguity, it is of great importance to have reliable methods available to validate that observed contrast actually stems from the cells of interest.

To verify the presence of iron at distinct tissue sites or inside cells, histological methods are often employed. The most popular and simple to implement approach is to perform optical microscopy of tissue slices that have been subject to Prussian blue staining. By this procedure iron appears blue in the images, and in combination with cell-specific stainings and exploiting the resolution of optical microscopy, iron can be attributed to distinct cells. However, Prussian blue staining is not specific for IONPs and thus does not provide information, whether the observed signal originates from administered IONPs, or whether IONPs are still intact or have been degraded by the organism. For a more precise visualization and localization of the IONPs, transmission electron microscopy (TEM) offers valuable additional information. Requiring more time-consuming slice preparation, TEM provides spatial resolutions that enable visualization of single particles and intracellular details. TEM thus allows for observing whether IONPs have been degraded, are present as single particles, or aggregated inside the cells. It can further be assessed whether IONPs have been released from the labeled cells or have been taken up by macrophages or other cells [14,18]. However, TEM does not distinguish between iron deposits of iron from IONPs and iron from other sources in the organism.

In the case that IONPs are not degraded in the body, this ambiguity may be resolved by methods that are sensitive for detection of intact particles. Over recent years, magnetic particle imaging (MPI) and a number of related methods have been developed [33–36]. These measure resonances in high-frequency electromagnetic fields, which are based on the interaction with superparamagnetic particles. With these methods, which even enable quantification of particles, signals are only detected if IONPs are still intact. While, in particular, MPI is capable of tracking labeled cells with high temporal resolution, it must be combined with MRI or other imaging methods to provide images of surrounding tissue [37]. Recently, a novel approach using subcellular diamond magnetometry was proposed to map the origins of MR contrast on a microscopic scale, which might in the future prove useful for IONP quantification [38].

To quantify local iron accumulation protocols using spectrophotometry, T_2-relaxometry or atomic absorption spectroscopy have been applied [30,39]. Also high-resolution MR microscopy of ex vivo tissue specimens has been used to follow up labeled cells and their fate after engraftment [40]. Yet, none of these methods is capable of separating administered from exogenous iron or of providing data on the physico-chemical state of IONPs.

To unambiguously identify administered IONPs or iron originating from the IONPs, IONPs have previously been spiked with different isotopes or elements. ^{59}Fe was used to add a radioactive signature to IONPs, mostly containing the most abundant stable isotope ^{56}Fe [41,42]. Using radiographic methods, tissue containing IONPs can be identified and thus confirm MRI detection of IONPs. Most recently, nonradioactive approaches have been proposed. IONPs have been doped with europium, which is detectable by fluorescence methods or mass spectrometry [43]. In particular elemental mass spectrometry using inductively coupled plasma (ICP-MS) enables ex vivo separation of different isotopes in tissue slices with high sensitivity and precision [44]. It was recently shown that IONPs containing the stable but rare isotope ^{57}Fe allowed the observation of the fate and metabolization of administered IONPs by combining MRI with laser ablation (LA) ICP-MS [14]. ^{57}Fe-containing SPIOs were found to be cleared quickly from the bloodstream, stored in liver and spleen, while iron was reused in normal iron metabolism after degradation of the particles. It was further shown that ^{57}Fe-SPIOs allowed for tracking labeled immune cells in a local inflammation model in mice, and image contrast could be separated from contrast due to endogenous iron.

In addition to the basic problem of assigning contrast to the applied particles, detection of cells of interest may be hampered by biological processes. In the case of labeled tumor cells, stem cells, or bacteria, cells may be attacked by the immune system and particles or whole cells taken up by phagocytizing cells. These cells may appear "positive" in the MR images and also show typical patterns of the label with specific validation methods such as MPI, LA-ICP-MS, or others. It may therefore be necessary to perform immunohistochemical stainings and correlate those data with MRI and physical validation procedures. Often also TEM is required to provide a conclusive picture of the fate of administered IONPs [14]. Only such a multimodal combination of imaging and analysis methods will finally identify a labeled cell as an unambiguous source of observed image contrast in MRI.

12.8 Summary

Starting from the first notions to use MR for imaging single cells, methodology has progressed impressively and as of today allows for highly specific tracking of moving cells in the organism. The use of superparamagnetic IONPs has evolved as the most efficient and sensitive method to label and detect cells by MRI. For successful single-cell tracking by MRI several experimental preconditions need to be fulfilled:

1. The cells of interest must be labeled efficiently to achieve a sufficiently high magnetic moment per cell.
2. Cells can be detected in T_2*-weighted images, which are obtained using gradient echo MR sequences. Long echo times enhance contrast generated by the labeled cells. However, long TE also increases image artifacts, requiring individual compromises and the use of medium TEs.
3. High magnetic fields enhance T_2* relaxation and are therefore favorable for cell tracking. However, single cells were also detected at field strengths as low as 1.5 T.
4. While in vitro detection of single cells might be achieved with standard methods even by the less experienced MR user, in vivo detection and longitudinal tracking of single cells is still a demanding task. Artifacts have to be minimized and, in particular, motion has to be suppressed and corrected for to resolve the delicate contrast changes evoked by single cells.

References

1 Kircher, M.F., Gambhir, S.S., and Grimm, J. (2011). Noninvasive cell-tracking methods. *Nature Reviews Clinical Oncology* 8 (11): 677–688.

2 Ahrens, E.T. and Bulte, J.W.M. (2013). Tracking immune cells in vivo using magnetic resonance imaging. *Nature Reviews Immunology* 13 (10): 755–763.

3 Errico, C., Pierre, J., Pezet, S. et al. (2015). Ultrafast ultrasound localization microscopy for deep super-resolution vascular imaging. *Nature* 527 (7579): 499–502.

4 Taruttis, A. and Ntziachristos, V. (2015). Advances in real-time multispectral optoacoustic imaging and its applications. *Nature Photonics* 9 (4): 219–227.

5 Ciobanu, L., Webb, A.G., and Pennington, C.H. (2003). Magnetic resonance imaging of biological cells. *Progress in Nuclear Magnetic Resonance Spectroscopy* 42 (3–4): 69–93.

6 Srinivas, M., Heerschap, A., Ahrens, E.T. et al. (2010). F-19 MRI for quantitative in vivo cell tracking. *Trends in Biotechnology* 28 (7): 363–370.

7 Colotti, R., Bastiaansen, J.A.M., Wilson, A. et al. (2017). Characterization of perfluorocarbon relaxation times and their influence on the optimization of fluorine-19 MRI at 3 tesla. *Magnetic Resonance in Medicine* 77 (6): 2263–2271.

8 Schmid, F., Holtke, C., Parker, D., and Boosting, F.C. (2013). F-19 MRI-SNR efficient detection of paramagnetic contrast agents using ultrafast sequences. *Magnetic Resonance in Medicine* 69 (4): 1056–1062.

9 Rho, J., Stares, E., Adams, S.R. et al. (2020). Paramagnetic fluorinated nanoemulsions for in vivo F-19 MRI. *Molecular Imaging and Biology* 22 (3): 665–674.

10 Schmidt, R., Nippe, N., Strobel, K. et al. (2014). Highly shifted proton MR imaging: Cell tracking by using direct detection of paramagnetic compounds. *Radiology* 272 (3): 785–795.

11 Matsumoto, Y. and Jasanoff, A. (2008). T2 relaxation induced by clusters of superparamagnetic nanoparticles: Monte Carlo simulations. *Magnetic Resonance Imaging* 26 (7): 994–998.

12 Vuong, Q.L., Gillis, P., and Gossuin, Y. (2011). Monte Carlo simulation and theory of proton NMR transverse relaxation induced by aggregation of magnetic particles used as MRI contrast agents. *Journal of Magnetic Resonance* 212 (1): 139–148.

13 Hak, S., Stenmark, S., Bjerkholt, F.F., and Haraldseth, O. (2015). Transverse relaxivity of iron oxide nanocrystals clustered in nanoemulsions: Experiment and theory. *Magnetic Resonance in Medicine* 74 (3): 858–867.

14 Masthoff, M., Buchholz, R., Beuker, A. et al. (2019). Introducing specificity to iron oxide nanoparticle imaging by combining ^{57}Fe-based MRI and mass spectrometry. *Nano Letters* 19 (11): 7908–7917.

15 Himmelreich, U. and Dresselaers, T. (2009). Cell labeling and tracking for experimental models using magnetic resonance imaging. *Methods* 48 (2): 112–124.

16 Modo, M.M.J.J. and Bulte, J.W.M. (ed.). (2007). *Molecular and Cellular MR Imaging*. Boca Raton, FL: CRC Press.

17 Balla, D.Z., Gottschalk, S., Shajan, G. et al. (2013). In vivo visualization of single native pancreatic islets in the mouse. *Contrast Media and Molecular Imaging* 8 (6): 495–504.

18 Dodd, S.J., Williams, M., Suhan, J.P. et al. (1999). Detection of single mammalian cells by high-resolution magnetic resonance imaging. *Biophysical Journal* 76 (1): 103–109.

19 Hinds, K.A., Hill, J.M., Shapiro, E.M. et al. (2003). Highly efficient endosomal labeling of progenitor and stem cells with large magnetic particles allows magnetic resonance imaging of single cells. *Blood* 102 (3): 867–872.

20 Shapiro, E.M., Sharer, K., Skrtic, S., and Koretsky, A.P. (2006). In vivo detection of single cells by MRI. *Magnetic Resonance in Medicine* 55 (2): 242–249.

21 Foster-Gareau, P., Heyn, C., Alejski, A., and Rutt, B.K. (2003). Imaging single mammalian cells with a 1.5 T clinical MRI scanner. *Magnetic Resonance in Medicine* 49 (5): 968–971.

22 Bowen, C.V., Zhang, X., Saab, G. et al. (2002). Application of the static dephasing regime theory to superparamagnetic iron-oxide loaded cells. *Magnetic Resonance in Medicine* 48 (1): 52–61.

23 Shapiro, E.M., Skrtic, S., and Koretsky, A.P. (2005). Sizing it up: Cellular MRI using micron-sized iron oxide particles. *Magnetic Resonance in Medicine* 53 (2): 329–338.

24 Hoehn, M., Kustermann, E., Blunk, J. et al. (2002). Monitoring of implanted stem cell migration in vivo: A highly resolved in vivo magnetic resonance imaging investigation of experimental stroke in rat. *Proceedings of the National Academy of Sciences of the United States of America* 99 (25): 16267–16272.

25 Stroh, A., Faber, C., Neuberger, T. et al. (2005). In vivo detection limits of magnetically labeled embryonic stem cells in the rat brain using high-field (17.6 T) magnetic resonance imaging. *Neuroimage* 24 (3): 635–645.

26 Heyn, C., Ronald, J.A., Mackenzie, L.T. et al. (2006). In vivo magnetic resonance imaging of single cells in mouse brain with optical validation. *Magnetic Resonance in Medicine* 55 (1): 23–29.

27 Heyn, C., Ronald, J.A., Ramadan, S.S. et al. (2006). In vivo MRI of cancer cell fate at the single-cell level in a mouse model of breast cancer metastasis to the brain. *Magnetic Resonance in Medicine* 56 (5): 1001–1010.

28 Mori, Y., Chen, T., Fujisawa, T. et al. (2014). From cartoon to real time MRI: In vivo monitoring of phagocyte migration in mouse brain. *Scientific Reports* 4: 6997.

29 Kirschbaum, K., Sonner, J.K., Zeller, M.W. et al. (2016). In vivo nanoparticle imaging of innate immune cells can serve as a marker of disease severity in a model of multiple sclerosis. *Proceedings of the National Academy of Sciences of the United States of America* 113 (46): 13227–13232.

30 Masthoff, M., Gran, S., Zhang, X. et al. (2018). Temporal window for detection of inflammatory disease using dynamic cell tracking with time-lapse MRI. *Scientific Reports* 8 (1).

31 Hoerr, V., Tuchscherr, L., Huve, J. et al. (2013). Bacteria tracking by in vivo magnetic resonance imaging. *BMC Biology* 11: 63.

32 Modo, M., Cash, D., Mellodew, K. et al. (2002). Tracking transplanted stem cell migration using bifunctional, contrast agent-enhanced, magnetic resonance imaging. *Neuroimage* 17 (2): 803–811.

33 Gleich, B. and Weizenecker, J. (2005). Tomographic imaging using the nonlinear response of magnetic particles. *Nature* 435 (7046): 1214–1217.

34 Bulte, J.W.M. (2019). Superparamagnetic iron oxides as MPI tracers: A primer and review of early applications. *Advanced Drug Delivery Reviews* 138: 293–301.

35 Espy, M., Matlashov, A., and Volegov, P. (2013). SQUID-detected ultra-low field MRI. *Journal of Magnetic Resonance* 228: 1–15.

36 Van De Loosdrecht, M.M., Draack, S., Waanders, S. et al. (2019). A novel characterization technique for superparamagnetic iron oxide nanoparticles: The superparamagnetic quantifier, compared with magnetic particle spectroscopy. *Review of Scientific Instruments* 90: 2.

37 Bulte, J.W.M., Walczak, P., Janowski, M. et al. (2015). Quantitative "hot spot" imaging of transplanted stem cells using superparamagnetic tracers and magnetic particle imaging (MPI). *Tomography* 1 (2): 91–97.

38 Davis, H.C., Ramesh, P., Bhatnagar, A. et al. (2018). Mapping the microscale origins of magnetic resonance image contrast with subcellular diamond magnetometry. *Nature Communications* 9 (1).

39 Dadashzadeh, E.R., Hobson, M., Henry Bryant, L. et al. (2013). Rapid spectrophotometric technique for quantifying iron in cells labeled with superparamagnetic iron oxide nanoparticles: Potential translation to the clinic. *Contrast Media and Molecular Imaging* 8 (1): 50–56.

40 Chen, T., Mori, Y., Inui-Yamamoto, C. et al. (2017). Polymer-brush-afforded SPIO nanoparticles show a unique biodistribution and MR imaging contrast in mouse organs. *Magnetic Resonance in Medical Sciences* 16 (4): 275–283.

41 Weissleder, R., Stark, D.D., Engelstad, B.L. et al. (1989). Superparamagnetic iron oxide: Pharmacokinetics and toxicity. *American Journal of Roentgenology* 152 (1): 167–173.

42 Freund, B., Tromsdorf, U.I., Bruns, O.T. et al. (2012). A simple and widely applicable method to ^{59}Fe-radiolabel monodisperse superparamagnetic iron oxide nanoparticles for in vivo quantification studies. *ACS Nano* 6 (8): 7318–7325.

43 Scharlach, C., Müller, L., Wagner, S. et al. (2016). LA-ICP-MS allows quantitative microscopy of europium-doped iron oxide nanoparticles and is a possible alternative to ambiguous Prussian blue iron staining. *Journal of Biomedical Nanotechnology* 12 (5): 1001–1010.

44 Reifschneider, O., Wentker, K.S., Strobel, K. et al. (2015). Elemental bioimaging of thulium in mouse tissues by laser ablation-ICPMS as a complementary method to heteronuclear proton magnetic resonance imaging for cell tracking experiments. *Analytical Chemistry* 87 (8): 4225–4230.

13

Imaging Biomarkers for Alzheimer's Disease Using Magnetic Resonance Microscopy

Alexandra Badea[1,2,3,4], *Jacques A. Stout*[3], *Robert J. Anderson*[1], *Gary P. Cofer*[1], *Leo L. Duan*[5], *and Joshua T. Vogelstein*[6]

[1] *Department of Radiology, Duke University Medical Center, Durham, NC, USA*
[2] *Department of Neurology, School of Medicine, Duke University, Durham, NC, USA*
[3] *Brain Imaging and Analysis Center, Duke University Medical Center, Durham, NC, USA*
[4] *Department of Biomedical Engineering, Pratt School of Engineering, Duke University, Durham, NC, USA*
[5] *Department of Statistics, University of Florida, Gainesville, FL, USA*
[6] *Department of Biomedical Engineering, Institute for Computational Medicine, Kavli Neuroscience Discovery Institute, Johns Hopkins University, Baltimore, MD, USA*

13.1 Introduction

13.1.1 Motivation

Understanding the brain and its functioning, manipulating, and even reverse engineering it are some of the major goals in neuroscience. A better understanding of how the brain's structure and connections give rise to motivation and behavior will lead to better therapies for psychiatric and neurological conditions, including neurodegenerative conditions such as Alzheimer's disease (AD). The difficulties in understanding this condition come from several sources. It is not yet clear what causes it, but the research community is coming close to defining a spectrum of risk factors, from amyloid and tau proteins, to inflammation, metabolic alterations, infections, injury, or toxic stressors. Thus, we may have not one disease for which to define biomarkers, such as amyloid and tau burden, but a family of diseases that lead to neurodegeneration and memory loss. Deviations from normal brain aging are clinically assessed through memory, problem-solving, and attention tests, patient- and family-declared symptoms, and by ruling out other causes, such as tumors, strokes, etc. But these symptoms are rarely apparent in early stages, may be due to normal aging, or may simply be nonspecific. It is thus difficult to observe the early development of multifactorial diseases such as AD, to understand why it first develops, or how to treat it before it reaches a debilitating stage. Part of modern research aims to understand the role of quantitative biomarkers that characterize the disease. Identifying biomarkers and their interactions may lead to early diagnosis, subgroup classification, and devising better treatments and methods to

Magnetic Resonance Microscopy: Instrumentation and Applications in Engineering, Life Science and Energy Research, First Edition. Edited by Sabina Haber-Pohlmeier, Bernhard Blümich, and Luisa Ciobanu.

measure their effectiveness. This is important since single-pronged approaches have not fully explained the etiology of the disease.

The hallmark AD biomarkers are extracellular amyloid deposits, and abnormal tau phosphorylation. These pathologies are accompanied by neurodegeneration and memory decline. Besides mutations in amyloid precursor protein (APP), and presenilin 1 and 2 (PS1, PS2), which explain ~5–10% of AD cases, known risk factors for late onset AD (which explain >90% of cases) are aging and an apolipoprotein E4 (APOE4) genotype. Measuring age- and genotype-associated changes will allow us to develop risk and disease progression models. Even more important is to understand what confers resilience to AD, be it cognitive reserve, protective molecular mechanisms against toxicity, or variants on how brain regions communicate, and possibly adapt their communication to protect the brain against pathologies, leading to successful aging.

Imaging, and in particular magnetic resonance imaging (MRI), can provide rich quantitative biomarkers based on morphometry, texture reflective of microstructure and relaxation properties, cerebral blood perfusion, metabolism, and even brain function and activation. MRI allows us to evaluate how these metrics change in various brain regions, providing multivariate biomarkers to understand multifactorial diseases such as AD. The relationships between concerted changes in different brain regions are central to connectomic analysis. But the genetics and environmental exposure of humans suffering from neurodegenerative conditions are complex, and difficult to control.

Mouse models offer powerful tools to study the effects of gene modifications, controlled environmental exposures, or preventive interventions, such as exercise. Individual gene modifications as well as interactions of several genes can thus be studied. There is a wide variety of mouse models of neurodegenerative disease: the Jackson Laboratory (https://www.jax.org) now distributes over 300 mouse strains designed to recapitulate various aspects of AD. These models can be used to reveal functional and structural changes accompanying pathologies associated with AD before its overt stages. Their easy-to-manipulate genomes can help reveal the cellular mechanisms responsible for these deficits. These mouse models have genes responsible for familial AD, abnormal amyloid and tau, or the three human APOE alleles, where APOE2 is associated with protection from AD, APOE3 is considered a control, and APOE4 a risk for late onset, sporadic AD.

In this chapter, we will examine the main AD biomarkers and how these are represented in imaging studies using small animal models. We will focus on MRI biomarkers including volume and texture properties related to iron metabolism, cellular density, and myelin properties, as well as from diffusion-weighted-derived parameters and tractography. We will examine how combining MRI biomarkers may increase the predictive values of image-based models for behavioral traits typical of AD patients.

13.1.2 Challenges Associated with Quantifying MRI Brain Biomarkers in Mouse Models

Compared with clinical human brain scans, acquired typically with voxels of $1 \times 1 \times 6\ mm^3$ and occasionally up to $1 \times 1 \times 1\ mm^3$ voxels, small animal MRI needs to use smaller voxel sizes. These voxels need to be ~15 times smaller as a linear dimension, or 3000 times smaller in volume, as indicated by the size ratio between a 16-cm-wide human brain, and 1-cm-wide mouse brain. This translates to a ~62.5-μm linear voxel size. To compensate for the loss in signal-to-noise ratio (SNR) due to the smaller voxel sizes, it has been necessary to develop dedicated imaging hardware and software. Animal studies carried out with smaller bore magnets and higher magnetic fields (3–21 T) than clinical studies (usually 1–7 T) provide increased magnetization and thus SNRs. Other steps toward improving SNR include stronger and faster gradient coils (400–3000 mT/m versus 10–200 mT/m gradient strength; and 640–20 000 T/m/s versus 10–700 T/m/s slew rate) [1,2]. This reduces the echo time and length of the gradient pulses, increasing the SNR and improving the point spread function/resolution. As a result, larger image matrices have become possible, requiring longer computing times and more memory.

Higher magnetic fields do, however, come with larger susceptibility variations at the interface of tissue and air, leading to signal drops and distortions. Relaxation parameters also change, resulting in longer T_1 and shorter T_2, which pose significant demands on pulse program parameters to maximize the signal [3].

Advances in high-field preclinical MRI make it well suited for phenotyping mouse models of neurological and psychiatric conditions [4–6]. Imaging technology developments have been reviewed before, e.g. in [7,8]. The next challenge is to incorporate quantitative morphometry [9,10], which is bias free or relies on anatomical priors [10–15]. Both approaches involve developing image analysis pipelines for single or multiple MR imaging protocols [16], relying on image registration and statistical analyses [17]. The results can reveal phenotypes associated with genotypes [18] that confer risk for disease [19–22], with age [23], or with interventions [24] to help understand brain plasticity [25]. These associations may lead to novel mechanistic insight into disease etiology and development.

13.1.3 Animal and Specimen Preparation for Magnetic Resonance Microscopy

The higher heart and respiration rates in the mouse versus human pose challenges to achieving high resolution without motion artifacts. For in vivo scanning, animals are usually maintained under isoflurane anesthesia (1.2 ± 0.5%), delivered though a nose cone or mask. The animals need to be physiologically supported throughout imaging, provided air and heating within the magnet bore to maintain heart (300–600 beats/min), respiratory rates (60–200 beats/min), and body temperature (36–38°C) within normal limits. To enhance contrast and speed up neuroimaging acquisitions, contrast agents may be introduced into the brain by opening the blood–brain barrier using, e.g. ultrasound, as one single injection, or implantable mini-pumps that slowly deliver small doses over longer periods of time, e.g. 0.5 μl/h, thus reducing toxicity and delivering a higher total amount of contrast, up to 180 mg/kg of manganese chloride tetrahydrate over 7 days [24]. Several studies have reported MR in awake small animals [26,27] but such experiments require dedicated restraining systems and short durations.

Ex vivo scanning removes the scan time limitations related to animal safety and welfare. Still, it is advantageous to devise efficient imaging strategies, and one way to help this process is to prepare specimens using "active staining" with T_1 shortening agents, like gadolinium. For this, animals are anesthetized to a surgical plane and perfused through the left ventricle, with outflow from the right atrium. Saline (0.9%) with heparin (0.1%) is used to flush out the blood, e.g. at a rate of 8 ml/min, for 5 min. Fixation is achieved by perfusing with a 10% solution of neutral buffered formalin phosphate containing 10% (50 mm) gadoteridol (ProHance, Bracco Diagnostics Inc., Monroe Township, NJ), at the same rate of 8 ml/min, and for the same 5 min duration. This reduces the spin lattice relaxation time (T_1) of tissue to ~100 ms, enabling faster acquisitions with higher SNR. Mouse heads are then stored in 10% formalin for 24 h. Finally, the specimens are transferred to a 0.01 M solution of phosphate buffered saline (PBS) containing 0.5% (2.5 mm) gadoteridol at 4°C for at least 7 days to rehydrate the tissue, or longer to ensure tissue stability. Extraneous tissue around the cranium is removed prior to imaging, and specimens are positioned in MRI-compatible tubes, immersed in perfluoropolyether (Galden Pro, Solvay, NJ) for susceptibility matching. Brains are left inside the cranium to avoid distortions or damage to regions such as the olfactory bulbs or the parafloculi of the cerebellum [9].

We will present MRI-based phenotypes from four mouse models. Two of these model familial AD [19,20]. The other two model the APOE-associated genetic risk for late onset AD [28,29], one of them on a background that better models the innate human immune system [30].

The first mouse model of AD has a double mutant form of the APP, encoding the Swedish and Indiana familial mutations (APPswe/ind), and its expression is controlled by the tetracycline-responsive transactivator [31]. This system allows temporal control of the expression of APPswe/ind. Thus we can examine animals in which the disease is initiated after the brain has fully matured [19].

The second model of familial AD is the CVN mouse, which expresses the Swedish, Dutch, and Iowa mutations of the human APP gene, on a murine nitric oxide synthase 2 knockout background (*mNos2*$^{-/-}$) [32] to better mimic human immunity. CVN mice show amyloid deposition (>6 weeks of age, particularly around blood vessels), hyperphosphorylated and aggregated mouse tau (24 weeks), deficits in memory and learning (24 weeks), altered microglial activity [33], and neuronal degeneration and loss (36 weeks) [34].

APOE-targeted replacement mice have been generated at the University of North Carolina (UNC), where the mouse APOE has been replaced with each of the three principal human APOE alleles: APOE2 – thought to confer protection against AD, APOE3 – as the control, and APOE4 – as the major genetic risk for late onset AD [28,29]. In APOE-HN mice, the human APOE genes have replaced the mouse APOE, and the human nitric oxide synthase 2 (NOS2) gene has replaced the mouse Nos2 gene (HuNOS2$^{+/+}$/mNos2$^{-/-}$; abbreviated HN). These mice better model the APOE4-associated risk in humans with late onset AD, and the response to stressors and inflammation.

13.1.4 AD Biomarkers

AD biomarkers center on hallmark pathologies such as amyloid beta (Aβ) and tau, as well as neurodegeneration. Blood and cerebrospinal fluid (CSF) biomarkers hold great promise for diagnosis of AD. CSF markers including β-amyloid(1-42), Aβ42/40 ratio, total tau, and phospho-tau-181 have high specificity and sensitivity [35], but require invasive procedures. Blood-based markers for Aβ42/40 ratios are easier to collect but challenging because brain-derived molecules are present at much lower concentrations in blood than in CSF. However, almost all these markers are limited in their ability to inform directly on brain structure and function, or the disconnections between brain regions.

Positron emission tomography (PET) and MRI have become the imaging modalities of choice to examine changes throughout the brain. Fluorodeoxyglucose or FDG-PET informs on metabolic changes but makes it difficult to differentiate among tau related disorders, e.g. frontotemporal degeneration (FTD), progressive supranuclear palsy (PSP), and cortico-basal degeneration [36,37]. An MRI-based method reflective of metabolic changes can rely on perfusion imaging using arterial spin labeling sequences such as arterial spin labeling (ASL), continuous arterial spin labeling (CASL), or pseudo continuous arterial spin labeling (pCASL) [38].

Aβ pathology is consistently detected in postmortem studies of AD patients, but also in normal elderly individuals [39]. Up to 30% of adults in their 70s, and >50% of adults in their 80s can have positive results [40]. Amyloid burden has been detected in [18 F]FBP (florbetapir) scans among cognitively normal patients who have hypertension and an APOE4 genotype [41].

While amyloid plaques can be observed due to the differences in relaxation properties relative to the surrounding healthy parenchyma, there are challenges associated with their small size. Magnetic resonance microscopy can detect amyloid deposits in mouse models with amyloid pathology [19] (Figure 13.1). One method to accelerate imaging and enhance contrast is to deliver MRI contrast across the blood–brain barrier, which is facilitated by sonicating microbubbles (Definity) injected into the tail vein of the animal [42]. The contrast agents (Gd or Mn) reduce the T_1 of the brain, and permit scanning at shorter TR. We have acquired in vivo images at 7.1 T, at $40 \times 40 \times 625\ \mu m^3$ resolution, using blood–brain barrier opening and microbubbles to allow

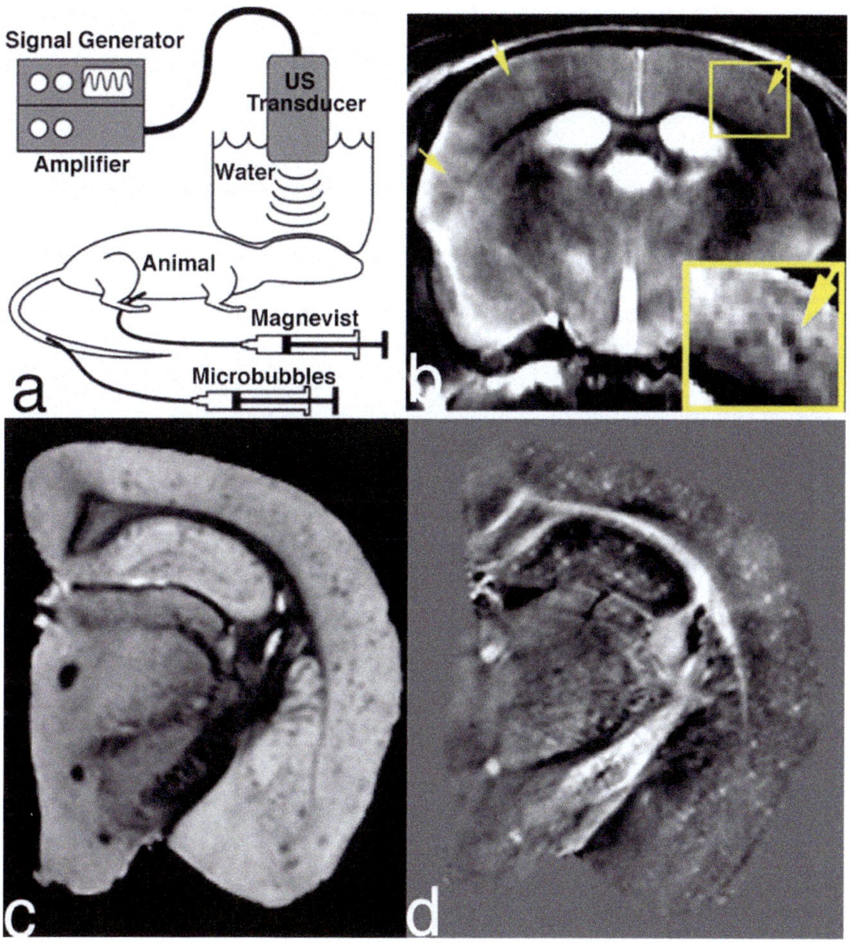

Figure 13.1 Opening the blood–brain barrier (a) allows for introducing magnetic resonance (MR) contrast agents into the brain parenchyma, and enhances contrast-to-noise ratios, allowing the detection of amyloid plaques in vivo (b). MR in vivo images were acquired at 7.1 T using a CPMG (Carr–Purcell–Meiboom–Gill) sequence with repetition time (TR) of 400 ms, echo train length of 8, min echo time (TE) = 7.5 ms, echo spacing of 6.5 ms, at a spatial resolution of $40 \times 40 \times 625\ \mu m^3$. Smaller plaques remain difficult to discern in vivo because of the partial volume effects. We thus imaged gadolinium-stained mouse brains at 9.4 T using a gradient echo sequence (GRE) with TE/TR 3.4 ms/50 ms, at 35.2 μm isotropic resolution. This allows the observation of smaller amyloid plaques throughout the brain. Quantitative susceptibility maps derived using STISuite [83] enhanced contrast, and characterized microstructural properties of brain tissue.

visualization of amyloid deposits in vivo (Figure 13.1a, b); however, smaller plaques are hard to discern because of partial volume effects. Without the time constraints of in vivo imaging, ex vivo MRI can achieve higher resolutions. We imaged actively stained mouse brain at 35.2 μm isotropic resolution, to detect significantly more, and smaller, amyloid plaques (Figure 13.1c), and to estimate quantitative susceptibility maps (Figure 13.1d).

Imaging tau is currently done using PET ligands, and MR-PET systems can come in handy. [18 F]AV-1451 binds to tau-positive inclusions in autopsy samples from subjects with neurocognitive disorders such as fronto-temporal dementia, progressive supranuclear palsy, and cortico-basal degeneration [43]. [18 F]Florbenazine (18 F-AV-133), shows lower uptake in Parkinson's relative to AD patients. [^{11}C]UCB-J [44] has detected synaptic loss in PSP and amyloid-negative cortico-basal syndrome (CBS) [45]. TDP-43 [46] has been added on the list of neurodegeneration markers, but it is performed postmortem. Chemical exchange saturation transfer and decreased amide proton transfer may offer an MRI-based method sensitive to hyperphosphorylated tau pathology [47].

Neurofilament light (NfL) is a biomarker of axonal degeneration, which can be imaged using diffusion-weighted imaging. CSF and plasma NfL levels are elevated in neurodegenerative disorders including AD [48], multiple sclerosis, stroke, traumatic brain injury, amyotrophic lateral sclerosis, and Parkinson disease [49]. An elevated plasma NfL is predictive of cognitive decline assessed by the Mini-Mental State Examination, and neuroimaging markers of neurodegeneration, e.g. cortical thickness, and diffusion MRI has similar effect sizes to CSF NfL [50]. However, it may be a nonspecific marker of degeneration.

Studies have linked αSyn to tau hyperphosphorylation, Aβ pathology [51], and the APOE4 allele, the latter being a major genetic risk factor for both AD and dementia with Lewy bodies [52]. Thus, newer studies are beginning to examine AD risk before overt cognitive decline, based on genetic risk such as that conferred by APOE4 alleles.

Inflammation and vascular disease may expand the biomarker framework and MRI can provide insight into these phenomena through blood–brain barrier integrity [53], angiography [54], cerebral blood perfusion [55], and quantitative susceptibility mapping [56]. Still, the majority of MRI studies have focused on mapping progression of neurodegeneration and microstructural changes either as direct indicators or as proxies for the above AD-related processes and these are the focus of the subsequent sections.

13.2 Morphometry

13.2.1 General Principles

Unlike fluid biomarkers, which are unable to convey spatial information, image-based biomarkers can help us to understand how disease affects the brain in different stages, by giving insight into the spatial distribution of brain atrophy, hypometabolism, and amyloid deposits. Multivariate imaging can also identify population subgroups with variable resistance to disease [57]. For example, La Joie et al. [58] revealed distinct subsets of regions, e.g. (i) in the hippocampus atrophy exceeded hypometabolism, even as Aβ load was minimal; (ii) in posterior association areas Aβ deposition was predominant, combined with high hypometabolism and lower atrophy; and (iii) in frontal regions Aβ deposition was maximal, whereas structural and metabolic alterations were low. Atrophy and hypometabolism were significantly correlated in the hippocampus and temporo-parietal cortex, whereas Aβ load was not significantly related to either atrophy or hypometabolism. These findings suggest differential involvement of region-specific pathological or protective mechanisms can be revealed by multivariate imaging biomarkers.

The effects of individual pathologies, and their interactions, as well as those of therapeutic interventions can be studied in mouse models, relative to age-matched controls raised in well-controlled experimental conditions.

Magnetic resonance microscopy of the mouse brain at 100 μm in vivo [55] and ~20–50 μm ex vivo [19] make it possible to observe changes in volume, shape, and texture – at the level of cellular layers. Cellular layers can be observed in a single mouse brain, as well as in the study-specific minimum deformation template (MDT) (Figure 13.2).

A single mouse brain specimen, stained with Gd, imaged at 21 μm resolution provides good definition for white matter tracts, subcortical nuclei, and cortical layers (Figure 13.2a). The MDT shows enhanced layer contrast, as a result of staining and good registration (Figure 13.2b). The MDT can be co-registered to a reference atlas [15] (Figure 13.2c). Atlas-based registration provides regional volume and texture changes [55].

To characterize morphometric changes due to amyloid or other pathologies typical of neurodegeneration requires mapping all of the subjects' images into a reference atlas space, or a study-specific MDT. We use Advanced Normalization Tools [59,60], and a symmetric atlas with 332 anatomical regions [15]. Regional properties are derived after first mapping the labeled atlas onto the MDT, then backpropagating the labels onto all subjects. This mapping also allows for voxel-wise comparisons, without a priori anatomical information. The determinant of the Jacobian matrix of the deformation fields mapping each individual to the MDT [61] directly encodes local volume information, after compensating for global changes. A high-performance computer enables efficient implementations of small animal multivariate brain analysis pipelines, e.g. small animal multivariate brain analysis (SAMBA) [55] (Figure 13.3).

SAMBA includes in its first preprocessing step: recentering, orientation setting, and enforcing a common reference space. All intra-subject contrasts are rigidly co-registered using Mutual Information (MI). Other tasks include bias field correction and skull-stripping [9]. In stages 2 and 3 all subjects are rigidly aligned to an atlas, then affinely to a target image. This can be one of the controls, or an unbiased average linear template created from all subjects. In stage 4, the template is iteratively created. The "MDT group" is typically either the controls or all subjects. Each cohort member is registered and warped to a template – initially just the affine template, then averaged to create an intermediate template. This process captures the optimal appearance of the group. Next, a "shape update" warp is created by averaging the inverse warps, and is in turn incrementally applied to the intermediate template. This process diffeomorphically moves the MDT toward the mean group shape. The resulting distance-minimized template becomes the target for the next iteration, and typically we use six iterations. In stage 5 all subjects are independently reregistered to the MDT (affine + diffeomorphic mapping). This minimizes bias toward the subjects in the MDT group. For nonlinear registrations between like contrasts the cross-correlation (CC) similarity metric is used, while MI is used for unlike contrasts.

Label sets are generated in stage 6 via affine and diffeomorphic registration between the MDT and a labeled brain. For each label and subject, SAMBA calculates the mean volume and other values based on contrast intensities. Study-wide regional statistics can be associated with predictors such as genotype, age, sex, and treatment. Stages 4, 5, and 6 are the computational bottlenecks of the pipeline due to the time requirements for diffeomorphic registration, and the iterative template creation.

In stage 7 SAMBA creates the Jacobian determinant images, then all contrasts are smoothed with a three-voxel sigma Gaussian kernel, before using SurfStat [62], ANTsR (https://antsx.

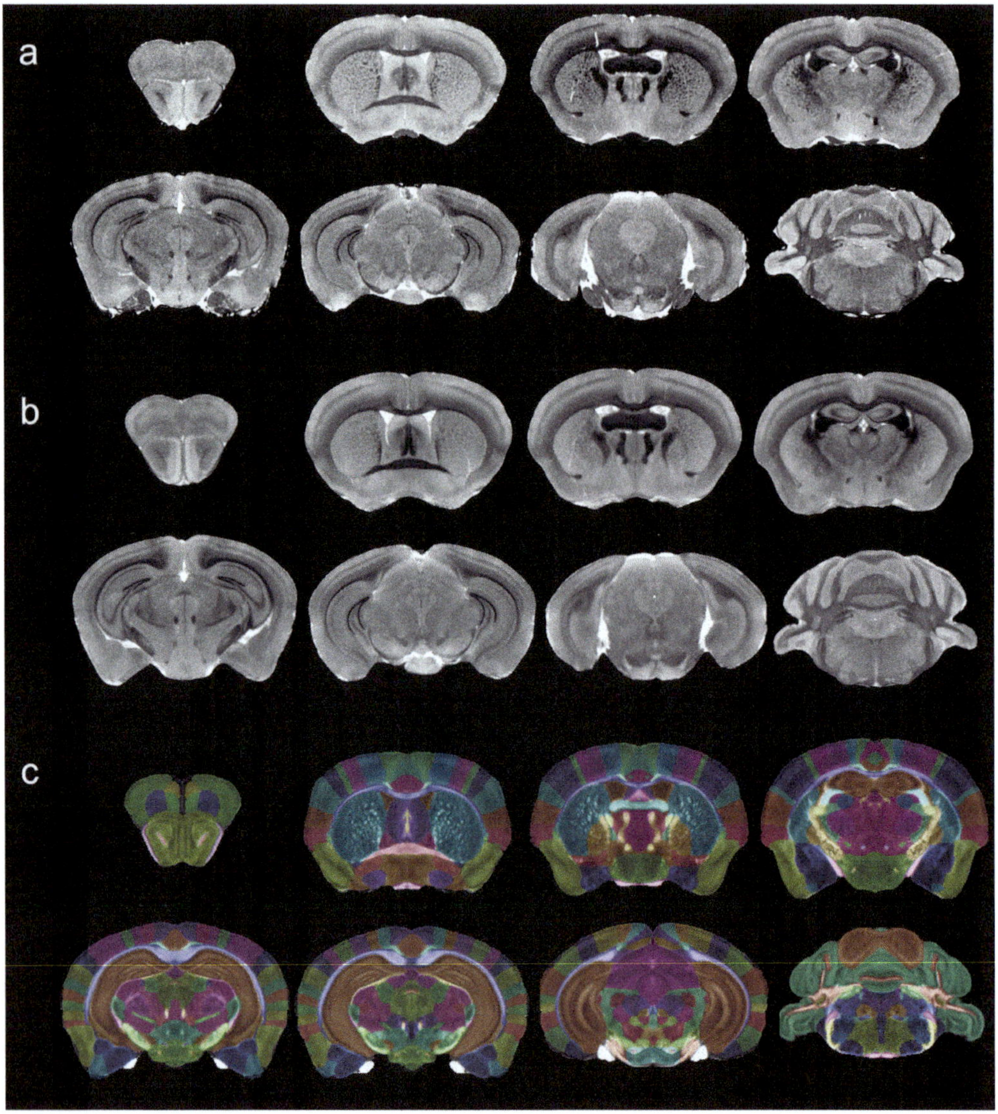

Figure 13.2 Morphometric studies assess group differences, and efforts are underway to identify differences at single-subject level. (a) Single mouse magnetic resonance (MR) image acquired at 9.4 T using a 3D T1w spin echo sequence with echo time (TE) 5.2 ms, repetition time (TR) 50 ms, 62.5 kHz bandwidth, field of view (FOV) = 11 × 11 × 22 mm^3. Fourier space was sampled asymmetrically at 384 × 384 × 768 locations, with zero filling on one side of the array to 512 × 512 × 1024. The receiver gain was increased at the periphery of Fourier space, to provide an expanded dynamic range and effectively amplify the higher frequencies during the phase-encoding steps. Images were reconstructed at 21.5 μm isotropic resolution, and scan time was 2 h 7 min (Johnson, Ali-Sharief et al. 2007). (b) Study-specific minimum deformation template (MDT). (c) Atlas reference providing anatomical priors to label the MDT, and individual subjects.

github.io/ANTsR), or FSL Randomise [63] to provide parametric or nonparametric voxel-based analysis in stage 8.

To optimize SAMBA we have evaluated the performance for its main stages [55]. The essential parameters for registration include: the gradient step size ("SyN" – ranging from 0.1 to 0.25, and 0.5

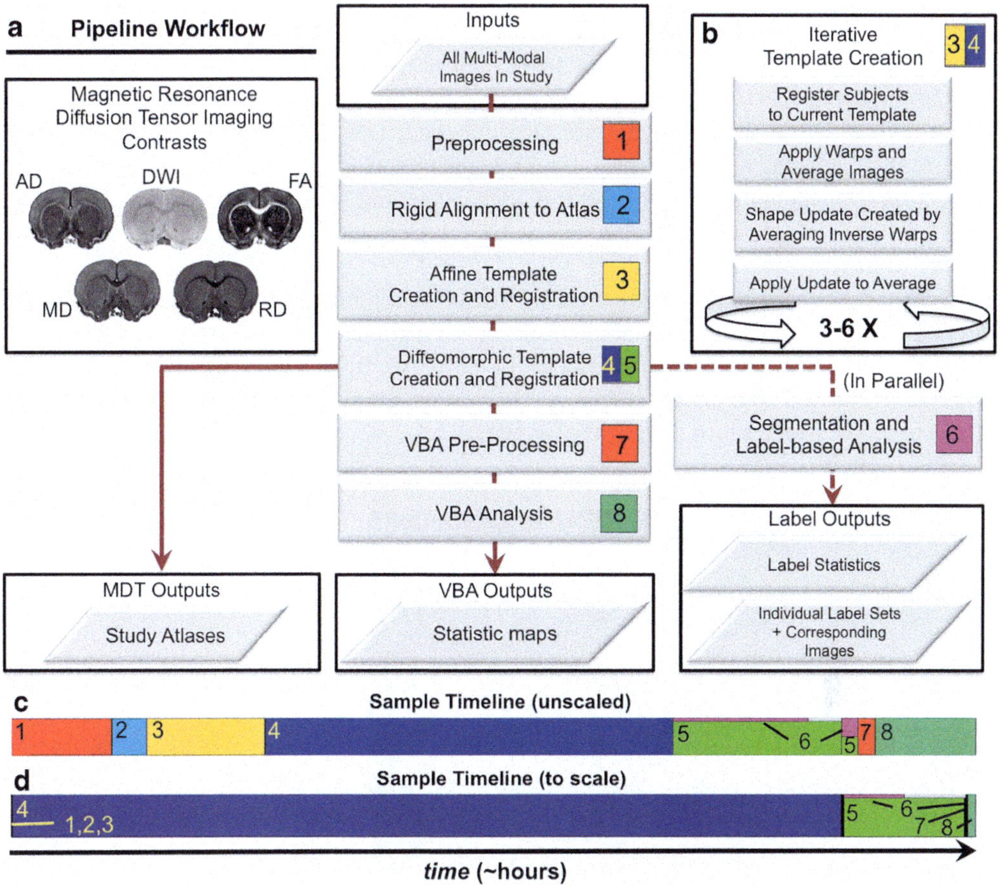

Figure 13.3 The small animal multivariate brain analysis (SAMBA) pipeline (a) processes one or multiple magnetic resonance contrasts through several stages. Stage 1 and 2 are preprocessing, and rigid alignment into a reference atlas space. Stages 3 and 4 iteratively create unbiased affine and diffeomorphic templates (b). Study-specific labeled templates (stages 4, 5) are generated in addition to statistical maps (stages 7, 8). Stage 6 produces regional labels and statistics. Stages 5 and 6 run in parallel, as seen in the unscaled timeline (c). The runtime is largely determined by the diffeomorphic registrations (stages 4, 5, 6), shown by the scaled timeline (d). [55] Anderson et al. (2019), figure 01[p.004]/ Springer Nature/CC BY 4.0.

voxels), a regularization parameter for the velocity/"*update*" field ("RegU"), and a regularization parameter for the *total* warp field ("RegT"). The smaller SyNs allowed us to examine the tradeoff between higher-quality results and runtime. RegU values were chosen as 3 and 5, for recovering large deformations in a reasonable timeframe. RegT values were 0 and 0.5 voxels. The parameter space of *MDT*(*SyN*, *RegU*, *RegT*) was $2 \times 3 \times 2 \times 2 = 24$ permutations. The runtimes ranged from 7.3 h for (*SyN*, *RegU*, *RegT*) set using controls for MDT – *Controls* (0.5, 5, 0) to 187 h (7.8 days) when using all subjects for MDT –*All*(0.1, 3, 0.5), running on 4 cluster nodes, with 32 central processing units (CPUs) each. The largest effect of changing any one parameter was due to SyN. SyN level set at 0.1 runs typically took 4× longer than SyN set at 0.25. Using RegU(5) instead of RegU(3) resulted in a ~25% reduction in runtime. A 30% difference came from choosing RegT(0.5), and a ~45% increase from creating an MDT from all subjects rather than just controls. The time penalty was high when SyN and RegU were small. Thus, better-performing parameter groups required longer runtimes.

13.2.2 In Vivo Morphometry

CVN mice and mNos2$^{-/-}$ controls, aged 14–18 months were imaged in vivo using a 7-T magnet (Bruker BioSpec 70/20 USR, Billerica, MA), using a gradient coil that can supply 450 mT/m, with a rise time of 110 μs, and a quadrature radiofrequency (RF) transmit–receive cryogenic coil (Figure 13.4). Imaging was done using a rapid acquisition with relaxation enhancement (RARE) protocol with a repetition time TR = 1.8 s, effective echo time TE = 28.3 ms, echo spacing 9.4 ms, RARE factor 8. Images were reconstructed at 100 μm isotropic resolution, and acquisition time was ~1 h. In vivo voxel-based analysis (VBA) detected volume loss (false discovery rate [FDR] 10%, cluster threshold 200 voxels) in areas including the septofimbrial complex, fornix, and hippocampus; in the ventral tegmental area, substantia nigra, and retorubral field; as well as the accessory olfactory areas, thalamic nuclei (anterior ventral, ventro-lateral, and reuniens), superior colliculus, and basal ganglia, with clusters extending over the internal capsule. Conversely, the lateral ventricles were enlarged [20].

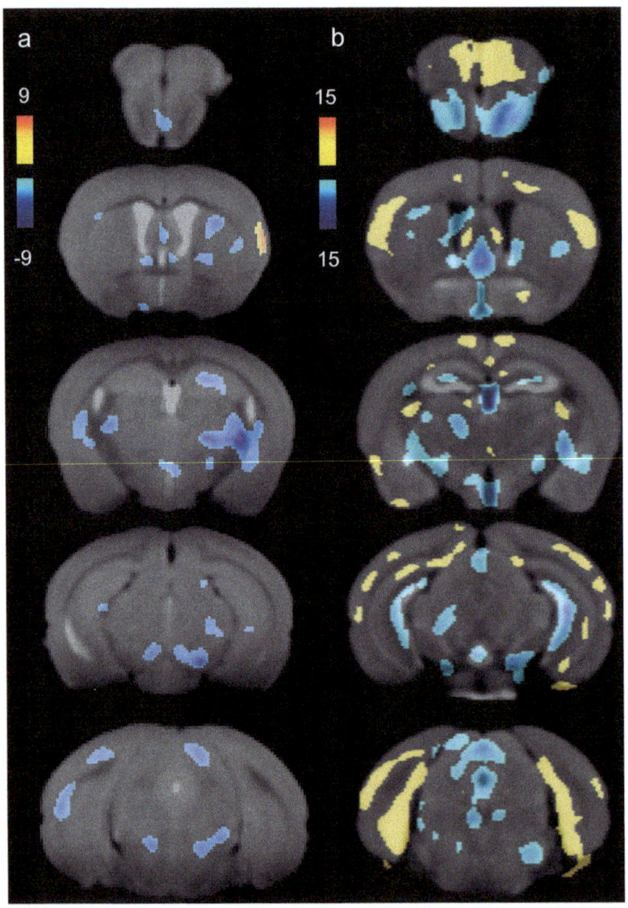

Figure 13.4 In vivo magnetic resonance imaging (MRI) detected atrophy in CVN mice relative to mNOS$_2^{-/-}$ controls (14–16 months) modeling middle-aged humans (a); manganese-enhanced MRI (MEMRI) detected differences earlier, at 12 months (b). These are more extensive, more symmetrical, and effect sizes are larger. The false discovery rate (FDR) threshold for unenhanced MRI was 0.11 [20]; and for MEMRI 0.02 [55]. [24] Badea et al. (2019), figure 5A[p.59]/with permission from Elsevier.

Introducing contrast agents into the brain using Alzet 1007D mini-pumps (Durect Corp., Cupertino, CA), containing 100 μl of 64 μg/μl $MnCl_2{*}4(H_2O)$ (Sigma-Aldrich, St. Louis, MO), dissolved in 100 mmol bicine (Sigma-Aldrich, St. Louis, MO) allowed for a faster acquisition with minimum echo time TE = 10.3 ms, effective echo time TEeff = 20.6 ms, repetition time TR = 150 ms, echo spacing 10.3 ms, RARE partitions 4, in 23 min time. These protocols revealed volume loss in similar areas, but changes were more symmetric and effects sizes were larger and survived more stringent FDR thresholding (0.02).

In vivo magnetic resonance microscopy at 100 μm isotropic resolution suggests that exogenous contrast increases the accuracy of registration, and identifies more voxels with significant differences in CVN models of AD relative to $mNos_2^{-/-}$ control mice [20,24].

13.2.3 Ex Vivo Morphometry

Given the higher imaging resolution, and enhanced contrast in ex vivo magnetic resonance microscopy, VBA provides better definition of areas with atrophy as well as the volume and texture (FA) properties of brain regions. We have examined the CVN model (Figure 13.5) of familial AD, and the APOE/HN model of late onset AD (Figure 13.6). CVN mice had significant volume loss relative to $mNos2^{-/-}$ controls in the hippocampus (6% smaller), including the CA1, the dentate gyrus (dorsal

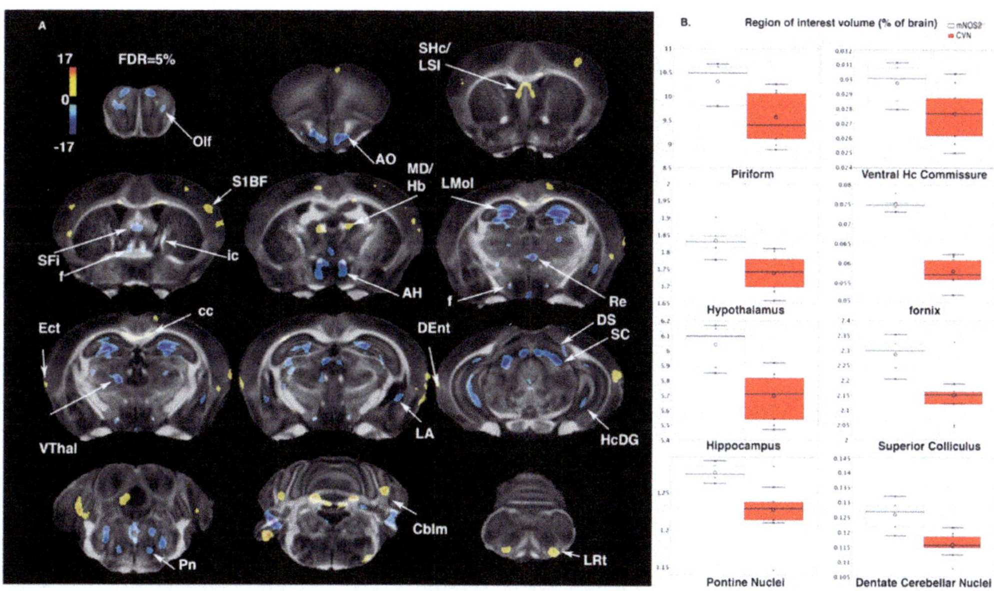

Figure 13.5 (A) Ex vivo diffusion tensor imaging (DTI) detected atrophy (blue) in CVN mice relative to controls (t maps, false discovery rate [FDR] = 5%) in the olfactory areas, preoptic, hypothalamus, lateral amygdala, hippocampal commissure, septofimbrial nucleus, septum, hippocampus, subiculum, ventral thalamic nuclei, superior colliculus, pons, the fornix, and internal capsule. Areas of hypertrophy (yellow) included the somatosensory cortex, corpus callosum, cerebellum, and the lateral reticular nucleus. (B) Regional volumes – the median (line), mean (circle), quantile (box), range (whiskers). AD: anterodorsal thalamic nuclei; AO: anterior olfactory; DEnt: dorsolateral entorhinal cortex; DS: dorsal subiculum; Ect: ectorhinal cortex; HcDG: dentate gyrus of hippocampus; LDDM: latero-dorsal thalamic nucleus (dorsomedial); LA: lateral amygdala; LRt: lateral reticular nucleus; LSI: lateral septal nucleus, intermediate part; LMol: lacunosum moleculare layer of the hippocampus; Olf: olfactory; Pn: pons; SC: superior colliculus; SFi: septofimbrial nucleus; cc: corpus callosum; f: fornix; ic: internal capsule. [20] Badea et al. (2016), figure 02[p.37]/with permission from Elsevier.

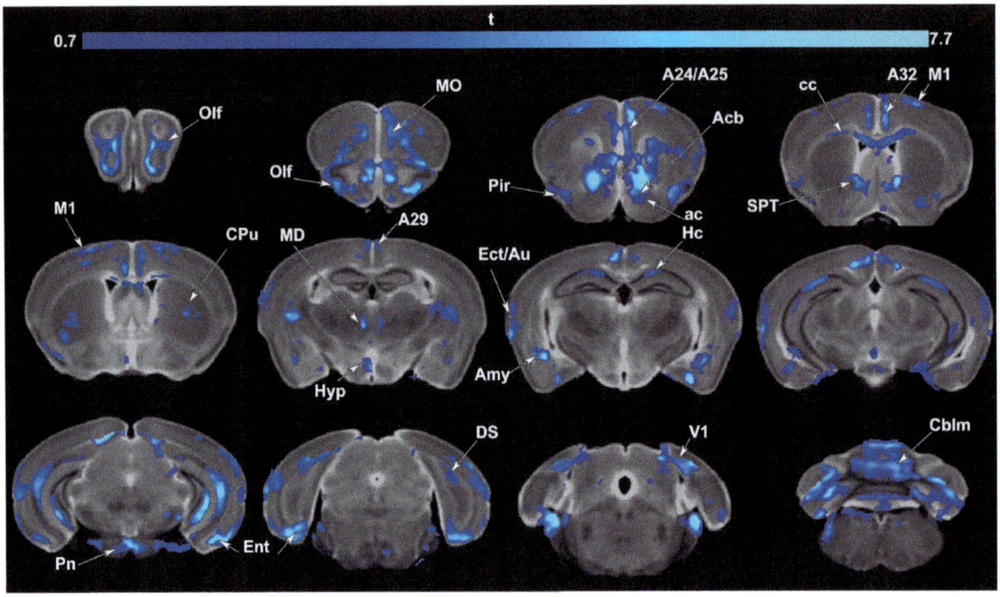

Figure 13.6 APOE4HN models of late onset Alzheimer's disease (AD) showed atrophy (blue) relative to APOE3HN mice in the olfactory/piriform (Olf, Pir) cingulate (A24,25,29, 32), sensory (Ect: ectorhinal, Au: auditory, V1: primary visual cortex) and motor cortex (M1), and the entorhinal cortex (Ent). Deep gray matter atrophy was found in the accumbens (Acb), caudate putamen (CPu), hippocampal formation (Hc, subiculum: DS), amygdala (Amy), thalamic nuclei (mediodorsal: MD), the cerebellum (Cblm), and pontine nuclei (Pn). White matter tracts, e.g. the anterior commissure (ac), and corpus callosum (cc) showed atrophy. Results show t maps, false discovery rate (FDR) corrected, using an initial cluster threshold of 0.05, and the whole brain as a mask. [30] Badea et al. (2019), figure 04[p.007]/Frontiers Media S.A./CC BY 4.0.

hippocampus), and CA3 in ventral hippocampus. Decreased volumes were seen in the subiculum, the hippocampal commissure, the olfactory areas, hypothalamus, lateral amygdala, septum, septofimbrial nucleus, ventral thalamic nuclei, superior colliculus, and pons. The largest reduction in white matter volume was in the fornix (20%), which agrees with findings in humans with AD.

These mouse models of familial [20] and sporadic AD [30] shared areas of vulnerability, e.g. local volume loss in the hippocampus and septum, while the models for sporadic AD had a stronger focus on cortical areas, the caudate putamen, and also the cerebellum.

We found the largest volume loss in APOE4HN mice relative to the APOE3HN controls to be ~10% for the cingulate cortex, the ventral intermediate entorhinal cortex, and the temporal association cortex. The accumbens, amygdalo-piriform transition area, and secondary visual cortex were 7% or smaller in APOE4HN mice. Finally, the cerebellar cortex, middle cerebellar peduncle, and pontine nuclei were ~4% smaller, while the piriform cortex was 2% smaller. VBA is thus sensitive to morphometric changes in mouse models of AD.

13.2.4 Challenges

A major issue with VBA is the long computational time. In its critical step, spatial normalization maps each subject to a common template. In the best scenario from a time perspective, one subject serves as the target template, and the process requires $(N-1)$ registrations. But this introduces a bias toward the selected subject. To eliminate bias it is best to construct a study-specific

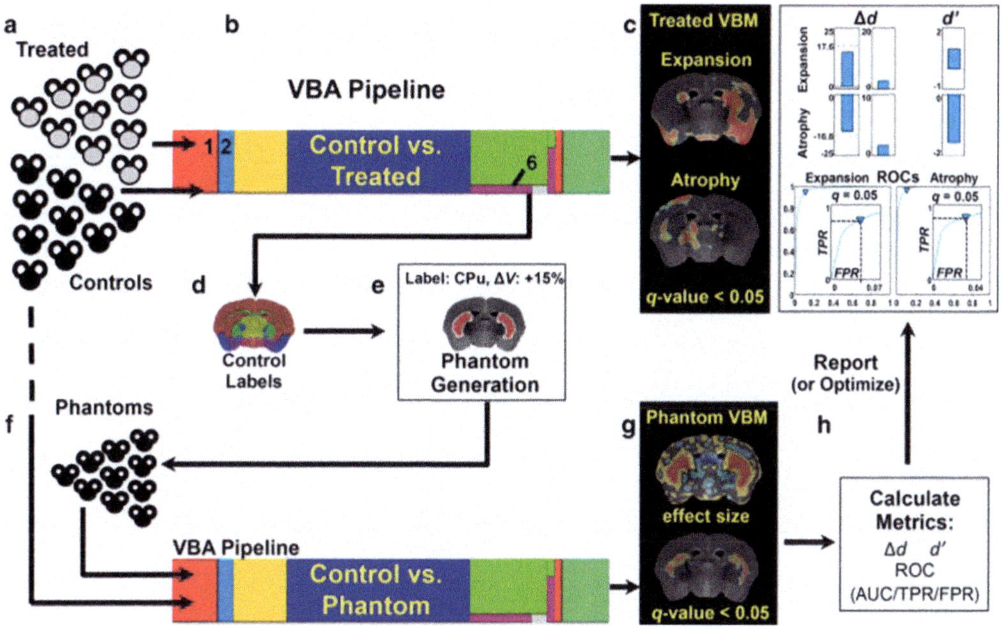

Figure 13.7 Voxel-based analysis (VBA) evaluation. Control and treated images (a) are fed into the VBA pipeline, initialized with a best-guess parameter set (b), to provide statistical results (c). Once automated labels are available for the control images in stage 6 (d), they provide input for phantom creation (e). The user can specify how much atrophy or hypertrophy to induce in phantoms. The pipeline is reinitialized with the control and phantom images (f). The results of the phantom VBA (g) help calculate evaluation metrics (h), reported alongside VBA results, or are used to optimize the registration or smoothing parameters. [55] Anderson et al. (2019), figure 02[p.07]/Springer Nature/CC BY 4.0.

MDT. This strategy requires around six iterations of pair-wise registrations between each subject and the target template, and a minimum of $3{*}N_{\mathrm{MDT}}$ jobs. Then all subjects need to be registered to the final MDT, for a total of $4{*}N_{\mathrm{MDT}}$ jobs. For a study consisting of 10 control (N_{C}) and 10 treated (N_{T}) mice, where only the controls are used to create the MDT we need $3{*}N_{\mathrm{C}}+(N_{\mathrm{C}}+N_{\mathrm{T}})=50$ jobs or ~30 weeks of CPU time if a single CPU is used. It is thus necessary to implement efficient computational strategies for MRI-VBA, and use high-performance compute clusters (HPC) to handle parallel image registrations.

VBA pipelines are sensitive to changes in the processing chain. Access to voxel-based morphometry (VBM) "phantoms" will enhance our ability to produce a quantitative validation. We have proposed synthetic phantom images, which can be tuned to the expected deformation in a study, to guide the selection of pipeline parameters, and estimate the accuracy of the VBA (Figure 13.7). We have also used manual segmentation in an example of a model of epilepsy injected in one hemisphere with kainic acid. The four regions included the left/right hippocampus (Hc) and left/right caudate putamen/striatum (CPu). The right Hc Dice values characterize how well atrophy was recovered. The left CPu functioned as a control, as it was minimally impacted by the kainic acid . We have evaluated different sets of registration parameters using Dice coefficients, distance metrics, areas under the curve, true positive, and false positive rates. RegT and SyN produced the largest effects in both Dice and phantom metrics. RegT paired t-tests featured the smallest p values and, in area under the curve (AUC) and true positive rate (TPR), effect sizes around two times greater than the closest values produced by SyN. RegU had a smaller but significant effect per the phantom metrics, which was not captured by the Dice coefficients. The choice of MDT group (just the controls, or all subjects) had a sizable effect per the Dice coefficients, particularly in the atrophied hippocampi.

The Dice values in the Hc (both left and right) ranged from 86.66% to 95.49%, with a mean of 92.58 ± 1.35%. For the CPu, these ranged from 88.89% to 95.20%, with a mean of 93.46 ± 1.26%.

Given the large differences obtained for different parameter sets (Figure 13.8), it is important to develop computational and experimental frameworks to promote quality assurance of VBA in neuroimaging, and facilitate the creation and communication of robust results.

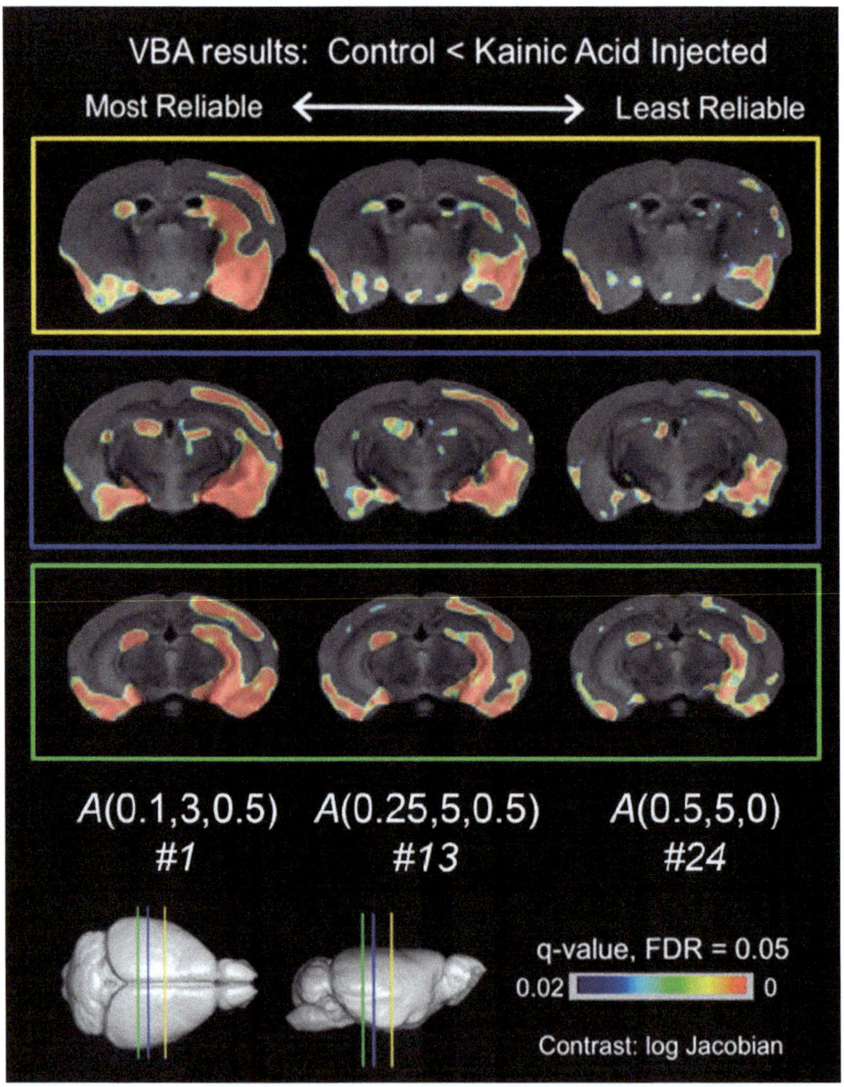

Figure 13.8 Voxel-based analysis (VBA) is sensitive to the choice of registration parameters including gradient step size (SyN), total warp field (RegT), and velocity/update field (RegU), and more accurate results may require longer processing times.

13.3 Diffusion-weighted Imaging and Analysis

Mouse models of familial AD provide useful tools for dissecting the relationship between multiple associated pathologies and white matter. Most prominently these pathologies have been linked to amyloid deposition [64], abnormal tau hyperphosphorylation [47,65], vascular damage [66], and the immune response [67]. MRI provides a translational approach that could provide early in vivo biomarkers, and a whole-brain perspective on such biomarkers. This allows the exploration and targeting of vulnerable networks, which may serve to propagate pathology from a seed region to other remote targets.

Microstructural changes are thought to precede gross neurodegenerative changes estimated with morphometry, which may also precede observable amyloid pathology [19] and cognitive loss. Diffusion-weighted imaging applied to mice modeling middle age (12-month-old mice correspond to 43-year-old humans, 18-month-old corresponding to 56-year-old humans, and 24-month-old to ~70-year-old humans; https://www.jax.org/news-and-insights/jax-blog/2017/november/when-are-mice-considered-old#) may give insight into the early sequence of events that lead to aging and AD-associated brain pathologies.

13.3.1 MR Estimates of Tissue Microstructure

Diffusion-weighted imaging models the water molecule diffusion in different compartments of brain tissues. Diffusion is relatively unconstrained in cerebrospinal fluid, but gets more restricted in more organized tissue, and can exhibit a preferential direction in well-organized white matter bundles. Axons limit diffusion in planes perpendicular to their main direction, an effect strengthened by the myelin sheets wrapping around them. By sensitizing our measurements with diffusion gradients of distinct orientations, we can calculate a diffusion tensor (D) in each voxel. Its eigenvalues denote the principal diffusion directions, and can be used to compute parameters reflecting how anisotropic the diffusion in one voxel is. Axial diffusivity (AD) is the primary eigenvalue, representing diffusivity along the primary axis of the diffusion tensor ellipsoid. Radial diffusivity (RD) is the mean of the secondary and tertiary eigenvalues, characterizing diffusivity perpendicular to the primary axis. Fractional anisotropy (FA) is calculated in each voxel based on the main eigenvalues, and ranges from 0 to 1; the higher the value, the more directional the local diffusion, and the more ordered the tissue. Thus, healthy white matter exhibits high anisotropy, characterized by the FA, higher AD, and lower RD than the gray matter. These parameters give insight into microstructural changes associated with pathology.

13.3.2 Microstructural Changes in Mouse Models of Familial AD

To quantify microstructural differences between CVN mice and age-matched controls we have used a modified 3D diffusion-weighted spin echo (SE) imaging protocol similar to that shown in Figure 13.9 . Fixed specimens were scanned using a 9.4-T, 8.9-cm vertical bore Oxford magnet, providing gradients up to 2000 mT/m (Resonance Research, Inc. Billerica, MA). The diffusion tensor imaging (DTI) protocol used 12 diffusion directions, TE/TR = 12 ms/100 ms, b ≈ 1595 s/mm^2, diffusion pulse width 4.04 ms, separation 5 ms, amplitude 90.94 G/cm. Images were reconstructed at 55 μm isotropic resolution, and scan duration was ~12 h.

The diffusion-weighted images were co-registered to the first, non-diffusion-weighted (B_0) image to correct for Eddy distortions, using an affine transform and advanced normalization tools (ANTs) [60]. Diffusion tensor calculation was performed using the Diffusion Toolkit (http://

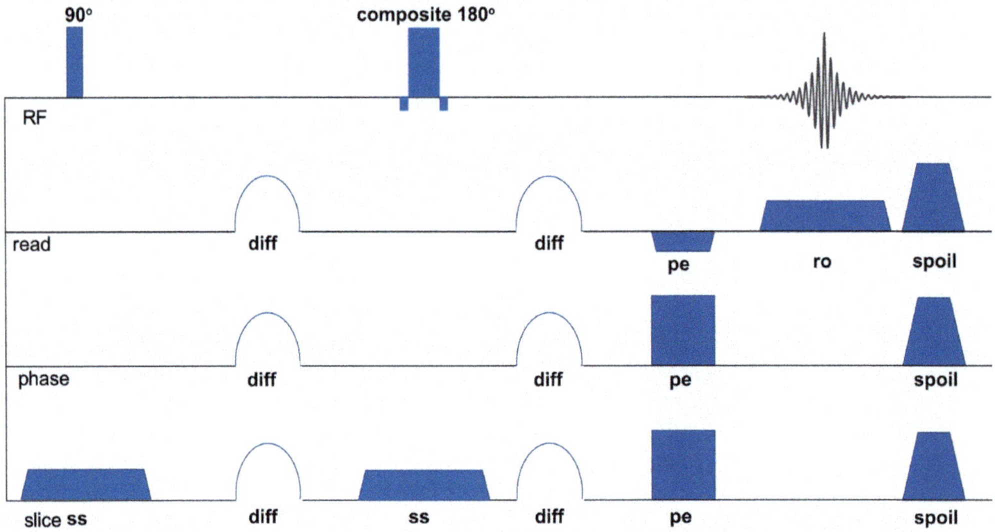

Figure 13.9 A pulse sequence for diffusion-weighted magnetic resonance imaging. Half sine gradients are used to minimize mechanical forces on the gradients since our protocols run for ~12 h.

trackvis.org). To evaluate regional differences in DTI parameters between CVN mice and controls, we used a segmentation pipeline [68] relying on ANTs to generate the MDT driven by the FA images, to map all DTI parametric images into the MDT space, and to label individual brains.

We used a reference atlas [69] to provide anatomical priors. We used MATLAB (MathWorks, Natick, MA) to compute region of interest (ROI) statistics for volume, FA, and electron microscopy parameters: axonal numbers, axonal diameters, and g-ratios.

The images mapped into the space of the MDT were smoothed with a 200-μm kernel. SPM 12 [70] was used for voxel-based statistics. Uncorrected statistics were considered significant at $p < 0.05$, and the minimum cluster size was 200 voxels. Multiple comparison corrections used false discovery rate and considered either: (i) all brain voxels, or (ii) voxels with FA > 0.3, to focus on white matter.

FA reductions (Figure 13.10) were localized largely to the fornix, a white matter tract that connects brain regions important in memory formation and storage [20]. This is consistent with FA changes in AD patients [71]. When relaxing the statistical threshold by constraining the analyses to white matter, we found additional areas with significant structural changes in the ventral hippocampal commissure, internal capsule, cerebral, and middle cerebellar peduncles.

The DTI results were validated using transmission electron microscopy (Figure 13.11). We used the Neural Networks toolbox in MATLAB to identify axons based on morphometric features including: area, eccentricity, equivalent diameter, Euler number, extent, filled area, major and minor axes length, orientation, perimeter, and solidity. The g-ratios were obtained by dividing distances from the axons' centroids to the nearest edge points in the distance maps, and located within 20% of the radius of the equivalent circles. We observed reduced axonal numbers in the CVN mice, relative to mNOS2$^{-/-}$ controls, but larger diameters, and g-ratios indicative of axonal, and possible myelin pathology. Our results suggest that connectivity is impaired in regions connected by the fornix, e.g. the septum, hippocampus, and hypothalamus.

Interestingly, while the CVN mouse model of familial AD had a clear focus on white matter tracts such as the fornix, the APOE4HN model of late onset AD had a more diffuse pattern of FA

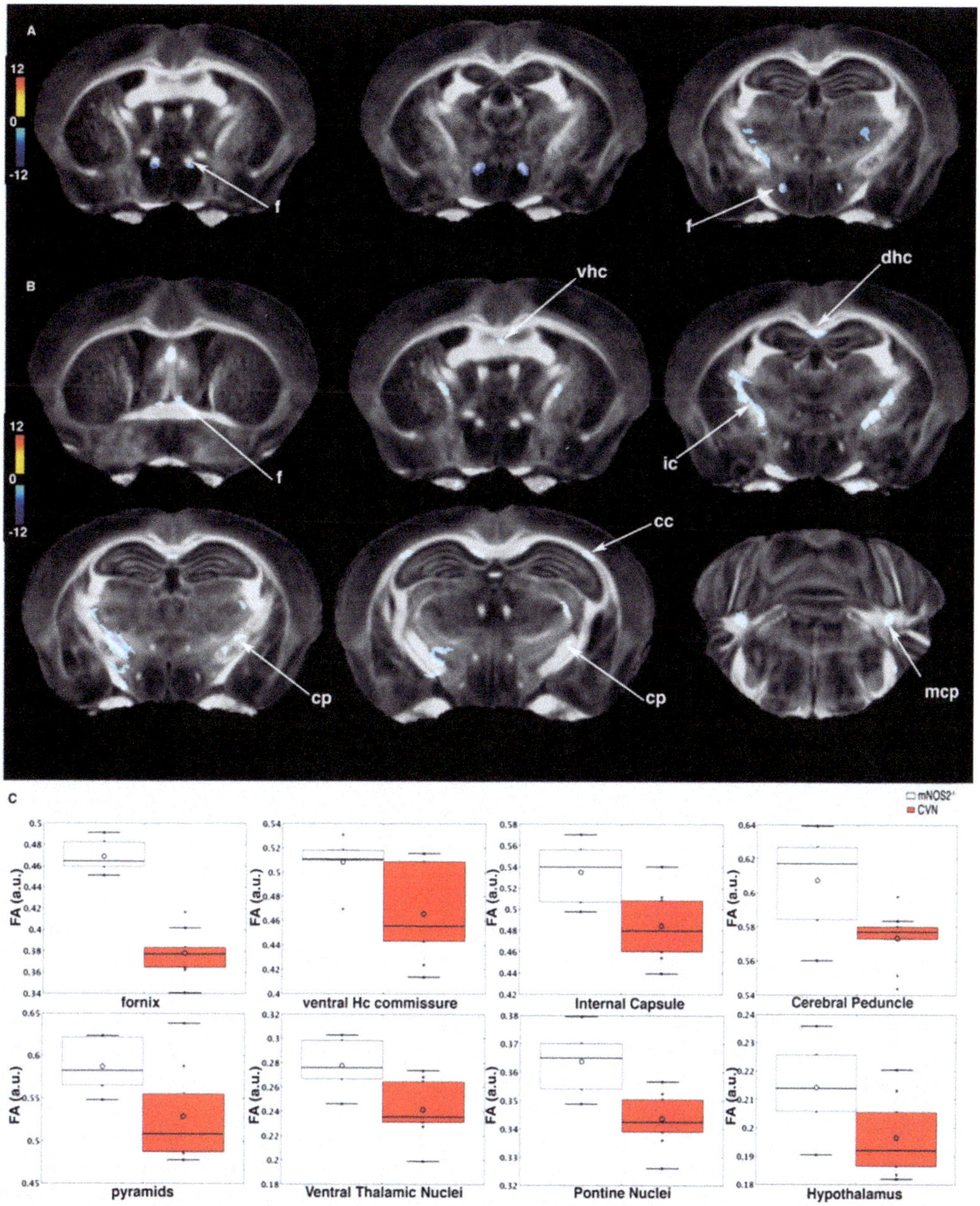

Figure 13.10 (A) Fractional anisotropy (FA) reductions in CVN mice were located in the fornix (20%), and thalamus (13%) (false discovery rate [FDR] = 0.05), using the brain as a mask. (B) Using a white matter mask (FA > 0.3) revealed significant reductions in the fornix, corpus callosum, internal capsule, cerebral, and middle cerebellar peduncle. (C) Region of interest (ROI) differences in FA. cc: corpus callosum; cp: cerebral peduncle; dhc (vhc): dorsal (ventral) hippocampal commissure; ic: internal capsule; f: fornix; mcp: middle cerebellar peduncle. [20] Badea et al. (2016), figure 03[p.42]/with permission from Elsevier.

changes (Figure 13.12) indicating that microstructural properties may change as a result of a different mechanism that starts earlier in vulnerable gray matter regions including the hippocampus.

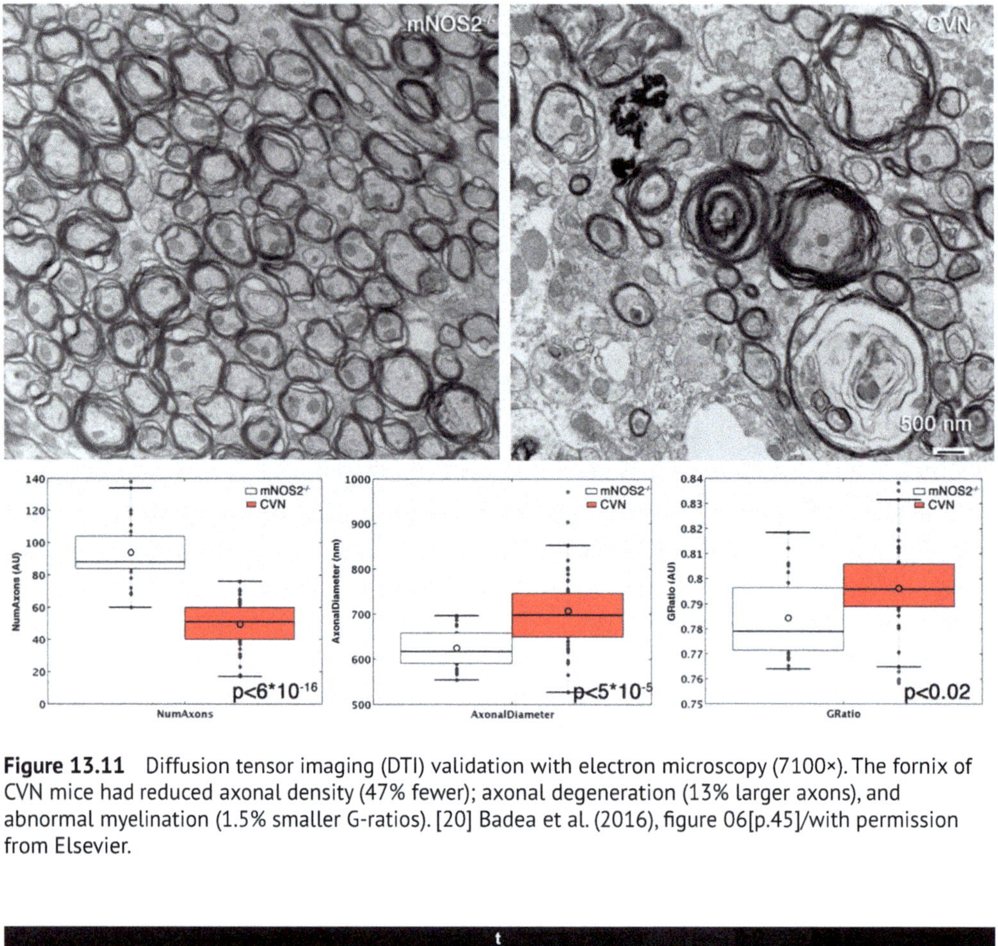

Figure 13.11 Diffusion tensor imaging (DTI) validation with electron microscopy (7100×). The fornix of CVN mice had reduced axonal density (47% fewer); axonal degeneration (13% larger axons), and abnormal myelination (1.5% smaller G-ratios). [20] Badea et al. (2016), figure 06[p.45]/with permission from Elsevier.

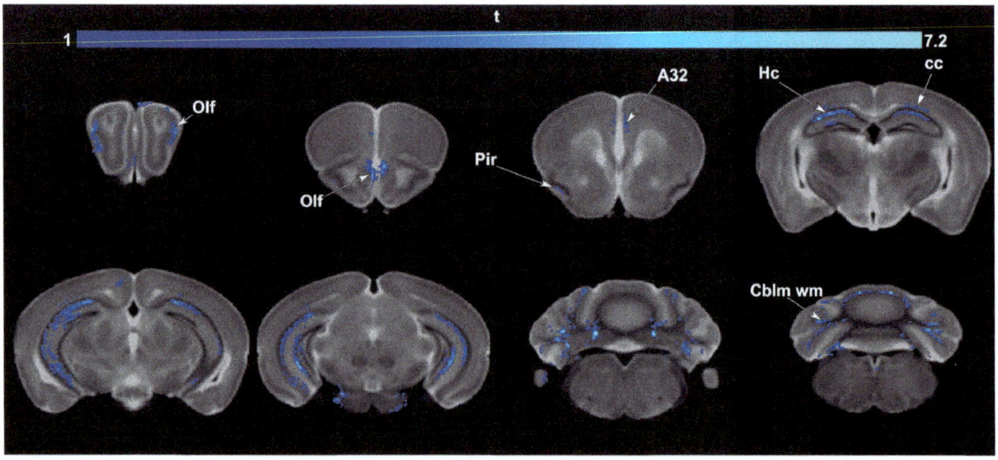

Figure 13.12 Fractional anisotropy (FA) analyses suggest vulnerable brain networks including the olfactory (Olf), piriform (Pir), cingulate cortex (A32), hippocampus (Hc), and the white matter of the corpus callosum (cc) and cerebellum (Cblm wm). Results are presented as t maps, false discovery rate (FDR) cluster-corrected using an initial cluster-forming threshold of 0.05, and the whole brain as a mask. [30] Badea et al. (2019), figure 04[p.007]/Frontiers Media S.A./CC BY 4.0.

13.3.3 Connectivity Changes in Mouse Models of Familial AD

Structural connectomes based on tractography are graphs where the association between a pair of nodes (i.e. atlas regions) is measured by the number of connecting streamlines. Adjacency matrices are generated based on tracts seeded in all atlas regions, as the number of fibers stemming from A that fall within B. The absolute connectivity depends on the number of voxels in the seed region, and the number of times each voxel is seeded for tractography. Connectivity estimates can be normalized by the volume of the seed region, to correct for differences in the total number of tractography seeds/structure. The result is a weighted, directed graph. We have examined global network differences and found age-related decrease of the nodal degree (expressing the probability of a node to share a link with k other nodes) in mice (Figure 13.13) of mixed CVN and mNos2$^{-/-}$ genotypes, in particular for interhemispheric connectivity.

Besides multivariate volume and FA changes between CVN and mNos2$^{-/-}$ controls, the latter predominantly located in the fornix, we have identified connectivity changes ($t = -2.831$, $p = 0.01$), and the subgraph accounting for those differences based on a distance between statistical distributions, such as the Wasserstein distance [72] (Figure 13.14).

These results revealed widespread network effects beyond those found using traditional voxel-based analyses.

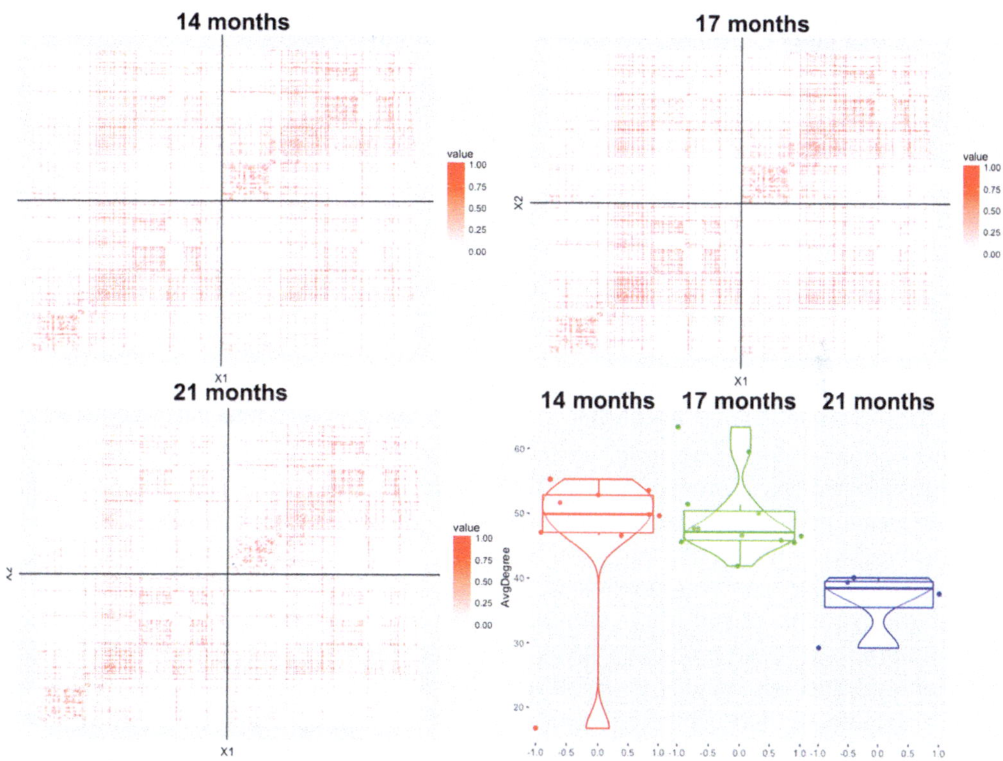

Figure 13.13 Mice show loss of connectivity with aging, as apparent from the binarized adjacency matrices. The 21-month-old mice had a lower average connectivity degree relative to the 17-month group (Cohen $d = 2.13$, $t = 3.69$, df = 14, $p = 0.001$).

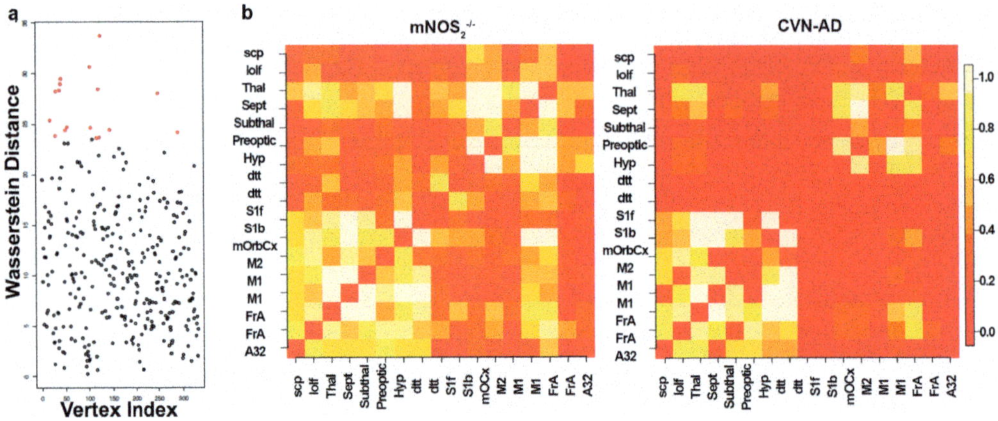

Figure 13.14 Subgraph identification based on the Wasserstein distance (a) of the nodal degree distribution. CVN mice showed lower connectivity in the identified subgraph (b). Scp = superior cerebral peduncles; lolf = lateral olfactory tract; Thal: rest of thalamus; Sept: septum; Subthal: subthalamic nucleus; Preoptic; dtt: dorsal tenia texta; Hypo: hypothalamus; S1f: primary somatosensory cortex, forelimb region; mOrbCx: medial orbital cortex; M1: primary motor cortex; M2: secondary motor cortex; FrA: frontal association cortex; A32: cingulate cortex area 32. Regions listed once are from the left hemisphere, those listed twice indicate the involvement of both hemispheres.

13.3.4 Accelerated Imaging Protocols Using Compressed Sensing

Image acquisitions for diffusion-weighted protocols benefit from a dense sampling in diffusion or $\boldsymbol{q}$-space, or a high angular sampling resolution. A regular, fully sampled Cartesian spin echo sequence like that used before to acquire data for 12 diffusion directions, and one non-diffusion-weighted acquisition needed 12 h, and though excellent at characterizing white matter tracts, is limited in its rotational stability and ability to capture gray matter to gray matter connectivity [73].

A simulation-based evaluation of different acquisitions, based on extracting subsets from a 120 diffusion table that required 10 days of total scan time, is shown in Figure 13.15. Our results suggest that 45 directions balance well the need for high angular resolution, time constraints, and scan costs [73], in order to retrieve connections of cortical regions.

We have sparsely sampled the $\boldsymbol{k}$-space to produce high-resolution diffusion-weighted images in one-quarter or one-eighth of the time. This compressed sensing approach [74] uses the Berkeley advanced reconstruction toolbox (BART) [75]). We have quantified reconstructed image errors (Figure 13.16), and demonstrated the similarity of connectomes between a fully sampled and a compressed sensing (CS)-sampled acquisition at 45 μm spatial resolution, up to a compression level of 8 (0.97 Spearman correlation).

For APOE4HN models of late onset AD, and APOE3HN controls, we have used diffusion-weighted imaging (DWI) protocols with an acceleration factor of 4. The DWI protocol used 46 diffusion-weighted acquisitions, 5 non-diffusion-weighted scans, and: TE = 12 ms, TR = 90 ms, BW = 125 kHz, $b \approx 4000$ s/mm^2, diffusion pulse width = 4 ms, separation 6 = ms, amplitude 130.67 G/cm. Spatial resolution was 55 μm isotropic. Scan time was 14 h.

To reconstruct tracts we have used a constant solid angle (Q-Ball) method implemented in Diffusion Imaging in Python (DIPY) [76]. For tract-based analyses we have used a relative peak ratio of 0.5, separation angle 25°, and 4 parallel compute threads. We used local tracking with 1

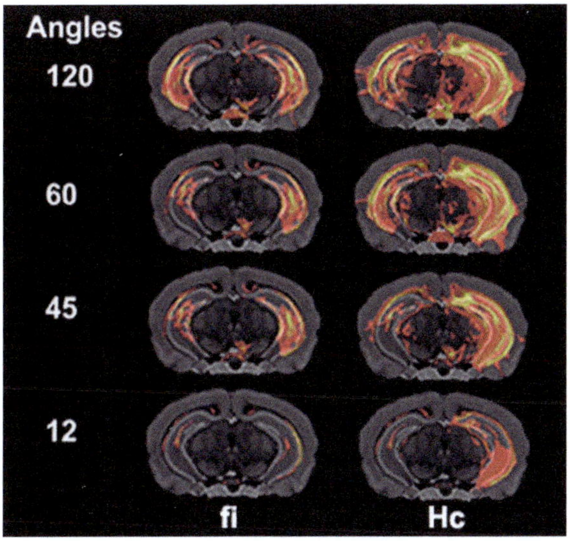

Figure 13.15 Simulations based on subsampling a 120 diffusion gradient table into 12, 45, 60 directions suggest that fimbria (fi) and hippocampal (Hc) projections need to be reconstructed from more than 45 angles.

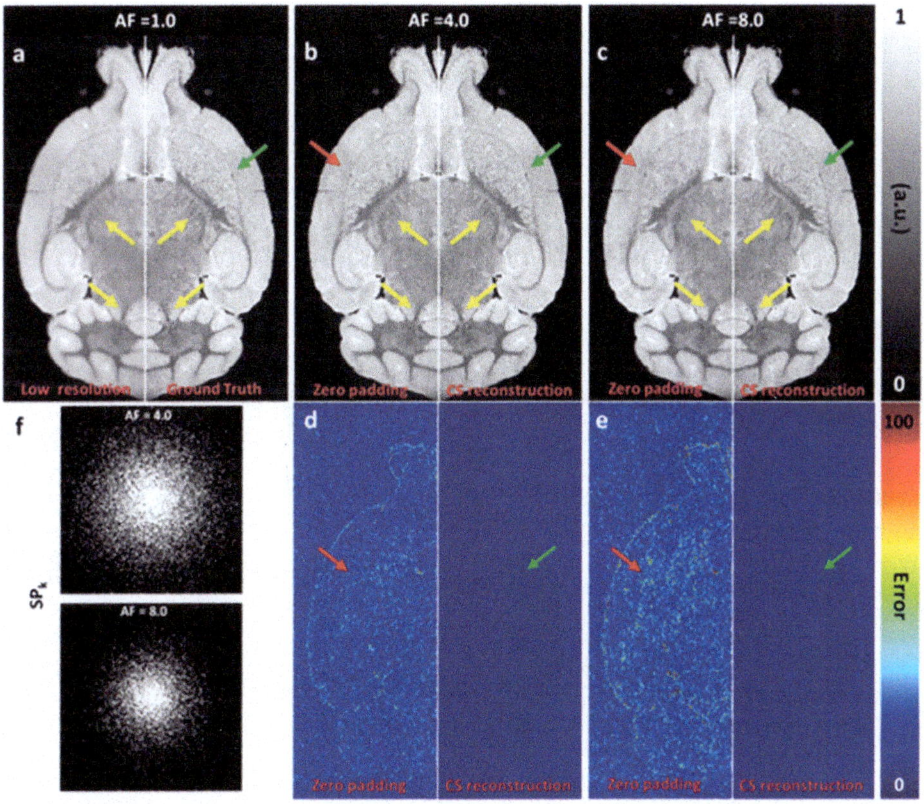

Figure 13.16 Simulations based on a fully sampled acquisition (a) indicate limited errors for acceleration factors of 4 (b) and 8 (c). The errors are substantially smaller for both sampling patterns (f), in the compressed sensing (CS) cases relative to zero filling (d, e). [74] Wang et al. (2018), figure 01[p.4326]/with permission from Springer Nature.

seed per voxel in the whole-brain mask, and 0.5 step size. We saved 10% of the 3 000 000 tracks, in trk files of ~1.5 GB.

We have used tensor network principal component analysis for connectomes (Figure 13.17) [77] to identify the top connections differentiating APOE4 versus APOE3 genotype carriers. The top 30 connected subnetworks included predominantly interhemispheric connections, and regions such as the hippocampus, piriform cortex, cerebellum, the caudate putamen, gigantoreticular nucleus, and the corpus callosum. Intrahemispheric connections pointed to a role for the cerebellum. The three most frequent structures included the piriform cortex, the hippocampus, and the cerebellum.

Tracts from individual brains were clustered based on a Euclidian distance metric minimization [78] then registered to a reference brain, before being clustered again for each specific population (APOE3HN and APOE4HN). Thus, interhemispheric connectivity between the hippocampus and piriform cortex was stronger in APOE3HN mice relative to APOE4HN mice. Figure 13.18 shows the top six largest bundles connecting the hippocampus to the piriform cortex, ranked according to their size. We compared the FA distribution along spatially matched subbundles between genotypes, and found that FA was nonuniformly distributed along the bundles. Together, our results suggest that APOE alleles (Figure 13.19) influence the brain connectivity, some connections showing deficits, some showing the reverse trend, suggestive of compensatory mechanisms.

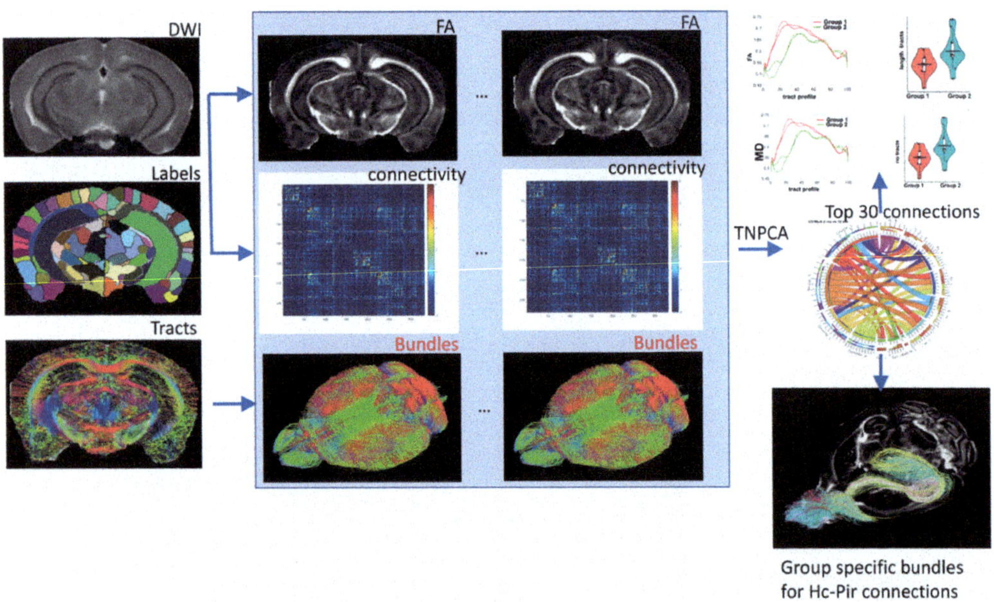

Figure 13.17 To identify vulnerable networks we co-registered the diffusion-weighted imaging (DWI) acquisitions, segmented the brains into 332 regions, reconstructed tracts, then connectomes. Tensor network principal components analysis (TNPCA) identified subgraphs discriminating two genotypes and the resulting selected connections between node pairs were analyzed for tract length and fractional anisotropy (FA) differences. The hippocampus–piriform Hc: hippocampus; Pir: piriform cortex connections are shown as an example. [30] Badea et al. (2019), figure 01[p.004]/Frontiers Media S.A/ Licensed under CC BY 4.0.

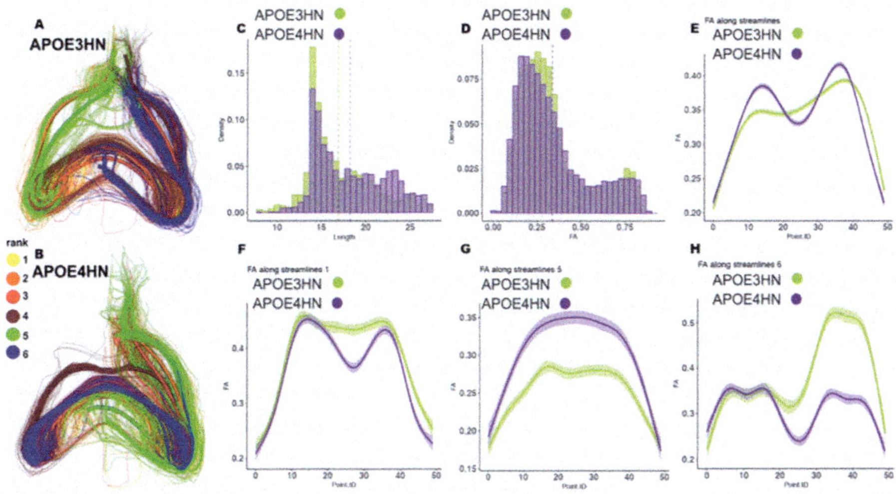

Figure 13.18 The six largest bundles of the piriform–hippocampal interhemispheric connections. Interhemispheric connections were stronger in APOE3HN relative to APOE4HN mice (A) and (B). Differences in distributions are shown for fiber length (C), and fractional anisotropy (FA) (D). The histogram densities indicate a slight shift toward longer length (C), but lower FA in APOE4HN mice (D). We established spatial correspondence through bundle registration, and observed differences along the bundle containing all connections between the piriform cortex and hippocampus were not uniform (E). We identified the top three subbundles accounting for the largest difference between the genotypes (F–H). FA was in general lower for APOE4HN mice in sections of two of these subbundles (F, H), and higher in one (G). >[30] Badea et al. (2019)/Frontiers Media S.A/Licensed under CC BY 4.0.

13.4 Integrating Multivariate Biomarkers Improves the Ability to Predict Cognition

To understand the mechanistic links between the anatomical and cognitive substrates of AD, it is helpful to conceptualize the brain as a network of interacting regions. Structural networks integrate effects occurring at different scales (synapse, cells, circuits), which in turn modulate the functional networks. Several networks have been mapped to the brain and characterized by distinct functional profiles, e.g. sensory perception, movement, attention, and cognition [79].

Approaches to map brain regions to behavioral descriptors have used voxel-based statistical analyses of deformation fields, structural and functional connectivity maps, vascular perfusion, or amyloid deposition and tau maps. But dichotomous strategies, aiming to separate data according to image features, do not necessarily explain the behavioral changes, or the biological processes underlying them. More recently predictive modeling approaches have been proposed to provide statistically relevant imaging correlates of memory changes [80]. We have used such strategies, and measures of atrophy, magnetic susceptibility, and manganese contrast uptake to identify vulnerable brain circuits responsible for behaviors typical of AD, such as the learning performance in a spatial navigation task.

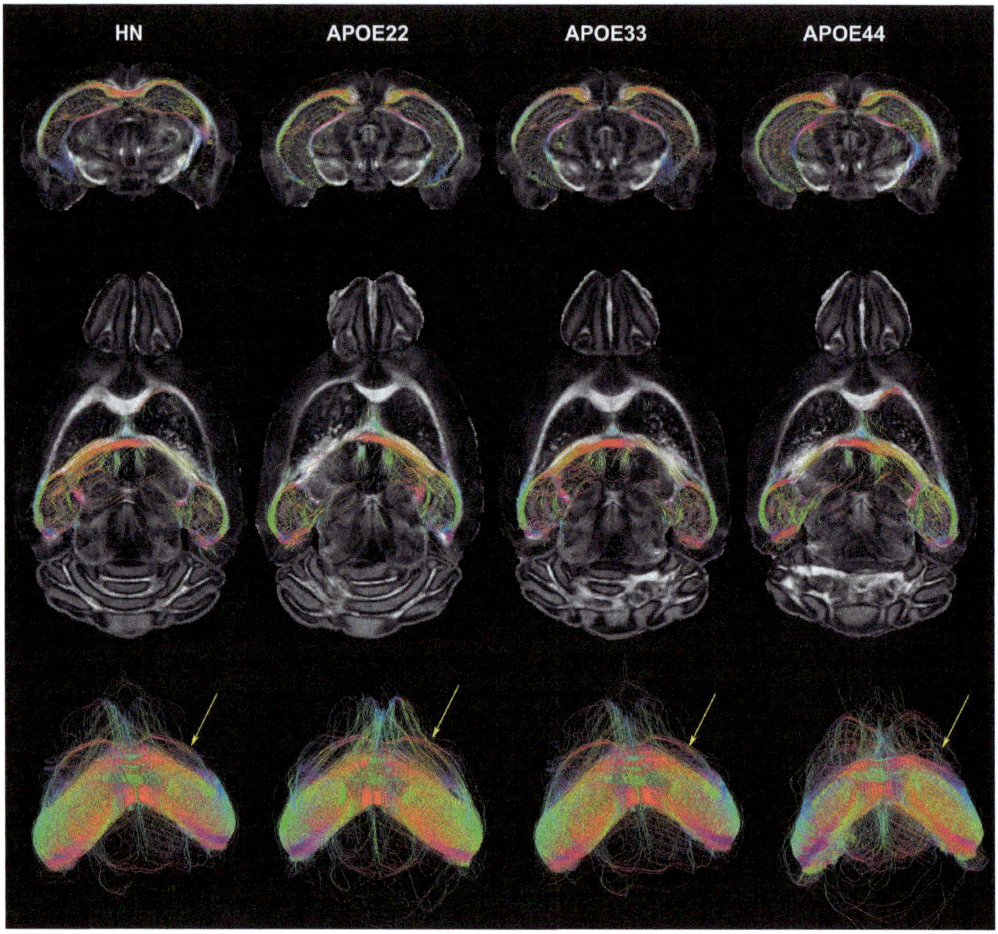

Figure 13.19 The three apolipoprotein E (APOE) alleles and the humanized nitric oxide synthase (NOS) background may influence the brain connectivity, in particular for subnetworks with a role in neurodegenerative conditions, such as those involving the hippocampus.

13.4.1 Integrative Analyses

Approaches involving multiple imaging measurements and dimensionality reduction help predict behavior dysfunction typical of AD, which in mice may be characterized as deficits in the Morris water maze test. To quantify morphometric and manganese uptake we used a RARE sequence with repetition time TR = 150 ms, min echo time = 10.3 ms, echo spacing = 10.3 ms, RARE partitions 4, field of view $2 \times 2 \times 1\ cm^3$, matrix $200 \times 200 \times 100$, 2 excitations, BW = 100 kHz, acquired in 23 min. To estimate quantitative susceptibility maps we used a multi-echo gradient recalled echo (mGRE) with field of view $1.92 \times 1.92 \times 0.9\ cm^3$, matrix $192 \times 192 \times 90$, 8 echoes with spacing of 5.5 ms, 1 excitation, min TE = 3.9 ms, TR = 100 ms, flip angle = 30°, BW = 62.5 kHz, respiratory gated, acquired in ~30–40 min. VBA identified significant differences in all three contrasts, and these regions were used to predict behavior.

We used eigenanatomy (Figure 13.20), a sparse dimensionality reduction method, to extract regions responsible for changes in morphometry (atrophy), signal intensity due to Mn^{2+} uptake

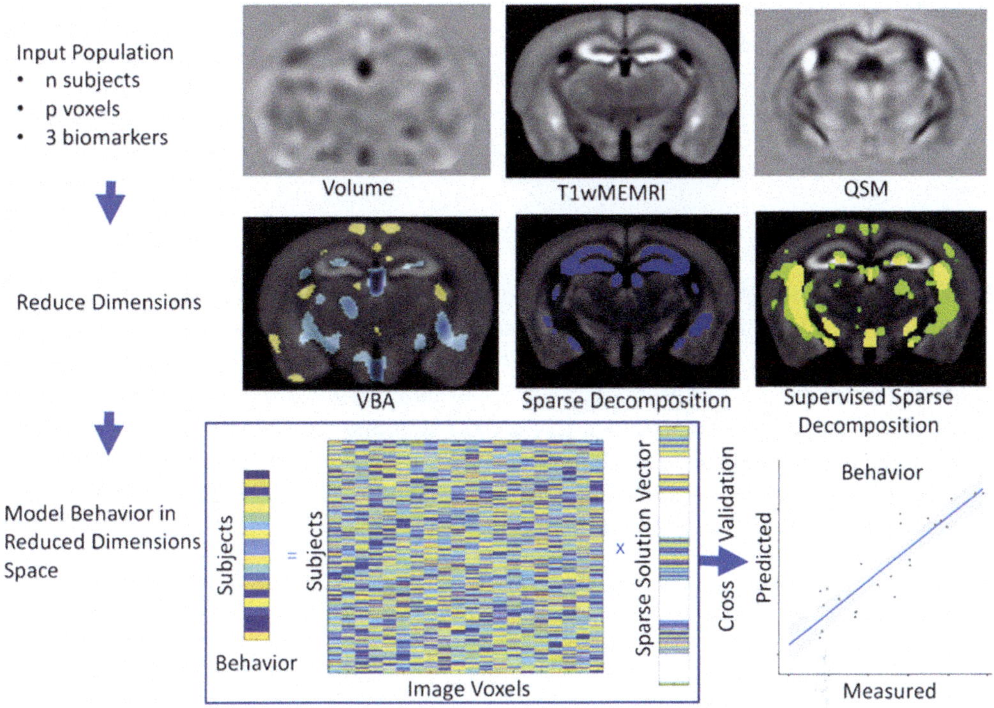

Figure 13.20 Predictive modeling approach in a reduced dimension space. Eigenanatomy identified brain areas with maximum covariance between subjects, producing a sparse decomposition (SD) for each imaging contrast. [24] Badea et al. (2019), figure 02[p.56]/with permission from Elsevier.

(reflective of brain activity), and magnetic susceptibility (reflective of altered iron homeostasis and conducive to oxidative stress and inflammation) [81]. We selected regions that appeared in one or more of the imaging contrasts. To predict cognitive dysfunction based on in vivo multivariate imaging markers, we used sparse canonical correlation analysis (SCCA) [60], which selects regions so as to maximize correlation among imaging and cognitive measures, in a supervised approach.

To identify brain networks associated with behavior, we used canonical correlation (CCA), the multivariate extension of correlation analysis. To find the linear projections of two random vectors, CCA maximizes the correlation between the two linear combinations of the variables in each dataset. Our datasets consist of imaging (X) and behavioral parameters (Y) for n subjects. The imaging parameter is an $n \times p$ multidimensional matrix, and the behavioral parameter is $n \times q$, where p is the number of voxels in X for each subject, and q is the number of behavioral parameters in Y for each subject. Due to the greater size of the imaging matrix compared with behavior, we used sparse CCA as a dimensionality reduction method to produce the solution vectors, x ($p \times 1$) and y ($q \times 1$), which act as weights on columns of X and Y [82]:

$$x^*, y^* = \underset{x,y}{argmax} \frac{xX^T Yy}{\| Xx \| \| Yy \|}; \text{ subject to } \sum_j \| x^j \|_1 \leq s,\ x^j \geq 0. \tag{13.1}$$

The x^* solution vector is subjected to the "L_1" norm, $\|.\|_1$, producing nonzero entries below the chosen sparsity threshold. A nonlinear gradient descent optimizer is used to provide the solution vectors. A cluster threshold and a smoothness constraint are enforced. As a result, sparse CCA produces solution vectors x^* and y^*, dimensioned as subsets of variables that maximize the correlation between imaging and behavior.

We performed a sparse regression between each imaging marker and behavior, measured as the swim distance required for mice to find a hidden platform in the water maze, using spatial cues.

$$\underset{x}{argmin}\frac{1}{2}\| Xx - y \|_2^2 + \frac{\lambda_1}{2}\| x \|_2^2 + \frac{\lambda_2}{2}\| \nabla x \|_2^2; \text{ subject to } \| x \|_1 \leq s. \tag{13.2}$$

Where s is the desired sparseness, λ_1 a ridge penalty that alleviates the problem of multicollinearity among regression predictor variables, λ_2 a smoothness penalty, and ∇ the gradient operator. This is solved through a projected gradient descent.

Both the imaging and behavioral data were split into training and testing sets in a ratio of 75:25. The sparse canonical correlation was initialized from each of the a priori selected image regions (1 eigenvector), using a medium prior (0.5). We used 5% sparseness, 250 voxels cluster threshold. No sparsity constraints were enforced on behavior. The projection vectors obtained from the training set model the relationship with behavior. These models were then used to predict swim distance for the testing set. To assess the quality of the predictive modeling we determined the relationship between the predicted and measured swim distance, using a fourfold cross validation scheme. The resulting eigenregions were used to fit a linear model to swim distance. The combined biomarkers modeling uses image concatenation to extend this approach (Figure 13.21).

The eigenregions based on volume revealed a network comprising the frontal pole, septum, medial thalamus, retrosplenial and cingulate cortex, amygdala, the CA1 area of hippocampus, and fimbria. We noted the involvement of sensory and motor cortices, and of the caudate putamen. The manganese-enhanced MRI (MEMRI) signal intensity-based decomposition revealed a network including olfactory areas, the cingulate cortex, bed nucleus of stria terminalis, substantia innominata, amygdala, and hippocampus. The quantitative susceptibility mapping (QSM) decomposition revealed involvement of the olfactory areas, septum, cingulate cortex, hippocampus, bed nucleus of stria terminalis, and substantia innominata. Iron-rich regions were also involved, such as the substantia nigra, caudate putamen, and globus pallidus.

By adding elements of supervision through the sparse canonical correlation of imaging and behavior, we observed additional areas. These included the fornix, dorsal thalamic nuclei, as well as the ventricles for the volume-based decomposition, the medial thalamic nuclei for the MEMRI-based decomposition, white matter such as the anterior commissure and corpus callosum (including areas below the motor cortex and S1) for the QSM-based decomposition, and more extensive areas on the ventral hippocampus and primary somatosensory cortex.

Specific brain areas common in the results for all three contrasts included the olfactory, cingulate cortex, and retrosplenial cortex, hippocampus, as well as the motor cortices, and septum. The substantia innominata and bed nucleus of stria terminalis were common to MEMRI and QSM.

Using a supervised approach and the combined regions from volume, MEMRI and QSM provided the lowest error among all the predictive models we have tested, a significant correlation of 0.94, explaining 88% of the variance in the swim distance. The tightest confidence intervals and largest adjusted R2 were obtained using the multivariate supervised approach. The integrative approach provided increased value when applied to a model of a multifactorial disease such as AD.

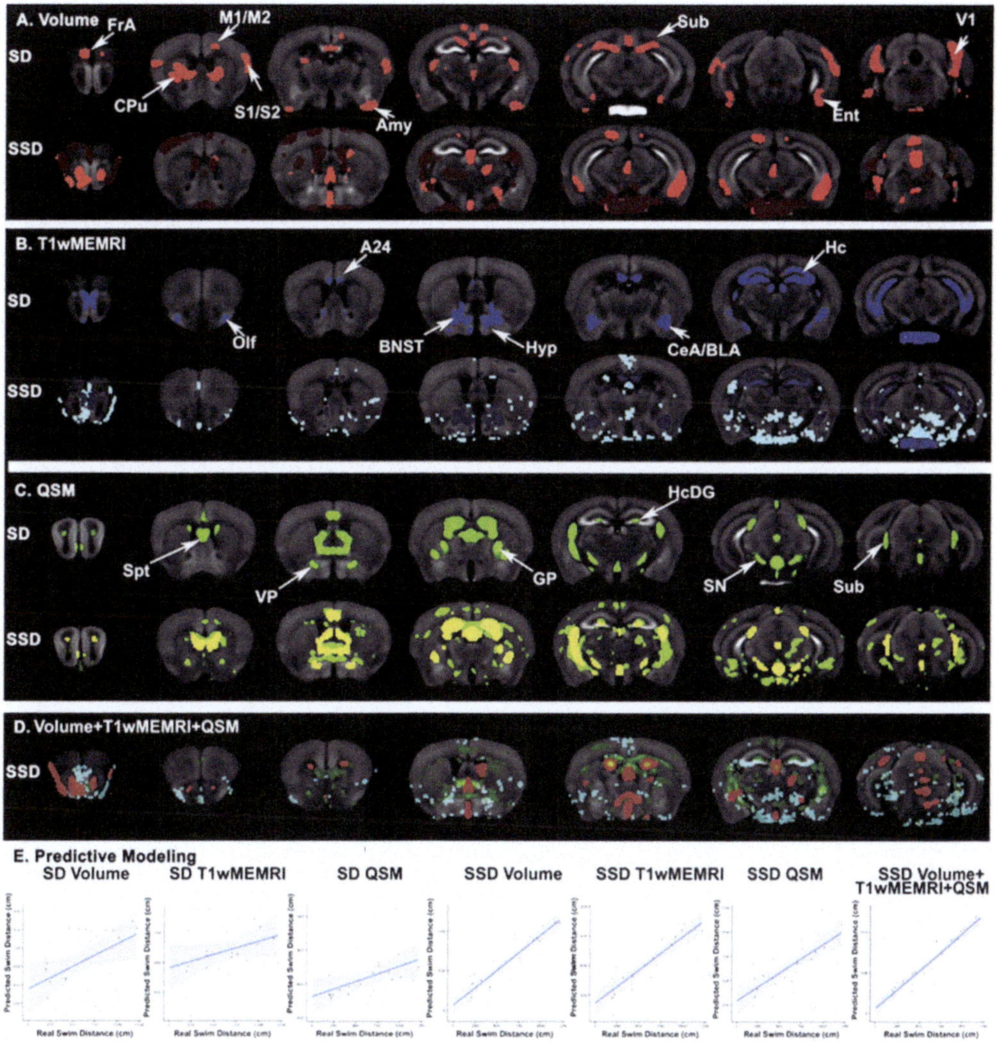

Figure 13.21 Using volume (A), manganese-enhanced magnetic resonance imaging (MEMRI) signal (B), quantitative susceptibility mapping (QSM) (C), or the combination of these biomarkers (D) as predictors revealed associations between the eigen regions and memory, estimated by swim distance on day 4 in the Morris water maze. For all contrasts, the first row shows networks identified by the sparse decomposition. The second row shows networks identified based on a behavior supervised decomposition. The predictive models for swim distance are shown with confidence intervals for each of the biomarkers, as well as for the combined approach (E). FrA: frontal association cortex; M1/M2: primary and secondary cortices; CPu: caudate putamen; S1/S2: primary and secondary cortices; Amy: amygdala; Sub: subiculum; Ent: entorhinal cortex; V1: primary visual cortex; Olf: olfactory areas; A24: cingulate cortex, area 24; BNST: bed nucleus of stria terminalis; Hyp: hypothalamus; CeA: central amygdaloid nucleus; BLA: basolateral amygdala; Hc: hippocampus; Spt: septum; VP: ventral pallium; GP: globus pallidus; HcDG: hippocampal dentate gyrus; SN: substantia nigra. [24] Badea et al. (2019), figure 06[p.61]/with permission from Elsevier.

Summary

Neurodegenerative conditions such as AD arise from multifactorial pathological processes, resulting in multiple biomarkers, some of which can be detected and measured objectively using neuroimaging. For example, volume, and image texture can give insight into neurodegeneration and microstructural changes based on diffusion or magnetic susceptibility, which may precede large atrophy. Not reviewed here are changes in cerebral blood perfusion, blood–brain barrier integrity, or functional connectivity, which may also be measured with magnetic resonance microscopy and be informative about events typical of prodromal AD. Integrative modeling is an important step toward better understanding of this complex disease etiology, as well as predicting its trajectory. Recent efforts to produce models for disease progression and response to treatment have shown promise in AD patients, and cognitively normal people at risk for AD. However, the genetic variability and differences in environmental conditions to which patients have been exposed make these studies difficult. Animal models provide attractive tools for conceptually advancing our understanding of complex pathological processes, and factor-to-factor interactions. Thus, animal models, imaging methods, and biomarkers hold promise to reveal insight into disease processes, and responses to therapeutic interventions.

Acknowledgments

We are grateful to the Center for In Vivo Microscopy (CIVM) and Brain Imaging and Analysis (BIAC) staff, and in particular to Dr Yi Qi for her creative input and solutions, and to Dr Cristian Badea for proofreading. Our research is sponsored by the National Institutes of Health through U24CA220245, RF1 AG057895, RF1AG070149, and R01 AG066184.

References

1 Tan, E.T., Lee, S.-K., Weavers, P.T. et al. (2016). High slew-rate head-only gradient for improving distortion in echo planar imaging: Preliminary experience. *Journal of Magnetic Resonance Imaging* 44 (3): 653–664.
2 Foo, T.K.F., Tan, E.T., Vermilyea, M.E. et al. (2020). Highly efficient head-only magnetic field insert gradient coil for achieving simultaneous high gradient amplitude and slew rate at 3.0T (MAGNUS) for brain microstructure imaging. *Magnetic Resonance in Medicine* 83 (6): 2356–2369.
3 De Graaf, R.A., Brown, P.B., McIntyre, S. et al. (2006). High magnetic field water and metabolite proton T1 and T2 relaxation in rat brain in vivo. *Magnetic Resonance in Medicine* 56 (2): 386–394.
4 Benveniste, H. and Blackband, S. (2002). MR microscopy and high resolution small animal MRI: Applications in neuroscience research. *Progress in Neurobiology* 67 (5): 393–420.
5 Natt, O., Watanabe, T., Boretius, S. et al. (2002). High-resolution 3D MRI of mouse brain reveals small cerebral structures in vivo. *Journal of Neuroscience Methods* 120 (2): 203–209.
6 Bock, N.A., Nieman, B.J., Bishop, J.B. et al. (2005). In vivo multiple-mouse MRI at 7 Tesla. *Magnetic Resonance in Medicine* 54 (5): 1311–1316.
7 Driehuys, B., Nouls, J., Badea, A. et al. (2008). Small animal imaging with magnetic resonance microscopy. *ILAR Journal* 49 (1): 35–53.

8 Badea, A. and Johnson, G.A. (2013). Magnetic resonance microscopy. *Studies in Health Technology and Informatics* 185: 153–184.

9 Badea, A., Ali-Sharief, A.A., and Johnson, G.A. (2007). Morphometric analysis of the C57BL/6J mouse brain. *NeuroImage* 37 (3): 683–693.

10 Johnson, G.A., Badea, A., and Jiang, Y. (2011). Quantitative neuromorphometry using magnetic resonance histology. *Toxicologic Pathology* 39 (1): 85–91.

11 Kovačević, N., Henderson, J.T., Chan, E. et al. (2005). A three-dimensional MRI atlas of the mouse brain with estimates of the average and variability. *Cerebral Cortex* 15 (5): 639–645.

12 Badea, A., Nicholls, P.J., Johnson, G.A. et al. (2007). Neuroanatomical phenotypes in the Reeler mouse. *NeuroImage* 34 (4): 1363–1374.

13 Ma, Y., Smith, D., Hof, P.R. et al. (2008). In vivo 3D digital atlas database of the adult C57BL/6J mouse brain by magnetic resonance microscopy. *Frontiers in Neuroanatomy* 2. https://doi.org/10.3389/neuro.05.001.2008.

14 Aggarwal, M., Zhang, J., and Mori, S. (2010). Magnetic resonance imaging-based mouse brain atlas and its applications. *Methods in Molecular Biology* 711: 251–270.

15 Calabrese, E., Badea, A., Cofer, G. et al. (2015). A diffusion MRI tractography connectome of the mouse brain and comparison with neuronal tracer data. *Cerebral Cortex* 25 (11): 4628–4637.

16 Ali, A.A., Dale, A.M., Badea, A. et al. (2005). Automated segmentation of neuroanatomical structures in multispectral MR microscopy of the mouse brain. *NeuroImage* 27 (2): 425–435.

17 Wu, T., Bae, M.H., Zhang, M. et al. (2012). A prior feature SVM-MRF based method for mouse brain segmentation. *NeuroImage* 59 (3): 2298–2306.

18 Badea, A., Johnson, G.A., and Williams, R.W. (2009). Genetic dissection of the mouse CNS using magnetic resonance microscopy. *Current Opinion in Neurology* 22 (4): 379.

19 Badea, A., Johnson, G.A., and Jankowsky, J.L. (2010). Remote sites of structural atrophy predict later amyloid formation in a mouse model of Alzheimer's disease. *NeuroImage* 50 (2): 416–427.

20 Badea, A., Kane, L., Anderson, R.J. et al. (2016). The fornix provides multiple biomarkers to characterize circuit disruption in a mouse model of Alzheimer's disease. *Neuroimage* 142: 498–511.

21 Badea, A., Ng, K.L., Anderson, R.J. et al. (2019). Magnetic resonance imaging of mouse brain networks plasticity following motor learning. *PLoS ONE* 14 (5): e0216596.

22 Yang, R., Walder-Christensen, K.K., Kim, N. et al. (2019). ANK2 autism mutation targeting giant ankyrin-B promotes axon branching and ectopic connectivity. *Proceedings of the National Academy of Sciences* 116 (30): 15262–15271.

23 Wadghiri, Y.Z., Blind, J.A., Duan, X. et al. (2004). Manganese-enhanced magnetic resonance imaging (MEMRI) of mouse brain development. *NMR in Biomedicine* 17 (8): 613–619.

24 Badea, A., Delpratt, N.A., Anderson, R. et al. (2019). Multivariate MR biomarkers better predict cognitive dysfunction in mouse models of Alzheimer's disease. *Magnetic Resonance Imaging* 60: 52–67.

25 Badea, A., Ng, K.L., Anderson, R.J. et al. (2019). Magnetic resonance imaging of mouse brain networks plasticity following motor learning. *PLoS ONE* 14 (5): e0216596.

26 Madularu, D., Mathieu, A.P., Kumaragamage, C. et al. (2017). A non-invasive restraining system for awake mouse imaging. *Journal of Neuroscience Methods* 287: 53–57.

27 Tsurugizawa, T., Tamada, K., Ono, N. et al. (2020). Awake functional MRI detects neural circuit dysfunction in a mouse model of autism. *Science Advances* 6 (6): eaav4520.

28 Sullivan, P.M., Mezdour, H., Aratani, Y. et al. (1997). Targeted replacement of the mouse apolipoprotein E gene with the common human APOE3 allele enhances diet-induced hypercholesterolemia and atherosclerosis. *Journal of Biological Chemistry* 272 (29): 17972–17980.

29 Johnson, L.A., Olsen, R.H., Merkens, L.S. et al. (2014). Apolipoprotein E-low density lipoprotein receptor interaction affects spatial memory retention and brain ApoE levels in an isoform-dependent manner. *Neurobiology of Disease* 64: 150–162.

30 Badea, A., Wu, W., Shuff, J.M. et al. (2019). Identifying vulnerable brain networks in mouse models of genetic risk factors for late onset Alzheimer's disease. *Frontiers in Neuroinformatics* 13: 72.

31 Jankowsky, J.L., Slunt, H.H., Gonzales, V. et al. (2005). Persistent amyloidosis following suppression of Abeta production in a transgenic model of Alzheimer disease. *PLoS Medicine* 2 (12): e355.

32 Colton, C.A., Vitek, M.P., Wink, D.A. et al. (2006). NO synthase 2 (NOS2) deletion promotes multiple pathologies in a mouse model of Alzheimer's disease. *Proceedings of the National Academy of Sciences* 103 (34): 12867–12872.

33 Kan, M.J., Lee, J.E., Wilson, J.G. et al. (2015). Arginine deprivation and immune suppression in a mouse model of Alzheimer's disease. *Journal of Neuroscience* 35 (15): 5969–5982.

34 Colton, C.A., Wilson, J.G., Everhart, A. et al. (2014). mNos2 deletion and human NOS2 replacement in Alzheimer disease models. *Journal of Neuropathology and Experimental Neurology* 73 (8): 752–769.

35 Humpel, C. (2011). Identifying and validating biomarkers for Alzheimer's disease. *Trends in Biotechnology* 29 (1): 26–32.

36 Suppiah, S., Didier, M.A., and Vinjamuri, S. (2019). The who, when, why, and how of PET amyloid imaging in management of Alzheimer's disease – Review of literature and interesting images. *Diagnostics (Basel)* 9 (2): PMC6627350.

37 Neary, D., Snowden, J., and Mann, D. (2005). Frontotemporal dementia. *Lancet Neurology* 4 (11): 771–780.

38 Hirschler, L., Munting, L.P., Khmelinskii, A. et al. (2018). Transit time mapping in the mouse brain using time-encoded pCASL. *NMR in Biomedicine* 31 (2). https://doi.org/10.1002/nbm.3855

39 Mormino, E.C., Kluth, J.T., Madison, C.M. et al. (2009). Episodic memory loss is related to hippocampal-mediated beta-amyloid deposition in elderly subjects. *Brain* 132 (Pt 5): 1310–1323.

40 Rowe, C.C., Ng, S., Ackermann, U. et al. (2007). Imaging beta-amyloid burden in aging and dementia. *Neurology* 68 (20): 1718–1725.

41 Rodrigue, K.M., Rieck, J.R., Kennedy, K.M. et al. (2013). Risk factors for β-amyloid deposition in healthy aging: Vascular and genetic effects. *JAMA Neurology* 70 (5): 600–606.

42 Bing, K.F., Howles, G.P., Qi, Y. et al. (2009). Blood-brain barrier (BBB) disruption using a diagnostic ultrasound scanner and definity in mice. *Ultrasound in Medicine & Biology* 35 (8): 1298–1308.

43 Rabinovici, G.D., Schonhaut, D., Baker, S. et al. (2015). IC-P-161: Tau PET with [18F]AV1451 in non-Alzheimer's disease neurodegenerative syndromes. *Alzheimer's & Dementia* 11 (7S_Part_2): P107–P9.

44 Nabulsi, N.B., Mercier, J., Holden, D. et al. (2016). Synthesis and preclinical evaluation of 11C-UCB-J as a PET tracer for imaging the synaptic vesicle glycoprotein 2A in the brain. *Journal of Nuclear Medicine* 57 (5): 777–784.

45 Holland, N., Jones, P.S., Savulich, G. et al. (2020). Synaptic loss in primary tauopathies revealed by [^{11}C]UCB-J positron emission tomography. *medRxiv* 24 January: 20018697.

46 James, B.D., Wilson, R.S., Boyle, P.A. et al. (2016). TDP-43 stage, mixed pathologies, and clinical Alzheimer's-type dementia. *Brain* 139 (11): 2983–2993.

47 Holmes, H.E., Colgan, N., Ismail, O. et al. (2016). Imaging the accumulation and suppression of tau pathology using multiparametric MRI. *Neurobiology of Aging* 39: 184–194.

48 Andersson, E., Janelidze, S., Lampinen, B. et al. (2020). Blood and cerebrospinal fluid neurofilament light differentially detect neurodegeneration in early Alzheimer's disease. *Neurobiology of Aging* 95: 143–153.

49 Khalil, M., Teunissen, C.E., Otto, M. et al. (2018). Neurofilaments as biomarkers in neurological disorders. *Nature Reviews Neurology* 14 (10): 577–589.

50 Preische, O., Schultz, S.A., Apel, A. et al. (2019). Serum neurofilament dynamics predicts neurodegeneration and clinical progression in presymptomatic Alzheimer's disease. *Nature Medicine* 25 (2): 277–283.

51 Crews, L., Tsigelny, I., Hashimoto, M. et al. (2009). Role of synucleins in Alzheimer's disease. *Neurotoxology Research* 16 (3): 306–317.

52 Twohig, D. and Nielsen, H.M. (2019). α-synuclein in the pathophysiology of Alzheimer's disease. *Molecular Neurodegeneration* 14 (1): 23.

53 Sweeney, M.D., Sagare, A.P., and Zlokovic, B.V. (2018). Blood-brain barrier breakdown in Alzheimer disease and other neurodegenerative disorders. *Nature Reviews in Neurology* 14 (3): 133–150.

54 Govaerts, K., Dresselaers, T., Van Leuven, F. et al. (2020). Quantitative assessment of age-associated alterations in brain vasculature in wild-type mice and in bigenic mice that model Alzheimer's disease. *Molecular Imaging and Biology* 22 (3): 578–586.

55 Anderson, R.J., Cook, J.J., Delpratt, N. et al. (2019). Small Animal Multivariate Brain Analysis (SAMBA) – A high throughput pipeline with a validation framework. *Neuroinformatics* 17 (3): 451–472.

56 Gong, N.J., Dibb, R., Bulk, M. et al. (2019). Imaging beta amyloid aggregation and iron accumulation in Alzheimer's disease using quantitative susceptibility mapping MRI. *NeuroImage* 191: 176–185.

57 Ossenkoppele, R., Lyoo, C.H., Sudre, C.H. et al. (2020). Distinct tau PET patterns in atrophy-defined subtypes of Alzheimer's disease. *Alzheimer's & Dementia* 16 (2): 335–344.

58 La Joie, R., Perrotin, A., Barré, L. et al. (2012). Region-specific hierarchy between atrophy, hypometabolism, and β-amyloid (Aβ) load in Alzheimer's disease dementia. *Journal of Neuroscience* 32 (46): 16265–16273.

59 Avants, B.B., Tustison, N.J., Song, G. et al. (2011). A reproducible evaluation of ANTs similarity metric performance in brain image registration. *Neuroimage* 54 (3): 2033–2044.

60 Avants, B.B., Cook, P.A., Ungar, L. et al. (2010). Dementia induces correlated reductions in white matter integrity and cortical thickness: A multivariate neuroimaging study with sparse canonical correlation analysis. *Neuroimage* 50 (3): 1004–1016.

61 Chung, M.K., Worsley, K.J., Paus, T. et al. (2001). A unified statistical approach to deformation-based morphometry. *Neuroimage* 14 (3): 595–606.

62 Worsley, K.J., Taylor, J.E., Carbonell, F. et al. (2009). SurfStat: A MATLAB toolbox for the statistical analysis of univariate and multivariate surface and volumetric data using linear mixed effects models and random field theory. *NeuroImage* 47: S102.

63 Jenkinson, M., Beckmann, C.F., Behrens, T.E. et al. (2012). FSL. *Neuroimage* 62 (2): 782–790.

64 Song, S.K., Kim, J.H., Lin, S.J. et al. (2004). Diffusion tensor imaging detects age-dependent white matter changes in a transgenic mouse model with amyloid deposition. *Neurobiology of Disease* 15 (3): 640–647.

65 Wells, J.A., O'Callaghan, J.M., Holmes, H.E. et al. (2015). In vivo imaging of tau pathology using multi-parametric quantitative MRI. *Neuroimage* 111: 369–378.

66 Pathak, A.P., Kim, E., Zhang, J. et al. (2011). Three-dimensional imaging of the mouse neurovasculature with magnetic resonance microscopy. *PLoS ONE* 6 (7): e22643.

67 Tan, J., Town, T., Paris, D. et al. (1999). Microglial activation resulting from CD40-CD40L interaction after beta-amyloid stimulation. *Science* 286 (5448): 2352–2355.

68 Badea, A., Gewalt, S., Avants, B.B. et al. (2012). Quantitative mouse brain phenotyping based on single and multispectral MR protocols. *Neuroimage* 63 (3): 1633–1645.

69 Johnson, G.A., Badea, A., Brandenburg, J. et al. (2010). Waxholm space: An image-based reference for coordinating mouse brain research. *Neuroimage* 53 (2): 365–372.
70 Friston, K.J., Worsley, K.J., Frackowiak, R.S. et al. (1994). Assessing the significance of focal activations using their spatial extent. *Human Brain Mapping* 1 (3): 210–220.
71 Perea, R.D., Rabin, J.S., Fujiyoshi, M.G. et al. (2018). Connectome-derived diffusion characteristics of the fornix in Alzheimer's disease. *NeuroImage Clinical* 19: 331–342.
72 Kolouri, S., Park, S., Thorpe, M. et al. (2017). Optimal mass transport: Signal processing and machine-learning applications. *IEEE Signal Processing Magazine* 34 (4): 43–59.
73 Anderson, R.J., Long, C.M., Calabrese, E.D. et al. (2020). Optimizing diffusion imaging protocols for structural connectomics in mouse models of neurological conditions. *Frontiers in Physics* 8: 88.
74 Wang, N., Anderson, R.J., Badea, A. et al. (2018). Whole mouse brain structural connectomics using magnetic resonance histology. *Brain Structure & Function* 223 (9): 4323–4335.
75 Lustig, M., Donoho, D., and Pauly, J.M. (2007). Sparse MRI: The application of compressed sensing for rapid MR imaging. *Magnetic Resonance in Medicine* 58 (6): 1182–1195.
76 Garyfallidis, E., Brett, M., Amirbekian, B. et al. (2014). DIPY, a library for the analysis of diffusion MRI data. *Frontiers in Neuroinformatics* 8: 8.
77 Badea, A., Ng, K.L., Anderson, R.J. et al. (2019). Magnetic resonance imaging of mouse brain networks plasticity following motor learning. *PLoS ONE* 14 (5): e0216596.
78 Garyfallidis, E., Brett, M., Correia, M. et al. (2012). QuickBundles, a method for tractography simplification. *Frontiers in Neuroscience* 6: 175.
79 Smith, S.M., Fox, P.T., Miller, K.L. et al. (2009). Correspondence of the brain's functional architecture during activation and rest. *Proceedings of the National Academy of Sciences* 106 (31): 13040–13045.
80 Avants, B., Dhillon, P., Kandel, B.M. et al. (2012). Eigenanatomy improves detection power for longitudinal cortical change. *Medical Image Computing and Computer-assisted Intervention: MICCAI International Conference on Medical Image Computing and Computer-Assisted Intervention* 15 (3): 206–213.
81 Jomova, K., Vondrakova, D., Lawson, M. et al. (2010). Metals, oxidative stress and neurodegenerative disorders. *Molecular and Cellular Biochemistry* 345 (1–2): 91–104.
82 Avants, B.B., Libon, D.J., Rascovsky, K. et al. (2014). Sparse canonical correlation analysis relates network-level atrophy to multivariate cognitive measures in a neurodegenerative population. *Neuroimage* 84: 698–711.
83 Liu, C., Li, W., Tong, K.A. et al. (2015). Susceptibility-weighted imaging and quantitative susceptibility mapping in the brain. *Journal of Magnetic Resonance Imaging* 42 (1): 23–41.

14

NMR Imaging of Slow Flows in the Root–Soil Compartment

Sabina Haber-Pohlmeier[1], *Petrik Galvosas*[2,3], *Jie Wang*[2,3], *and Andreas Pohlmeier*[4]

[1] *Institut für Wasser- und Umweltsystemmodellierung (IWS), Universität Stuttgart, Stuttgart, Germany*
[2] *School of Chemical and Physical Sciences, Victoria University of Wellington, Wellington, New Zealand*
[3] *MacDiarmid Institute for Advanced Materials and Nanotechnology, Wellington, New Zealand*
[4] *IBG-3, Research Center Jülich, Jülich, Germany*

14.1 Introduction

The growing global population needs a significant increase in food production, preferably plant-based, against a background of increasing stress due to climate change concomitant with a loss of acreage. At the same time, the food production increase must be accompanied by greater sustainability for optimal use of natural resources such as soil and water. Generally, plant productivity is strongly linked to water and nutrient uptake by the root system, which opens up opportunities for improving it as an alternative strategy besides genetic modifications. The aim is to use the valuable resource of water more efficiently by optimizing root–soil interactions, which requires as a first step a more thorough understanding of the flow processes that occur in the participating compartments of soil and plants. Due to the sensitive nature of the interaction of plant roots with soil, the first prerequisite is the noninvasive and three-dimensional observation of flow processes under the most natural conditions possible. Generally, flow of water through a compartment is a result of the interplay between the difference in pressure or water potential and the ability of the structures in the compartment to transport water, often parameterized by permeability or hydraulic conductivity. Consequently, observed flow velocities can be linked to water potential gradients, and thus they are the basis of the further development and parameterization of model concepts for soil–plant interaction. These concepts, in turn, are a prerequisite for the development of strategies which can contribute to improved and more sustainable plant productivity. In this chapter, we will first briefly highlight the transport processes from the perspective of soil physics and plant physiology, in order to use this as a basis to discuss the measurement of flow velocities in roots and soil with magnetic resonance imaging (MRI) and the associated challenges. We would also like to draw the interested reader's attention to Chapter 11, which, among other things, reports on transport processes in the above-ground parts of plants.

14.2 Transport in the Soil–Plant–Atmosphere Continuum

Figure 14.1 sketches schematically the main compartments related to flow processes in the soil–plant system: atmosphere, leaves, stem, root, and soil. Flow processes take place predominantly in the transport system of roots, stem, and leaves, where the xylem tissue is responsible for upward transport of water and the phloem tissue for downward transport of nutrients. Both

Magnetic Resonance Microscopy: Instrumentation and Applications in Engineering, Life Science and Energy Research, First Edition. Edited by Sabina Haber-Pohlmeier, Bernhard Blümich, and Luisa Ciobanu.

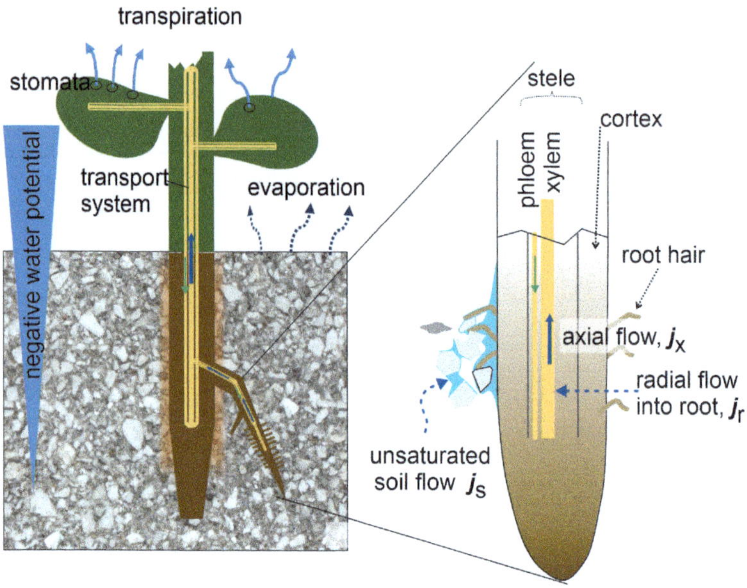

Figure 14.1 Water flow in the soil–plant–atmosphere continuum (SPAC). Left: water moves along the negative water potential gradient from the soil compartment to the atmosphere while passing soil, rhizosphere, roots, shoot, and leaves. Right: schematic root anatomy showing the cortex and the stele region with xylem and phloem vessels, flow paths into the root (j_r) through the soil (j_s) and inside the root (j_x).

tissues are closely adjacent in the vascular bundles. The actual driver for flow is the transpiration from the leaves into the atmosphere, a complex interplay of solar irradiation, air humidity, wind speed, and opening/closing of stomata on the leaf surface. Eventually, the huge difference in water potential of −70 to −95 MPa between the liquid phase in the leaves and vapor in the unsaturated atmosphere causes transpiration [1]. As a result, the water potential decreases in the leaves' mesophyll so that more water is dragged out of the leaves' xylem. This process continues further in the xylem in the stem and finally in the roots where the water potential decreases and the difference from the relatively high water potential in the soil leads to the root water uptake from the soil compartment [1]. However, this simplified picture ignores multiple interactions and hormonal signaling inside the plants. It focuses on the fact that the entire process represents a continuum in which all parts are in a mutual dynamic equilibrium and no part can be considered in isolation [1–3]. Consequently, it is termed the soil–plant–atmosphere continuum concept (SPAC), as depicted in Figure 14.1.

With respect to flow velocities, the lowest values are found in the soil compartment for the unsaturated flow of water in soil ($\boldsymbol{j}_s$). Values range from close to zero in the matrix of clayey material to the order of a few micrometers per second in sandy soils [4]. Approaching the roots, the radial flow velocity $\boldsymbol{j}_r$ is defined as the transfer of water from the soil zone directly at the root surface to the xylem transport system in the stele region in the center of the roots. Noteworthy is that this is a simplified approach since it summarizes the complex regulation processes in one parameter [5].

Due to the large area of the absorbing root surface, radial flow velocities remain small, i.e. generally below 10 μm/s, and they may vary over a broad range depending on ambient conditions [6,7]. Water is axially transported further in hierarchically organized root systems from the outer lateral roots to the taproot and stem. Consequently, flow velocities may increase to values of several tens of micrometers per second. The highest xylem flow velocities occur in the stem with values in the order of tenths of millimeters per second to millimeters per second as observed in smaller plants such as cucumber or tomato [8,9] and well-irrigated trees [1].

14.3 Root–Soil Diagnostics

14.3.1 Geophysical Methods

Direct observation of root water uptake is difficult due to the opaque nature of soil. Therefore, most investigations have determined root water uptake indirectly by placing water potential sensors (potentiometers) or water content-sensitive time-domain reflectometry probes (TDR) in the soil around the roots. Miniaturization may open a route for targeted probing in the vicinity of roots in soil outdoors [10]. Geoelectrical methods such as electrical resistivity tomography (ERT) and electrical impedance tomography (EIT) measure the spatial distribution of electrical properties as a response of injected current into the soil or into the plant itself [11–13]. However, the derivation of information such as water or solute content is performed by inversion techniques and requires great computational effort. From the analysis of these data, flow processes can be expressed in terms of a temporal series of water content changes. For example, ERT allows resolution given by the distance of the observation electrodes placed in the ground. It is usually in the order of few centimeters to several meters, so ERT addresses the scales from a single plant to the field.

14.3.2 X-ray Computer Tomography and Neutron Imaging

X-ray computer tomography (XCT) is a method with the highest resolution down to some microns. The contrast of XCT is highly sensitive to the solid matrix, but water, air, and roots are difficult to segment and require elaborate image processing methods [14]. At the same time, it requires relatively long measuring times since the object is rotated in the beam line through many angles, taking a transmission image at each one. The 3D image is finally reconstructed offline from all transmission images, which typically takes several minutes to hours. Thus, Peng et al. were able to monitor water infiltration fronts in a double porosity model composed of sintered clay spheres with a spatial resolution of 63 μm in three dimensions [15].

In contrast to X-ray CT, neutron radiography in two dimensions and neutron tomography in three dimensions are highly sensitive to all hydrogen atoms in a sample and thus convenient for imaging root water processes [16,17], especially in synergistic combination with MRI [18,19]. Flow velocities are not directly mappable but can be assessed in combination with tracer tracking such as D_2O, either in the soil compartment or in the roots themselves, since the neutron scattering cross-section of deuterium is significantly lower than that of hydrogen. Thus, regions containing H_2O appear dark in the transmission images and regions with higher D_2O content result in higher transmission and appear bright.

Zarebanadkouki et al. [20] grew white lupin in 1-cm-thick rhizotrons (2D) where the soil was compartmentalized in some centimeter-wide quadrants by inserting capillary barriers that are 1-cm broad strips of coarse sand. After a few weeks of growth 3 ml D_2O was injected into a defined compartment and the increase of transmitted intensity was imaged (see Figure 14.2).

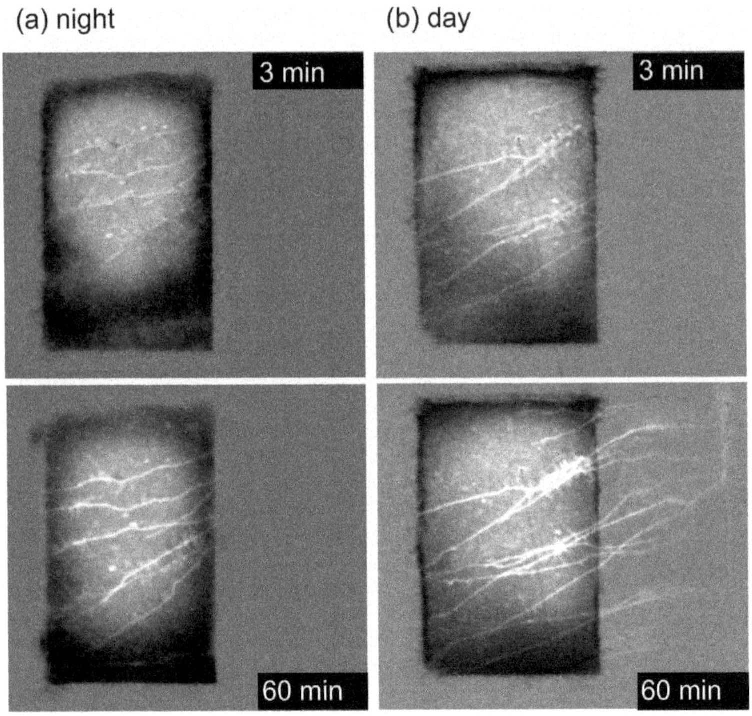

Figure 14.2 Neutron radiography of water uptake in root systems of white lupins, 3 and 60 min after injection of D_2O. (a) Under low transpiration conditions at night. (b) With full illumination of 300 μmol/m^2/s and consequently high transpiration. The rectangles are defined soil compartments into which D_2O was injected at time-point zero and which were separated from the neighboring compartments. [20] Zarebanadkouki et al. (2012), figure 03 [p.007] / with permission from John Wiley & Sons, Inc.

Roots took up D_2O rapidly during day and night phases, but transported it quickly into the hydraulically disconnected neighboring compartment only during the day (Figure 14.2b, bottom). By fitting a transport model they calculated radial flows $\boldsymbol{j}_{\mathrm{r}}$ with an average of 0.17 μm/s and axial volume flows between 6 and 8 × 10^{-6} cm^3/s [20]. Most recently, the group extended a comparable approach for the investigation of water uptake by maize roots [21].

14.3.3 MRI in Combination with Tracer Tracking

MRI, on the other hand, provides the option of using tracers for assessing flow in the soil system even at very low concentrations by exploiting the increase in contrast caused by shortening of relaxation times [22], an effect called paramagnetic relaxation enhancement (PRE) [23–25]. With this approach, we were able to track and quantitatively evaluate accumulation of a T_1-sensitive tracer on the roots of lupine in sand in 3D. The aim of this method is to indirectly deduce the important physiological parameter of hydraulic conductivity K_{r} [26,27] by inverse modeling, i.e. by fitting a fully coupled soil–plant model to the experimentally observed enrichment patterns around roots in their natural environment. The experimental backbone for this study were MRI images of the root system of a three-week-old lupine grown in sand, yielding a time series of 3D tracer concentration maps. We had chosen $Na_2GdDTPA$ as contrast agent since it is chemically inert, does not adsorb at the soil particles, and was not taken up during the experimental period. Its solution concentration was monitored by T_1 mapping using an inversion-recovery (IR) preparation in combination with a spin echo multi-slice pulse sequence (SEMS).

The sensitivity as well as the accessible T_1 range could be enhanced by normalization of the IR sensitive image series to a reference image [22]. Figure 14.3a shows a maximum intensity projection image obtained by a fast spin echo multi-slice (FSE) sequence and (b) depicts three selected slices for further analysis. In (c) we exemplarily show $GdDTPA^{2-}$ concentration maps after five days of irrigation from above with a 1-mM tracer solution. The root density was higher in the top layers leading to more pronounced advective enrichment. With increasing depth the enrichment pattern became increasingly diffuse due to the counteraction between advective enrichment and dispersive spreading.

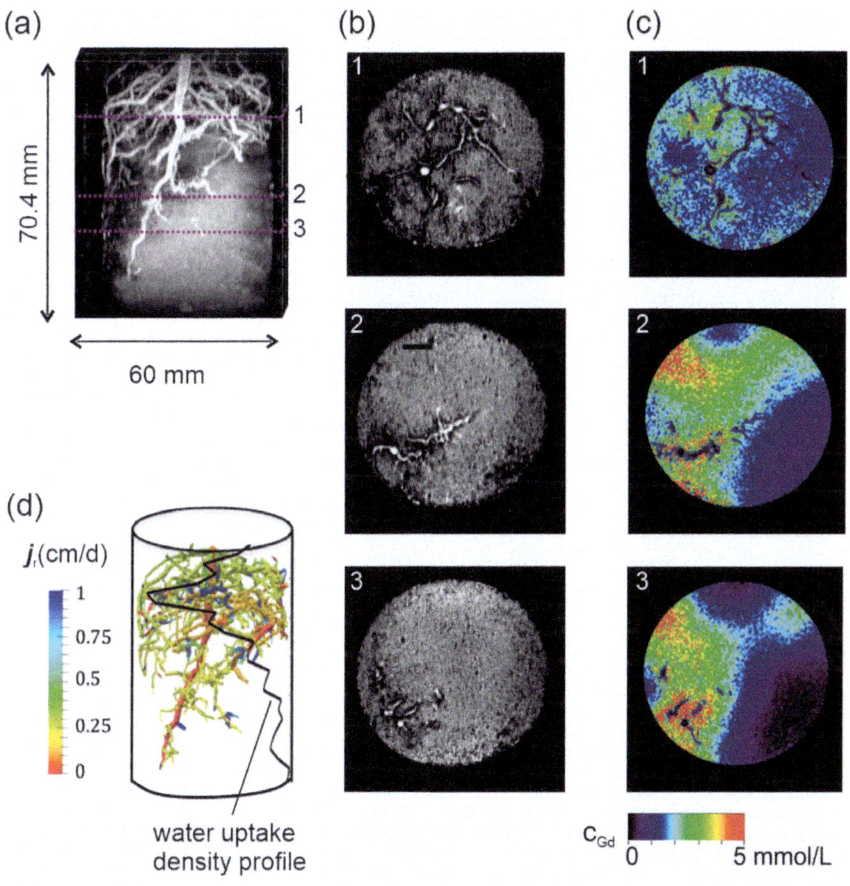

Figure 14.3 Determination of root water uptake rates by tracer enrichment. (a) fast spin echo multi-slice (FSE) image of the root system. (b) Axial cross-sections at three selected depths of the roots. (c) Tracer concentration maps after five days of enrichment determined from spin echo multi-slice pulse sequence (SEMS) T_1 maps. The resolution is 270 μm × 270 μm with a matrix size of 256 × 256 pixels. (d) Three-dimensional radial flow density map, obtained by fitting a fully coupled soil–root model. (a)–(c) [22] Pohlmeier et al., (2017), figure 06 [p.7477] / with permission from John Wiley & Sons, Inc. (d) Modified with kind permission by Creative Commons CC-BY license from Koch et al. 2019 [7].

The fully coupled soil–root model [27] was set up with the root system architecture shown in Figure 14.3a and initiated with transpiration, percolation, and irrigation data. Since hydraulic conductivity depends also on root age, the authors parameterized different classes of roots and fitted the model to the experimentally observed tracer concentration distributions. Thus, Koch et al. determined values for the radial and axial hydraulic conductivities [7] in good agreement with ranges of values reported in the literature for roots of lupine and wheat [28]. These data together with the actual transpiration data and soil water content yielded local radial flow velocities around lateral roots to the order of 0.4 cm/day (see Figure 14.3d).

14.3.4 MRI Velocity Mapping

14.3.4.1 Above-ground Parts of Plants

Beyond applications of tracers, MRI has the potential to directly map local velocities and diffusion coefficients [29]. One typically measures the MRI signals using several $\boldsymbol{q}$-steps in pulsed field gradient (PFG) echo experiments [30] and obtains the displacements as the propagators by voxel-by-voxel Fourier transformation along the $\boldsymbol{q}$-coordinate. This approach returns displacements as spatially resolved coherent motion (flow maps) and random motion (diffusion maps). In the above-ground ("green") compartments of plants the high volumetric water content (θ= 0.7 to 0.9 cm^3/cm^3) and the relatively high flow velocities provide favorable conditions for magnetic resonance flow imaging. On the other hand, the proximity of phloem and xylem transport vessels may lead to an overlay of counter-directed flows in single voxels yielding partial volume effects. This makes a quantitative measurement difficult and is still challenging.

In an early work, Xia and Callaghan successfully imaged the axial flows in a *Stachys sylvatica* stem with a maximum average velocity of 45 μm/s using a multiple pulsed gradient spin echo (PGSE) approach, see Figure 14.4 [31,32]. Later Köckenberger et al. continued similar investigations of a castor bean with comparable in-plane resolution of 47 μm [33,34]. They developed a subtraction technique to overcome the problem of a small fraction of moving water on a background of a large fraction of stationary water [35].

They confirmed for the first time a circulation mechanism for water exchange between xylem and phloem even when the plant did not transpire, which was already proposed by Münch [36].

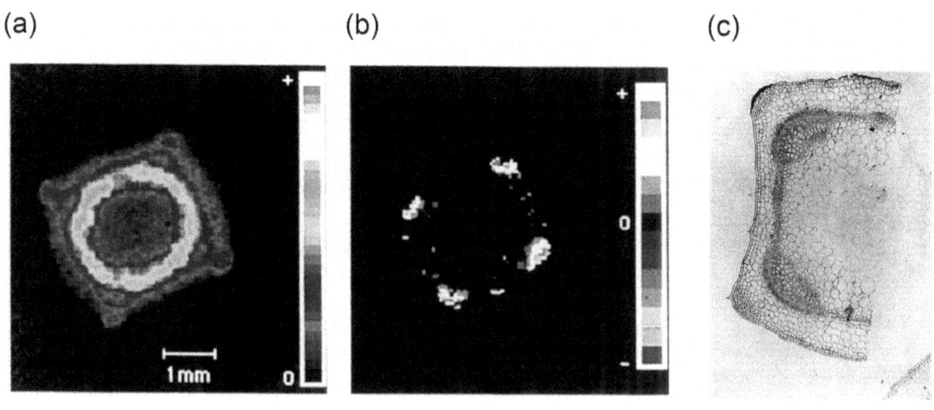

Figure 14.4 MRI of a *Stachys sylvatica* stem. (a) Water proton density image at 60 MHz with an in-plane resolution of 75 μm × 75 μm. (b) Velocity image. (c) Optical micrograph of a half stem. [32] Xia et al. (1993), figure 01 & 02 [p.172 & 173] / with permission from Springer Nature.

Kuchenbrod et al. compared flow imaging based on a phase contrast technique under the assumption of a parabolic flow profile with a Gd tracer tracking velocimetry. The volume flows from MRI and integral transpiration measurements for maize corresponded satisfactorily and showed different behavior during night and daytime [37].

Around the same time, Van As and coworkers further developed flow-imaging methods as well as dedicated hardware for the application to transport processes in plants [29,38–40]. The basic propagator method [41] was combined with a turbo spin echo imaging module to accelerate image acquisition time down to several minutes [42,43]. Data processing can now be performed model-free for the flow profile within a single pixel. Thus it is now possible to differentiate between small changes in flow velocities connected to day and night transpiration differences [8]. With this method, the authors could also differentiate between stationary, only diffusing, water pools and coherent flow motion, for instance, in stems of tomato, tobacco, castor, and maize with velocities in the order of a few hundred micrometers per second [8,9,44].

MRI velocity mapping can also be used outdoors for the investigation of living trees in situ. For that purpose, Kose et al. developed a 0.2 T mobile MRI system and measured xylem and phloem flow velocities in a young zelkova tree. The resolution of the flow velocity is in the submillimeter per second range at a spatial resolution of 1 mm, which is due to the large stem diameter [45]. Although this is still inferior to typical resolutions of magnetic resonance microscopy, it opens the route for higher resolution of on-site MRI of plants so that they can be investigated in their natural environment.

14.3.4.2 Below-ground parts of plants

The need for high spatial resolution with velocity mapping in plants motivated the group of Jakob to employ an 11.75 T magnet for flow imaging of French bean roots under high illumination intensity [46]. Here, the imaging protocol consisted of a spin echo where multi-slice features were not implemented. The imaging module followed a velocity-encoding filter consisting of one PFG pair within a stimulated echo scheme. The velocity map, Figure 14.5, was obtained by fitting the signal equation to a set of eight $\boldsymbol{q}$-steps according to [47] resulting in a flow velocity resolution of 0.2 mm/s. They could demonstrate that the application of ammonium salt led to a reduction in the number of functional xylem vessels and, subsequently, to a reduction of the xylem flow velocities. Since the roots were grown in aquaculture without soil, internal magnetic field gradients did not impede the imaging process.

Thus MRI can measure flows in isolated or hydroponically grown roots, but it has been postulated that hydraulic properties of hydroponically grown roots are different from those of soil-grown roots, and, furthermore, root water uptake is affected by water distribution patterns in the soil matrix [20]. It still remains a challenge to map flow velocities in and around roots in their natural environment, i.e. in soil, due to the following experimental conditions:

1) Flow imaging in soils due to water uptake by plant roots differs from more commonly known flow mapping in packed beds and rocks because of the inherently low flow velocities in the order of several micrometers per second.
2) In root systems, contrary to shoots and stems, flow occurs in all three directions, not only predominantly in the axial (z) direction.
3) One should be careful of long total measurement times since flow patterns may change due to root growth and water uptake. Therefore, we used a multi-slice sequence with interleaved acquisition scheme.

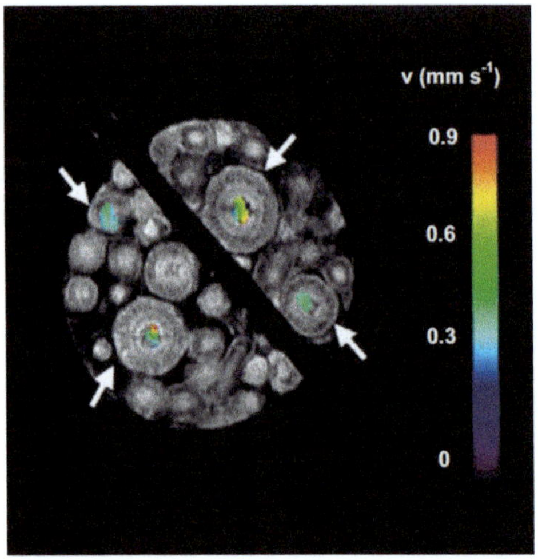

Figure 14.5 Cross-sectional image of root bundles of French bean grown in aquaculture at B_0 = 11.75 T. The grayscale image is the turbo spin echo image with a resolution of 19 μm × 19 μm and an echo time of t_E = 9.1 ms. Overlaid in color is the corresponding velocity image obtained by pulsed field gradient magnetic resonance imaging using a stimulated echo followed by spin echo imaging. [46] Schulze-Till et al. (2009), figure 02 [p.312] / with permission from Springer Nature.

4) Short T_1 may impose restrictions on the observation time and consequently limit the velocity resolution, since the T_1 decay should not be significantly shorter than the z-storage period of the stimulated echo pulse sequence
5) Internal magnetic field gradients, also termed internal magnetic fields, are abundant in many natural porous media. They are frequently caused by paramagnetic impurities in the minerals such as iron or manganese oxides but occur also at interfaces between materials of different magnetic susceptibilities, for instance at a solid–liquid interface or at a gas–liquid interface of air bubbles [48,49]. They scale with the main magnetic field strength, which counterbalances the general push toward higher fields for better resolution [50] and improved signal-to-noise ratios. Noteworthy is that they accelerate transverse relaxation and contribute to so-called cross terms in PFG pulse sequences.

To address this set of experimental conditions, efforts have been made to develop new MRI protocols. The influence of internal magnetic field gradients can be reduced by using bipolar field gradient pulses as proposed by Cotts [51]. This becomes apparent by analyzing the equation for the nuclear magnetic resonance (NMR) signal simultaneously influenced by diffusion and flow:

$$S(t) = S(0)\exp\left\{-D\gamma^2\left[A_{\text{appl}}\left(t,\boldsymbol{G}_{xyz}^2\right) + A_{\text{cross}}\left(t,\boldsymbol{g}_0\boldsymbol{G}_{xyz}\right) + A_{\text{int}}\left(t,\boldsymbol{g}_0^2\right)\right]\right\}$$
$$\times\exp\{\mathrm{i}\gamma\,\boldsymbol{v}[\boldsymbol{C}_{\text{appl}}\left(t,\boldsymbol{G}_{xyz}\right) + \boldsymbol{C}_{\text{int}}\left(t,\boldsymbol{g}_0\right)]\}, \tag{14.1}$$

where $\boldsymbol{C}_{\text{appl}}(t,\boldsymbol{G}_{xyz})$ and $\boldsymbol{C}_{\text{int}}\ (t,\boldsymbol{g}_0)$ describe the influence of PFGs and internal magnetic gradients[1] on the NMR signal under flow with velocity $\boldsymbol{v}$. Further, A_{int} is the contribution to the signal

1 We deviate from both Cotts' and Callaghan's notations by using $\boldsymbol{g}_0$ and $\boldsymbol{G}_{xyz}$ for the internal (background) and applied pulsed field gradients, respectively.

attenuation due to diffusion, which is controlled by the temporally constant, voxel-averaged internal magnetic gradient $\mathbf{g}_0$ alone. A_{appl} is the term contributing to the echo attenuation due to diffusion in the PFGs ($\boldsymbol{G}_{xyz}$) applied in the three Cartesian directions x, y, z; and D is the translational diffusion coefficient, which is assumed to be constant and isotropic. The internal magnetic gradients contribute to a so-called cross term $A_{cross}(t,\mathbf{g}_0\boldsymbol{G}_{xyz})$ and the quadratic term $A_{int}(t,\mathbf{g}_0^2)$ and thereby increase the diffusional decay of the signal $S(t)$. Note that the influence of $A_{int}(t,\mathbf{g}_0^2)$ can be eliminated by normalizing the signal using PFGs to the signal without PFGs [52,53]. It is further noteworthy (i) that there is no cross term influencing the phase shift due to coherent motion (flow), and (ii) that the influence of the internal magnetic gradients by $\boldsymbol{C}_{int}$ on the phase shift can be ignored by measuring a series of different flow-encoding gradients $\boldsymbol{G}_{xyz}$ [51,52]. Furthermore, we developed and employed a pulse sequence where the imaging parts, i.e. the slice selection, phase encoding, and the stimulated echo module were nested into the bipolar PFG pairs [52]. This allowed minimizing of the T_2-sensitive periods between the 90° and 180° pulses and, consequently, prevented unnecessary T_2 decay, compared with a sequence design where the entire imaging part follows the displacement encoding.

We tested the performance of this so-called 13-interval stimulated echo multi-slice imaging (13-interval STEMSI) sequence by studying slow plug flow of $CuSO_4$-doped water in sand. The sand contained small amounts of iron and manganese so that transverse relaxation times were between 20 and 160 ms [52]. The experiments reported in the following were conducted in a vertical 9.4 T Bruker wide-bore magnet, equipped with a 1.5 T/m gradient system (Micro2.5) and a 1.5 cm-wide resonator for a Larmor frequency of 400 MHz. The accuracy, taken as the scattering of the z-velocities around the average value was ~40 μm/s at a spatial resolution of 140 μm. This was confirmed by comparison with the applied external flow and by the scattering of the values for one transverse component, shown in Figure 14.6b on the bottom subplot.

In a next step, we inserted a suction cup in a sand pack as the simplest possible model for root water uptake, since it stimulates radial flow patterns as expected for flow to roots (see Figure 14.1). After initial saturation, a steady-state flow was established by adjusting the height difference of a reservoir outside the magnet, indicated by Δh, while average flow velocity was measured volumetrically via the outflow by collecting the water outside of the magnet. Different experiments were performed with average irrigation flow rates j [54] between 0.15 mm/s and 0.035 mm/s. This resulted in pore flow velocities v_p, defined as $v_p = j/\theta$, between 0.41 and 0.097 mm/s for volumetric water content of saturated sand of $\theta = 0.36\ cm^3/cm^3$.

For all measurements, the flow velocity vector field was determined by successive applications of the flow-encoding gradients $\boldsymbol{G}_{xyz}$ in the three Cartesian coordinates in two $\boldsymbol{q}$-steps. Further MRI parameters were set as follows: field of view (FOV) = 18 mm × 18 mm, matrix size = 128 × 128 resulting in an in-plane resolution of 140 μm × 140 μm, acquisition bandwidth = 100 kHz, 7 axial slices of 2.0 mm thickness, two $\boldsymbol{q}$-steps in the Cartesian directions with $\boldsymbol{G}_{max} = 0.559$ T/m, $\Delta = 30$ ms, and $\delta = 0.2$ ms. The phase cycle was identical to that reported in [52]. Figure 14.7 shows the flow patterns obtained for the lowest average flow rate. In the top slice we observed plug flow with an average pore flow velocity of 110 μm/s, top row in Figure 14.7b and Figure 14.7c, top. It correlates well to the applied pore flow velocity of $v_p = 97$ μm/s.

The pore flow velocity was slightly higher as expected near the wall due to the locally enhanced hydraulic conductivity caused by packing faults at the walls. The suction cup induced a funnel-shaped flow pattern as can be seen in Figure 14.7b, middle row. The z-component pointed downward for the most part, whereas the y-components point toward the center, made obvious by blue colors (negative velocities) on the left and red colors (positive velocities) on the right side. We

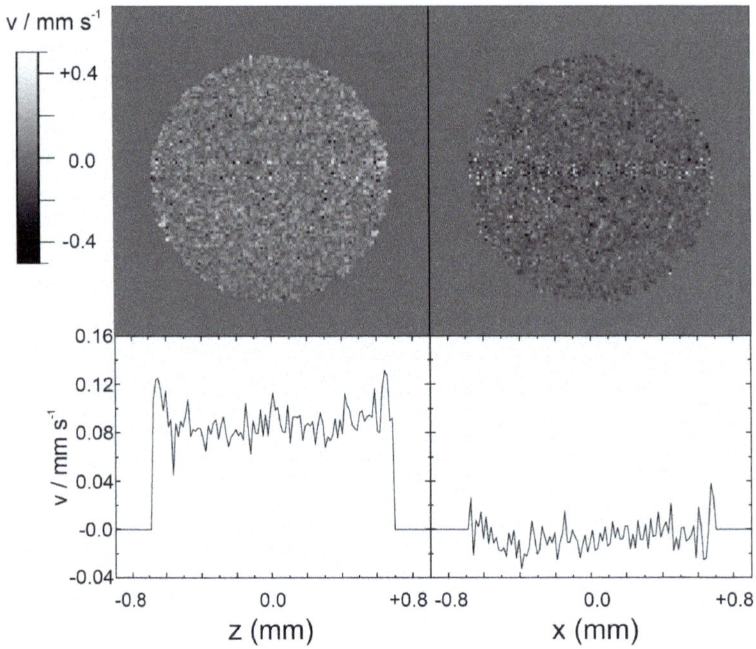

Figure 14.6 Flow patterns in a homogeneous sand pack. Top: axial (a) and one radial (b) component of the velocity vector in an axial slice. Bottom: profile lines through the maps. [52] Spindler et al. (2011), figure 02 [p.219] / with permission from Elsevier.

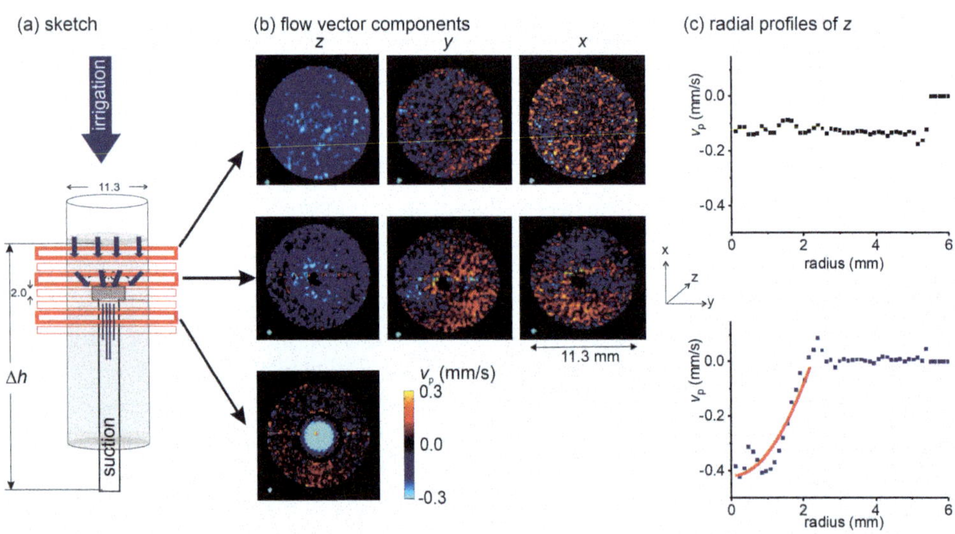

Figure 14.7 Results for the heterogeneous flow phantom. (a) Three slices were selected (see highlighted red rectangles). (b) *z*-, *y*-, and *x*-components of the flow vector. (c) Radial profiles through slice 1 (top), showing plug flow pattern and through slice 6 (bottom), showing a flow pattern through the capillary resembling a parabolic shape, red curve.

found analogous patterns for the x-components rotated by 90°. In the capillary below the suction cup the flow resembles a parabolic flow pattern within uncertainties, whereas the net flow outside was zero (Figure 14.7c, bottom). To visualize the 3D flow field, the flow components in three directions were combined into flow vectors in Figure 14.8. Generally, the vectors point to the center of the suction cup with higher magnitudes (red), combined with a vortex-like pattern, tentatively caused by asymmetric injection from the top. As a first conclusion, one can state that with this pulse sequence it is possible to measure the three Cartesian components of the flow vector field at slow velocities.

Finally, we investigated flow velocities in a living root system in soil, which required further optimizations. Typical values of T_1 are around one second for the water in plants and acceleration by addition of $CuSO_4$ like it was done for the suction cup experiment and in [52], leads to plant damage. Therefore, imaging an entire root system by, e.g. 20 slices, 3 Cartesian directions, a repetition time of $T_R = 5T_1$, and 32 accumulations for improving signal-to-noise ratios would result in total measurement times of several days with the existing 13-interval STEMSI. During such periods, significant changes due to root growth could change the flow patterns and lead to uncertainties in the interpretation of the data. Therefore, the acquisition time needed to be considerably shortened by implementing true multi-slice velocity encoding in interleaved slice selection mode and by replacing all hard radiofrequency pulses with three-lobe sinc-shaped slice-selective pulses according to the concepts of the stimulated echo acquisition mode (STEAM) [55], as shown in Figure 14.9. This redesigned sequence is termed as 13-interval STEAM-MSI. The black gradient lobes along $\boldsymbol{G}_x$, $\boldsymbol{G}_y$, and $\boldsymbol{G}_z$ directions are readout (frequency-encoding), phase encoding, and slice selection gradients, respectively, and depict the actual imaging part together with the radiofrequency pulses. The red bipolar gradient pairs, schematically shown for the x-direction, encode displacement.

Five points have to be considered for the application of this modified pulse sequence: (i) The PFG pairs should be centered between the $\pi/2$ radiofrequency pulses to minimize artifacts from internal field gradients [56]. (ii) The integral of the slice refocusing gradient should be 50.6% of that of the slice-encoding gradient because of nonlinear Bloch equations [57]. (iii) The observation (mixing) time Δ should be sufficiently long to achieve the high resolution in velocity encoding necessary for slow flows but it should be shorter than or equal to T_1. In

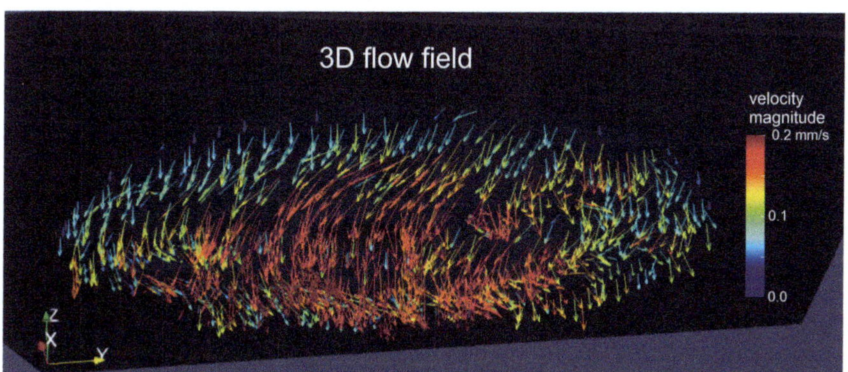

Figure 14.8 Three-dimensional flow field for the second highlighted slice in Figure 14.7, constructed from the three Cartesian components.

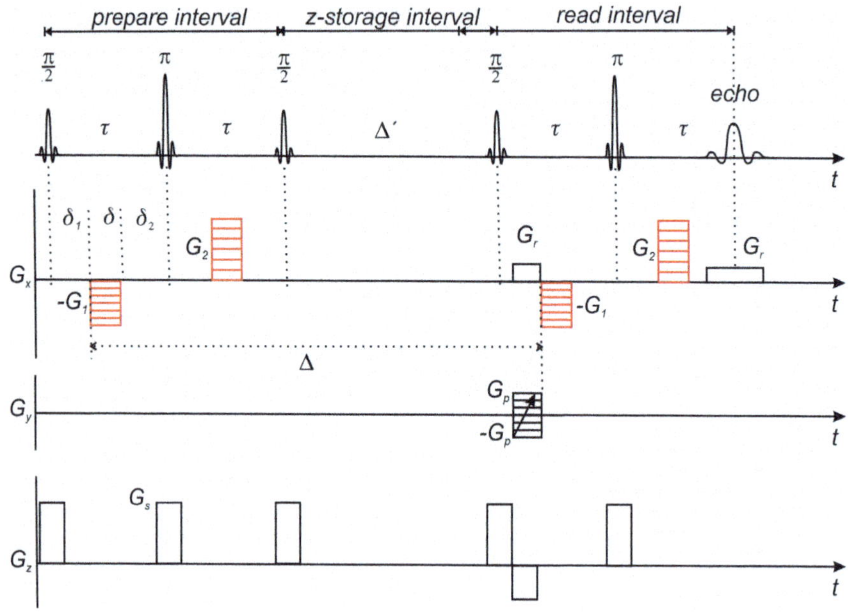

Figure 14.9 Modified 13-interval stimulated echo acquisition mode multi-slice imaging (13-interval STEAM-MSI) in interleaved slice selection mode with STEAM acquisition scheme. The flow-encoding gradient G_x is exemplarily applied in the x-direction. Please note the unequal flow-encoding gradient pairs of strengths G_1 and G_2, length δ, and separation Δ for encoding and decoding of displacement. For simplicity, only the x-direction is shown, but they are applicable in all Cartesian directions. Note that the imaging part is nested into the pulsed field gradient scheme by placing the slice selection gradient G_s and phase-encoding gradient G_p upstream of the second bipolar pair. All radiofrequency pulses are weak and slice-selective.

practice, the long T_1 of water in roots allowed the observation time to be set as $\Delta = 300$ ms. (iv) It is pointed out that, when measuring slow velocities, one runs the risk of not being able to differentiate displacements due to flow from the displacements due to diffusion processes that are always taking place. Thus, to enhance the influence of the signal from flow (proportional to Δ) as opposed to diffusion (proportional to the square root of Δ) we chose long observation times Δ and small $\boldsymbol{q}$-values, see also [44]. (v) The pulse sequence allows slice selection in interleaved mode so that the acquisition scheme can iterate through all required slices during one repetition time t_R without extending the total acquisition time. We tested the performance of the modified sequence by using different flow phantoms, where we determined parabolic flow profiles in capillaries, the neck of the plant cuvette, and plug flow in sand packs analogous to Figure 14.7. Most notably we established that coherent displacements due to flow can be determined despite the root mean square displacement due to diffusion being much larger (see [58] for details).

With this knowledge, we applied the sequence for the investigation of flow processes in the root system of broad bean (*Vicia faba*) in natural sand. A first plant (plant A) was used to establish the validity of the procedure by monitoring the total water uptake of the plant, which was measured via the flow through the neck at the bottom of the cuvette. A 6-day-old seedling was replanted into a 20 mm × 90 mm MRI cuvette, filled with medium sand, and grown for 3 more days. Then the cuvette was placed directly into the scanner and connected at the bottom to an outer water reservoir to sustain steady-state water content while the plant was transpiring. The average water level was adjusted to the bottom end of the soil column by varying the height difference Δh outside (see Figure 14.10a), thus maintaining water supply to the root system due to capillary forces in the soil. In this steady state, the flow velocity maps were acquired by the 13-interval STEAM-MSI pulse sequence after a further 6 days (days after seeding (DAS) = 15) in unsaturated states under continuous illumination and ventilation from top to keep the transpiration rate constant (the actual transpiration rates will vary depending on plant age, environmental and experimental conditions, and individual plant properties). The sample was scanned with the following parameters: FOV = 25 mm × 25 mm, matrix size = 128 × 64 (in-plane resolution 195 μm × 390 μm), 21 slices with a thickness of 2.0 mm, and 8 $\boldsymbol{q}$-steps in the Cartesian directions with $G_{max} = 0.28$ T/m, $\Delta = 300$ ms, and $\delta = 0.8$ ms. Finally, the velocity maps were calculated slice-wise by Fourier transform of $\boldsymbol{q}$-space (after zero-filling to 1024) into displacement space and extracting the maximum position of the propagator. Recently, Williamson et al. [59] have shown that the velocity resolution for phase contrast flow imaging improves with the inverse of the signal-to-noise ratio. Using their results we determined velocity uncertainty for the results reported here in the order of ± 1 μm/s. For further analysis, we assumed an error of ± 1 velocity steps after zero-filling, i.e. 1.2 μm/s (see [58] for more details).

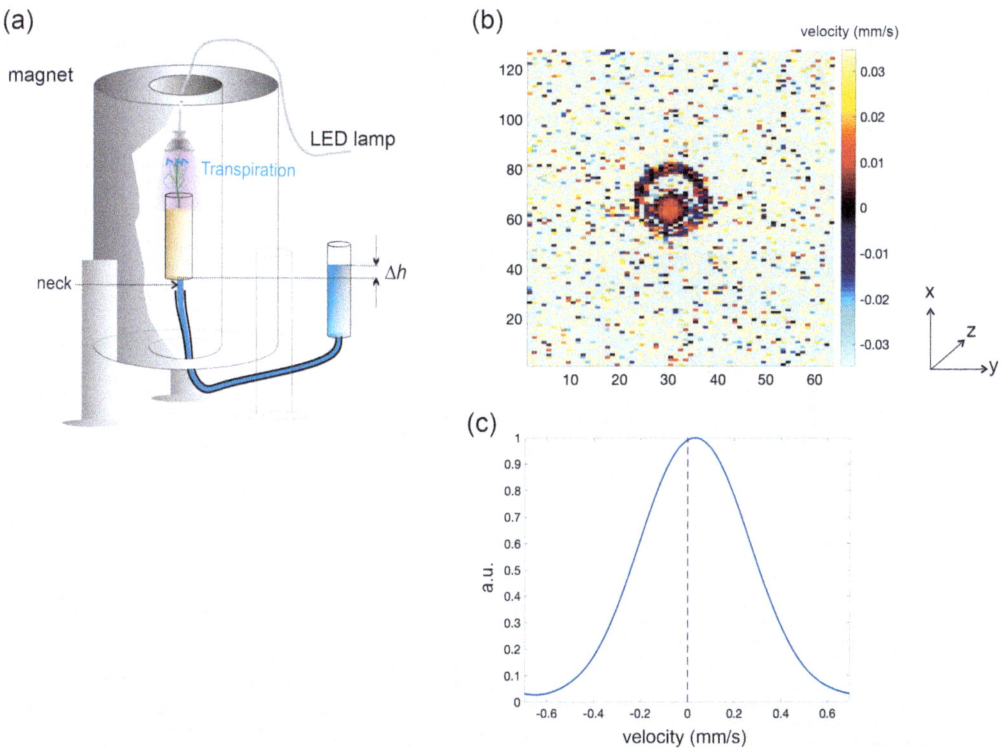

Figure 14.10 Test of the performance of the stimulated echo acquisition mode multi-slice imaging (STEAM-MSI) sequence at the example of the flow in the neck at the bottom of the plant cuvette. (a) Schematic setup. The height drop in the external water reservoir was 1.8 mm, yielding a consumption of 3.7 ± 1 ml/day. (b) Image of an axial slice showing flow velocity in the z-direction. (c) Flow propagator of one voxel inside the neck. The difference of the peak maximum position to the line at zero is the displacement due to coherent motion, rescaled in velocity units [58].

The sketch in Figure 14.10a depicts the position of the plant inside the magnet where the transpiration is promoted by intense illumination and mild ventilation. The neck at the bottom of the cuvette was connected via a silicon tube to the outer reservoir to sustain water flow in unsaturated state, adjusted by the hydraulic height difference h. The flow velocity map in Figure 14.10b shows an axial cross-section through the neck at the lower end of the cuvette (see bottom part of Figure 14.10a). The outer ring is the proton signal from the silicon tube with no net flux and a spatial off-center shift in frequency-encoding direction due to a chemical shift artifact. The velocities in the inner spot, i.e. inside the neck, are entirely positive compared with the random distributions in the outer layer. Determining the volume flow by integration over the neck area delivers 3.1 ± 0.8 ml/day. This compares well with the volumetric flow determined by the drop of the water level in the external water reservoir by 1.8 mm, yielding a consumption of 3.7 ± 1 ml/day. This agreement suggests that despite the very wide propagator, its displacement can be detected accurately enough to measure even such small velocities.

Before further analyzing velocity images of the roots we will first discuss the morphological details of young roots to be expected in MRI images, since transport processes inside roots are strongly correlated with them. It is known that the cell differentiation in epidermis, cortex, endodermis, and stele is not yet so pronounced in young roots as in stems, but will have started already a few millimeters away from the root tip in the root hair zone [1]. To compare features visible in MRI with the real structures in young roots, we performed optical microscopy of roots of the same species and cultivar as used for the velocity maps, see Figure 14.11. The MRI image was recorded for a 14-day-old roots system in unsaturated sandy soil using a T_2-weighted multi-slice rapid acquisition with relaxation enhancement (RARE) sequence with $t_E = 4.5$ ms, a RARE factor of 8, and a slice thickness of 1 mm for 50 slices.

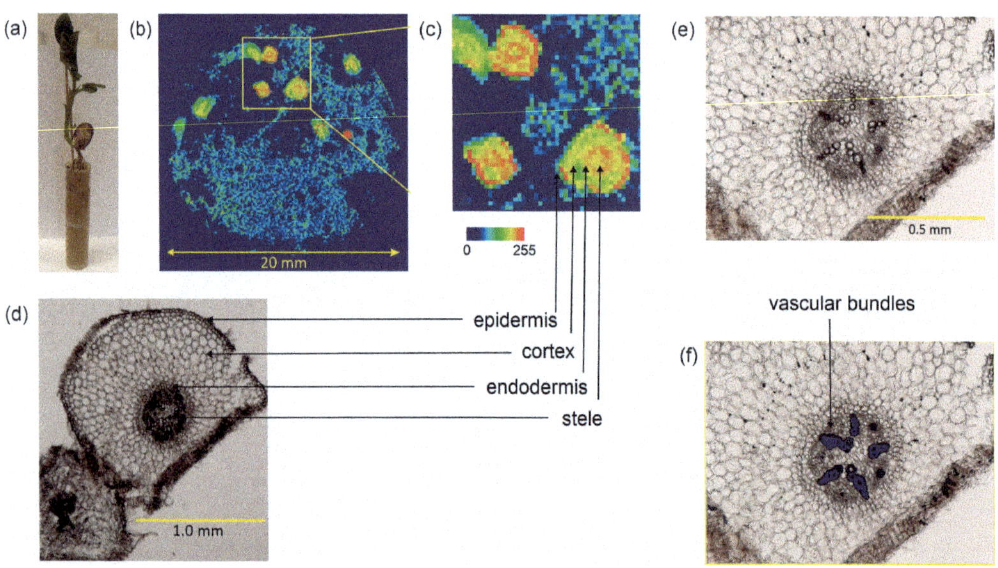

Figure 14.11 Root anatomy of *Vicia faba*. (a) Photo of the plant in the magnetic resonance imaging (MRI) cuvette. (b) Cross-section of an MRI rapid acquisition with relaxation enhancement (RARE) image and (c) magnification with 0.12 mm resolution. (d)–(f) Optical micrographs of an excerpted root obtained with a Zeiss Axio microscope, courtesy of Dr. Cordula Vogel, TU Dresden, Germany. (f) Blue stains show the lignified cell walls of the vascular bundles.

The matrix size was 256 × 256 resulting in an in-plane resolution of 117 μm × 117 μm and a total measuring time of 8 min 32 s. Optical micrographs were performed with a resolution of 0.92 μm (Figure 14.11d) and 0.23 μm (Figure 14.11e and Figure 14.11f). Comparing the micrographs with the MRI cross-section slices, one can clearly identify the epidermis, the cortex, the endodermis, and the stele area, indicated by the arrows between Figure 14.11c and Figure 14.11d. By the blue staining of cell walls containing lignin, one can see in the stele area the vascular bundles (xylem and phloem) in which upward and downward transport depending on water potential and nutrient gradients take place (Figure 14.11f).

With this information in mind, one can return to the discussion of the flow-sensitive MRI images. A second plant (plant B) was scanned and processed with similar parameters as used for plant A. However, only four $\boldsymbol{q}$-steps were used, $\boldsymbol{G}_{\max}$ was reduced to 0.13 T/m, and δ was set to 1ms. The four-dimensional raw data (x, y, slice, displacement-encoding) were zero-filled to a matrix size of 128 × 128 and filtered in $\boldsymbol{k}$-space using a 2D sinebell filter resulting in an in-plane resolution of 195 μm × 195 μm. The plant was planted in the cuvette cell 7 days after seeding, placed in the magnet system after another 15 days, and measured after a further 3 days in the MRI system, i.e. 25 days after seeding. The magnitude image of slice 3 in Figure 14.12b clearly shows that the root system spread across the entire cuvette cross-section, with some vertical roots

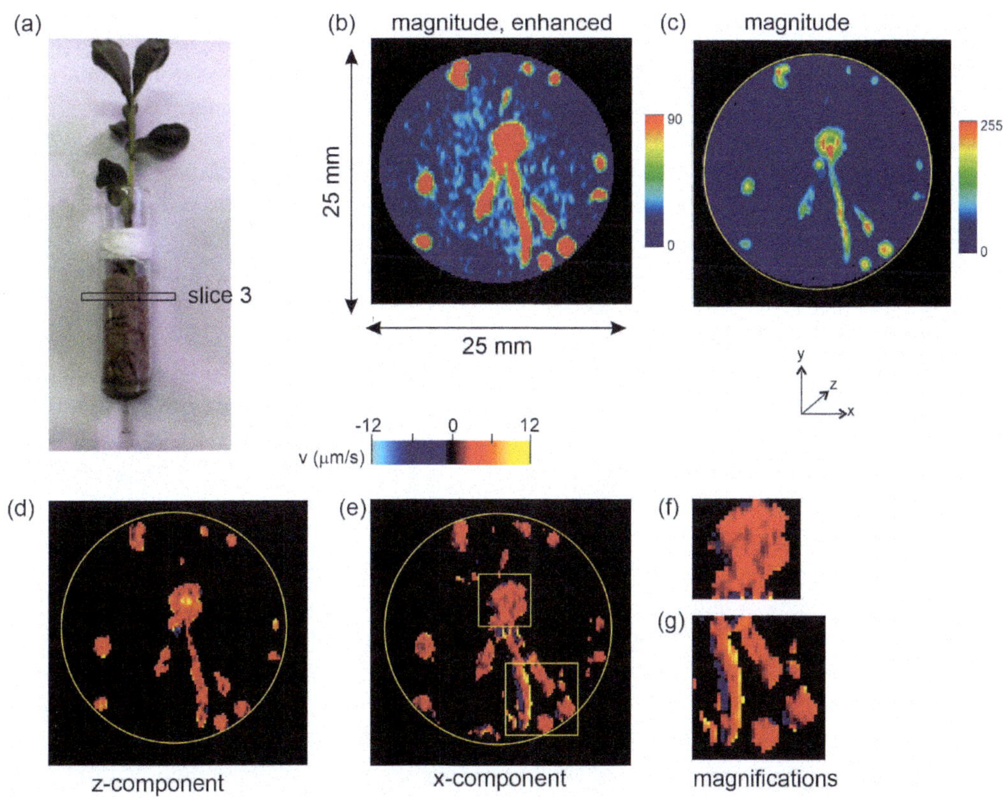

Figure 14.12 Flow measurements in the root system of *Vicia faba* grown in unsaturated sand using 13-interval stimulated echo acquisition mode multi-slice imaging (STEAM-MSI) in interleaved mode. (a) Position of the imaging slice. (b, c) Magnitude images at different contrast settings. (d, e) Maps of the z- and x-velocity components. (f, g) Magnified image regions. Please note that for the x-direction "yellow" denotes positive velocities pointing from right to left, and blue denotes negative velocities pointing from left to right.

near the wall, a taproot in the center, and some lateral roots grown approximately in-plane, as seen for the one pointing at 5:30 on the clock. In addition to the roots, soil moisture is also visible, not spread evenly around all roots but more around the central roots. The morphological details in the roots become more visible when the entire magnitude range is displayed (Figure 14.12c). The endodermis enclosing the stele manifests itself as a bright ring in the taproot and bright spots in the peripherical, axial lateral roots (red to yellow color). The cortex appears with a lower intensity in the green to light blue color range.

The calculated velocity distributions along the z-gradient direction (Figure 14.12d) and along the x-gradient direction (Figure 14.12e) match the positions of the roots in the magnitude image. The flow velocities in the individual roots were different. The taproot exhibits the highest value in z-direction (axial) of 8 ± 1.2 μm/s in the stele region, decreasing to a mean of 3 ± 1.2 μm/s in the cortex. The positive flow direction in the stele of the taproot indicates that it can be assigned to the upward flow in the xylem vessels of the vascular bundles.

The x-components of the taproot are more uniform and slow with an average of 4 ± 1.2 μm/s (Figure 14.12e). A possible explanation is a tilt of the taproot. Due to the natural growth the taproot was not exactly vertical but shifted by about 3 pixels per slice, which corresponds to a small tilt angle of about 17°. Correspondingly, the expected additional velocity component in the x-direction is about 2.4 ± 1.2 μm/s, if the z-velocity is 8 μm/s. Furthermore, in the magnified image in Figure 14.12f, one can identify negative velocities on the left side (blue spots) and positive x-components on the right side (yellow spots) in the cortex indicating radial flow directed from the root surface to the stele region (Figure 14.12f). These patterns are expected for a young taproot, where water is taken up by the root hairs and epidermis and transported through the cortex toward the endodermis. After passing the endodermis it is transported rapidly upward in the vascular bundles. Figure 14.13a shows this schematically. Therefore, finite horizontal flow velocities are expected in the root cortex, whereas transport proceeds only axially in the stele in the above-ground stem.

The observed patterns are different for the lateral root pointing to 5:30 on the clock. The z-components are very low (Figure 14.12d), whereas considerable difference in the x-components occurred with negative directions on the left side (Figure 14.12g, blue strip) combined with positive directions on the right side (Figure 14.12g, yellow strip). This disparity indicates horizontal flow in the cortex of the lateral root directed from the root surface to the center. This is illustrated schematically by Figure 14.13b, where the root has grown out of the paper plane in the y-direction. In addition, one must still consider that the lateral root is also inclined more than the taproot within the image plane. This means that the flow along the root axis holds components in the z-direction and x-direction (Figure 14.13c).

Our measurements went a step further than reported in the literature so far. Schulze-Till et al. determined flow velocity maps of older French bean roots in hydroponic culture, where relatively high flow velocities in the range of 400 μm/s occurred only in the vascular bundles at a position close to the shoot explaining the observed high axial velocities [46]. In contrast, the plant investigated in our study was much less developed with only five leaves and the root system (not shown here) did not express a clear hierarchy, i.e. with several parallel strands so that the entire volume flow was distributed over several individual roots. The group of Carminati used neutron radiography for D_2O tracer plume detection and analysis of water movement in white lupin root systems grown in sand [20,60]. By inverse modeling, they estimated axial volume flow rates of about 8×10^{-6} cm^3/s. By normalization of these rates on the stele cross-section area one can calculate the corresponding flow velocities $\boldsymbol{j}_x$. The radius of the stele is given as 46% to that of the entire

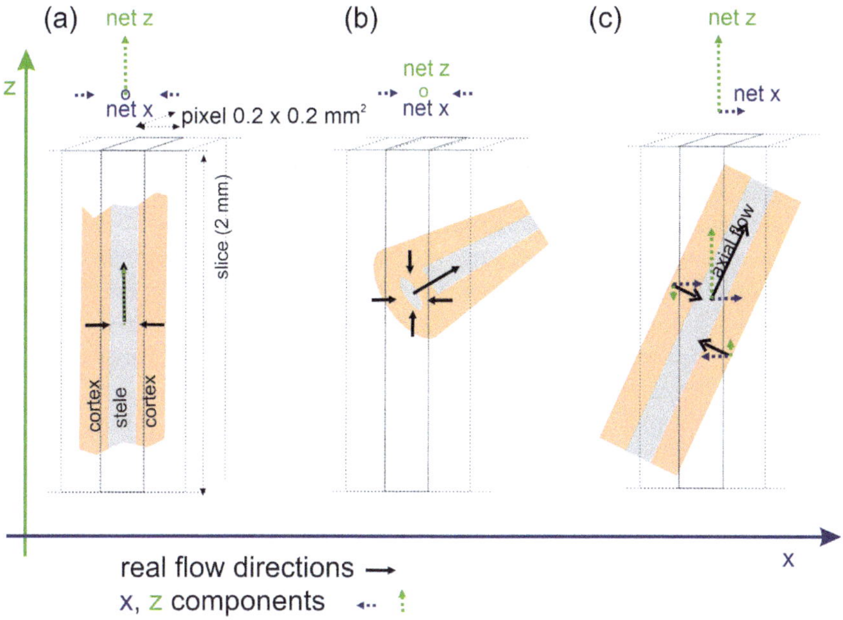

Figure 14.13 Sketch of possible flow vector compositions in anisotropic slice-selective voxels. The top row shows the observable *z*- and *x*-components for the voxels. Orange and gray regions indicate the root cortex and stele regions. (a) Vertical root with *z*-flow in the stele and horizontal flow in the cortex. (b) Horizontal root in the *y*-direction (out-of-paper) with net positive and negative *x*-components in the cortex voxel. (c) Flow vector decomposition for a tilt angle of a root of 25°.

root, which is about 0.25 mm, see Figure 14.2 [20]. This leads to an axial flow velocity of 190 μm/s, which is higher than the axial flow velocity reported here. The most obvious reason is the much thicker root of the broad bean where the stele region has a diameter of about 1 mm. Since flow velocities scale with the cross-section of the stele for a given volume flow, the distribution of the volume flow in lupine on a hypothetical stele with a fourfold diameter would lead to a flow velocity of 12 μm/s, which is in the same order as the axial velocity of 8 μm/s determined in this work. In a recent work, Zarebanadkouki et al. could differentiate between apoplastic and cell-to-cell pathway velocities with inverse modeling of neutron tomography data and they found relatively high radial velocities of 15–30 μm/s for the apoplastic pathway [61]. This value is in the same order as the radial flow velocity of 12 μm/s, determined here using velocity-MRI, see also Figure 14.12g. This suggests the hypothesis that the values we have measured here are caused by the apoplastic path transport of water during daylight.

14.4 Conclusions

Water uptake and the associated nutrient uptake are central to plant growth and thus yield. It is a particular challenge to investigate and understand roots in their natural environment because hydraulic properties can change depending on conditions and flow velocities are very slow. In this chapter, we have shown that the main advantage of MRI over neutron radiography is the ability to directly measure flow velocities.

NMR and MRI have been adapted and applied to plant research over decades and much understanding of plant physiology has resulted from this. However, the application of magnetic resonance techniques to root systems in natural sand and soil remains challenging due to magnetic field inhomogeneities caused by the soil and the resulting reduction of relaxation times and S/N ratios. Moreover, flows are distributed in the root system, not concentrated in a single stem and may not point in well-defined directions. Despite these difficulties, it is possible to measure flow velocities in 3D in multi-slice mode under realistic conditions in lateral roots and taproots and link them to the radial and axial flows as demonstrated in this chapter.

However, the uncertainties of these velocities are still significant. A possible way out could be the use of ultrahigh field strength with the advantage of higher S/N ratio and higher spatial resolution. This was proved for multi-slice multi-echo imaging of roots with a resolution of 13 μm× 13 μm × 13 μm at a field strength of 22 T [62], and the combination with a displacement-encoding module will be the next task to be undertaken. The combination of high resolution with low flow velocity mapping will allow the flows in different tissues of the root to be differentiated and the elucidation of how plants adapt their root hydraulics as a reaction to soil and climate conditions.

References

1 Nobel, P.S. (2005). *Physicochemical and Environmental Plant Physiology*. Amsterdam: Elsevier Science & Technology.

2 Philip, J.R. (1966). Plant water relations – Some physical aspects. *Annual Review of Plant Physiology* 17: 245–26810.1146/annurev.pp.17.060166.001333.

3 Kirkham, M.B. (2005). *Principles of Soil and Plant Water Relations*. Burlington, MA: Elsevier.

4 Marshall, T.J., Holmes, J.W., and Rose, C.W. (1999). *Soil Physics*. Cambridge, UK: Cambridge University Press.

5 Steudle, E. and Petersen, C.A. (1998). How does water get through roots? *Journal of Experimental Botany* 49: 775–788.

6 Zarebanadkouki, M., Kroener, E., Kaestner, A. et al. (2014). Visualization of root water uptake: Quantification of deuterated water transport in roots using neutron radiography and numerical modeling. *Plant Physiology* 166 (2): 487–499. 10.1104/pp.114.243212.

7 Koch, A., Meunier, F., Vanderborght, J. et al. (2019). Functional-structural root-system model validation using a soil MRI experiment. *Journal of Experimental Botany* 70 (10): 2797–2809. doi:10.1093/jxb/erz060.

8 Scheenen, T.W.J., Heemskerk, A.A.D. et al. (2002). Functional imaging of plants: A nuclear magnetic resonance study of a cucumber plant. *Biophysical Journal.* 82: 481–492.

9 Windt, C., Vergeldt, F., De Jager, P.A. et al. (2006). MRI of long-distance water transport: A comparison of the phloem and xylem flow characteristics and dynamics in poplar, castor bean, tomato and tobacco. *Plant, Cell and Environment* 29: 1715–1729.

10 Green, S.R., Kirkham, M.B., and Clothier, B.E. (2006). Root uptake and transpiration: From measurements and models to sustainable irrigation. *Agricultural Water Management* 86 (1–2): 165–176. 10.1016/j.agwat.2006.06.008.

11 Vanderborght, J., Huisman, J.A., van der Kruk, J. et al. (2013). Geophysical methods for field-scale imaging of root zone properties and processes. In: *Soil-Water-Root Processes: Advances in*

Tomography and Imaging Vol. 61. (eds. S.H. Anderson and J.W. Hopmans), 247–282. Madison, WI: SSSA Special Publications.

12 Cimpoiasu, M.O., Kuras, O., Pridmore, T. et al. (2020). Potential of geoelectrical methods to monitor root zone processes and structure: A review. *Geoderma* 365: 114232. 10.1016/j.geoderma.2020.114232.

13 Ehosioke, S., Nguyen, F., Rao, S. et al. (2020). Sensing the electrical properties of roots: A review. *Vadose Zone Journal* 19: 29. 10.1002/vzj2.20082.

14 Gao, W., Schluter, S., Blaser, S. et al. (2019). A shape-based method for automatic and rapid segmentation of roots in soil from X-ray computed tomography images: Rootine. *Plant and Soil* 441 (1–2): 643–655. 10.1007/s11104-019-04053-6.

15 Peng, Z., Duwig, C., Delmas, P. et al. (2015). Visualization and characterization of heterogeneous water flow in double-porosity media by means of X-ray computed tomography. *Transport in Porous Media* 110 (3): 543–564. 10.1007/s11242-015-0572-z.

16 Kardjilov, N., Hilger, A., Manke, I. et al. (2016). CONRAD-2: The new neutron imaging instrument at the Helmholtz-Zentrum Berlin. *Journal of Applied Crystallography* 49: 195–202. 10.1107/s1600576715023353.

17 Esser, H.G., Carminati, A., Vontobel, P. et al. (2010). Neutron radiography and tomography of water distribution in the root zone. *Journal of Plant Nutrition and Soil Science* 173 (5): 757–764. 10.1002/jpln.200900188.

18 Haber-Pohlmeier, S., Tötzke, C., Lehmann, E. et al. (2019). Combination of magnetic resonance imaging and neutron computed tomography for three-dimensional rhizosphere imaging. *Vadose Zone Journal* 18: 180166. 10.2136/vzj2018.09.0166.

19 Oswald, S., Tötzke, C., Haber-Pohlmeier, S. et al. (2015). Combining neutron and magnetic resonance imaging to study the root-soil interface. *Physics Procedia* 69: 237–243.

20 Zarebanadkouki, M., Kim, Y.X., Moradi, A.B. et al. (2012). Quantification and modelling of local root water uptake using neutron radiography and deuterated water. *Vadose Zone Journal* 11: 3. 10.2136/vzj2011.0196.

21 Hayat, F., Zarebanadkouki, M., Mohsen, A. et al. (2020). Quantification of hydraulic redistribution in maize roots using neutron radiography. *Vadose Zone Journal* 19: 1. 10.1002/vzj2.20084.

22 Haber-Pohlmeier, S., Vanderborght, J., and Pohlmeier, A. (2017). Quantitative mapping of solute accumulation in a soil-root system by magnetic resonance imaging. *Water Resources Research* 53: 7469–7480. 10.1002/2017WR020832.

23 Clore, G.M. and Iwahara, J. (2009). Theory, practice, and applications of paramagnetic relaxation enhancement for the characterization of transient low-population states of biological macromolecules and their complexes. *Chemical Reviews* 109 (9): 4108–4139. 10.1021/cr900033p.

24 Helm, L. (2006). Relaxivity in paramagnetic systems: Theory and mechanisms. *Progress in Nuclear Magnetic Resonance Spectroscopy* 49 (1): 45–64. 10.1016/j.pnmrs.2006.03.003.

25 Furtado, F., Gavosas, P., Stallmach, F. et al. (2011). Paramagnetic relaxation enhancement (PRE) as a tool for probing diffusion in environmentally relevant porous media. *Environmental Science & Technology* 45 (20): 8866–8872. 10.1021/es2010946.

26 Doussan, C., Pages, L., and Vercambre, G. (1998). Modelling of the hydraulic architecture of root systems: An integrated approach to water absorption – Model description. *Annals of Botany* 81 (2): 213–223.

27 Javaux, M., Schröder, T., Vanderborght, J. et al. (2008). Use of a three-dimensional detailed modeling approach for predicting root water uptake. *Vadose Zone J* 7 (3): 1079–1088.

28 Bramley, H., Turner, N.C., Turner, D.W. et al. (2009). Roles of morphology, anatomy, and aquaporins in determining contrasting hydraulic behavior of roots. *Plant Physiology* 150 (1): 348–364. 10.1104/pp.108.134098.

29 Van As, H., Homan, N., Vergeldt, F.J. et al. (2012). Soil-plant-atmosphere continuum studied by MRI. In: *Encyclopedia of Magnetic Resonance* (ed.R.K. Harris) 1, 633–644. New York: John Wiley & Sons.

30 Callaghan, P.T., Eccles, C.D., and Xia, Y. (1988). NMR Microscopy of dynamic displacements – k-space and q-space imaging. *Journal of Physics E* 21: 820–822.

31 Xia, Y. and Callaghan, P.T. (1992). One-shot velocimetry microscopy – NMR imaging of motion using a single phase-encoding step. *Magnetic Resonance in Medicine* 23 (1): 138–153. 10.1002/mrm.1910230115.

32 Xia, Y., Sarafis, V., Campbell, E.O. et al. (1993). Noninvasive imaging of water flow in plants by NMR microscopy. *Protoplasma* 173 (3–4): 170–176. 10.1007/bf01379005.

33 Köckenberger, W., Pople, J.A., Xia, Y. et al. (1997). A non-invasive measurement of phloem and xylem water flow in castor bean seedlings by nuclear magnetic resonance microimaging. *Planta* 201: 53–63.

34 Köckenberger, W., De Panfilis, C., Santoro, D. et al. (2004). High resolution NMR microscopy of plants and fungi. *Journal of Microscopy – Oxford* 214: 182–189. 10.1111/j.0022-2720.2004.01351.x.

35 Callaghan, P.T., Köckenberger, W., and Pope, J.M. (1994). Use of difference propagators for imaging of capillary-flow in the presence of stationary fluid. *Journal of Magnetic Resonance Series B* 104 (2): 183–188. 10.1006/jmrb.1994.1074.

36 Münch, E. (1930). *Stoffbewegungen in Der Pflanze*. Jena, Germany: Gustav Fischer Verlag.

37 Kuchenbrod, E., Kahler, E., Thurmer, F. et al. (1998). Functional magnetic resonance imaging in intact plants – Quantitative observation of flow in plant vessels. *Magnetic Resonance Imaging* 16 (3): 331–338. 10.1016/s0730-725x(97) 00307-x.

38 MacFall, J.S. and Van As, H. (1996). Magnetic resonance imaging of plants. In: *Nuclear Magnetic Resonance in Plant Biology* (eds. Y. Shachar-Hill and P.E. Pfeffer), 33–76. Rockville, MD: Plant Physiology Society of America.

39 Van As, H. and van Dusschoten, D. (1997). NMR methods for imaging of transport processes in micro-porous systems. *Geoderma* 80 (3–4): 389–403.

40 Van As, H., Scheenen, T., and Vergeldt, F.J. (2009). MRI of intact plants. *Photosynthesis Research*102 (2–3): 213–222. 10.1007/s11120-009-9486-3.

41 Callaghan, P.T. (1991). *Principles of Nuclear Magnetic Resonance Microscopy*. Oxford, UK: Oxford University Press.

42 Scheenen, T.W.J., van Dusschoten, D., de Jager, P.A. et al. (2000). Quantification of water transport in plants with NMR imaging. *Journal of Experimental Botany* 51 (351): 1751–1759.

43 Scheenen, T.W.J., van Dusschoten, D., de Jager, P.A. et al. (2000). Microscopic displacement imaging with pulsed field gradient turbo spin-echo NMR. *Journal of Magnetic Resonance* 142 (2): 207–215.

44 Scheenen, T.W.J., Vergeldt, F.J., Windt, C.W. et al. (2001). Microscopic imaging of slow flow and diffusion: A pulsed field gradient stimulated echo sequence combined with turbo spin echo imaging. *Journal of Magnetic Resonance* 151 (1): 94–100.

45 Terada, Y., Horikawa, Y., Nagata, A. et al. (2020). Dynamics of xylem and phloem sap flow in an outdoor Zelkova tree visualized by magnetic resonance imaging. *Tree Physiology* 40: 290–304. 10.1093/treephys/tpz120.

46 Schulze-Till, T., Kaufmann, I., Sattelmacher, B. et al. (2009). A H-1 NMR study of water flow in *Phaseolus vulgaris* L. roots treated with nitrate or ammonium. *Plant and Soil* 319 (1–2): 307–321. 10.1007/s11104-008-9871-2.

47 Rokitta, M., Peuke, A.D., Zimmermann, U. et al. (1999). Dynamic studies of phloem and xylem flow in fully differentiated plants by fast nuclear-magnetic-resonance microimaging. *Protoplasma* 209 (1–2): 126–131.

48 Duschl, M., Galvosas, P., Brox, T.I. et al. (2015). In situ determination of surface relaxivities for unconsolidated sediments. *Water Resources Research* 51 (8): 6549–6563. 10.1002/2014wr016574.

49 Stingaciu, L.R., Pohlmeier, A., Blümler, P. et al. (2009). Characterization of unsaturated porous media by high-field and low-field NMR relaxometry. *Water Resources Research* 45: W08412.

50 Krug, J.R., van Schadewijk, R., Vergeldt, F.J. et al. (2020). Assessing spatial resolution, acquisition time and signal-to-noise ratio for commercial microimaging systems at 14.1, 17.6 and 22.3 T. *Journal of Magnetic Resonance* 316: 106770. 10.1016/j.jmr.2020.106770.

51 Cotts, R.M., Hoch, M.J.R., Sun, T. et al. (1989). Pulsed field gradient stimulated echo methods for improved NMR diffusion measurements in heterogeneous systems. *Journal of Magnetic Resonance* 83 (2): 252–266. 10.1016/0022-2364(89)90189-3.

52 Spindler, N., Galvosas, P., Pohlmeier, A. et al. (2011). NMR velocimetry with 13-interval stimulated echo multi-slice imaging in natural porous media under low flow rates. *Journal of Magnetic Resonance* 212 (1): 216–223. 10.1016/j.jmr.2011.07.004.

53 Galvosas, P., Stallmach, F., and Karger, J. (2004). Background gradient suppression in stimulated echo NMR diffusion studies using magic pulsed field gradient ratios. *Journal of Magnetic Resonance* 166 (2): 164–173. 10.1016/j.jmr.2003.09.012.

54 SSSA (2008). *Glossary of Soil Science Terms*. Madison, WI: Soil Science Society of America.

55 Frahm, J., Merboldt, K.D., Hanicke, W. et al. (1985). Stimulated echo imaging. *Journal of Magnetic Resonance* 64 (1): 81–93. 10.1016/0022-2364(85)90033-2.

56 Sorland, G.H., Hafskjold, B., and Herstad, O. (1997). A stimulated-echo method for diffusion measurements in heterogeneous media using pulsed field gradients. *Journal of Magnetic Resonance* 124 (1): 172–176. 10.1006/jmre.1996.1029.

57 Brown, R.W., Cheng, Y.-C.N., Haake, E.M. et al. (2014). *Magnetic Resonance Imaging: Physical Principles and Sequence Design*. Hoboken, NJ: Wiley-Blackwell.

58 Wang, J., Haber-Pohlmeier, S., Pohlmeier, A. et al. (2022). Imaging slow flow in plant roots in preparation.

59 Williamson, N.H., Komlosh, M.J., Benjamini, D. et al. (2020). Limits to flow detection in phase contrast MRI. *Journal of Magnetic Resonance Open* 2–3: 100004. 100004.10.1016/j.jmro.2020.10000

60 Zarebanadkouki, M., Kim, Y.X., and Carminati, A. (2013). Where do roots take up water? Neutron radiography of water flow into the roots of transpiring plants growing in soil. *New Phytologist* 199 (4): 1034–1044. 10.1111/nph.12330.

61 Zarebanadkouki, M., Trtik, P., Hayat, F. et al. (2019). Root water uptake and its pathways across the root: Quantification at the cellular scale. *Scientific Reports* 9: 11. 10.1038/s41598-019-49528-9.

62 van Schadewijk, R., Krug, J.R., Webb, A. et al. (2021). MRM microcoil performance calibration and usage demonstrated on *Medicago truncatula* roots at 22 T. *JoVE* (167): e61266. 10.3791/61266.

15

Magnetic Resonance Studies of Water in Wood Materials

Bruce J. Balcom[1] *and Minghui Zhang*[2]

[1] *University of New Brunswick, Fredericton, Canada*
[2] *Inner Mongolia Agricultural University, Hohhot, Inner Mongolia, China*

15.1 Introduction

Wood plays a vital role in our world as a structural material, a packaging material, an energy source, and a decorative material. Wood is easy to process, has a high strength-to-weight ratio, and is a good thermal and electrical insulator. It is usually esthetically pleasing in color and texture, and it is a renewable resource. Wood materials are typically good acoustic insulators and of course it is a major source of cellulose for paper and cardboard. At the same time, wood has drawbacks. It is hygroscopic and dimensional change can result from changes in hydration. Natural defects are common, and the material is subject to deterioration. Significant research in wood science and technology has been undertaken to employ these materials more advantageously. Research areas include water and wood relationships, drying, wood chemistry, manufacturing, mechanical properties, preservation, and many others. Among these areas, water and wood relationships are among the most important since >90% of all problems with the material involve water [1]. Water in the pore space has a great influence on material dimensions, mass, and density, as well as thermal, elastic, mechanical, electrical, and transport properties [2]. Magnetic resonance (MR) techniques are powerful tools to investigate water and wood relationships. The MR signal amplitude depends on the quantity of water. The MR signal lifetimes depend on the environment and dynamics of water interacting with the wood fiber matrix. In this chapter, we focus on water behavior in wood materials, in particular the interactions that lead to different MR signal lifetimes.

15.2 Basic Wood Anatomy

Wood is a natural porous material with a hierarchal structure, with water residing in the cell walls and in the pore space. MR techniques probe water states in the material, but also reveal information on the pore space. We now consider basic wood anatomy to better understand the behavior of water in the material [3–5].

Trees, the wood source, are divided into two broad classes: softwood and hardwood. Softwoods are from the *Gymnospermae* subdivision, with seeds that are not enclosed in the ovary of the flower. Softwoods are characterized by needle-like or scale-like leaves. They are commonly known as evergreens because most softwood species remain green all year. Hardwoods belong to

Magnetic Resonance Microscopy: Instrumentation and Applications in Engineering, Life Science and Energy Research, First Edition. Edited by Sabina Haber-Pohlmeier, Bernhard Blümich, and Luisa Ciobanu.

the *Angiospermae* subdivision, with seeds that are enclosed in the ovary of the flower. Hardwoods are generally characterized by broad leaves that are seasonal. The cellular structures of hardwoods and softwoods have important differences. Hardwoods have a characteristic cell type termed a vessel element, with size on the order of 100 μm, that exist as pores in the fiber of the wood. Softwoods lack this structure, the basic cells are tracheid cells. The tracheid cells, described further in the next section, are the primary source of porosity in softwoods. In this chapter we focus on MR/magnetic resonance imaging (MRI) measurements of softwoods.

15.2.1 Macrostructural Wood Anatomy

Both softwoods and hardwoods have growth rings that reveal regular growth of xylem/wood. Each growth ring consists of earlywood and latewood layers. The earlywood is grown in the early part of the growth season and is characterized by a wider band that is lighter in color and more porous. Earlywood has larger-diameter cells that are shorter, with cell walls that are thinner. Latewood is grown in the later part of the growth season and is characterized by narrower bands, darker color, and higher density. Latewood has smaller-diameter cells that are longer, and cell walls that are thicker. The variable cell size in earlywood and latewood results in different MR signal lifetimes for water in the cell interior.

Figure 15.1 shows a light microscope view of the cellular structure of a softwood [6]. The image shows the cell walls and the large open cavity in the tracheid cells, which is the lumen. The lumen width in earlywood softwood is a few tens of microns. Characteristic density changes are visible in the latewood and earlywood. The majority of the water in a fresh softwood is held in the lumen.

In cross-section (perpendicular to the tree trunk), the tree has a heartwood interior and a sapwood exterior. Heartwood is darker in visual appearance and all cells in the heartwood are dead.

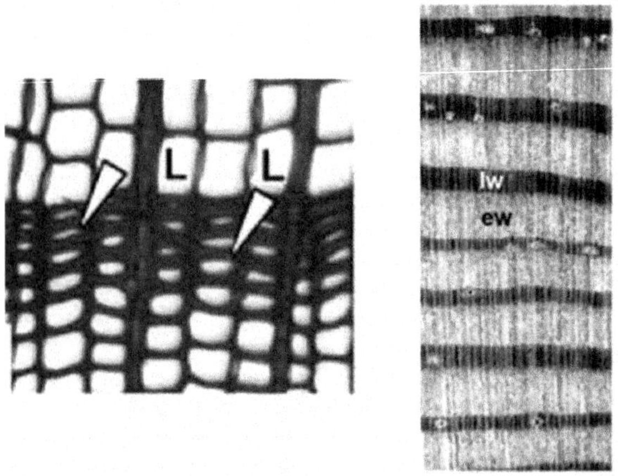

Figure 15.1 Cross-sectional light microscope view (left) showing the lumen (L) and cell walls (arrowheads) of a softwood. The lumen width in typical softwood, earlywood, is a few tens of microns. The lumen height is orders of magnitude longer. The right-hand image represents a hand-lens view of growth rings, each composed of earlywood (ew) and latewood (lw). The light spots present in the latewood are resin canals. [6] Ross et al. (2010), figure 03.03[p.66]/USDA Forest Service/Public Domain.

The water content in the heartwood is usually low. Sapwood is usually lighter colored than heartwood. There are living parenchyma cells in sapwood. The water content in the sapwood is usually very high, easily above 100% (as defined in Section 15.3.2).

Figure 15.2 shows the macroscopic character of a tree in cross-section. Based on the cellular arrangement, wood has three nominal directions: the radial direction (from pith to bark); the tangential direction (perpendicular to the rays and substantially parallel to the growth rings); and the longitudinal direction (parallel to tree trunk or wood grain). Macroscopically wood also has three nominal surfaces: the radial surface; the tangential surface; and the cross-sectional surface (transverse surface) as shown in Figure 15.2.

15.2.2 Microstructural Wood Anatomy

Wood comprises small cells with each cell having a precise structure including tiny openings, membranes, and layered cell walls. Cell walls are principally composed of cellulose, hemicellulose, and lignin. Cellulose is not incorporated into the cell wall as individual molecules. Rather, it forms intricately arranged clusters or strands of molecules. Hemicellulose somewhat selectively interacts with cellulose through hydrogen bonding and chemically links to lignin macromolecules.

The wood cell wall consists of a primary wall and a secondary wall while the wood cell has a hollow center called the cell cavity or lumen (Figure 15.1). The lumen cavities are the dominant pore space of the softwood. The secondary wall is the main part of the cell wall. Cellulose molecules in the secondary wall are embedded in a matrix and provide a structurally reinforcing network. This matrix is also composed of hemicellulose and lignin "sheets." Pit regions are areas of the cell wall that lack a secondary wall. Normally pits appear as a matched pair between two adjoining cells. Pit pairs are the primary avenue of water transport between cells. Water transport is of course fundamental to drying.

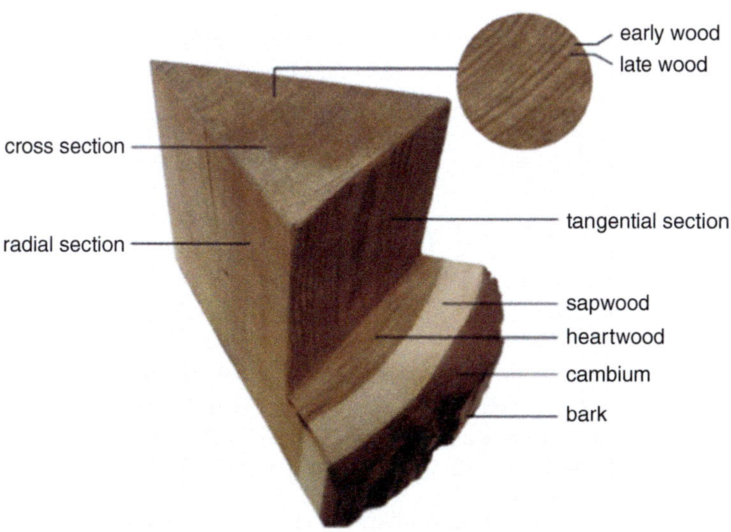

Figure 15.2 Macroscopic structure of a tree in cross-section, and in radial and tangential section.

15.3 MR Studies of Water States in Wood

The importance of water in wood is well known and various techniques have been used to study water in wood. Among these techniques, MR has proved to be one of the more informative methods.

15.3.1 Bound and Free Water vs. Cell Wall and Cell Lumen Water in Wood

The fiber saturation point (FSP) is a fundamental feature of the relationship between water and wood in wood science. Although the concept of the FSP is of long standing, proposed more than 100 years ago, an exact definition is elusive. Tiemann defined the FSP as the moisture content (MC) at which the cell wall is saturated with water, but no moisture is present in the cell lumen [7]. However, Stamm believed that the FSP should be defined as the critical MC at which wood physical properties change abruptly [8]. Still other opinions about FSP were reviewed by Engelund [9]. Different methods for FSP measurement yield different results [8]. Despite the range of opinion found in the literature, the Tiemann definition is the most commonly accepted.

There are two principal locations for water in wood: one is in the cell wall. In the wood science literature this water is often called bound water. The second location is the cell lumen, which is the void space/pore space in the cellular material. In the wood science literature, water in the cell lumen is often called free water [4].

One of the major goals of this review chapter is to change the description of bound and free water in wood science [10]. The wood science consideration of bound and free water is at odds with the MR description, as we further develop in this section. The MR description is physically correct and is used to understand the behavior of a broad range of porous media.

Most wood science researchers already believe water in the cell wall exists in two forms; namely water that may freeze and water that will not freeze. These two components can be quantified using differential scanning calorimetry [11,12]. Other studies have shown that water that freezes in cellulosic materials is similar in properties to free water [13]. There is strong evidence that at least two general types of water exist in the cell walls of wood. The evidence suggests that one is more motionally restricted than the other. If this is correct, it means that cell wall water should not universally be termed bound water. It would be better in fact to refer to motionally restricted water (bound water) and water that is less motionally restricted (free water). This result conforms to the MR understanding of water behavior in many porous materials. We often encounter materials where there is a bound water layer or environment and a bulk-like water environment. Water in these two environments is often in rapid exchange.

In the case of water in wood, from an MR point of view, there are four different water environments. Two correspond to water in the cell wall; two correspond to water in the cell lumen. We enumerate and discuss the four water environments in the next few paragraphs. (i) We consider first motionally restricted water in the cell wall. Stamm [3] employed the sigmoidal adsorption curve proposed by Brunauer, Emmett, and Teller [14], the BET equation, to calculate the average number of molecular layers of water at fiber saturation. Stamm reported 6.5 layers for sitka spruce (*Picea sitchensis*) and 7.5 layers for sugar maple (*Acer saccharum*). The moisture needed to form a monomolecular surface layer within the amorphous regions of cellulose corresponds to a MC of about 5% [3,15]. BET theory considers the cell wall water to be a surface phenomenon with a strongly adsorbed monolayer. (ii) Additional water in the cell wall will be less motionally restricted water. The water content at the FSP (when cell lumen water is gone but cell wall water remains) is approximately 30%. Nonbound water dominates in the cell wall although both types of water exist in significant quantity. MR measurement of wood as a function of MC is readily able to determine the FSP, as illustrated by the experiments shown in Figure 15.3 [10].

The cell lumen of Figure 15.1 are open cavities/pores that are full of water at saturation. There must be two different types of water in the lumen. One is the bulk lumen water (iii), which is very mobile and truly bulk-like in behavior. The other (iv) is the small fraction, which adheres to the lumen walls. The observed MR relaxation rate from the water in the lumen (conventionally called free water) is the weighted average rate of the surface water and bulk water relaxation according to Wong [16] and Lamason [10].

In MR, relaxation describes specific processes by which nuclear magnetization in nonequilibrium states returns to equilibrium. Different time constants characterize the longitudinal (spin–lattice) and transverse (spin–spin) relaxations. T_1 (spin–lattice relaxation time) is a measure of the time required for a spin system to return to thermal equilibrium with its surroundings (lattice). T_2 (spin–spin relaxation time) governs the decay of transverse magnetization due to spin–spin interactions. T_2* is the time constant that describes the decay of transverse magnetization in a free induction decay experiment. It combines decay due to fundamental T_2 decay and decay due to magnetic field inhomogeneities [17]. The MR signal typically decays exponentially with T_2 and T_2*, which are related to the type of water environment in wood. It is more common to measure spin echoes and T_2 lifetimes as opposed to T_1 or T_2* lifetimes due to the ease of experiment and sensitivity of the experiment.

In the case of rapid exchange of water between two environments (e.g. bound surface water in the lumen and bulk water in the pore space), one observes an average relaxation rate.

$$\frac{1}{T_2} = \frac{n}{T_{2s}} + \frac{m}{T_{2b}} \quad n + m = 1 \tag{15.1}$$

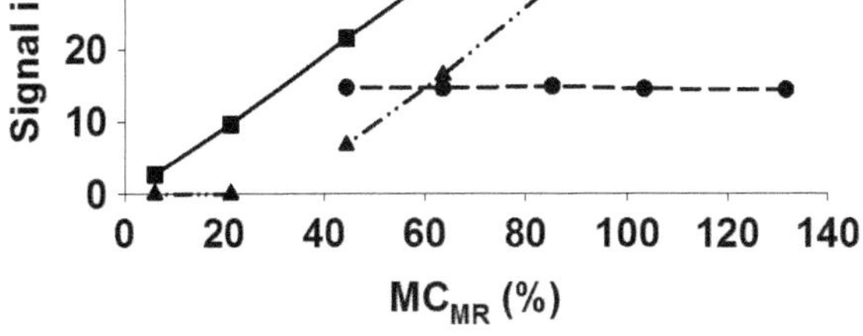

Figure 15.3 Plot of ^{1}H magnetic resonance-determined total water content (squares), lumen water content (triangles), and cell wall water content (circles) in a black spruce sample as a function of moisture content (MC). Cell wall water content is constant above 40% MC. Lumen water content is zero at low MC. The intersection of the cell wall water content and total water content is the fiber saturation point (FSP), estimated to be 33% MC. Reproduced with minor modifications from *Wood Material Science and Engineering* 2015, 10, 86–93 [10].

where $1/T_{2s}$ is the surface relaxation rate and $1/T_{2b}$ is the bulk relaxation rate. The variables n and m indicate the relative quantities of water in the surface and bulk environments.

Typically, in the lumen, there is more bulk water than surface water when the wood is green, but the surface relaxation rate is much greater than the bulk relaxation rate. A different ratio of bound and free water in the cell lumen of different physical sizes results in different MR lifetimes [18]. During drying the bulk-like free water content of the lumen decreases but the surface bound water content does not. This also results in a change of the weighted average T_2 of the lumen water [19]. Similarly, the lumen water T_2 from earlywood and latewood differs since the lumen cell sizes are different [20].

We can reasonably anticipate exactly the same behavior (Equation 15.1) with cell wall water where the two environments are more evenly matched in quantity compared with lumen water. Changes in either the population ratio or dynamics of either population would result in a change in weighted average T_2 lifetime.

According to BET theory, the cell wall water may be divided into monomolecular and polymolecular water [21]. Monomolecular water has strong hydrogen bonds with the fiber matrix and is considered bound water. Polymolecular water has very weak interactions with the fiber matrix and is motionally less restricted. The different ratios of bound and free water (MR description) in the cell wall versus cell lumen result in MR lifetimes that differ. The cell wall water T_2 is typically on the order of 1 ~ 10 ms while the cell lumen water T_2 is on the order of 100 ms. MR measurement of a water-saturated wood sample will measure water in both environments, and biexponential T_2 decays are usually observed. Figure 15.4 shows an MR biexponential decay of cell wall and cell lumen water signal as a function of MC in a black spruce sapwood sample [10].

There have been many studies of water in wood with MR techniques since the 1970s. Following the existing wood science literature, researchers assigned the short lifetime signal to bound water in the cell wall and free water in the cell lumen. As discussed earlier the water of these two environments should not be termed bound and free water because each environment has both bound and free water. In the remainder of this chapter we will speak of cell wall water and cell lumen water, not bound and free water.

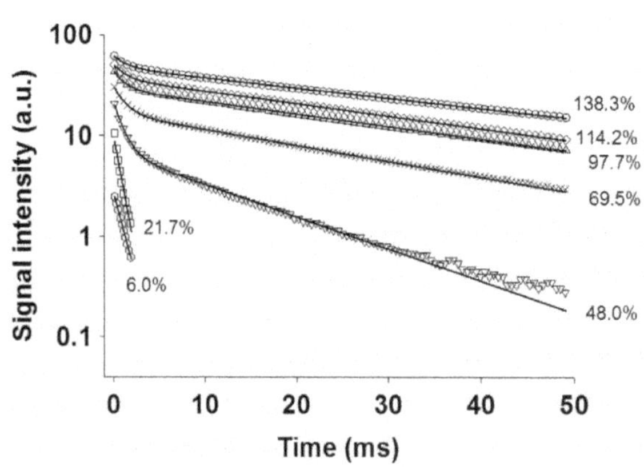

Figure 15.4 Semilog plot of ^{1}H MR signal decay from a sample of black spruce sapwood as a function of moisture content (MC). The observed decay is clearly biexponential until the MC falls below the fiber saturation point (FSP). MC is labeled to the right of each decay. Reproduced from *Wood Material Science and Engineering* 2015, 10, 86–3 [10].

15.3.2 MR Studies of MC in Wood

Wood is a hygroscopic and anisotropic material whose properties are deeply influenced by the presence of water. Measurements of MC of wood are therefore of considerable importance. MC is an important concept in wood science since it is related to other important definition such as FSP and equilibrium moisture content (EMC), and is also related to the physical and mechanical properties of wood. As mentioned earlier, the FSP is usually around 30% MC. The EMC depends on local climate, most notably the average relative humidity. In northwest China the EMC is 7 ~ 9% MC. Interactions between wood and water influence all steps of the production chain and the final product quality. For example, MC variations in wood cause unequal shrinkage. Excessive moisture increases the cost of transportation and decreases the amount of available thermal energy by hindering combustion [6].

In the wood products industry, MC is expressed as the amount of water in proportion to the oven-dry wood weight. The most widely accepted method to measure MC is to weigh the wet sample and then dry it in an oven for 24 h at 103 ± 2°C to remove all water. The oven-dried sample is then reweighed. More information about this process can be found in the method of ASTM-D4442-07 [22]. This definition of MC results in MCs of fully saturated green wood that are often above 100%. This simply means there is a greater mass of water in the original sample than the mass of wood fiber.

The gravimetric measurement is destructive, it involves a measurement time on the order of a day, and there is a practical limit to the size and number of pieces that can be measured. Volatile constituents are removed during the drying process, which causes errors in MC calculations. Gravimetric methods of this type are nevertheless commonly employed in the industry and are considered to be the gold standard.

If wood has been treated or impregnated with chemicals, oven-drying moisture measurements may be inaccurate. An impregnant that is volatile at oven temperatures will evaporate during oven drying, and the resulting weight loss can be misinterpreted as being due to evaporated water. An impregnant that is nonvolatile will remain in the wood and so increase the apparent oven-dry weight of the wood. For treated wood, the distillation method may be more accurate than oven drying for measuring MC. In this method, the wood material is chipped, then an organic solvent is added, and the mixture heated. Water will evaporate from the mixture but may be condensed and collected in a closed system with the quantity of water measured directly. The distillation method is more accurate than the oven-drying method for species that naturally contain large amounts of volatile materials other than water [22].

As with gravimetric methods, distillation is a time-consuming and destructive measurement. In addition, it requires considerable experimental skill. These problems reinforce the need for alternate methods to measure MC. Electrical moisture meters were developed for rapid wood MC measurement. However, their accuracy is low for samples with MC >25%. Resistance meters are calibrated for a particular species and manufacturers advise correction factors to apply for other species [2,23]. With electrodes driven into the wood, this method is destructive for some wood applications (such as veneers and furniture). MC measurement can also be determined with microwave moisture meters. The spatial resolution in microwave systems is on the order of several centimeters, which precludes a thorough examination of drying mechanisms in wood. Also, there is significant attenuation of the microwave signal at high frequencies. The attenuation is significant for green wood and MC measurements are not possible with microwaves in green wood. The power needed to overcome the attenuation is so high that it would heat or even burn the wood.

The MR signal is proportional to the amount of ^{1}H nuclei. In wood materials we usually assume that the wood fiber ^{1}H signal lifetime is very short lived, and is therefore not observable in ordinary time-domain MR measurements or images. This means that the water ^{1}H signal, if properly calibrated, will provide a high-quality MC measurement. MR measurement of bulk MC is independent of wood species. The first research papers exploiting this idea were from Japan [24,25]. MC measurement with MR is fast, robust, and nondestructive with benchtop MR instruments. The MR signal decays reported in Figure 15.4 were calibrated according to MC.

However, bulk MR methods in benchtop instruments are often not applicable to wood science problems because they are whole-sample measurements of small samples. Portable unilateral MR instruments offer the prospect of local MC measurement. Casieri et al. [26] employed mobile MR as a nondestructive and noninvasive tool for water content analysis and moisture fraction determination in wood. Senni et al. [27] employed a portable MR sensor for in situ noninvasive determination of MC of wooden works of art and painted wood. Portable unilateral MR was employed by Dvinskikh et al. [28] to study MC profiles and uptake kinetics in wood materials. Pourmand et al. [29] evaluated the moisture protective properties of wood coating by a portable MR sensor. Most of the sensors employed by these researchers have a maximum depth of penetration of 3–5 mm.

One of the most successful unilateral magnets, designed and fabricated in the University of New Brunswick (UNB) MRI Centre, is a lightweight three-magnet array [30]. It offers open access and portability. The unilateral magnet has a homogeneous magnetic field of 1 cm^3 volume located approximately 1.3 cm away from the top of the magnet. This permits measurements beyond the surface and near surface, and beyond most bark layers and cambium, in log samples. The importance of measurement below the surface, to avoid the surface bark, is illustrated in Figure 15.5 [31]. Employing the three-magnet array, Lamason and coworkers have measured local MC, the local cell wall and lumen water content and they have monitored freezing and thawing of lumen water in wood materials [18]. Most recently, in combination with MRI, they have examined drying in logs and developed an empirical method to estimate the rate of log drying [10].

15.3.3 MR Studies of Water Migration in Wood

Drying is a near-ubiquitous processing step in wood materials. Drying is however energy intensive, and an inappropriate drying regime risks damaging or destroying the material. Putative drying mechanisms are complicated, but some features are straightforward. We note

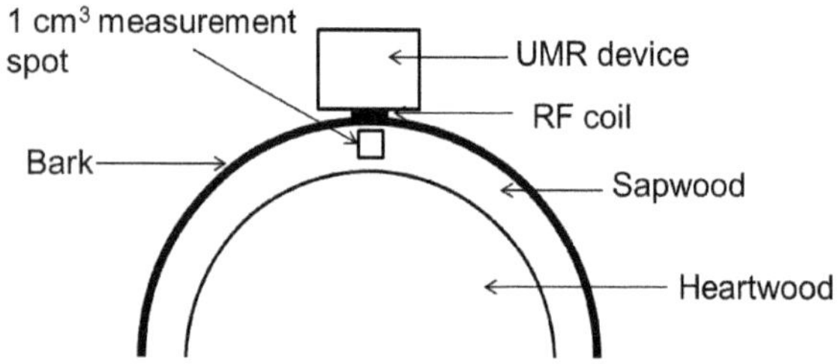

Figure 15.5 Schematic diagram of wood log moisture content measurement in sapwood. The sensitive spot of the three-magnet array unilateral nuclear magnetic resonance (UMR) device is displaced from the surface, past the bark, into the sapwood, where the moisture contest is high in a freshly cut log. Reproduced from *Forest Products Journal* 2017, 67, 266–274 [31].

that drying to the EMC involves removal of all lumen water in softwoods and a significant reduction in the cell wall MC as well.

Drying involves water migration and results in MC change and drying stress. The loss of cell wall and cell lumen water in wood during drying is usually studied independently, because it is difficult to examine these two water environments simultaneously with traditional methods. Time-domain MR techniques are widely used in wood-drying studies because they are quick, nondestructive, and quantitative, and most importantly they can easily distinguish among water states based on T_2 relaxation times [32–41]. The cell wall water, in a more motionally restricted environment, has T_2 lifetimes that are substantially shorter lived than cell lumen water as reported in Section 15.3.1 [10,20]. Because the amplitude of the magnetization is proportional to the number of ^{1}H nuclei, different water components may be quantified from the signal integrals. Figure 15.6 shows spatially resolved MC in the sapwood of a drying black spruce log measured with the UNB three-magnet array unilateral MR device [31]. This study showed clearly that lumen water is lost before cell wall water during drying. The unilateral nuclear magnetic resonance (UMR) measurement was spatially resolved down the length of the log but with a finite measurement volume.

Bulk measurements of drying are limited since the bulk measurement does not report spatially resolved water content or water behavior. However, this approach yields rapid results, within a few minutes. So, with caveats it can be used to study water migration online during wood drying.

Zhang et al. [19,32] studied different wood-drying methods with MR T_2 measurements. They found that different drying methods show different water migration behaviors in wood. Water migration in conventional drying can be divided into two phases: one is a linear function of time while the other is an exponential function with time. As anticipated, the amount of cell wall water does not change significantly above the FSP during drying.

Drying with microwaves is characterized by a fast drying rate, and there is a linear relation between MC and microwave drying time. The drying rate of cell lumen water is about 2.7 times more

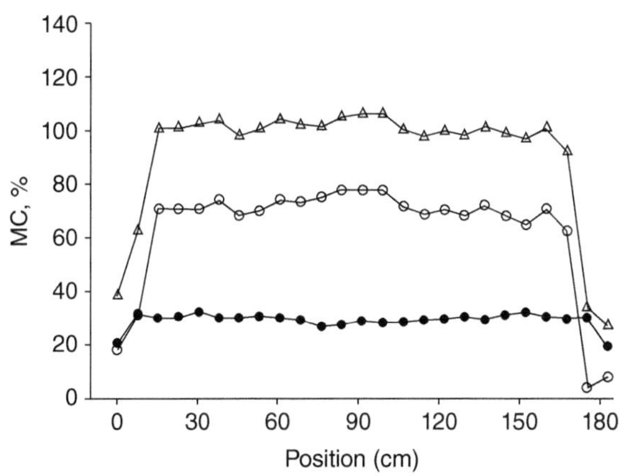

Figure 15.6 Total moisture content (MC) (triangle), lumen water MC (open circle), cell wall water MC (closed circle) as a function of length along a drying black spruce log, one month after commencement of drying. Drying is principally from the cut ends at 0 cm and 180 cm. The measurement is made by translating a unilateral nuclear magnetic resonance three-magnet array along the surface and making a local through-bark measurement in the sapwood. The initial total MC was uniform at 130% at time 0. The cell wall water MC is constant until the local total water content reaches fiber saturation point (FSP). Reproduced from *Forest Products Journal* 2017, 67, 266–274 [31].

rapid than that of cell wall water. The spin–spin relaxation time (T_2) revealed that most of the water was lumen water situated in smaller pores. The irregular T_2 signal amplitudes of cell lumen water indicated that fractional water in smaller pores was transferred into bigger pores during drying.

MRI is the natural method to study drying in wood materials because spatial resolution is obvious in an image. The first studies focused on conventional MRI techniques to image wood anatomy and structure [42–44]. Later researchers employed MRI to study water distribution in wood [45–52]. Most MRI studies consider relatively wet wood because conventional MRI techniques are not quantitative at low MCs. The imaging quality is poor compared with other imaging methods and one cannot resolve structure on the scale of the lumen. Spatial resolution of millimeters or submillimeters is most common.

The image quality achieved tends to depend on MC. Imaging MC in wood below 30% MC, when there is no lumen water, is very challenging. Spatial resolution of the evolving MC during drying is critical to understand drying mechanisms and drying rates. Menon did one-dimensional MRI measurements of water distribution in wood with MC 180% [45]. The SPRITE (single point ramped imaging with T_1 enhancement) MRI measurement has been applied to measure wood water content even at low MC [46–50]. Figure 15.7 shows that the local image intensity from SPRITE MRI images of low-MC wood samples directly correlates with MC [50].

SPRITE is a pure phase-encode MRI measurement that does not require echo formation. It employs ramped magnetic field gradients and brief-duration broadband radiofrequency pulses, which permit rapid image acquisition in cases where the signal lifetimes are short, as in the case of low water content wood materials [48,50]. For a centric scan SPRITE image employing a radiofrequency excitation pulse with uniform radiofrequency flip angle (α), the signal intensity denoted by S, at any point in the image, is determined by the following equation [51]:

$$S = S_0 e^{-\frac{t_p}{T_2^*}} \sin\alpha, \tag{15.2}$$

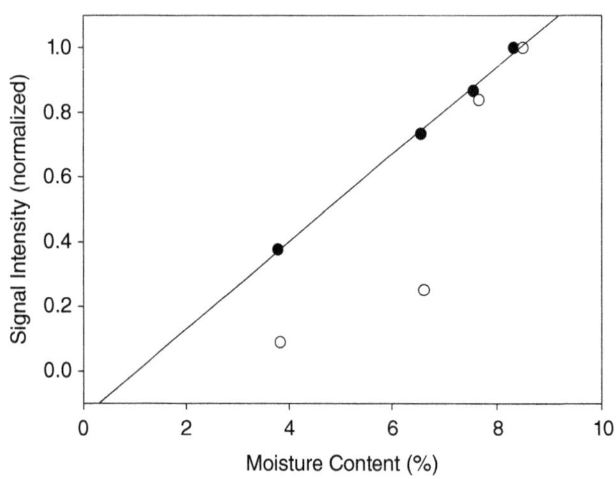

Figure 15.7 Local image intensity versus moisture content (MC) in yellow birch measured with SPRITE MRI (single point ramped imaging with T_1 enhancement magnetic resonance imaging) (closed circles) and a conventional spin echo MRI measurement (open circles). The SPRITE measurement yields a local image intensity that is linear with known MC. Reproduced from *Canadian Journal of Forest Research* 2011, 41, 2158–2162 [50].

where S_0 is the signal intensity at $t = 0$. S_0 is proportional to the quantity of observable water. Variable t_p is the phase-encode time and T_2^* is the effective transverse signal lifetime. T_2^* is a function of both the intrinsic transverse relaxation time, T_2, and the local magnetic field inhomogeneity ΔB_0 as shown in Equation 15.3:

$$\frac{1}{T_2^*} = \frac{1}{T_2} + \gamma \Delta B_0, \tag{15.3}$$

where γ is the nuclear gyromagnetic ratio. Generally, for water in wood, $t_p \ll T_2^*$ meaning that the SPRITE signal intensity does not suffer from T_2^* attenuation and the signal intensity is directly proportional to liquid water content. SPRITE has made quantitative low-MC wood imaging possible. However, the SPRITE method cannot directly distinguish cell wall and cell lumen water. The method measures total water content spatially resolved. Figure 15.8 shows profiles from 3D SPRITE MRI images, in the sapwood, of drying black spruce showing total water content variation during drying [31].

Xiao and Balcom [52] have recently developed a method to spatially resolve cell wall water and cell lumen water distribution, at high and low wood MC. They succeed, not by spatial resolution of the cell wall and lumen, but rather by analysis of the spatially resolved signal lifetimes. The spatial resolution achieved averages over cell walls and the cell lumen. This development is critical for wood-drying and modeling studies because it permits monitoring space-resolved cell wall water and cell lumen water migration. This method combines ultrashort echo time (UTE) imaging with multiple spin echo refocusing to generate a series of images with T_2 weighting. The T_2 decay curves for each image voxel can be extracted and multiple T_2 relaxation components are quantitatively evaluated. Figure 15.9 shows results of their work discriminating cell wall water and cell lumen water in black spruce test samples.

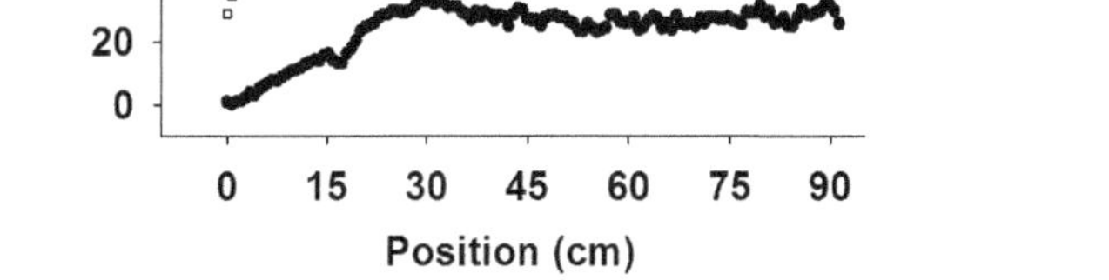

Figure 15.8 Moisture content measurement spatially resolved in the sapwood of a drying black spruce log. The 1D profile is assembled from a composite series of 3D SPRITE MRI (single point ramped imaging with T_1 enhancement magnetic resonance imaging) images. The profile reveals water content in one-half of the drying log. Drying is principally through a fresh-cut end at 0 cm. The profiles reveal a diffusive drying front near the cut end and uniform drying along the full length of the log. Reproduced from *Forest Products Journal* 2017, 67, 266–274 [31].

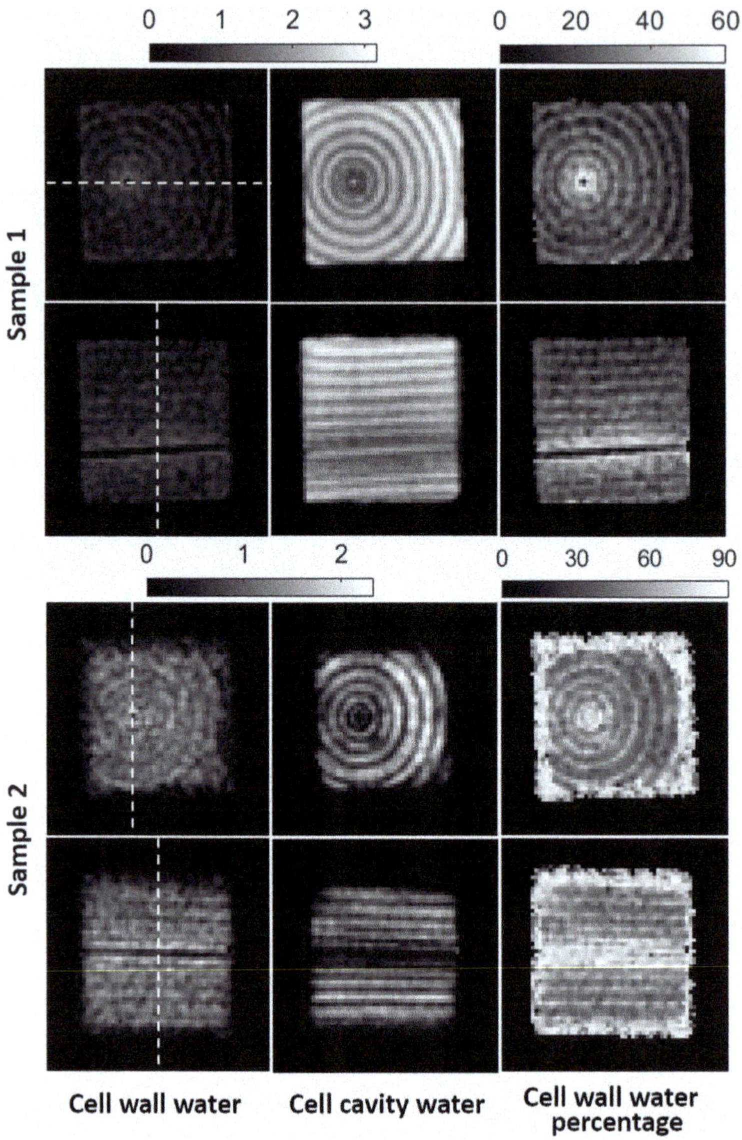

Figure 15.9 2D slices from 3D maps of the cell wall water content (first column), cell cavity water content (second column), and cell wall water percentage (third column) of two black spruce samples, 2.2 cm × 2.2 cm × 2.2 cm. The ring pattern observed in the images is due to the growth rings of the samples. Sample 1 was fully water saturated. Sample 2 was dried for 1 h at 80°C and ambient relative humidity. Note, as expected, the cell wall water percentage increased as lumen water evaporated in sample 2. Transverse and longitudinal slices are displayed from the 3D images of both samples. The dashed lines indicate the position of the relevant slices. [52] Xiao D et al. (2019), figure 07[p.22]/with permission from Elsevier.

15.4 MR Studies of the Pore Space in Wood

Wood is a natural porous material. It contains pores/voids, which may range from nanometer scale to millimeter scale. The pore space in wood is important for mass migration, catalyst, and energy storage material development. MR is a powerful tool to study the pore space of materials since T_1 and T_2 values are related to pore size and pore occupancy.

15.4.1 Relaxation Time Distribution for Wood Pore Studies

Brownstein and Tarr combined classical diffusion ideas in biological cells with MR relaxation measurements. Their results showed the cell size and shape are related to the multiexponential MR relaxation decay curve [53]. A continuous distribution of relaxation times instead of the limited sum of discrete T_2 values is better explained by a continuous distribution of the wood cell sizes [34]. Araujo et al. developed a diffusion model for compartmentalized water in wood based on spin–spin relaxation and used the diffusion model to estimate the size of the cell lumen distribution directly from the CPMG (Carr–Purcell–Meiboom–Gill) decay curves [54].

Simple models of MR relaxation in the pore space show that both the T_1 and T_2 distribution reflect the pore size distribution according to Equations 15.4 and 15.5 [55]:

$$\frac{1}{T_1}=\frac{1}{T_{1b}}+\rho_1\left(\frac{S}{V}\right)_{\text{pore}}, \tag{15.4}$$

$$\frac{1}{T_2}=\frac{1}{T_{2b}}+\rho_2\left(\frac{S}{V}\right)_{\text{pore}}+\frac{1}{T_{2\text{diffusion}}}, \tag{15.5}$$

where T_{1b} is the bulk T_1 relaxation time; T_{2b} is the bulk T_2 relaxation time of the pore fluid; ρ_1 is the T_1 surface relaxivity (T_1 relaxing strength of the grain surfaces), while ρ_2 is the T_2 surface relaxivity (T_2 relaxing strength of the grain surfaces); $(S/V)_{\text{pore}}$ is the ratio of pore surface to fluid volume; and $T_{2\text{diffusion}}$ is a diffusion effect that is manifest when internal magnetic field gradients exist.

If $T_2\text{b} \gg T_2$ and the third term in Equation 15.5 is very small, $1/T_2$ is approximately equal to $\rho_2(S/V)_{\text{pore}}$. It is easier and more accurate to determine pore sizes (S/V) with T_1 methods, but the T_1 measurement time is much longer than for T_2. However, the surface relaxation rate is related to many factors [55]. Employing standard samples with known and specified pore sizes is a good way to calibrate pore size measurement [56]. We note however that Figure 15.1 shows that the cell lumen size is remarkably constant in this hierarchal material. The major difference in cell lumen size is between earlywood and latewood.

15.4.2 MR Cryoporometry Studies of Water in Wood Pores

MR cryoporometry is an effective way to determine pore size information, according to the Gibbs–Thomson equation [57]. The principle for this method is that liquid within small pores has depressed melting or freezing temperatures compared with large pores and bulk fluid [12,58]. Kekkonen et al. employed this method to determine the pore size distribution of thermally modified wood [59]. In wood materials the Gibbs–Thomson equation can be simplified with the assumption that all pores are cylindrical [60,61]. Gao et al. applied the method to determine swollen wood cell pore size distribution [62]. Lamason employed simpler freezing and thawing methods to differentiate cell wall water and cell lumen water [18]. The cryoporometry measurement cannot be applied to online measurements. It is more appropriate as a research measurement.

15.5 Conclusions

MR/MRI techniques are powerful research tools for wood science. They play an important role in distinguishing water states in wood, water migration during drying and sorption, and exploring wood pore structure. It is likely that portable and desktop MR measurements will become increasingly important in near line applications; however, true online measurements are unlikely due to practical limits on MR sensitivity and the high-throughput requirements of industry. High-value-added wood products are the most likely avenue for increased utilization of MR in the industry. High-quality data, MR and MRI, will be of great use to help guide and optimize energy-intensive drying protocols.

Acknowledgments

This work was supported by awards from the China Scholarship Council (CSC no. 201508150005) and National Natural Science Foundation of China (31860185, 31160141, and 30800866). The MRI Research Centre at the UNB gave tremendous technical support to the second author during his sabbatical year in Canada (Dr. Bryce MacMillan, Dr. Dan Xiao, and Dr. Clevan Lamason).

References

1 Hoadley, R.B. (2000). *Understanding Wood: A Craftsman's Guide to Wood Technology*. Newtown, CT: Taunton Press.
2 Skaar, C. (1988). *Wood–Water Relations*. Heidelberg, Germany: Springer.
3 Stamm, A.J. (1964). *Wood and Cellulose Science*. New York: The Ronald Press Company.
4 Panshin, A.J. and Zeeuw, C.D. (1980). *Textbook of Wood Technology*. New York: McGraw-Hill.
5 Bowyer, J.L., Shmulsky, R., and Haygreen, J.G. (2007). *Forest Products and Wood Science: An Introduction*. Oxford, UK: Blackwell Publishing.
6 Ross, R.J. (2010). *Wood Handbook: Wood as an Engineering Material*. Centennial edn. General technical report FPL; GTR-190. Madison, WI: USDA Forest Service, Forest Products Laboratory.
7 Tiemann, H.D. (1906). *Effect of Moisture Upon the Strength and Stiffness of Wood*. USDA Forest Service, Bulletin 70. Washington, DC: Government Printing Office.
8 Stamm, A.J. (1971). Review of nine methods for determining the fiber saturation points of wood and wood products. *Wood Science* 4: 114–128.
9 Engelund, E.T., Thygesen, L.G., Svensson, S. et al. (2013). A critical discussion of the physics of wood-water interactions. *Wood Science and Technology* 47: 141–161.
10 Lamason, C., MacMillan, B., Balcom, B. et al. (2015). Water content measurement in black spruce and aspen sapwood with benchtop and portable magnetic resonance devices. *Wood Material Science & Engineering* **10 (1): 86–93.**

11 Nakamura, K., Hatakeyama, T., and Hatakeyama, H. (1981). Studies on bound water of cellulose by differential scanning calorimetry. *Textile Research Journal* 51 (9): 607–613.
12 Berthold, J., Rinaudo, M., and Salmeń, L. (1996). Association of water to polar groups; estimations by an adsorption model for ligno-cellulosic materials. *Colloids and Surfaces A: Physicochemical and Engineering Aspects* 12 (2–3): 117–129.
13 Zelinka, S.L., Lambrecht, M.J., Glass, S.V. et al. (2012). Examination of water phase transitions in loblolly pine and cell wall components by differential scanning calorimetry. *Thermochimica Acta* 53: 9–45.
14 Brunauer, S., Emmett, P.H., and Teller, E.J. (1938). Adsorption of gases in multilayers. *Journal of the American Chemical Society* 60: 309–319.
15 Walker, J.C. (2006). *Primary Wood Processing Basic Wood Chemistry and Cell Wall Ultrastructure*. Dordrecht, the Netherlands: Springer.
16 Lucatorto, T., Graef, M.D., and Wong, P.Z. (1999). *Methods in the Physics of Porous Media*. San Diego, CA: Academic Press.
17 Liang, Z.P. and Lauterbur, P.C. (2000). *Principles of Magnetic Resonance Imaging – A Signal Processing Perspective*. New York: IEEE Press.
18 Lamason, C., MacMillan, B., Balcom, B. et al. (2014). Examination of water phase transitions in black spruce by magnetic resonance and magnetic resonance imaging. *Wood and Fiber Science* 46 (4): 1–14.
19 Zhang, M.H., Wang, X.M., and Gazo, R. (2013). Water states in yellow poplar during drying studied by time-domain nuclear magnetic resonance. *Wood and Fiber Science* 45 (4): 423–428.
20 Menon, R.S., Mackay, A.L., Hailey, J.R. et al. (1987). An NMR determination of the physiological water distribution in wood during drying. *Journal of Applied Polymer Science* 33 (4): 1141–1155.
21 Lamason, C. (2017). *Log Moisture Content Determination with Quantitative Magnetic Resonance*. Fredericton: University of New Brunswick. 2017.
22 ASTM Standard D4442-07 (2007). *Direct Moisture Content Measurement of Wood and Wood-base Materials*. West Conshohocken, PA: ASTM International.
23 James, W.L. (1988). *Electric Moisture Meters for Wood, Forest Products Laboratory, General Technical Report*. Madison, WI: USDA Forest Service.
24 Nanassy, A.J. (1973). Use of wide-line NMR for measurement of moisture content in wood. *Wood Science* 5 (3): 187–193.
25 Nanassy, A.J. (1976). True dry-mass and moisture content of wood by NMR. *Wood Science* 9 (2): 104–109.
26 Casieri, C., Senni, L., Romagnoli, M. et al. (2004). Determination of moisture fraction in wood by mobile NMR device. *Journal of Magnetic Resonance* 171 (2): 364–372.
27 Senni, L., Casieri, C., Bovino, A. et al. (2009). A portable NMR sensor for moisture monitoring of wooden works of art, particularly of paintings on wood. *Wood Science and Technology* 43 (1–2): 167–180.
28 Dvinskikh, S.V., Furo, I., Sandserg, D. et al. (2011). Moisture content profiles and uptake kinetics in wood cladding materials evaluated by a portable nuclear magnetic resonance spectrometer. *Wood Material Science and Engineering* 6: 119–127.
29 Pourmand, P., Wang, L., and Dvinskikh, S.V. (2011). Assessment of moisture protective properties of wood coatings by a portable NMR sensor. *Journal of Coatings Technology and Research* 8 (5): 649–654.
30 Marble, A.E., Mastikhin, I.V., Colpitts, B.G. et al. (2007). A compact permanent magnet array with a remote homogeneous field. *Journal of Magnetic Resonance* 186 (1): 100–104.

31 Lamason, C., MacMillan, B., Balcom, B.J. et al. (2017). Magnetic resonance studies of wood log-drying processes. *Forest Products Journal* 67 (3–4): 266–274.
32 Li, X.Y., Gao, Y.L., Zhang, M.H. et al. (2017). Water migration in poplar wood during microwave drying studied by time domain nuclear magnetic resonance (TD-NMR). *Holzforschung* 71 (11): 881–887.
33 Quick, J.J., Hailey, J.R., and MacKay, A.L. (1990). Radial moisture profiles of cedar sapwood during drying: A proton magnetic resonance study. *Wood and Fiber Science* 22 (4): 404–412.
34 Araujo, C.D., Mackay, A.L., Hailey, J.R. et al. (1992). Proton magnetic resonance techniques for characterization of water in wood: Application to white spruce. *Wood Science and Technology* 26 (2): 101–113.
35 Hartley, I.D., Kamke, F., and Peemoeller, H. (1994). Absolute moisture content determination of aspen wood below the fiber saturation point using pulsed NMR. *Holzforschung* 48 (6): 474–479.
36 Almeida, G., Gagné, S., and Hernández, R.E. (2007). A NMR study of water distribution in hardwoods at several equilibrium moisture contents. *Wood Science and Technology* 41 (4): 293–307.
37 Elder, L.G. and Elder, T. (2009). Moisture in untreated, acetylated, and furfurylated Norway spruce monitored during drying below fiber saturation using time domain NMR. *Wood and Fiber Science* 41 (2): 194–200.
38 Hoffmeyer, P., Engelund, E.T., and Thygesen, L.G. (2011). Equilibrium moisture content (EMC) in Norway spruce during the first and second desorptions. *Holzforschung* 65 (6): 875–882.
39 Zhang, M.H., Li, X.Y., Zhou, Y.J. et al. (2014). Water status change in wood drying studied by time-domain NMR. *Scientia Silvae Sinicae* 50 (12): 109–113.
40 Wang, Z., Wang, X.M., and Chen, Z.J. (2018). Water states and migration in Xinjiang poplar and Mongolian Scotch pine monitored by TD-NMR during drying. *Holzforschung* 72 (2): 113–123.
41 Xu, K., Lu, J.X., Gao, Y.L. et al. (2017). Determination of moisture content and moisture content profiles in wood during drying by low-field nuclear magnetic resonance. *Drying Technology* 35 (15): 1909–1918.
42 Wang, P.C. and Chang, S.J. (1986). Nuclear magnetic resonance imaging of wood. *Wood and Fiber Science* 18 (2): 308–314.
43 Hall, L.D., Rajanayagam, V., Stewart, W. et al. (1986). Magnetic resonance imaging of wood. *Revue Canadienne de Recherche Forestière* 16 (2): 423–426.
44 Hall, L.D. and Rajanayagam, V. (1986). Evaluation of the distribution of water in wood by use of three dimensional proton NMR volume imaging. *Wood Science and Technology* 20 (4): 329–333.
45 Menon, R.S., Mackay, A.L., Flibotte, S.G. et al. (1989). Quantitative separation of NMR images of water in wood on the basis of T_2. *Journal of Magnetic Resonance* 82 (1): 205–210.
46 Balcom, B.J., MacGregor, R.P., Beyea, S.D. et al. (1996). Single-point ramped imaging with T_1 enhancement (SPRITE). *Journal of Magnetic Resonance A* 123 (1): 131–134.
47 Hameury, S. and Sterley, M. (2006). Magnetic resonance imaging of moisture distribution in pinus sylvestris L. exposed to daily indoor relative humidity fluctuations. *Wood Material Science & Engineering* 1 (3–4): 116–126.
48 MacMillan, B., Schneider, M.H., Sharp, A.R. et al. (2002). Magnetic resonance imaging of water concentration in low moisture content wood. *Wood and Fiber Science* 2: 276–286.
49 Stenström, S., Bonazzi, C., and Foucat, L. (2011). Magnetic resonance imaging for determination of moisture profiles and drying curves. *Modern Drying Technology* 2: 91–142.

50 MacMillan, B., Veliyulin, E., Lamason, C. et al. (2011). Quantitative magnetic resonance measurements of low moisture content wood. *Canadian Journal of Forest Research* 41: 2158–2162.

51 Halse, M., Goodyear, D.J., MacMillan, B. et al. (2003). Centric scan SPRITE magnetic resonance imaging. *Journal of Magnetic Resonance* 165 (2): 219–229.

52 Xiao, D. and Balcom, B.J. (2019). Ultra-short echo time imaging with multiple echo refocusing for porous media T_2 mapping. *Journal of Magnetic Resonance* 299: 33–41.

53 Brownstein, K.R. and Tarr, C.E. (1979). Importance of classical diffusion in NMR studies of water in biological cells. *Physical Review A. General Physics* 19 (6): 2446–2453.

54 Araujo, C.D., Mackay, A.L., Whittall, P.K. et al. (1993). A diffusion model for spin-spin relaxation of compartmentalized water in wood. *Journal of Magnetic Resonance, Series B* 101 (3): 248–261.

55 Coates, G.R., Xiao, L., and Prammer, M.G. (2000). *NMR Logging Principles and Applications.* Houston, TX: Gulf Publishing Company.

56 Zhao, Z.H., Liu, W.J., and Zhang, M.H. (2020). Measurement of pore sized microporous-mesoporous materials by time-domain nuclear magnetic resonance. *BioResources* 15 (1): 1407–1418.

57 Aksnes, D.W. and Kimtys, L. (2004). ^{1}H and ^{2}H NMR studies of benzene confined in porous solids: Melting point depression and pore size distribution. *Solid State Nuclear Magnetic Resonance* 25 (1–3): 146–152.

58 Kärenlampi, P.P., Tynjälä, P., and Ström, P. (2005). Phase transformations of wood cell wall water. *Journal of Wood Science* 51 (2): 118–123.

59 Kekkonen, P.M., Ylisassi, A., Telkki, V.V. et al. (2014). Absorption of water in thermally modified pine wood as studied by nuclear magnetic resonance. *Journal of Physical Chemistry C* 118 (4): 2146–2153.

60 Park, S., Venditti, R.A., Jameel, H. et al. (2006). Changes in pore size distribution during the drying of cellulose fibers as measured by differential scanning calorimetry. *Carbohydrate Polymers* 66 (1): 97–103.

61 Zauer, M., Kretzschmar, J., Grossmann, L. et al. (2014). Analysis of the pore-size distribution and fiber saturation point of native and thermally modified wood using differential scanning calorimetry. *Wood Science and Technology* 48: 177–193.

62 Gao, X., Zhuang, S., Jin, J. et al. (2015). Bound water content and pore size distribution in swollen cell walls determined by NMR technology. *Bioresources* 10 (4): 8208–8224.

Part IV

Applications in Energy Research

16

In Situ Spectroscopic Imaging of Devices for Electrochemical Storage with Focus on the Solid Components

Elodie Salager

CNRS, CEMHTI UPR3079, Université d'Orléans, Orléans, France
Réseau sur le Stockage Electrochimique de l'Energie (RS2E), CNRS FR3459, Amiens, France

16.1 Introduction

Energy storage becomes increasingly crucial in the transition from fossil energy to renewable energies, which are intrinsically intermittent. Electrochemical storage is one of the currently exploited answers and has received a lot of attention already. The corresponding devices, mainly lithium-ion batteries (LiBs) and electrochemical double-layer capacitors (EDLCs), are omnipresent in today's life. Mobile applications range from small portable electronics to electrified transport, and stationary storage is increasingly important. The ultimate goal is to obtain devices with high power (fast charge or delivery of energy), high capacity (autonomy), and long lifetime. Power and capacity are difficult to optimize jointly, so that the emphasis is put on one or the other characteristics depending on the intended use.

While a full understanding of the mechanisms of storage is essential to devise new materials for enhanced energy storage and power, the parasitic chemical reactions or morphological evolutions undergone in the longer term (after multiple cycles) often relate to capacity loss and therefore device lifetime. In addition the LiB and EDLC energy-storing materials are highly reactive and prompt to reach metastable (electro)chemical states. Unsurprisingly there is a strong demand for characterization techniques that can follow in real time their global evolution and ideally the evolution of each of their components, as demonstrated by several reviews on that subject [1–3].

Magnetic resonance microscopy (MRM), usually defined as magnetic resonance imaging (MRI) techniques with resolution better than ~100 μm [4], fits in the limits of the spatial requirements for these devices. Its relatively low sensitivity compared with other techniques is counterbalanced by its nonionizing characteristics and the low deposited energy that does not degrade the highly reactive constituents.

In this chapter we concentrate on in situ MR spectroscopic microscopy of the contents of fully functional electrochemical storage devices, for which a complete nuclear magnetic resonance (NMR) spectrum is correlated to a specific voxel in the image. The heterogeneity and complexity of the electrochemical devices, however, call for using more constraints than just the spin density to interpret the images obtained during operation. The spectroscopic information is an asset for such studies, thanks to the strong sensitivity of the NMR spectrum to the chemical environment. The resulting spectroscopic images can also be regarded as localized spectra. The goal of this

Magnetic Resonance Microscopy: Instrumentation and Applications in Engineering, Life Science and Energy Research, First Edition. Edited by Sabina Haber-Pohlmeier, Bernhard Blümich, and Luisa Ciobanu.

chapter is not to review exhaustively the in situ MR field for electrochemical storage, as a number of reviews have been published in recent years [5–10]. Instead, we introduce, using selected examples, the specific challenges for in situ and operando MR of electrochemical devices and highlight for solids the opportunities offered by the full exploitation of the combination of the two MR modalities: spectroscopy and imaging.

16.2 Electrochemical Storage

This part aims at giving a rapid overview of the electrochemical storage devices. Further details are available in reference books [11–14].

16.2.1 Lithium-ion Batteries

LiBs rely on reversible reduction-oxidation (redox) electrochemical reactions occurring in the two electrodes to store and deliver energy [15,16]. On discharge, the reduction at the positive electrode captures electrons that are released by the oxidation at the negative electrode. Those electrons are exchanged through the load (phone, computer, car, etc.) and provide current to it. On charge, the charger provides energy to the battery to perform the inverse redox reactions (Figure 16.1a). In commercial LiBs these reactions are associated with reversible intercalation of lithium ions (i.e. without major change of the structure) in the active material, hence their name.

Commercial LiBs use layered transition metal oxides as the positive electrode, such as nickel, manganese, and cobalt lithium oxides (NMC, $Li(Ni_yMn_zCo_{1-y-z})O_2$) or Li-rich NMC (with extra Li ions in the transition metal sites) [17]. Graphite is classically used as a negative electrode, with the possibilities of silicon–graphite blends being extensively explored [18,19]. The electrodes are electrically insulated by a porous separator, and the electrolyte – usually a salt dissolved in carbonate solvents with various additives – soaks the separator and the electrodes. The requirements for transport electrification have triggered a large body of work to enhance the existing technology, in terms of capacity (amount of stored energy per charge), power (rate of charge), coulombic efficiency (longer lifetime of the battery), and sustainability, but there are still fundamental challenges to meet to cover a larger range of applications [20–22].

Lately a significant effort has been made to remove the liquid – flammable – solvent by using a solid. The main issues are the lower conductivity and chemical reactivity with the electrodes [23]. Other technologies including Na, K, and multivalent ions are also enthusiastically explored in post-Li-ion or all-solid-state batteries [24,25].

16.2.2 Electrochemical Double-Layer Capacitors

EDLCs correspond to a different usage, with smaller autonomy but high power (fast delivery of energy). In EDLCs energy is not stored through faradic processes but through capacitive mechanisms [26]: the electronic charge accumulates on the electrodes and it is charge-compensated by the organization of the ions in the liquid electrolyte (formation of a double layer) and adsorption on the surface of the electrodes (Figure 16.1b). The most common electrodes are nanoporous carbons with controlled porosity. The electrolyte is usually a salt in an organic

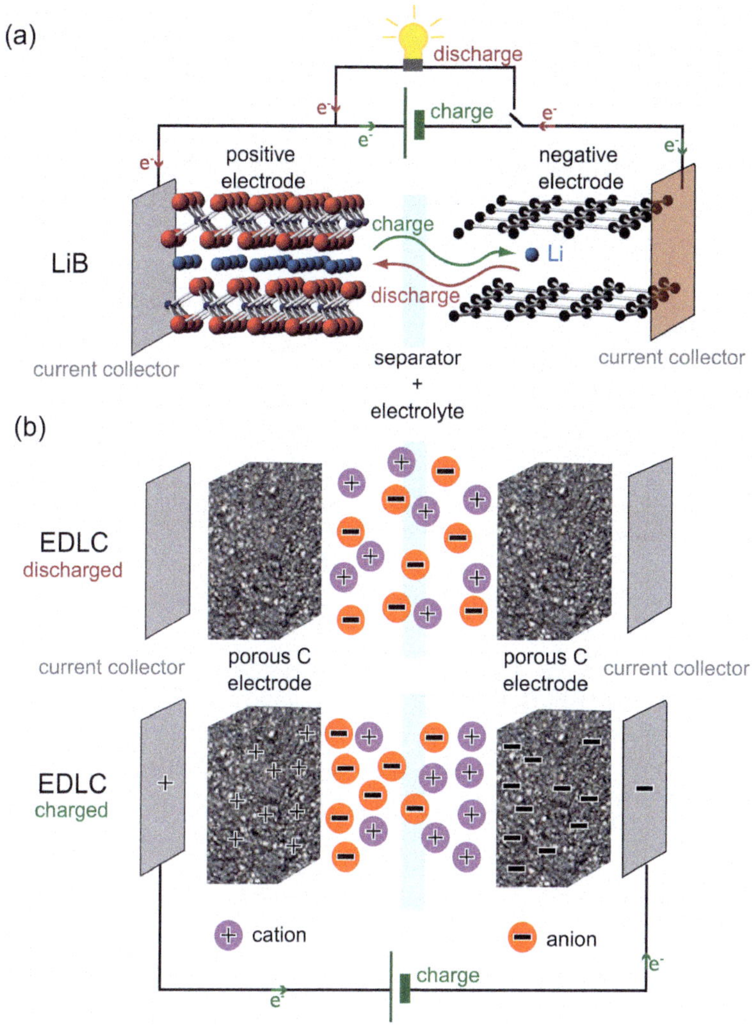

Figure 16.1 Schematics of lithium-ion battery (LiB) and electrochemical double-layer capacitor (EDLC) electrochemical mechanisms. (a) The LiB concept relies on redox reactions: exchange of electrons through the electrical circuit and exchange of Li ions through the separator. (b) The EDLC is based on the capacitive behavior of each electrode–electrolyte interface. The capacitance is obtained through accumulation of charges in the nanoporous carbon electrodes and organization of the ions in the electrolyte. Energy is released through the load, which permits the exchange of charges.

solvent to enlarge the voltage range and hence the capacitance. The capacitive nature of the energy storage process allows for faster energy storage and delivery (higher power) than in LiBs, but the stored charge is lower. EDLCs have however met a large range of applications, in which power and light weight are strong requirements while an autonomy of minutes is acceptable. The challenge relies on storing more energy, through carefully designed electrolytes and fine control of the electrode–electrolyte interface, or using hybrid devices such as pseudocapacitors and Li-ion capacitors [26–30].

16.2.3 Electrochemical Characterization

A large range of electrochemical characterizations is available for LiBs and EDLCs [31,32]. The electrochemical performance of new electrode composites, electrolyte compositions, or chemistries is usually characterized first through multiple charge–discharge cycles. In practice a galvano-potentiostat controls precisely the current or the voltage applied to the battery or EDLC and measures the cell voltage or current response, respectively. Experimentally, a cable links the galvano-potentiostat (and its electronics) to the electrochemical device, using small BNC (Bayonet Neill–Concelman) or clip terminations. Frequency-dependent impedance measurements also provide information on the limiting factors for the electrochemical device.

For batteries, the capacity (amount of charge stored during charge, and returned during discharge) is a key parameter of the electrochemical characterization, as well as the coulombic efficiency (efficiency of the charge–discharge process). Batteries are therefore usually charged and discharged in a galvanostatic mode.

EDLCs' stored energy and power depend on the maximum allowed voltage, on the accessible surface area of the electrode, and on the average distance between the adsorbed ions and the electrode surface. Cyclic voltammetry is the common characterization tool for EDLC; the current is ramped between 0 V and the maximum allowed by the electrolyte and returned to its initial value.

Two electrochemical procedures are also relatively common, and especially interesting for in situ studies, as they correspond to longer times and allow for (pseudo)equilibrium of the device. In the first procedure – open circuit (OC) – no current is applied to mimic resting stages. The voltage may be followed to determine its equilibrium value and the time needed to reach it. This process is also applied in the galvanostatic or potentiostatic intermittent titration technique (GITT or PITT) to determine the kinetic overpotentials and equilibrium potentials in galvanic cells [31,33]. The second procedure, often termed "constant-voltage" (CV) or "floating," maintains the voltage at a chosen value to characterize the decreasing current. In EDLCs this allows the stabilization of the electrode/electrolyte potentials and is sometimes used as a way to mimic aging due to multiple cycles of charge–discharge [27].

16.2.4 Electrochemical Devices in Practice

The LiB or supercapacitor is a multi-component device [15] as shown in Figure 16.2, including:

- an airtight casing: stainless steel for cylindrical (Figure 16.2a) and coin cells (Figure 16.2b), aluminized plastics for prismatic (Figure 16.2c) or pouch cells (Figure 16.2d) with electrical contacts to enable charge and discharge;
- a positive electrode and a negative electrode;
- an electrolyte, electronically insulating but ionically conducting.

The electronic insulation is provided in the case of liquid electrolytes by a microporous membrane (separator), porous to ions to ensure ionic conduction.

The electrodes each contain the electrochemically active material (redox or double-layer), a binder for mechanical strength and processability, and a conductive additive for better electronic conductivity. Note that, contrary to LiBs, standard EDLCs contain similar positive and negative electrodes.

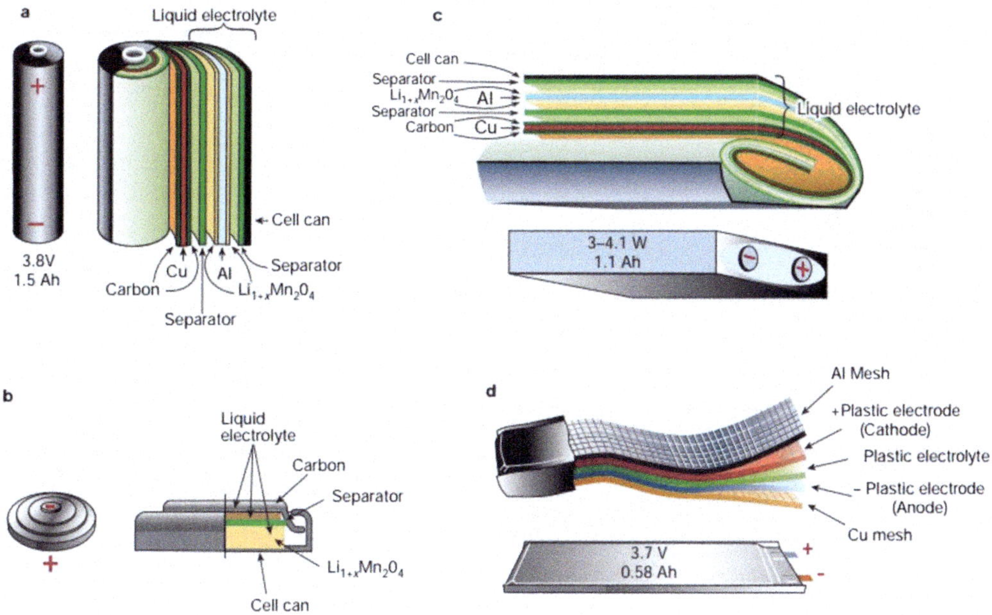

Figure 16.2 Lithium-ion battery devices. (a) Cylindrical cell, (b) coin cell, (c) prismatic cell, and (d) pouch cell. Reproduced with permission from [15].

16.3 Opportunities and Challenges for in Situ MR Spectroscopic Imaging of Electrochemical Storage Devices

Let us first define three terms used in this chapter: ex situ, in situ, and operando. This specific terminology was developed to discriminate three different ways of approaching the characterization of an electrochemical device. In situ originates from Latin and means "in the natural, original, or appropriate position" [34]. In our case, an in situ measurement is performed on the fully operational device, but not necessarily functioning during the measurement. Operando, Latin for "operating," is a subgroup of in situ studies and was devised to indicate the characterization of the electrochemical cell while it is operating, i.e. while it has current flowing through. Both terms are commonly opposed to ex situ, used for "out of the natural position" and referring to studies for which the components are extracted from the electrochemical cell, thereby destroying a new device at each state of charge of interest.

16.3.1 General Experimental Setup

The experimental setup includes the elements necessary for the MR measurement – supraconducting magnet, magnetic field gradient, radiofrequency resonator, and the circuit to the corresponding electronics (pulse programmer, transmitter, radiofrequency amplifiers, receiver, digitizer, etc.) [35,36], but also the electrical circuit to connect the electrochemical cell to the galvano-potentiostat (Figure 16.3).

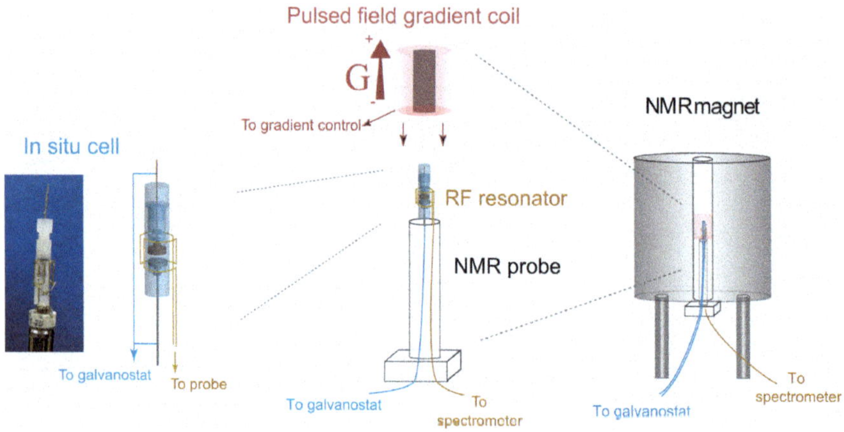

Figure 16.3 General experimental setup of an in situ magnetic resonance (MR) experiment. Electrochemical cell and probe designs for MR spectroscopic imaging are presented in Section 16.4.

16.3.2 Opportunities

The highly heterogeneous nature of the electrochemical devices results in a complex overall MR signal. The wide range of constitutive materials in the device (metals, paramagnetic oxides, conductive powders, liquids, microporous separator membranes, etc.) results in distortions of the in situ MR images that are difficult to identify. Strongly shifted peaks may also create artifacts in the images, and magnetic susceptibility blurring at the interfaces is a common issue.

Reversing the approach, from the in situ MR spectroscopy point of view, assignment of the peaks in the MR spectrum to the various components of the battery is essential to extract relevant information about its evolution. This is challenging due to the numerous interfaces at the microscopic scale that result in shifts and broadenings (mainly magnetic susceptibility effects [37,38]).

But this complexity is also an opportunity, as it contains precious information about the contents of the device. The combination of the spectroscopic and spatial dimensions enables easier exploitation of that information. Figure 16.4a shows the spectrum of a LiB containing a paramagnetic positive electrode (broad component in the baseline of the 1D spectrum, not detected in chemical shift 1D image, Figure 16.4b and c), a negative lithium electrode, and a standard electrolyte, LiPF6 in EC:DMC (1:1). The signal for the electrolyte spans the zone −5 ppm to 15 ppm, a range 10 times larger than the ^{7}Li chemical shift range for diamagnetic compounds. In this case the chemical information is negligible compared with other effects in the cell. In Figure 16.4b and c, localization adds an extra dimension to expand the spectrum and helps with efficient assignment of the spectra. The negative shifts can be assigned to electrolyte out of the electrochemical core, above the copper and below the aluminum current collectors.

The spectroscopic information is often reduced to the chemical shift, considered as the most relevant spectroscopic information thanks to its sensitivity to the chemical environment. The NMR spectrum provides in reality more than just the shift to the image. Several other characteristics (width, shape, relaxation, diffusion, connectivity maps) can be correlated to each peak and their evolution followed as the device is charged and discharged. These characteristics are also critical to exploit and correct distorted images.

In short, operando spectroscopic imaging or localized spectroscopy is a multidimensional measurement, in which the image dimension can be in principle 1D (profile), 2D (slice or

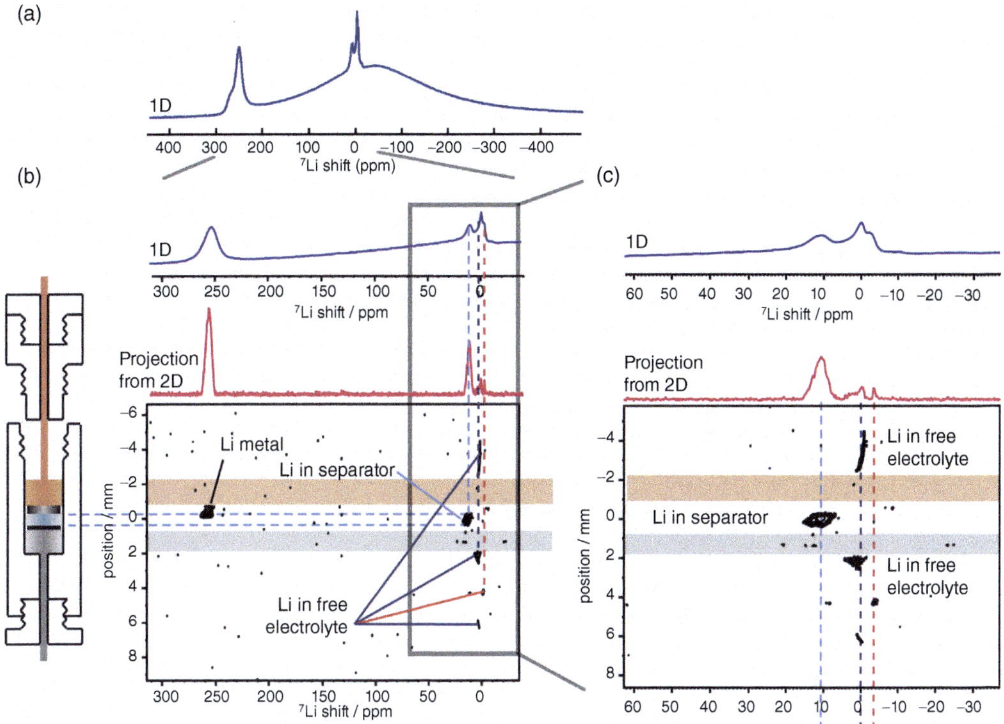

Figure 16.4 In situ ^{7}Li magnetic resonance (MR) data of a $Li_2Ru_{0.75}Sn_{0.25}O_3$-Li battery with LP30 electrolyte. (a) Spectrum, (b) chemical shift 1D image, and (c) close up on the electrolyte part. Adapted with permission from [39]. Copyright (2014) American Chemical Society.

projection), or 3D, while the spectroscopic information for each slice/pixel/voxel is in the form of a 1D spectrum, or a series of 1D spectra used to extract other parameters such as relaxation times (T_1, T_2, T_2*), self-diffusion coefficients, or even n-dimensional spectra.

An important feature of operando or stop-and-go in situ measurements is the evolution of the contents of the electrochemical device upon charge and discharge, as it stores and releases energy. Following the *evolution* of the device brings time as an additional dimension, which is useful to analyze and discriminate components compared with a single spectroscopic image. A last group of dimensions is therefore formed by the electrochemical properties (state of charge, impedance, galvanostatic or potentiostatic intermittent titration, etc.). The only limit to dimensionality is the time required to record data with a satisfying signal-to-noise ratio.

16.3.3 Experimental Challenges

The first in situ measurements were performed using a homemade design with a toroid cavity, with a Cu foil having a dual functionality: current collector and NMR detector [40]. In this pioneering study, rotating frame NMR imaging was used to obtain the spectra as a function of distance from the center of a radial cell and detect lithium plating on a carbon electrode. This approach however prevented operando measurement, as no charging current could flow in the detector during the measurements.

Innovative strategies are currently explored by Jerschow and coworkers, with the aim of keeping the commercial aluminum casing. In a first step, the commercial electrochemical device was studied without any modification, using pure MRI. The state of the device was observed indirectly by distortions in the image of a liquid filling a surrounding reservoir [41–43]. This original approach allowed for fast screening of defective cells, with measurement times as short as 12 s for a two-dimensional image. Spectroscopy was not possible however, so the chemical origins of cell failure could not be tracked. Instead, the cell image was compared with a premade database of functional and dysfunctional cells to identify the source of failure. Key features were identified using statistical tools, such as principal components analysis (PCA). Another creative approach has been to use the casing of the device as the resonator, and promising NMR spectra of the inside of an aluminized battery pouch cell were obtained [44]. This second approach has not yet provided spectroscopic imaging data.

The most popular approach, which was used for MR spectroscopic imaging, was pioneered by Letellier et al. in 2003 for LiBs [45] and fully exploited by Grey et al. [5]. The principle is to separate physically the electrochemical and MR circuits. One of the key advantages of that approach is to enable operando measurements, i.e. with current flowing in the cell during the MR measurement. Additionally the device can be transferred from or to an electrochemical bench for preliminary or additional charge and discharge.

Two main practical challenges are faced for in situ MR of electrochemical devices. The first challenge comes from the electrical circuit running between the galvano-potentiostat and the device, through the MR magnet and radiofrequency resonator. Coupling to the RF resonator brings parasitic signals that alter significantly the signal-to-noise ratio of the MR experiment, even without current running. A careful choice of shielded electrical cables and low-pass filters (cutoff frequency well below the Larmor frequency of the nuclei of interest) efficiently reduces the noise pickup at the Larmor frequency and restores a better signal-to-noise ratio.

The second challenge is the low penetration of the radiofrequency field through conductors: less than 10 μm in aluminum for a radiofrequency carrier frequency of 77 MHz, corresponding to ^{7}Li in a 4.7-T magnet. Stainless steel and aluminized casings shield the contents of the device from orthogonal time-dependent radiofrequency fields, inducing eddy currents and blind zones near the current collectors [46]. Letellier's approach and its derivatives are associated with the design of a specific geometry for the electrochemical cell to prevent issues in radiofrequency shielding from the cell, keeping the probe largely unmodified – apart from the radiofrequency resonator in some cases. These designs are described further in Section 16.4.

16.3.4 Quantitative Measurements

Another issue, often neglected, is the difficulty of performing properly quantitative measurements. These are challenging as several factors act on the signal intensity, with varying strength as the device is charged and discharged. First, part of the signal may be broadened beyond detection for some states of charge, changing the total intensity of the spectrum. Changes in the conductivity or the charge of the constituents modify the quality factor of the measuring circuit (hence the overall intensity of the spectrum). The NMR relaxation properties also vary with the state of charge and must be determined prior to the operando measurement or on the fly, resulting in a loss of time resolution. The last issue is the limited penetration of the radiofrequency wave in the bulk of metallic conductors, which results in a heterogeneous radiofrequency field-strength distribution and excitation of the spins in the metallic electrodes. Metallic transitions upon lithiation or delithiation are relatively common in positive and negative electrodes (e.g. $LiCoO_2$, graphite). However, concentrations profiles have been obtained through analysis of spectroscopic images [47–49].

16.3.5 Time Resolution and Sensitivity

In situ and operando MR of electrochemical storage devices is demanding, especially in terms of spatial and temporal resolution: battery components are usually around or less than 100 μm thick and the target for electric car batteries is set to charging in as fast as 15 min. Time resolution is an important point to take into account, as the MR measurement is an average of the battery states over the recording time. In particular, spectroscopic imaging measurements are relatively slow, so during the course of recording a nonnegligible evolution of the state of charge may occur. The device may be left in open circuit to avoid this "temporal blurring" and the MR data recorded only after stabilization of the electrochemical state – this "stop-and-go" in situ approach is not strictly operando but provides relevant snapshots of the battery state without opening it.

Space and time resolution are ultimately limited by sensitivity. As a result, in situ MR spectroscopic imaging is usually limited to sensitive isotopes, such as ^{1}H, ^{19}F, ^{31}P and ^{7}Li. For the solid components, the dimensionality is limited to spectroscopic 1D profiles or slice-selective measurements, using a compatible electrochemical cell geometry.

16.4 Electrochemical Cell and Probe Designs for in Situ Spectroscopic Imaging

Performance of the MR spectroscopic imaging is intimately linked to the probe design: pulsed field gradient (PFG) strength and match between the sample holder and the radiofrequency resonator. Electrochemical cell and probe designs have been explored enthusiastically to increase sensitivity.

Most of the work performed for in situ MR spectroscopic imaging follows Letellier's pouch cell approach, with variations in the design and geometry of the dedicated electrochemical cell. Three main families of setup have been successfully used to study LiBs and EDLCs: pouch cells, Swagelok cells, and plastic capsule cells (Figure 16.5). Improvement of the designs is still ongoing work and numerous variations in each design have been published; in the following only the electrochemical cells that have been used for spectroscopic imaging are reported, with reference to the seminal work of the corresponding design.

Commercial microimaging probes with 3D-PFG, or probes meant for diffusion measurements (1D-PFG), are commonly employed. The 3D-PFG are relatively moderate (up to 3 T/m) while in the 1D-PFG design a stronger pulsed-field gradient is made available (30 T/m) by sacrificing the two other directions.

16.4.1 Pouch Cells

The first pouch cells for in situ NMR spectroscopy had an aluminized blue plastic film sealed as a pouch around the battery. The electrode and separator films were plasticized and laminated onto copper grids that came out of the pouch (Bellcore technology [53,54], Figure 16.5a).

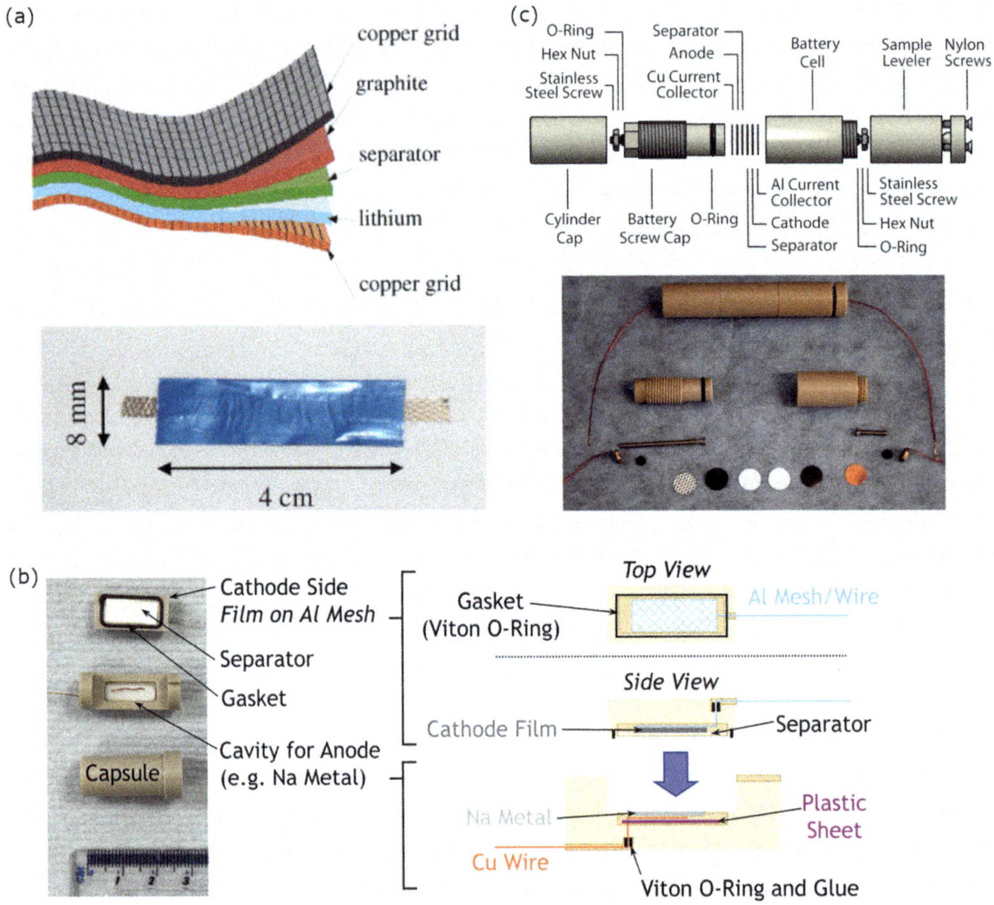

Figure 16.5 Electrochemical cells for in situ magnetic resonance (a) pouch, (b) capsule, and (c) Swagelok. Reproduced with permission from [50–52], respectively.

Chandrashekar et al. were the first to use this design in 2012 for in situ chemical shift-2D imaging on symmetric Li-Li cells [55] placed in an NMR tube. They used a simple polyester film as the casing; the removal of the aluminum barrier lowered the gas tightness of the casing but enhanced the MR sensitivity. Lithium dendrite formation could be characterized in situ, using the chemical shift information to discriminate bulk metallic lithium from mossy and dendritic lithium.

16.4.2 Plastic Capsule/Cartridge Cells

A plastic "cell capsule" was designed in the Grey and Jerschow groups [51] (Figure 16.5b). This design uses rectangular electrodes sitting between two polyetheretherketone (PEEK) half-cylinders. Air tightness is ensured by compression of a rectangular sealing gasket, and pressure is provided using chocks that press the electrochemical core inside the casing. The first use of the plastic cell capsule for spectroscopic 2D-imaging was performed by Ilott et al.

on EDLCs in 2014 with a polytetrafluoroethylene (PTFE) cell [56]. It enabled the separation of the spectroscopic signatures of the ions in the two similar electrodes in in situ and operando modes.

16.4.3 Swagelok Cells

A cylindrical polypropylene housing (Swagelok design) was pioneered in 2011 by Poli et al. for in situ NMR spectroscopy but it was not described precisely [57]. It inspired the development in 2013 by Tang et al. of a PEEK cell for in situ stray-field MR imaging (STRAFI), an approach using the strong fringe field of a magnet that unfortunately does not allow for spectroscopic measurements [52,58]. The electrode disks are perpendicular to the cylinder axis and sandwich a separator soaked with liquid electrolyte, with several O-rings ensuring air tightness. The main advantage of that design compared with the two others is the better control of compression via the screw cap (Figure 16.5c). The first report of spectroscopic imaging with this family of cells was shortly afterwards by Salager et al. [39]. Their cell was machined in polychlorotrifluoroethylene (PCTFE), a transparent plastic enabling easier positioning of the battery's electrochemical core in the probe. In this study in situ chemical shift-1D imaging enabled the localization of the signals for the electrolyte and metallic lithium, including those distorted by the conductors in the electrochemical cells.

16.4.4 Promising Designs

A promising design from Balcolm and coworkers was recently published. This cartridge design, close to the pouch cell geometry [59], is made of plastic (garolite/PC board) plates and sealed by compressing a silicon rubber gasket with screws running through the plastics. This cartridge is inserted in a parallel plate resonator within a homemade probe, using commercial PFG (max 0.3 T/m). ^{7}Li T_1–T_2 correlations identified the electrolyte located in different environments (separator, electrode porosity) and in the graphite electrode. Up to now only T_1–T_2 maps have been published, without further spectroscopic information, owing to the low homogeneity (1 ppm) and intensity of the magnetic field used in the study (2.4 T).

Another promising design, similar to a jelly roll, was proposed by Freytag et al. [60,61]. The current collectors were replaced by cellulosic paper and allowed very slow charging of the battery (one charge in 100 h). The radiofrequency penetration is optimum and the cell could be spun in the magnetic field, resulting in better spectral resolution. The design still needs improvement, as lithium dendrites should not form at such low charging rates, but the concept may enable spectroscopic imaging using rotating PFG synchronized with magic angle spinning [62].

16.5 Classical Chemical Shift Imaging for the Solid Parts of Batteries

The electrochemical device components are extremely heterogeneous in terms of spectroscopic NMR characteristics: the liquid electrolyte, narrow with long signal lifetime; and the solid components, broad and usually short signal lifetime. The field of pure MR imaging has been rapidly expanded to image the solid components of a battery, using for example spin or gradient echoes for the narrowest signals (mobile electrolyte) and single point imaging (SPI) and derivatives [8–10,63–65] for the broader signals (solid electrodes).

16.5.1 Signal Lifetime

The spectroscopic imaging approach is usually optimized for a given component of the battery. A constraint for chemical shift imaging (CSI) is that the signal lifetime of the material under study must be longer than the manipulation of magnetization for the chemical shift image formation.

The initial studies with in situ CSI focused on the liquid electrolyte, thanks to better sensitivity. Indeed, the liquids in the EDLC or battery are the easiest to record, mainly due to their narrow peaks, high intensity, and long lifetime. CSI of the electroactive ions (detected with ^{11}B and ^{1}H) provided atom-specific information, localized at the positive or the negative electrode [56]. CSI was also performed to correct 1D profiles for the study of polarization in liquid electrolytes at various temperatures [48]. Dynamics (diffusion and relaxation) is also the subject of active research: recent examples of slice-selective self-diffusion coefficients or slice-selective relaxation measurements applied to the liquid electrolyte of batteries can be found in [47,49,66,67]. Solids have faster transversal relaxation so that considerably shorter and stronger gradient pulses are needed, as will be discussed in Section 16.6.

16.5.2 Phase-encoded Chemical Shift 1D Imaging Sequence

A classical phase-encoded sequence used for MR spectroscopic 1D imaging is displayed in Figure 16.6. This version of CSI [68] is based on a spin echo, with the free induction decay recorded in the direct dimension. The spatial dimension is encoded as a phase evolution in the first half of the echo, obtained by applying a PFG. The PFG amplitude is ramped at every increment in the indirect dimension to sample the reciprocal space:

$$G_n = \frac{n}{L} G_{\max}, \tag{16.1}$$

with n varying in 2 L steps from $-L$ to $L-1$.

The spatial dimension is recovered by Fourier transform against the reciprocal space vector k_n, defined as:

$$k_n = \frac{1}{2\pi} \int_{t=0}^{tp} \gamma G_n f_G(t)\, dt, \tag{16.2}$$

with t_p the PFG pulse duration, γ the gyromagnetic ratio, and $f_G(t)$ the normalized temporal shape of the PFG pulse.

In the case of a perfect rectangular gradient pulse, $f_G(t) = 1$ between 0 and t_p, and

$$k_n = \frac{1}{2\pi} \gamma G_n\, t_p. \tag{16.3}$$

The field of view (FOV) is then given by:

$$\text{FOV} = \frac{1}{\Delta k} = \frac{2\pi}{\gamma \Delta G\, t_p}, \tag{16.4}$$

with Δk the increment in reciprocal space vector and ΔG the increment in the gradient amplitude.

Replacing ΔG with G_{max}/L in Equation 16.4 results in:

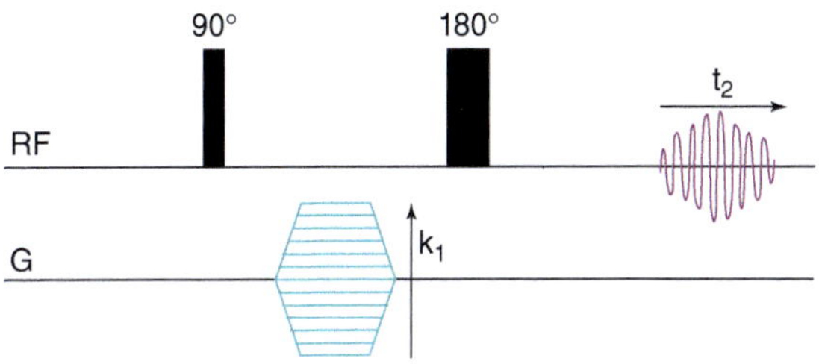

Figure 16.6 Chemical shift 1D imaging sequence. A Hahn-echo is performed with a pulsed field gradient (PFG) in the first evolution period (in cyan) to phase-encode position. The spatial dimension is encoded as a phase evolution, obtained by increment of the PFG strength for each new slice.

$$\mathrm{FOV} = \frac{2\pi L}{\gamma G_{\max}\, t_{\mathrm{p}}}. \tag{16.5}$$

The slice thickness is controlled by the gradient duration and maximum amplitude:

$$S = \frac{\mathrm{FOV}}{2L} = \frac{\pi}{\gamma G_{\max}\, t_{\mathrm{p}}}. \tag{16.6}$$

A smaller slice thickness can be obtained by an increase in $G_{\max}$, but in order to keep the FOV identical it must be accompanied by an increase in the number of encoding steps L. Better resolution therefore requires longer recording times: more encoding steps are needed and sensitivity is lowered due to the smaller number of atoms within each slice. Sensitivity is again the limiting factor.

16.5.3 Metallic Electrode

The signals of metallic electrodes have shorter characteristic lifetimes, but classical CSI techniques can still be applied provided that short echo times are used – the PFG strength is increased to shorten the PFG pulse duration and keep reasonably small slice thickness. Note that the FOV may be distorted for extremely short delays by nonideal PFG shapes and eddy currents from the last radiofrequency pulse.

The peak positions are shifted from the diamagnetic parts (Knight shift), making them easier to identify in the spectrum and reducing the need for imaging. Small variations in morphology impact slightly the shift [69], so that CSI can be useful to localize newly formed dendrites or mossy structures.

Although the Knight-shifted signal of metallic lithium is easily separated spectrally from the other constituents, it has been a great playground for in situ MR studies, thanks to its reasonable transversal relaxation and sensitivity of the shift to morphology.

Chandrashekar at al. implemented the first CSI measurements on a symmetric Li-Li cell in a seminal paper [55]. They demonstrated indisputably that the deposited mossy/dendritic metallic lithium had a slightly different shift compared with bulk lithium in Figure 16.7. Such assignment had been proposed in a previous in situ MR spectroscopic-only study [70].

Later Chang et al. [71] performed CSI in a Li-Li symmetric cell with widely spaced electrodes (flooded cell) to discriminate the signals of the two electrodes and identify various types of lithium microstructures from their shift. Their formation could be linked to lithium depletion in

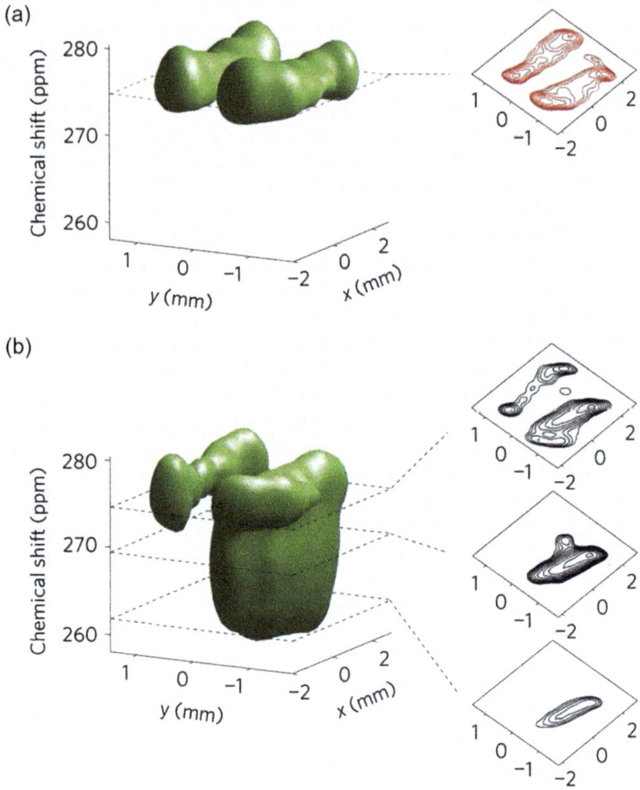

Figure 16.7 Chemical shift imaging of the metallic Li resonance in the pristine (a) and after polarization (b). The two flat electrodes are parallel to the *xz*-plane. In (b), dendritic lithium is evidenced by the signal in the 260–275 ppm range, lower than the Knight shift of the bulk electrode. Reproduced with permission from [55].

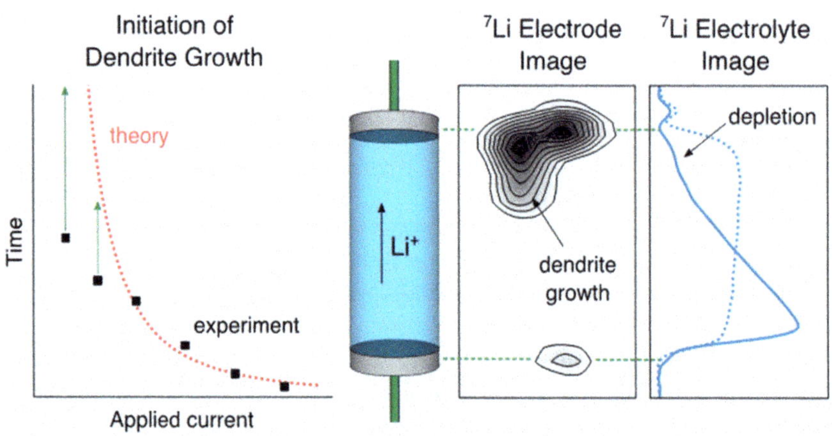

Figure 16.8 Left: plot showing the time (capacity) at the onset of dendritic growth (black squares) and Sand's model for simulating the evolution (dotted line). Center: flooded electrochemical cell. Right: chemical shift imaging of metallic Li and 1D profile of the electrolyte. Adapted with permission from [71]. Copyright (2015) American Chemical Society.

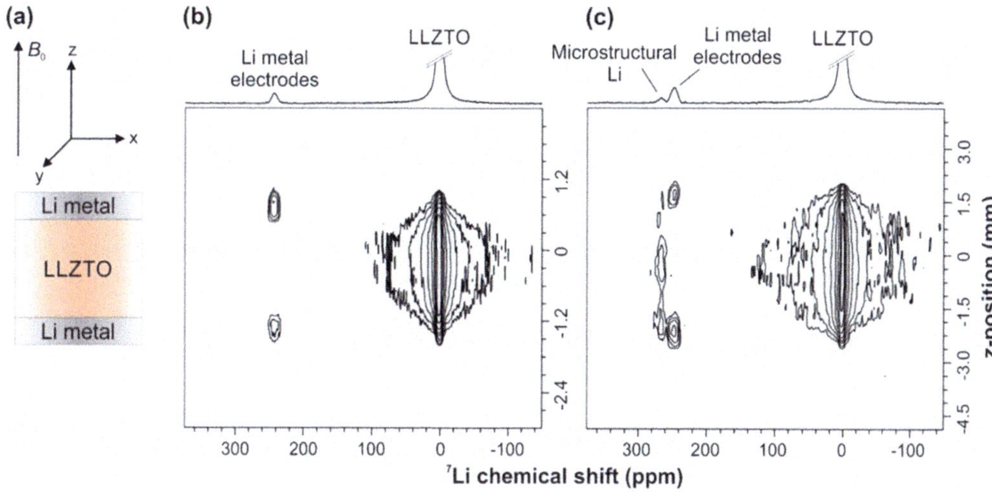

Figure 16.9 (a) Schematic of a symmetrical Li–LLZTO–Li cell depicting the orientation with respect to the external magnetic field, B_0, during the ^{7}Li chemical shift imaging experiment. ^{7}Li chemical shift images of (b) pristine and (c) cycled Li–LLZTO–Li symmetrical cells. The ^{7}Li nuclear magnetic resonance signal of LLZTO is significantly more intense than that of Li metal and is truncated as indicated by hashes. Marbella et al. (2019). ^{7}Li NMR chemical shift imaging to detect microstructural growth of lithium in all-solid-state batteries. Reproduced from [75], published under a CC-BY licence.

the electrolyte near the lithiating electrode in fast charging mode, thanks to an in situ 1D profile of the electrolyte in the cell. Importantly, this study demonstrated experimentally the validity of Sand's model in that case (Figure 16.8). Metallic sodium does not necessitate CSI as the shift of metallic Na is so different from that of other Na-containing compounds that they can be discriminated by using the limited bandwidth of the probe [65,72–74].

16.5.4 Solid Electrolyte

In situ chemical shift 1D imaging was performed by Marbella et al. [75] to identify dendritic growth in a garnet-type solid electrolyte (LLZTO), promising for all-solid-state LiBs. In Figure 16.9c, the ^{7}Li signal for the electrolyte appears near 0 ppm while additional signal located inside the separator appears at the shift of microstructural Li after polarization. It is ascribed to metallic Li dendrites forming across the solid separator.

16.6 A Specific Sequence for Paramagnetic Electrodes: S-ISIS

The most difficult part to image in a LiB is the paramagnetic electrode, containing unpaired electrons localized around the transition metal atoms. Dipolar and Fermi-contact nucleus–electron interaction [76] result in strong shifts and broadenings, so that the typical transverse relaxation times of those solids (T_2') are shorter than 100 μs. Classical CSI approaches based on spin echoes become inefficient: the signal of the paramagnetic electrodes is absent from the chemical shift image (Figure 16.4). The echo time cannot be shortened enough, notably because the strong PFG needed for imaging those solids cannot be switched on faster than 1 ms.

16.6.1 1D-ISIS Sequence for Localized Spectroscopy

Despite these constraints, Tang et al. managed to record in situ spectroscopic 1D images of a full battery, based on a frequency-encoding sequence [77]. The spectroscopic signatures of all the components inside the battery were obtained, including the two solid paramagnetic electrodes, $Li_{1-x}CoO_2$ and $Li_{4+y}Ti_5O_{12}$. This was achieved by combining the strongest 1D-PFG available (24 T/m) with a carefully designed pulse sequence for localized spectroscopy (Figure 16.10). This pulse sequence was initially designed for localized spectroscopy in living organisms and dubbed image selected in vivo spectroscopy or ISIS [78]. Note that the ISIS approach is compatible with 3D, but the resolution would be small, as the commercial three-axes PFG are currently limited to 3 T/m, one order of magnitude smaller than the strongest commercial one-axis PFG (30 T/m).

First, the PFG is turned on and left to stabilize for at least 1 ms. Then a selective 180° radiofrequency pulse inverts the spins in a slice. Finally, the gradient is switched off and the NMR spectrum is acquired. The key point is to switch the PFG off while the spins are experiencing longitudinal relaxation (~10 times slower than transversal relaxation). As shown in Figure 16.11b, the resulting NMR spectrum contains contributions from both the inverted slice and the noninverted slices. A reference spectrum (Figure 16.11a) is subtracted in order to retrieve the

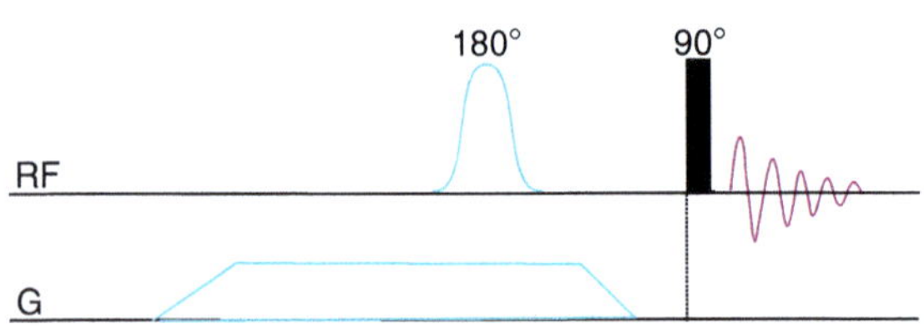

Figure 16.10 Image selected in vivo spectroscopy (ISIS) pulse program for 1D localized spectroscopy. Adapted from original work published under a CC-BY-4.0 license in [77].

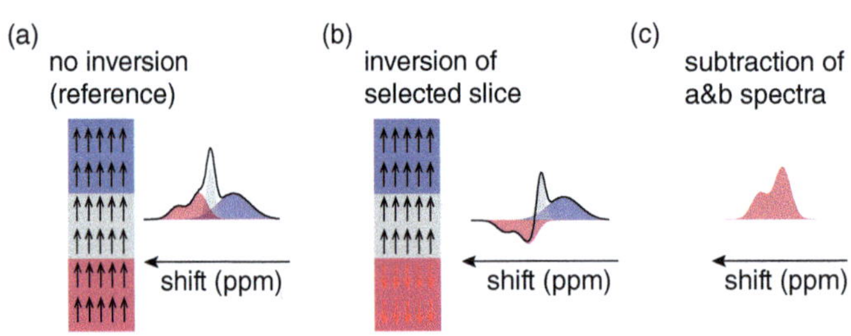

Figure 16.11 Schematic principle of 1D-ISIS (one-dimensional image selected in vivo spectroscopy) using three horizontal slices. (a) The reference spectrum is recorded without the inversion pulse. (b) An inversion, selective for the pink slice, is performed prior to the free induction decay (FID) recording. The resulting spectrum contains the contribution of the noninverted spins in the two other slices and that of the inverted spins in the pink slice. (c) Subtraction of the spectra obtained in (a) and (b) provides the spectrum of the pink slice. Adapted from original work published in [77] under a CC-BY-04 license.

slice spectrum (Figure 16.11c). This reference spectrum is recorded without the inversion pulse (but keeping all delays the same).

16.6.2 Slice Selection

The ISIS sequence relies on frequency encoding, as defined in Equation 16.7:

$$\nu(z) = \frac{\gamma}{2\pi} B_0 + \frac{\gamma}{2\pi} zG, \tag{16.7}$$

with $\nu(z)$ the frequency of the spins at z distance from the origin of the PFG, γ the gyromagnetic ratio, B_0 the main static magnetic field, and G the PFG strength.
The slice center z is then defined by:

$$z = \frac{2\pi \upsilon_{\text{pulse}}}{\gamma G} - \frac{B_0}{G}, \tag{16.8}$$

with υ_{pulse} the RF pulse frequency.

Note that this expression is only valid if the shift from the Larmor frequency due to the other interactions (chemical, paramagnetic, Knight, susceptibility shift, etc.) is negligible compared with the PFG. If not, slightly displaced images are overlapping that result in a degraded spatial selectivity. The strongest PFG strength should therefore be aimed at.

The peak width of a solid is controlled by a mixture of homogeneous and heterogeneous broadenings, as defined by Maricq and Waugh [79]. The minimal width that can be excited is controlled by the intrinsic width [35] – the set of frequencies that cannot be excited separately, even if the selective pulse was infinitely selective. As a consequence, the smallest achievable slice thickness is:

$$\Delta z = \frac{2\pi \Delta \upsilon_{\text{intrinsic}}}{\gamma G}, \tag{16.9}$$

with $\Delta\upsilon_{\text{intrinsic}}$ the intrinsic width of the spectrum.

16.6.3 Scanning ISIS

Tang et al. went one step further and used the ISIS technique in a scanning mode, dubbed S-ISIS [77]. They recorded the ^{7}Li spectra of all the slices successively so as to reconstruct the full spectroscopic image, as shown in Figure 16.12. The contributions of the two electrodes, overlapping in a single 1D spectrum could be separated and slices in each electrode could be extracted.

There is an additional constraint on the slice thickness from the in situ measurements. As the device is heterogeneous and contains several materials with variable intrinsic width, the optimal spatial resolution depends on the location in the device. This makes the interpretation intractable unless the localized intrinsic width is measured (ex situ) for all states of charge of interest. This option however would reduce time resolution, which is already limited, or it would require analyzing destructively tens of batteries per cycle.

In Tang's work the choice was to avoid control by the intrinsic width and ensure the slice thickness was known at all times. The RF pulse bandwidth was chosen so that the intrinsic

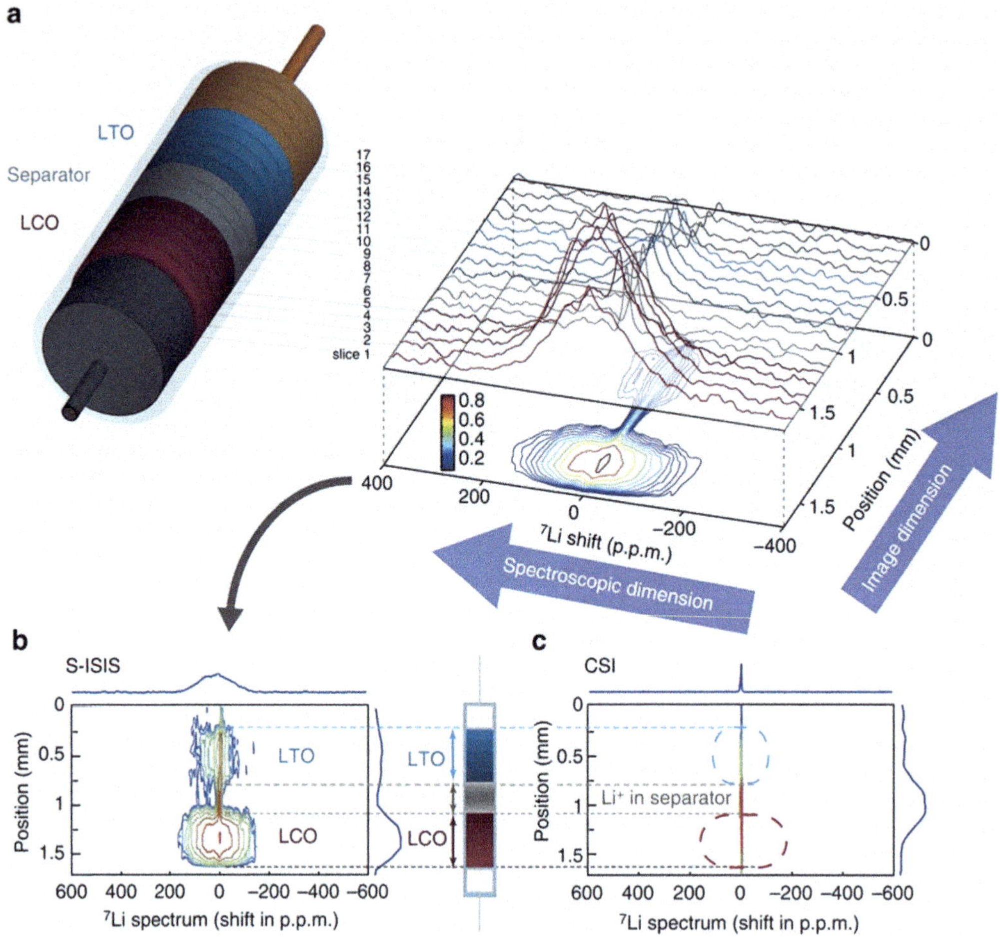

Figure 16.12 ^{7}Li scanning image selected in vivo spectroscopy (S-ISIS) of a $LiCoO_2/Li_4Ti_5O_{12}$ (LCO/LTO) battery. (a) Series of spectra obtained for horizontal slices of thickness 100 μm in the battery. The spectroscopic image is reconstructed from the series of spectra and correlates spectrum and position. The corresponding contour map is shown below. (b, c) Comparison of the nuclear magnetic resonance spectroscopic images (contour maps) for the LCO/LTO battery obtained with (b) the S-ISIS method and (c) the standard chemical shift imaging. Reproduced from [77], published under a CC-BY 4.0 license.

width of the peaks was negligible in comparison – the slice thickness did not depend any more on the intrinsic width of the spectrum, but only on the radiofrequency pulse bandwidth:

$$\Delta z = \frac{2\pi \Delta \upsilon_{\text{pulse}}}{\gamma G}, \tag{16.10}$$

with $\Delta \upsilon_{\text{pulse}}$ the radiofrequency pulse bandwidth.

Experimentally, this radiofrequency pulse bandwidth was set larger than the largest homogeneous width (38 kHz) obtained in all the materials throughout the battery charge and discharge cycles. The PFG strength was set to the maximum available: 24 T/m to ensure a spatial resolution of 100 μm, well controlled over the whole course of the in situ study, but worse than what could have been obtained at other states of charge.

16.6.4 S-ISIS Study of Lithiation Fronts in the Solid Electrodes of the Battery

A direct application of this 1D-S-ISIS sequence is the analysis of thick electrodes. Such electrodes are studied for the increase in capacity and hence autonomy, but they usually suffer from kinetic limitations. Visualizing and analyzing the lithiation fronts inside the electrodes provides useful information to identify the origin of the limitations. The redox reaction indeed requires a triple point, where the electrons, lithium ions, and transition metal ions must be present. Sluggish kinetics can be due to low percolation of lithium, or electrons, in the material. By construction of the cell, electrons are provided from the current collector side, while lithium ions originate from the electrolyte. The spatial distribution of states of charge obtained with 1D-S-ISIS for a given capacity indicated limitation from the electrolyte side, with a slow percolation of the lithium ions inside the electrode bulk.

16.7 Electron Paramagnetic Resonance Spectroscopic Imaging for Electrochemical Storage

Spectroscopic imaging has also been applied in the sister field of electron paramagnetic resonance (EPR). At room temperature, magnetic field gradients cannot be pulsed fast enough, so EPR imaging is performed by raster ray encoding [80–82]. Recently, Dutoit et al. published an application of spectroscopic imaging to characterize the morphology of lithium particles in a Li-Li battery [83]. In Figure 16.13, the comparison of the 2D images and the 1D-spectroscopic images identifies and locates submicrometric particles deposited on or stripped from the metallic Li electrodes during positive and negative polarization, respectively.

16.8 Conclusions

Since the first "modern" characterization by in situ MR spectroscopy of a LiB in 2003, the field of in situ MR of electrochemical devices has expanded to various probe and electrochemical cell designs. MR microscopy has been included into the MR toolbox to obtain localized information. When required, for example when the signals arising from several components of interest in the electrochemical device overlap, correlation of spatial and spectral information (spectroscopic imaging) provides added value. Specific methods have been introduced that allow reaching below 100 μm resolution for the most challenging solid components. Contrast in the images or localization of the spectra using the additional spectroscopic, electrochemical, and time dimensions still has its best days ahead.

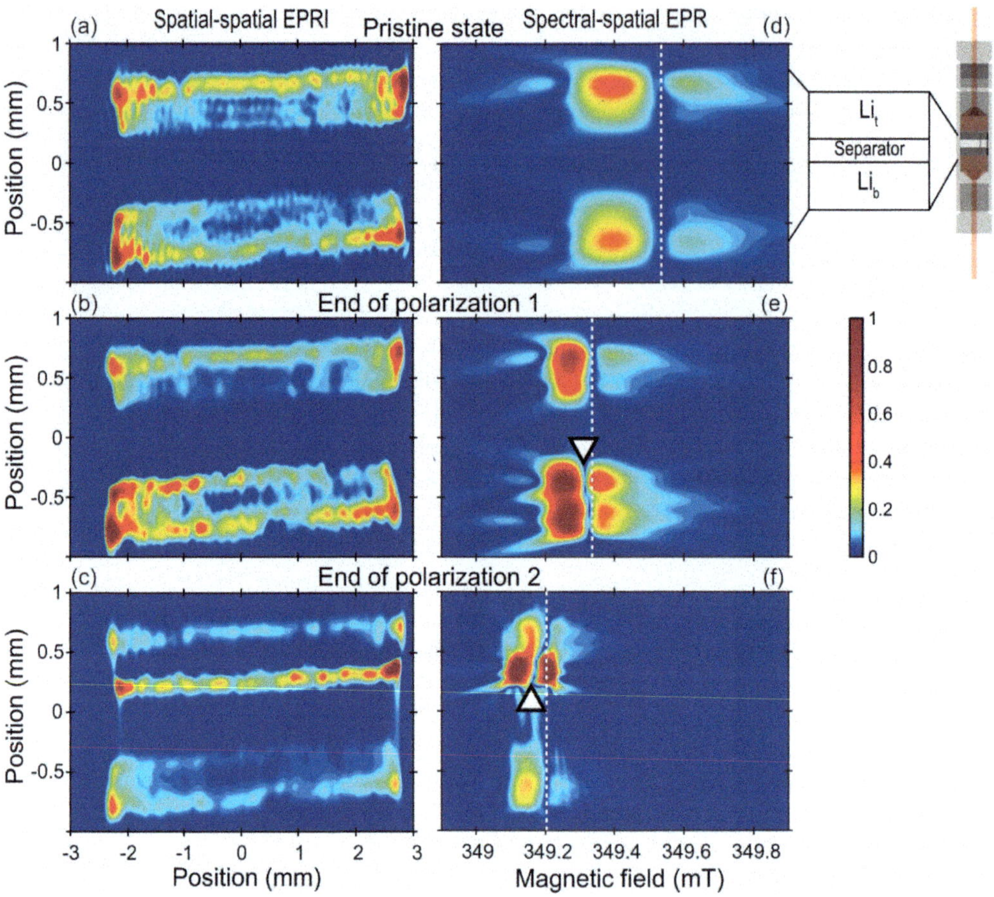

Figure 16.13 2D X-band electron paramagnetic resonance (EPR) images of a Li-Li electrochemical cell (a–c) 2D images; (d–f) "spectral-spatial" EPR images between 1D profile and spectrum. Reproduced from [83] Dutoit et al. (2021), figure 03/Springer Nature/CC BY 4.0.

References

1 Harks, P.P.R.M.L., Mulder, F.M., and Notten, P.H.L. (2015). In situ methods for Li-ion battery research: A review of recent developments. *Journal of Power Sources* 288: 92–105.

2 Grey, C.P. and Tarascon, J.M. (2017). Sustainability and in situ monitoring in battery development. *Nature Materials* 16 (1): 45–56.

3 Liu, D., Shadike, Z., Lin, R. et al. (2019). Review of recent development of in situ/operando characterization techniques for lithium battery research. *Advanced Materials* 31 (28): 1806620.

4 Köckenberger, W. and Granwehr, J. (2010). MR microscopy. In: *Encyclopedia of Spectroscopy and Spectrometry*, (ed. JC Lindon), 2e, 1654–1662. Oxford, UK: Academic Press.

5 Pecher, O., Carretero-González, J., Griffith et al. (2017). Materials' methods: NMR in battery research. *Chemistry of Materials: A Publication of the American Chemical Society* 29 (1): 213–242.

6 Borzutzki, K. and Brunklaus, G. (2017). Magnetic resonance imaging studies of the spatial distribution of charge carriers. In: *Annual Reports on NMR Spectroscopy*, 115–141. London: Academic Press.

7 Vyalikh, A., Köhler, T., Zakharchenko, T. et al. (2018). Magnetic resonance spectroscopy approaches for electrochemical research. *Physical Sciences Reviews* 3 (10): 20170155.

8 Mohammadi, M. and Jerschow, A. (2019). In situ and operando magnetic resonance imaging of electrochemical cells: A perspective. *Journal of Magnetic Resonance* 308: 106600.

9 Krachkovskiy, S., Trudeau, M.L., and Zaghib, K. (2020). Application of magnetic resonance techniques to the in situ characterization of Li-ion batteries: A review. *Materials* 13 (7): 1694.

10 Liu, X., Liang, Z., Xiang, Y. et al. (2021). Solid-state NMR and MRI spectroscopy for Li/Na Batteries: Materials, interface, and in situ characterization. *Advanced Materials* 2005878.

11 Wakihara, M. and Yamamoto, O. (1998). Lithium Ion Batteries: Fundamentals and Performance, 261. Weinheim, Germany: Wiley-VCH.

12 Conway, E.B. (1999). Electrochemical Supercapacitors, 698. New York: Springer.

13 Moseley, P.T. and Garche, J. (2014). Electrochemical Energy Storage for Renewable Sources and Grid Balancing. Amsterdam: Elsevier.

14 Beard, K.W. (2019). Linden's Handbook of Batteries, 5e. New York: McGraw-Hill Education.

15 Tarascon, J.-M. and Armand, M. (2001). Issues and challenges facing rechargeable lithium batteries. *Nature* 414 (6861): 359–367.

16 Armand, M. and Tarascon, J.-M. (2008). Building better batteries. *Nature* 451 (7179): 652–657.

17 Choi, J.U., Voronina, N., Sun, Y.-K. et al. (2020). Recent Progress and perspective of advanced high-energy Co-less Ni-rich cathodes for Li-ion batteries: Yesterday, today, and tomorrow. *Advanced Energy Materials* 10 (42): 2002027.

18 Asenbauer, J., Eisenmann, T., Kuenzel, M. et al. (2020). The success story of graphite as a lithium-ion anode material – Fundamentals, remaining challenges, and recent developments including silicon (oxide) composites. *Sustainable Energy & Fuels* 4 (11): 5387–5416.

19 Li, P., Kim, H., Myung, S.-T. et al. (2021). Diverting exploration of silicon anode into practical way: A review focused on silicon-graphite composite for lithium ion batteries. *Energy Storage Materials* 35: 550–576.

20 Armand, M., Axmann, P., Bresser, D., Copley, M. et al. (2020). Lithium-ion batteries – Current state of the art and anticipated developments. *Journal of Power Sources* 479: 228708.

21 Grey, C.P. and Hall, D.S. (2020). Prospects for lithium-ion batteries and beyond – A 2030 vision. *Nature Communications* 11 (1): 6279.

22 Zhang, H., Li, C., Eshetu, G.G. et al. (2020). From solid-solution electrodes and the rocking-chair concept to today's batteries. *Angewandte Chemie International Edition* 59 (2): 534–538.

23 Randau, S., Weber, D.A., Kötz, O. et al. (2020). Benchmarking the performance of all-solid-state lithium batteries. *Nature Energy* 5 (3): 259–270.

24 Tian, Y., Zeng, G., Rutt, A. et al. (2021). Promises and challenges of next-generation "beyond Li-ion" batteries for electric vehicles and grid decarbonization. *Chemical Reviews* 121 (3): 1623–1669.

25 Hasa, I., Mariyappan, S., Saurel, D. et al. (2021). Challenges of today for Na-based batteries of the future: From materials to cell metrics. *Journal of Power Sources* 482: 228872.

26 Shao, H., Wu, Y.-C., Lin, Z. et al. (2020). Nanoporous carbon for electrochemical capacitive energy storage. *Chemical Society Reviews* 49 (10): 3005–3039.

27 Slesinski, A., Fic, K., and Frackowiak, E. (2018). New trends in electrochemical capacitors. In: *Advances in Inorganic Chemistry* (eds. R. van Eldik and W. Macyk), 247–286. London: Academic Press.
28 Lin, Z., Goikolea, E., Balducci, A. et al. (2018). Materials for supercapacitors: When Li-ion battery power is not enough. *Materials Today* 21 (4): 419–436.
29 Ding, J., Hu, W., Paek, E. et al. (2018). Review of hybrid ion capacitors: From aqueous to lithium to sodium. *Chemical Reviews* 118 (14): 6457–6498.
30 Simon, P. and Gogotsi, Y. (2020). Perspectives for electrochemical capacitors and related devices. *Nature Materials* 19 (11): 1151–1163.
31 Yang, X. and Rogach, A.L. (2019). Electrochemical techniques in battery research: A tutorial for nonelectrochemists. *Advanced Energy Materials* 9 (25): 1900747.
32 Dugas, R., Forero-Saboya, J.D., and Ponrouch, A. (2019). Methods and protocols for reliable electrochemical testing in post-Li batteries (Na, K, Mg, and Ca). *Chemistry of Materials: A Publication of the American Chemical Society* 31 (21): 8613–8628.
33 Bard, A.J. and Faulkner, L.R. (2001). Electrochemical Methods: Fundamentals and Applications, 2e. New York: Wiley.
34 Collins English Dictionary. London: HarperCollins Publishers
35 Callaghan, P.T. (2003). Principles of Nuclear Magnetic Resonance Microscopy. Oxford, UK: Oxford University Press.
36 Keeler, J. (2010). Understanding NMR Spectroscopy, 2e . Chichester, UK: John Wiley and Sons Ltd. 526.
37 Trease, N.M., Zhou, L., Chang, H.J. et al. (2012). In situ NMR of lithium ion batteries: Bulk susceptibility effects and practical considerations. *Solid State Nuclear Magnetic Resonance* 42: 62–70.
38 Zhou, L., Leskes, M., Ilott, A.J.et al. (2013). Paramagnetic electrodes and bulk magnetic susceptibility effects in the in situ NMR studies of batteries: Application to Li1.08Mn1.92O4 spinels. *Journal of Magnetic Resonance* 234: 44–57.
39 Salager, E., Sarou-Kanian, V., Sathiya, M. et al. (2014). Solid-state NMR of the family of positive electrode materials Li2Ru1−ySnyO3 for lithium-ion batteries. *Chemistry of Materials: A Publication of the American Chemical Society* 26 (24): 7009–7019.
40 Gerald, II R.E., Sanchez, J., Johnson, C.S. et al. (2001). In situ nuclear magnetic resonance investigations of lithium ions in carbon electrode materials using a novel detector. *Journal of Physics: Condensed Matter* 13 (36): 8269.
41 Ilott, A.J., Mohammadi, M., Schauerman, C.M. et al. (2018). Rechargeable lithium-ion cell state of charge and defect detection by in-situ inside-out magnetic resonance imaging. *Nature Communications* 9 (1): 1776.
42 Romanenko, K. and Jerschow, A. (2019). Distortion-free inside-out imaging for rapid diagnostics of rechargeable Li-ion cells. *PNAS* 116 (38): 18783–18789.
43 Mohammadi, M., Silletta, E.V., Ilott, A.J. et al.(2019). Diagnosing current distributions in batteries with magnetic resonance imaging. *Journal of Magnetic Resonance* 309: 106601.
44 Benders, S., Mohammadi, M., Klug, C.A. et al. (2020). Nuclear magnetic resonance spectroscopy of rechargeable pouch cell batteries: Beating the skin depth by excitation and detection via the casing. *Scientific Reports* 10 (1): 13781.

45 Letellier, M., Chevallier, F., Clinard, C. et al. (2003). The first in situ 7Li nuclear magnetic resonance study of lithium insertion in hard-carbon anode materials for Li-ion batteries. *Journal of Chemical Physics* 118 (13): 6038.
46 Serša, I. and Mikac, U. (2018). A study of MR signal reception from a model for a battery cell. *Journal of Magnetic Resonance* 294: 7–15.
47 Krachkovskiy, S.A., Bazak, J.D., Werhun, P. et al. (2016). Visualization of steady-state ionic concentration profiles formed in electrolytes during Li-ion battery operation and determination of mass-transport properties by in situ magnetic resonance imaging. *Journal of the American Chemical Society* 138 (25): 7992–7999.
48 Bazak, J.D., Krachkovskiy, S.A., and Goward, G.R. (2017). Multi-temperature in situ magnetic resonance imaging of polarization and salt precipitation in lithium-ion battery electrolytes. *Journal of Physical Chemistry C* 121 (38): 20704–20713.
49 Bazak, J.D., Allen, J.P., Krachkovskiy, S.A. et al. (2020). Mapping of lithium-ion battery electrolyte transport properties and limiting currents with in situ MRI. *Journal of the Electrochemical Society* 167 (14): 140518.
50 Letellier, M., Chevallier, F., and Morcrette, M. (2007). In situ 7Li nuclear magnetic resonance observation of the electrochemical intercalation of lithium in graphite; 1st cycle. *Carbon* 45 (5): 1025–1034.
51 Pecher, O., Carretero-González, J., Griffith, K.J. et al.(2016). Materials' methods: NMR in battery research. *Chemistry of Materials* 29: 213–242.
52 Tang, J.A., Dugar, S., Zhong, G. et al. (2013). Non-destructive monitoring of charge-discharge cycles on lithium ion batteries using 7Li stray-field imaging. *Scientific Reports* 3: 2596.
53 Tarascon, J.-M., Gozdz, A.S., Schmutz, C. et al. (1996). Performance of Bellcore's plastic rechargeable Li-ion batteries. *Solid State Ionics* 86: 49–54.
54 Chevallier, F., Letellier, M., Morcrette M. et al. (2003). In situ 7Li-nuclear magnetic resonance observation of reversible lithium insertion into disordered carbons. *Electrochemical and Solid-State Letters* 6 (11): A225.
55 Chandrashekar, S., Trease, N.M., Chang, H.J. et al. (2012). 7Li MRI of Li batteries reveals location of microstructural lithium. *Nature Materials* 11 (4): 311–315.
56 Ilott, A.J., Trease, N.M., Grey, C.P., and Jerschow, A. (2014). Multinuclear in situ magnetic resonance imaging of electrochemical double-layer capacitors. *Nature Communications* 5: 4536.
57 Poli, F., Kshetrimayum, J.S., Monconduit, L., and Letellier, M. (2011). New cell design for in-situ NMR studies of lithium-ion batteries. *Electrochemistry Communications* 13 (12): 1293–1295.
58 Tang, J.A., Zhong, G., Dugar, S. et al. (2012). STRAFI NMR probe for material imaging of quadrupolar nuclei. *Journal of Magnetic Resonance* 225: 93–101.
59 Aguilera, A.R., MacMillan, B., Krachkovskiy, S. et al. (2021). A parallel-plate RF probe and battery cartridge for 7Li ion battery studies. *Journal of Magnetic Resonance* 325: 106943.
60 Freytag, A.I., Pauric, A.D., Krachkovskiy, S.A. et al.(2019). In situ magic-angle spinning 7Li NMR analysis of a full electrochemical lithium-ion battery using a jelly roll cell design. *Journal of the American Chemical Society* 141 (35): 13758–13761.
61 Freytag, A.I., Pauric, A.D., Jiang, M. et al. (2019). 7Li and 29Si NMR enabled by high-density cellulose-based electrodes in the lithiation process in silicon and silicon monoxide anodes. *Journal of Physical Chemistry C* 123 (18): 11362–11368.

62 Yon, M., Sarou-Kanian, V., Scheler, U. et al. (2017). Solid-state 31 P and 1 H chemical MR micro-imaging of hard tissues and biomaterials with magic angle spinning at very high magnetic field. *Scientific Reports* 7 (1): 8224.

63 Romanenko, K., Forsyth, M., and O'Dell, L.A. (2014). New opportunities for quantitative and time efficient 3D MRI of liquid and solid electrochemical cell components: Sectoral fast spin echo and SPRITE. *Journal of Magnetic Resonance* 248: 96–104.

64 Chien, P.-H., Griffith, K.J., Liu, H. et al. (2020). Recent advances in solid-state nuclear magnetic resonance techniques for materials research. *Annual Review of Materials Research* 50 (1): 493–520.

65 Bray, J.M., Doswell, C.L., Pavlovskaya, G.E. et al. (2020). Operando visualisation of battery chemistry in a sodium-ion battery by 23 Na magnetic resonance imaging. *Nature Communications* 11 (1): 2083.

66 Krachkovskiy, S.A., Pauric Allen, D., Halalay, I.C. et al. (2013). Slice-selective NMR diffusion measurements: A robust and reliable tool for in situ characterization of ion-transport properties in lithium-ion battery electrolytes. *Journal of Physical Chemistry Letters* 4 (22): 3940–3944.

67 Balbierer, R., Seegert, P., Herberger, S. et al. (2021). Investigation of transverse relaxation rate distribution via magnetic resonance imaging: Impact of electrode formation. *Energy Technology* 9: 2000579.

68 Maudsley, A.A., Hilal, S.K., Perman, W.H. et al. (1983). Spatially resolved high resolution spectroscopy by "four-dimensional" NMR. *Journal of Magnetic Resonance* 51 (1): 147–152.

69 Ilott, A.J., Chandrashekar, S., Klöckner, A. et al. (2014). Visualizing skin effects in conductors with MRI: Li MRI experiments and calculations. *Journal of Magnetic Resonance* 245: 143–149.

70 Bhattacharyya, R., Key, B., Chen, H. et al. (2010). In situ NMR observation of the formation of metallic lithium microstructures in lithium batteries. *Nature Materials* 9 (6): 504–510.

71 Chang, H.J., Ilott, A.J., Trease, N.M. et al. (2015). Correlating microstructural lithium metal growth with electrolyte salt depletion in lithium batteries using 7Li MRI. *Journal of the American Chemical Society* 137 (48): 15209–15216.

72 Bayley, P.M., Trease, N.M., and Grey, C.P. (2016). Insights into electrochemical sodium metal deposition as probed with in situ 23Na NMR. *Journal of the American Chemical Society* 138 (6): 1955–1961.

73 Xiang, Y., Zheng, G., Liang, Z. et al. (2020). Visualizing the growth process of sodium microstructures in sodium batteries by in-situ 23 Na MRI and NMR spectroscopy. *Nature Nanotechnology* 15 (10): 883–890.

74 Rees, G.J., Jolly, D.S., Ning, Z. et al. (2021). Imaging sodium dendrite growth in all-solid-state sodium batteries using 23Na T2-weighted magnetic resonance imaging. *Angewandte Chemie* 133 (4): 2138–2143.

75 Marbella, L.E., Zekoll, S., Kasemchainan, J. et al. (2019). 7Li NMR chemical shift imaging to detect microstructural growth of lithium in all-solid-state batteries. *Chemistry of Materials* 21: 2762–2769.

76 Pell, A.J. and Pintacuda, G. (2015). Broadband solid-state MAS NMR of paramagnetic systems. *Progress in Nuclear Magnetic Resonance Spectroscopy* 84–85: 33–72.

77 Tang, M., Sarou-Kanian, V., Melin, P. et al. (2016). Following lithiation fronts in paramagnetic electrodes with in situ magnetic resonance spectroscopic imaging. *Nature Communications* 7: 13284.

78 Ordidge, R.J., Connelly, A., and Lohman, J.A.B. (1986). Image-selected in vivo spectroscopy (ISIS). A new technique for spatially selective NMR spectroscopy. *Journal of Magnetic Resonance* 66 (2): 283–294.

79 Maricq, M.M. and Waugh, J.S. (1979). NMR in rotating solids. *Journal of Chemical Physics* 70 (7): 3300–3316.

80 Wandt, J., Marino, C., Gasteiger, H.A. et al. (2015). Operando electron paramagnetic resonance spectroscopy – Formation of mossy lithium on lithium anodes during charge–discharge cycling. *Energy & Environmental Science* 8 (4): 1358–1367.

81 Sathiya, M., Leriche, J.-B., Salager, E. et al. (2015). Electron paramagnetic resonance imaging for real-time monitoring of Li-ion batteries. *Nature Communications* 6 : 6276.

82 Niemöller, A., Jakes, P., and Eichel R-A, G.J. (2018). EPR imaging of metallic lithium and its application to dendrite localisation in battery separators. *Scientific Reports* 8 (1): 1–7.

83 Dutoit, C.-E., Tang, M., Gourier, D. et al. (2021). Monitoring metallic sub-micrometric lithium structures in Li-ion batteries by in situ electron paramagnetic resonance correlated spectroscopy and imaging. *Nature Communications* 12 (1): 1410.

17

Magnetic Field Map Measurements and Operando NMR/MRI as a Diagnostic Tool for the Battery Condition

Stefan Benders and Alexej Jerschow

New York University, New York, NY, USA

17.1 Batteries – Essential Components with Complex Geometries

Rechargeable batteries form a key technology for providing portable energy and are used in many applications such as mobile computing, communications, electric mobility, and consumer devices [1–3]. In addition, rechargeable energy storage is a central element in the use of alternative energy, and also in grid-level intermittent storage. Over the past 20 years, battery technology has evolved very quickly, leading to the availability of lightweight and potent cells today. Among the different classes of rechargeable cells, lithium-ion cells have emerged as especially favorable due to their high energy content per weight and also per volume and their favorable voltage range. Their popularity led to the development of many battery types and forms. The success of lithium-ion batteries also derives from the ability to reversibly shuttle lithium ions between the cathode and the anode. Apart from the desired electrochemical reactions, additional, unwanted side reactions can happen that can lead to the degradation of the electrolyte, gas evolution, and the buildup of lithium microstructures (mossy lithium or lithium dendrites) on the electrodes. All these processes not only degrade capacity, but also can pose severe safety risks.

Different cell geometries have been established for commercial use. Typically, one can find two design principles: (i) the most common types are stacked rectangular types, where multiple layers of cathode, separator, and anode are stacked above each other; and (ii) rolled cells, where one long strip of cathode–separator–anode layers is arranged up into a "jelly-roll." These exist either in a flat form (many cell-phone cells) or in cylindrical cells (e.g. the Tesla 2170 and 4680 cells). Larger cells employed in electrical cars, for example, may have more complex or hybrid geometries.

The manufacturing of large quantities of these cells is challenging and defects can lead to major safety problems and severe limitations in their applications. For this reason, battery diagnostics is essential to rule out manufacturing errors, which can pose serious risks to the public as unexpected fires can occur. Many diagnostic techniques such as electrical impedance spectroscopy (EIS) provide global information on the state of the whole cell but do not provide detailed information, especially on the spatial features [4–6]. There are only a few techniques with access to spatial resolution. Examples are localized EIS, computed tomography, or acoustic methods [7–13]. All of these techniques are accompanied by specific challenges such as limited imaging contrast or limitations of access to information from the battery's interior. Therefore,

Magnetic Resonance Microscopy: Instrumentation and Applications in Engineering, Life Science and Energy Research, First Edition. Edited by Sabina Haber-Pohlmeier, Bernhard Blümich, and Luisa Ciobanu.

additional, complementary techniques are desirable to provide insights into the processes of commercial batteries, in particular. Recently, magnetic resonance imaging (MRI) techniques have been developed that have allowed the study of commercial and prototype cells, and have the potential to provide detailed and spatially resolved analyses that can help to establish sorely needed rechargeable cell diagnostics.

17.2 Magnetic Resonance Techniques Applied to Research Cells

Nuclear magnetic resonance (NMR) and MRI are excellent techniques for the study of complex reactions, flow, and electrochemical processes. In situ studies including the observation of chemical transitions via the chemical shift, the quantification of multinuclear spectroscopic signatures, and measurements at multiple resolution scales and with different dimensionality have contributed to enhanced understanding of electrochemical processes.

In the electrochemical field, MRI has been applied extensively, thereby laying the foundation for imaging of batteries. Davenport et al. used the relaxation rates of a solution next to a zinc wire to visualize its corrosion. The results showed marked changes in relaxation behavior during the process, which can be very localized [14]. These observations are based on structural changes associated with zinc ion–water complexes. Corrosion processes of copper electrodes can be followed in a similar way [15]. The dissolution of copper into solution can be visualized at high resolution by using indirect detection, i.e. correlating copper ion concentration to the ^{1}H R_1 relaxation rate. These experiments show a heterogeneous dissolution process. Similar to that, reactions in a Zn–Ti cell could be observed by following characteristic relaxation changes [16].

When ions flow in solution during an MRI measurement, they are likely to be subject to the so-called magnetohydrodynamic effect. This effect occurs when the direction of ion flow is not parallel with the magnetic field resulting in a Lorentz force, and can result in high velocities as well as increased kinetics of electrochemical reactions [17–24].

More generally, transport properties of liquids in electrochemical systems have been studied for several systems, with special interest for Nafion. Codd et al. investigated the solvent mobility in Nafion membranes [25] and Baker et al. employed relaxation parameters to investigate diffusion along these under electrical potential [26]. Fuel cells are a special system of interest, warranted by their potential use in electromobility and have been investigated thoroughly [27,28], even in operation [29–31].

There have been many NMR-based studies on separate components of the battery, exploring the electrolyte, the electrodes, or even the solid–electrolyte interface (SEI). For example, a recent study by Hope et al. employed Overhauser-dynamic nuclear polarization (DNP) to reveal species in the SEI [32]. Their approach allowed the enhancement of the signals of species close to the metallic lithium since they were in the range of spin diffusion. In particular, not all species experienced the same enhancement.

For assembled battery systems, Chandrashekar et al. have demonstrated the ability to perform phase encoding as well as chemical shift imaging (CSI) in a model in situ lithium–metal cell [33]. In that work, a lithium bag cell was studied and the spatial changes in microstructural lithium observed, in particular the formation of dendrites with charge was followed. Since the radiofrequency radiation cannot penetrate through thick layers of lithium due to the skin depth being just a few micrometers, only microstructured lithium is observable. It is essential to align the bulk metal plates with the $\boldsymbol{B}_1$ field to optimize the observable signal [34].

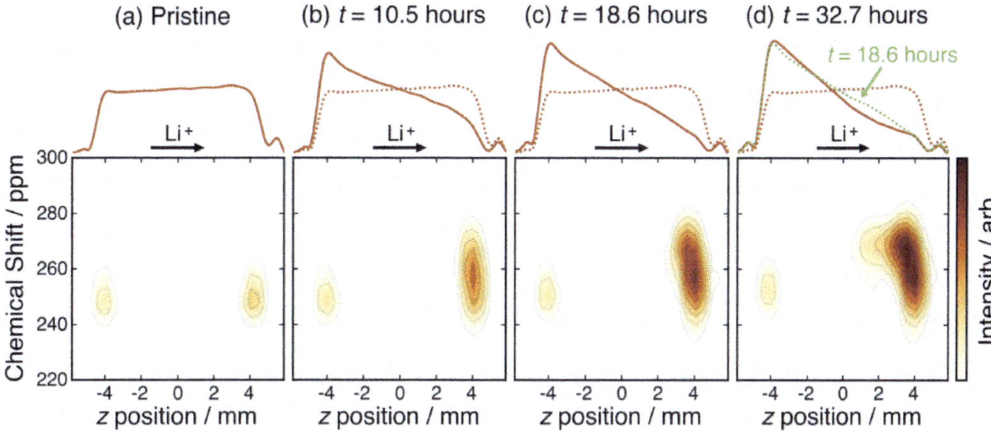

Figure 17.1 Magnetic resonance imaging (MRI) time series showing the evolution of the ^{7}Li electrolyte concentration profile (top) and the ^{7}Li chemical shift image of the metal (bottom) for the cell charged at 0.76 mA/cm^2. Reprinted (adapted) with permission from Chang et al. [35].

Following the characteristic shift of lithium dendrites allowed Chang et al. to follow dendrite growth in a lithium–lithium cell [35]. Here, the cell consisted of two lithium-metal disks with electrolyte in between. Figure 17.1 shows the (electrolyte) concentration profiles (top) at different times during charge with a current density of 0.76 mA/cm^2 as well as the CSI images for the metallic components (bottom). Similar concentration profiles were also observed in studies focused on the electrolyte [36–40]. A chemical shift of 250 ppm is typical of ^{7}Li metal perpendicular to $\boldsymbol{B}_0$. The metal positions seem to overlap with the electrolyte, which can be attributed to the low resolution of the CSI and the uneven surfaces. With increasing charge time, notable changes set in. In (b), after 10.5 h, the electrolyte signal shifts its maximum intensity to the negative electrode. Additionally, signals with chemical shifts of 260–270 ppm can be observed at the positive electrode, which grow in intensity for charge step (c) at 18.6 h, and finally, at 32.7 h of charge (d) the 270 ppm signal also extends 2–3 mm in the direction of the negative electrode. Mossy lithium microstructures can be assigned to a shift of 260 ppm while dendritic lithium appears at 270 ppm. Therefore, dendrite growth can be observed.

While these results demonstrate promising tools in detecting lithium dendrite formation, a more detailed three-dimensional image is desirable. To that end, good temporal and spatial resolution is required, but generally cannot be achieved due to signal-to-noise considerations. Ilott et al. have shown that this microstructure growth can be assessed by indirect imaging of the electrolyte and reconstructing the dendrite growths from the missing signal [41]. This approach allows the growth process of dendrites up to cell shorting to be followed (Figure 17.2).

Recently, the development and use of sodium batteries has received considerable interest, especially due to their cost and sustainability. While these batteries offer many advantages, there are technological hurdles to overcome concerning the materials used in these cells. Bray et al. employed ^{23}Na MRI to observe an in situ sodium battery [42]. The ^{23}Na spectrum exhibits two species: a metallic Na species at 1131 ppm, and an electrolyte species with two peaks at 0 and 4 ppm and the localization of species can clearly be observed. After one cycle of the cell, the picture changes. The electrolyte peak splits into four peaks (1, 0, −5, −7 ppm) hinting at carbon entering the microporous carbon electrode and degradation of the electrolyte.

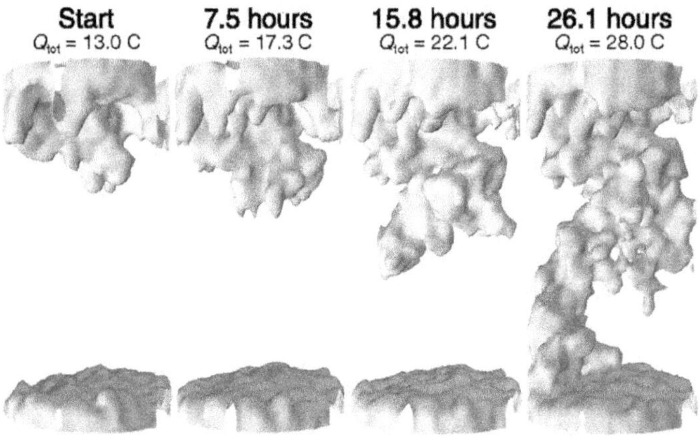

Figure 17.2 In situ ^{1}H 3D Fast Low Angle SHot (FLASH) imaging results from the electrochemical cell, with each 3D image acquired in 16 min 40 s. The cell was charged at 50 μA (0.16 mA/cm^2) for 72 h before these measurements and then charged at 160 μA (0.51 mA/cm^2) for the times specified, with the total charge applied also given in Coulombs. The three-dimensional segmented images were subject to an additional Gaussian filter to smoothen the visualization. [41] Ilott et al. (2016), figure 01 [p.002] / with permission from National Academy of Sciences.

The cell was discharged at a current of 150 mA/g to promote galvanostatic plating. Following this process, additional metallic sodium peaks appear, which can be associated with newly deposited sodium at different angles with respect to $\boldsymbol{B}_0$. This finding hints at the formation of dendritic structures, which can also be seen in the spatially resolved maps. After plating with a higher specific current of 600 mA/g, the images change drastically, as the metallic signals dominate the spectrum and the images/profiles. In that study the indirect imaging technique [41] via the electrolyte also revealed the formation of microstructures after plating.

17.3 The Effect of Conductive Materials on Radiofrequency Propagation

It is well known that conductive materials have a strong influence on the propagation, distribution, and distortion of radiofrequency irradiation. The origin of this influence can be traced to the electrodynamic boundary conditions, or perhaps more fundamentally, to the induced field generated by the motion of charged particles.

In the following magnetic field penetration discussion, we follow the description provided in Ref. [34]. The time-harmonic Maxwell equations in a homogeneous and isotropic region of space can be described by:

$$\nabla \times \boldsymbol{H} + ik\boldsymbol{E} = 0, \tag{17.1}$$

$$\nabla \times \boldsymbol{E} + ik\boldsymbol{H} = 0 \tag{17.2}$$

$$\nabla \times \boldsymbol{E} = 0, \tag{17.3}$$

$$\nabla \times \boldsymbol{H} = 0, \tag{17.4}$$

where k is the wave number, $\boldsymbol{H}$ is the magnetic field, and $\boldsymbol{E}$ the electric field. Note that these expressions are represented in a dimensionless form for convenience. At the surface of a perfect electronic conductor, the following boundary conditions would have to be observed:

$$\boldsymbol{n} \times \boldsymbol{E}^{\text{tot}} = 0, \tag{17.5}$$

$$\boldsymbol{n} \times \boldsymbol{H}^{\text{tot}} = 0. \tag{17.6}$$

Here $\boldsymbol{n}$ is the vector normal to the conductor surface. These expressions illustrate that the magnetic field can have no component normal to the surface, while the electric field has no component parallel to the surface. These extreme examples illustrate that significant field perturbations are generated immediately upon the introduction of a conducting surface into the region. This effect is clearly demonstrated by ^{7}Li images obtained from a region with one lithium plate as shown in Figure 17.3. As is seen there, near-zero NMR signal is obtained for perfect normal incidence, and a very good signal is found when the incidence is parallel. At the edges, one observes the strongest signals when the plate is perpendicular and a weaker signal when it is parallel. A similar study was performed by Serša et al., in which plastic cubes with conductive coating were exposed to different field orientations of $\boldsymbol{B}_0$ and $\boldsymbol{B}_1$ [43]. The optimal orientation of the conductors was parallel to both fields, and it was particularly important to choose an arrangement parallel to $\boldsymbol{B}_1$. Furthermore, the readout gradient should be set to be parallel to the electrodes as well. Additional improvements could be achieved by the use of pure phase encoding methods, which can suppress susceptibility effects but may still be susceptible to the radiofrequency-induced eddy currents.

These considerations are typically used when designing an experiment with electrochemical cells: in particular, it is desirable to minimize distortions and to maximize the magnetic fields in the vicinity of conductors. It is clear, however, that such choices cannot always be made, and the resulting magnetic field distortions need to be taken into account when a quantitative analysis of the results is desired. For example, it is also clear that more gradual and nuanced changes to the field distribution can be observed when the conductor is not perfect.

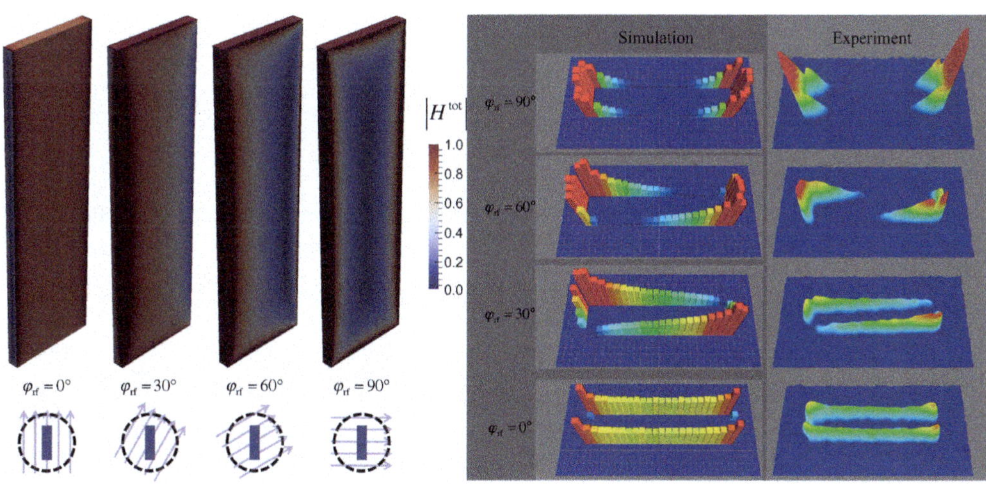

Figure 17.3 Left: plots of the calculated field at the metal surface, when the direction of the applied field is rotated. Right: results from simulations and experiment corresponding to rotations of 0° to 90° from bottom to top. Adapted from Ilott et al. [34].

Some of the early in situ MRI work described the observation of ^{7}Li signals originating from within a conducting surface [33,44]. For example, when quantifying the growth of lithium dendrites on the anode surface, one can use the signal from the surface region of the electrode as a reference.

While a field parallel to the surface can enter the conductor, it cannot travel very far. Induced fields from the motion of conducting particles generate an opposing field, which ultimately leads to the extinction of the field over a distance scale commonly referred to as the skin depth.

Assuming an infinite plane boundary, the radiofrequency field entering a conducting region can be written as:

$$B_1(\boldsymbol{r}) = B_{10}\, \boldsymbol{e}^{(-\beta \boldsymbol{n}\cdot\boldsymbol{r})} \boldsymbol{e}^{i\alpha \boldsymbol{n}\cdot\boldsymbol{r} - i\omega t}, \tag{17.7}$$

with $\boldsymbol{B}_{10}$ the radiofrequency field at the surface of the conductor, $\boldsymbol{n}$ the propagation direction, $\boldsymbol{r}$ the location vector, and α and β the real and imaginary parts of the wave vector $\chi = \alpha + i\beta$, defined by [45] as:

$$\begin{Bmatrix} \alpha \\ \beta \end{Bmatrix} = \sqrt{\mu\varepsilon}\,\frac{\omega}{c}\left[\frac{1}{2}\sqrt{1+\left(\frac{2\sigma}{\nu\varepsilon}\right)^2} \pm \frac{1}{2}\right]^{\frac{1}{2}}, \tag{17.8}$$

with the plus and minus sign used for α and β, respectively. Here, ε is the dielectric constant of the conductor, μ the permeability and σ the conductivity of the conductor, and c the speed of light in a vacuum. For a good conductor $\alpha \sim \beta = 1/\delta$, where δ is the skin depth given as

$$\delta = \sqrt{\frac{1}{\pi\mu\nu\sigma^2}}. \tag{17.9}$$

Using these expressions in Equation 17.7, one observes a depth dependence of both amplitude and phase, and the field decays exponentially with the distance from the surface, as given by

$$\widetilde{B}_1(\boldsymbol{r}) = B_{10}\boldsymbol{e}^{-\beta z}\boldsymbol{e}^{i\beta z}, \tag{17.10}$$

with $z = \boldsymbol{n}\cdot\boldsymbol{r}$. If such a field is used to excite a signal from spins residing within the conductor, one can determine the excitation flip angle as

$$\alpha_P(r) = \gamma\tau|B_{10}|e^{-\beta z} = \alpha_P(0)e^{-\beta z}, \tag{17.11}$$

if a pulse of duration τ is used.

The principle of reciprocity [46,47] can subsequently be used to relate the excitation field to the detection field [48]. These considerations allow establishing an expression for the detected signal s of the form

$$s = -2i\omega M_0 \frac{B_{10}^2}{|B_{10}|}\int_{z=0}^{\infty} e^{2i\beta z}e^{-\beta z}\sin[\alpha_P(z)]dz. \tag{17.12}$$

A particularly striking consequence of the principle of reciprocity is the phase doubling. One may understand the effect of phase doubling as originating from the fact that whatever phase lag is incurred in the forward (excitation) wave propagation has to arise in the backward (detection) propagation of the wave. Equation 17.12 can then be used, for example, to quantify the amount of lithium from the observed signal. Further uses of the known radiofrequency field behavior include the possibility of performing subsurface microscopy and imaging, and the determination of particle sizes [49].

This discussion also indicates that it would seem impossible to penetrate radiofrequency irradiation into a fully conductively enclosed container, as it would seem that no field should survive beyond the skin depth (typically tens of micrometers). Since most commercial batteries are of this kind – fully enclosed by a conductor, or at least have very small spacing between electrodes – it would seem that such devices are off limits for NMR or MRI study. Recently, however, it was demonstrated that the radiofrequency penetration into such devices is possible under certain conditions (related to driving opposite sides of the cell out of phase), which we describe toward the end of this chapter [50].

We first point our attention to the body of work on inside-out MRI or io-MRI, which uses as its principle the fact that information about the samples and devices can be obtained without the need to directly access the inside compartment.

17.4 MRI on Commercial Batteries

While experiments on model cells provide important information and can lead to more insights into processes associated with charging and discharging, commercial batteries exhibit distinct features worth exploring for potential applications. In particular, many cell designs fail in the scaling up process from model coin cells to the intermediate and commercial scale. One reason for the widespread use of model cells is definitely the inability of radiofrequency radiation to penetrate through the battery casing, commonly made from stainless steel or aluminum. Therefore, other means of analyzing the processes in the battery have been developed. More recently, information on important battery characteristics has been assessed with the io-MRI approach [51]. Employing this methodology, information about the magnetic field around the battery is obtained and monitored as a function of electrochemical properties. The first experiments of this kind were presented by Ilott et al., in which battery state of charge (SOC) as well as defects could be detected noninvasively [51].

A 3D-printed holder containing detection medium can be employed as a measurement cell (Figure 17.4). The holder can be adapted to the cell dimensions and the detection medium is usually doped water to enable rapid imaging. By using a multiple echo time gradient echo approach (Fast Low Angle SHot, FLASH [52]) focused on the detection medium, the field around the commercial cell can be acquired.

Since the signal phase is proportional to the echo time as well as the field, the desired information can be accessed by linear regression of the observed signal phase as a function of echo time. One of the challenges in this approach is that the phase of the signal is subject to wrapping, which needs to be handled appropriately to enable full information recovery, especially in areas of steep changes of the field [53]. The field change around the cell is associated with the magnetic susceptibility distribution within the battery, which can evolve due to lithiation state of the cathode or changes in the geometry or defects. For example, common cathode materials such as LCO (Li_xCoO_2) and NMC ($Li_xNi_yMn_yCo_{1-2y}O_2$) are well characterized with respect to their paramagnetic magnetic susceptibility changes in response to changes of lithiation state [54,55]. In addition, graphite, a commonly used anode material, exhibits strong changes in the diamagnetic susceptibilities. Therefore, not only is the SOC available as an indirect observable, but also its distribution in the material [56].

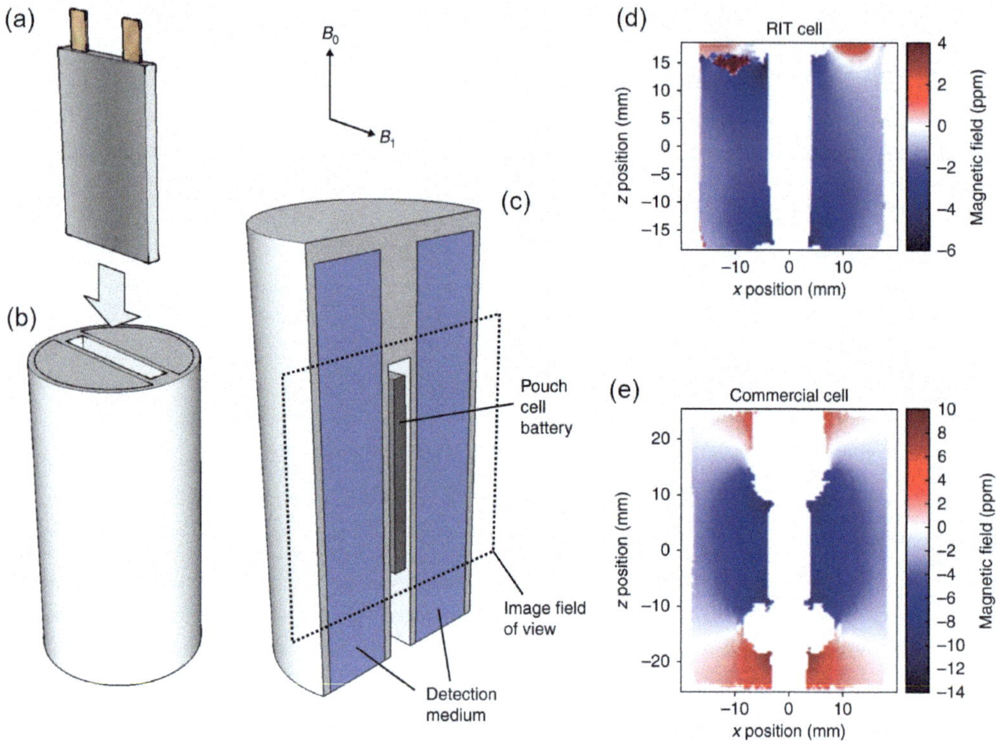

Figure 17.4 Sample placement and holder in the inside-out experiment and field maps obtained from such an experiment. Adapted from Ilott et al. [51].

The right side of Figure 17.4 shows exemplary field maps for two different cells. These were referenced to an empty holder, i.e. the field of an empty holder is subtracted to remove background gradients from $\boldsymbol{B}_0$ inhomogeneities, for example. The measured field at different SOCs of the cell changes significantly mostly due to the different amounts of lithium within the cathode matrix (Figure 17.5a). This information, paired with knowledge about the material composition, allows the calculation of the susceptibility of the cathode as an average value (Figure 17.5b, c). Obtaining a localized susceptibility map from such data is desirable, and could be obtained by regularized magnetic field inversion techniques [57,58].

For quality control purposes, the identification of defects is of particular importance. It was subsequently verified that the approach of measuring magnetic field changes in response to changes of the magnetic susceptibility or its distribution produces data that allow the identification or classification of a range of battery cell defects (Figure 17.6). For example, a folded electrode can easily be identified by its typical pattern (Cell 07/08), a missing part leads to a higher than expected field (Cell 09/10), and foreign parts also leave patterns (Cell 11/12). A large dataset of such defects for different cells can enable rapid identification of a range of defects, and it was also shown for a smaller dataset with principal component analysis (PCA). The imaging time can range from a few seconds to several minutes, depending on resolution and dimensionality.

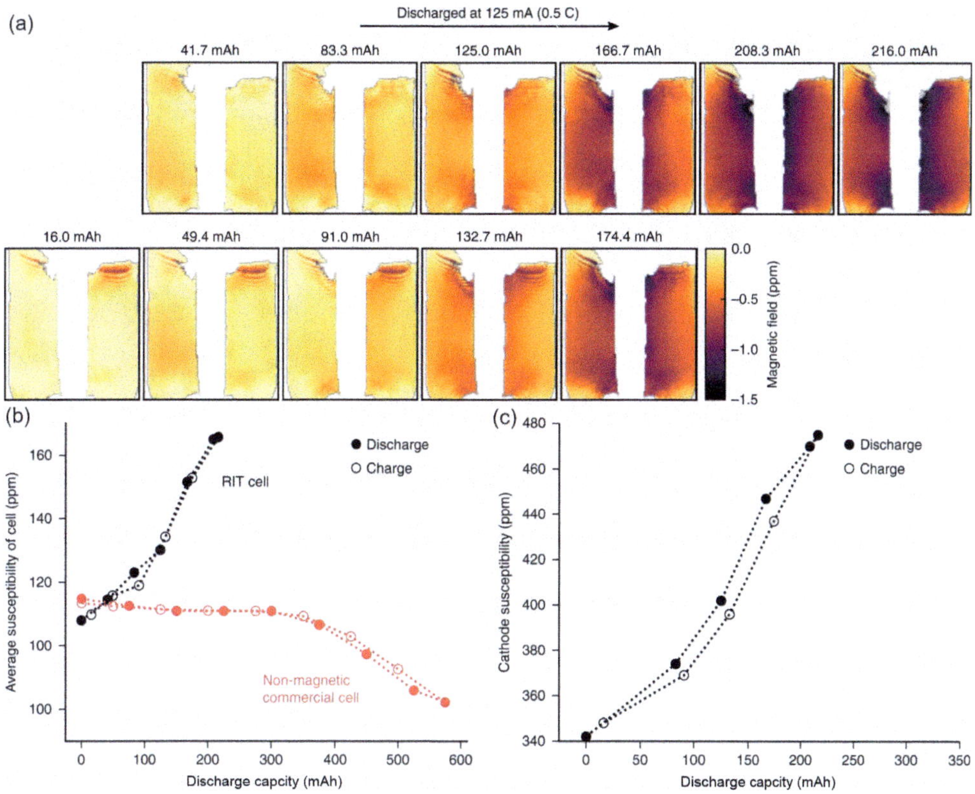

Figure 17.5 (a) Series of magnetic field maps taken at intervals during discharge and then charge of the cell. The plots are labeled by the discharge capacity of the cell at each step. The magnetic field maps are referenced to the field map produced by the fully charged cell. A cell produced by the Rochester Institute of Technology (RIT) was used for this purpose, and the susceptibility increased upon discharge. (b, c) Fitted magnetic susceptibilities of the cell during the discharge/charge. For simplicity, the cell is assumed to have a uniform volume susceptibility in the bottom left graph. Error bars are smaller than the size of the points. Knowing the cell construction, one can, however, also determine the average susceptibility of the cathode material in the bottom right graph. The susceptibility values are given in ppm indicating a factor of 10^{-6}. Adapted from Ilott et al. [51].

While producing images is often desirable, this is by no means always necessary. The induced magnetic fields produced by the cell materials can be encoded in a free-induction decay (FID) measurement without the application of any background magnetic field gradients. If the goal is the classification of different cells, then this approach could provide an extremely fast diagnostic method.

As demonstrated recently, the field changes caused by susceptibility are clearly visible in spectral changes in a range of cells [59]. These spectra contain information of the susceptibility within the cell as an integral measurement. The advantage of an FID measurement is of course its simplicity and speed. Since no magnetic field gradients are required, the technique could be implemented with minimal instrumentation. Spectra for a battery at different SOC points are shown in Figure 17.7 (left) where a shift in pattern is observed clearly. While one could perhaps argue that the overall changes appear small, a comprehensive study with a large number

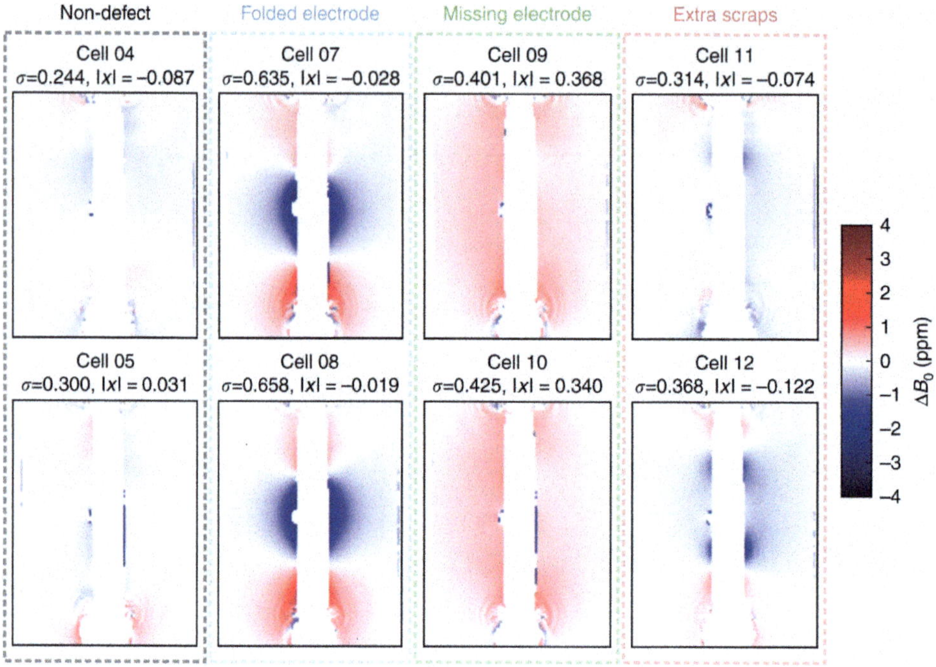

Figure 17.6 Magnetic field maps for the defective cells, with the mean and standard deviation indicated (taken over all the voxels in each image). The fields are given relative to one of the nondefect cells. Adapted from Ilott et al. [51].

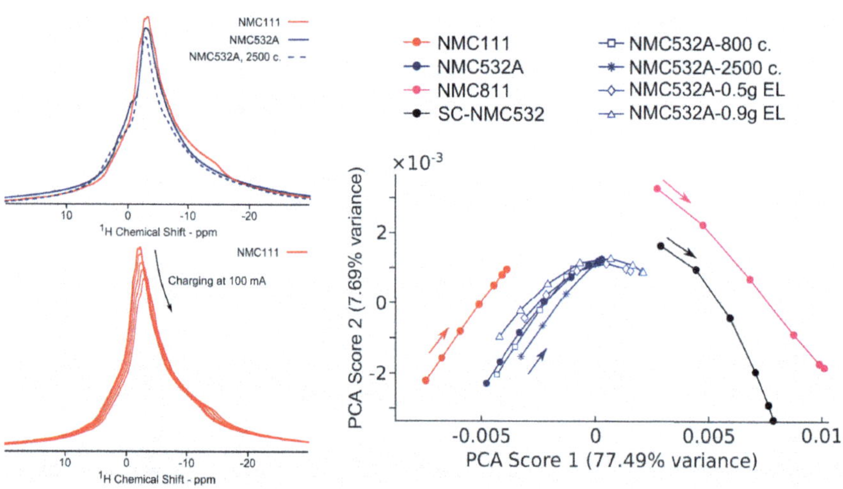

Figure 17.7 Left, top: ^{1}H nuclear magnetic resonance (NMR) spectra taken at 2.5 V for cells containing, as cathode: NMC111 (red), NMC532A (solid blue), and NMC532A having already undergone 2500 cycles prior to the NMR analysis (dashed blue). Bottom: ^{1}H NMR spectra taken at intervals during discharge of the cell containing NMC111 as cathode. The black arrow shows the direction of shift of the spectra as the cell undergoes discharge. Right: Principal component analysis performed on the ^{1}H NMR spectra of all the cells. Each arrow indicates the direction of the trend(s) of the corresponding color as the cell undergoes discharge. Adapted from Pigliapochi et al. [59].

of cells made with different cathode materials, different electrolyte amounts, and at different SOC has shown that a model-free PCA approach delivers unique classification power. In Figure 17.7 (right) it is shown that cells follow unique trajectories on the PCA plot (when using the two strongest principal components against each other) for each cathode material. In addition, the trajectories are traversed based on the SOC in a well-defined and unique manner for each cell type.

One further advantage of the FID approach is that it allows for straightforward capturing of the short T_2* components. In the gradient echo approach discussed earlier, there is a minimum echo time related to gradient switching and encoding requirements. During this time, signal and information related to high susceptibility components can be extracted. In the regions affected by high susceptibility materials, there is a large variation of offset fields within the measurement volume, represented by ΔB. This value is related to T_2* via $\frac{1}{T_2^*} = \frac{1}{T_2} + \frac{1}{T_2'}$ with $\frac{1}{T_2'} \alpha \gamma \Delta B$. This aspect can explain the particularly strong classification power observed in the FID approach. While valuable information can be gained with a gradient echo, in some cases, batteries contain significant amounts of ferro- or ferrimagnetic components, which leads to distortions and limits the ability to observe the paramagnetic changes that are of interest for the susceptibility measurements. The cells used in Figure 17.7 contained nickel tabs, and hence it was determined that the presence of such potentially problematic components did not hamper the ability to perform cell classification. Overall, the FID approach [59] represents an ultrafast nondestructive cell classification technique, which can operate well with minimal hardware and in the presence of strong background magnetic field gradients. Along similar lines, ferro- or ferrimagnetic components would create image distortions that could help with proper defect localization. Due to the rapid signal decay in areas of these components, techniques providing spatial resolution have to be performed with low encoding times, i.e. the time of applied gradients, which is not available with the FLASH technique presented earlier.

Single point imaging with T_1 enhancement (SPRITE) features low encoding times, relatively fast speed, and high stability. SPRITE is a fully phase-encoded technique employing continuous gradient switching to save time and reduce noise and vibrations. Compared with a regular gradient echo technique, strongly magnetic features as often found in smartphone batteries can be assessed (Figure 17.8) [60]. This approach is essential, if changes in battery behavior or safety are sought in proximity of such areas. A distinct feature of the SPRITE approach is the ability to sample multiple points similar to an FID in one scan. This approach enables the reconstruction of fields from these points due their different phase linked with their different evolution time. Since each of these points correlates to a different field of view caused by a longer time integral of the gradient, a *z*-chirp transformation has to be employed for reconstruction [60].

In Figure 17.8, the left side shows FLASH images, and image distortions and even "holes" are clearly visible in areas close to ferromagnetic materials/edges. On the right side, the SPRITE method provides very clear signal in these areas, albeit with reduced resolution. The ability to image short T_2* components with the SPRITE technique also enables the usage of a different sample holder constituted of a solid-state detection medium, which improves the reliability of the approach, especially in a high-throughput scenario. This approach was demonstrated in [61]. In addition, with this method, a gel-like detection medium can be used to wrap around devices with irregular shapes, which would also be immune to convection effects or leakage.

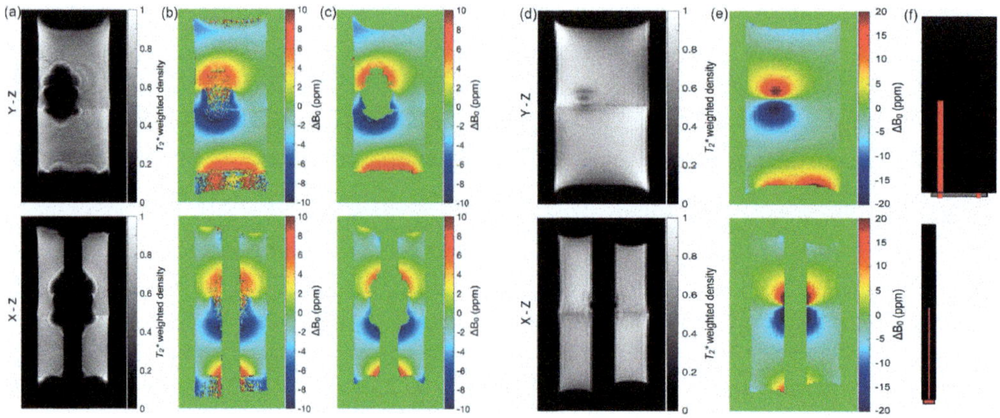

Figure 17.8 Left (a–c): Slices through a 3D Fast Low Angle SHot (FLASH) image (iPhone 5 battery in holder) with spin density, unmasked field map, and masked field map. Right (d–e): SPRITE (single point imaging with T_1 enhancement) images of the same battery holder as spin density and unmasked field maps. The rightmost panel depicts the current collector and leads responsible for some of the high susceptibility. Reprinted (adapted) from Romanenko et al. [60].

17.5 Expanding the Method to Current Density Sensing

A crucial parameter for battery performance is the distribution of electrical current during charge and discharge at different states of charge. Since electrical current is always associated with a magnetic field, this effect can be measured in principle by acquiring magnetic field maps around the device while it is being charged or discharged. A challenge in measuring these effects is the magnitude of the generated field, since for typical battery capacities it is lower than or in the range of the field perturbation caused by the magnetic susceptibility. Therefore, it is crucial to minimize or remove contributions from the background field. The background field changes with SOC, but if measurements are performed separately with and without current flow, the differential effect can be determined. To enhance the robustness of the measurement, one may also perform the field measurements for the different echo time experiments such that for each current experiment the state of the battery is returned to the initial state before the experiment. Furthermore, minimizing the time of each experiment further decreases the influence of background field evolution during an experiment.

Figure 17.9 shows the magnetic field maps obtained for the field attributed to electrical currents. These maps and histograms were generated from field maps during current flow, with the background field during the rest period subtracted [62]. Interesting symmetries are observed between the fully charged and the almost fully discharged state: for example, if a cell near full charge is being charged, the magnetic field histogram appears bimodal. This finding could be a consequence of the fact that near full charge, it becomes more difficult to reach efficient charge storage locations, and hence the current distribution becomes less uniform. By contrast, if the cell is being discharged near full charge, it is easy to draw current from many places in the cell, and hence the histogram appears to indicate a more ordered distribution. The opposite is found for the case of near full discharge: when attempting to discharge further, one sees a bimodal field distribution, again an indication for the difficulty in localizing potential charge sources. When charging in this state, it is easy to find recipients for charge, and hence a more orderly

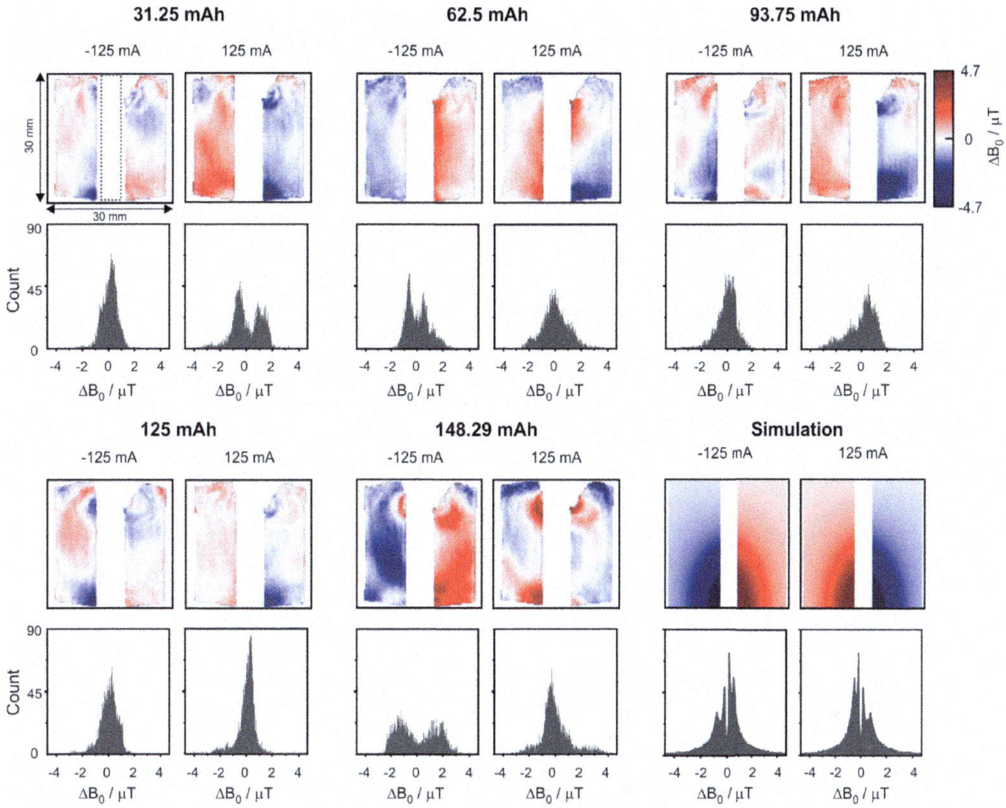

Figure 17.9 Magnetic field maps and histograms during discharge (negative current) and charge (positive current) at different depth of discharge (DOD) as indicated. The maps were obtained after subtracting a reference image from the resting period between discharging and charging at each DOD. A constant background field was further subtracted to reduce effects of magnet field drifts. The bottom right panel shows a simulation. The battery leads in these measurements are located at the bottom of each map. The discharge capacity is indicative of the starting point of each step, and discharge measurements were performed before charge measurements. The imaging dimensions and the location of the cell (dotted rectangle) are indicated in the top left image. Adapted from Mohammadi et al. [62].

distribution is found (monomodal magnetic field distribution). Clearly, battery defects would lead to changes of such patterns. For example, defects in electrode coating, scratches, or separator blockage would lead to significant perturbations in current flow, and consequently in the magnetic fields observed. Hence it is anticipated that these techniques could become sensitive reporters of normal and abnormal charge storage behavior in cells, and hence could represent a powerful diagnostic technique.

The current sensing approach can be combined with SPRITE imaging for the purposes of performing this technique in the presence of sizable magnetic background fields from ferro- or ferrimagnetic components. Figure 17.10 shows results from such measurements performed on iPhone batteries. The results show the possibility of obtaining distortion-free magnetic field maps providing observations of electrical current-induced effects. Notably, a symmetry between charge and discharge was observed, but cell damages also became clearly reflected in the observed magnetic field distributions.

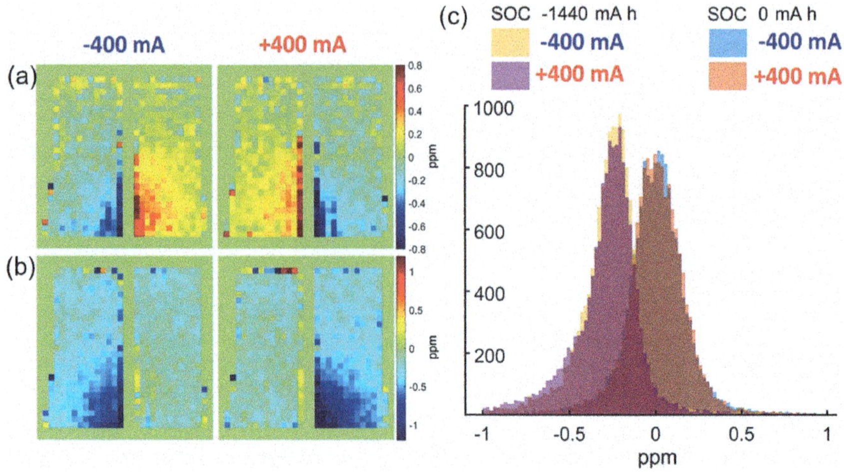

Figure 17.10 Longitudinal *xz*-slices through 3D DCF maps of the iPhone 8 battery (±400 mA) at (a) SOC = 0 mAh (fully charged) and (b) SOC = −1440 mAh. (c) DC-induced magnetic field distributions of the operating iPhone 8 battery. Reference images: intact cell (open circuit) at given states of charge. [61] Romanenko et al. (2020), figure 05 [p.2110] / with permission from American Chemical Society.

17.6 Mapping Oscillating Magnetic Fields Around Battery Cells

Spatial maps of the fields generated by direct current during charge and discharge are important measures in the characterization of batteries, since they are essential in their immediate practical applications. More dynamic processes within batteries can be probed with alternating current. A common technique, called EIS, analyzes the frequency-dependent impedance response of an electrical circuit. In this technique, alternating currents are applied over a given frequency range, and the complex impedance recorded [63,64]. Fields of its application range from material research to the investigation of biological tissue [65–67]. With regard to batteries, EIS can aid in characterizing cell performance, and properties such as the nature of the solid–electrolyte interface (SEI) and porosity of the electrode material [4–6,68–70]. The frequency range of 0.1–10 Hz is particularly interesting, since charge-transfer processes in batteries can be observed in this region [71]. Although there exist implementations that introduce spatial resolution to this technique, this approach still requires a direct localized contact with the electrical components [12,13]. Of course, for the nondestructive analysis of devices such as commercial batteries, this procedure is inapplicable due to the inability to access the interior. Therefore, accessing the field generated by alternating currents via MR techniques could be a promising strategy to gain further insights. While there exist techniques that can be used to obtain such information with precise synchronization of parts of the AC wave to the pulse sequence [72–76], a less restricted approach is desirable. Inspiration can be drawn from the medical field, where Yang et al. and Kim et al. have proven feasibility of less demanding approaches [77,78]. Focusing on electronic components, Benders et al. have developed the io-TRACING method [79]. This method involves less restrictive synchronization than other techniques and uses gating of the AC to the battery in certain parts of the battery, allowing to access frequencies from 0.1 Hz to 2 kHz. Especially the lower end of that range was difficult to access with MRI sequences in the past, while being important for charge-transfer processes [71]. The method maps the intensity proportional to the oscillating field outside of the battery in a detection medium. This intensity can be converted into fields with careful calibration, but also the changes in the pattern reveal important information. In Figure 17.11, field intensity maps for two different cells in a slice parallel to the battery are shown for various frequencies. These can potentially be converted to accurate field maps by calibration. Two batteries are investigated in this study, a PowerStream (PS) commercial battery and a Rochester Institute of Technology (RIT) prototype cell. For the PS battery, the images have a zero node, which changes dimensions with frequency. In addition, a pattern change can be seen at the top at very low frequencies, while the RIT cell undergoes more changes with frequency. Lock-in and phase-sensitive detection could ultimately also enable the recovery of the phase information of the field.

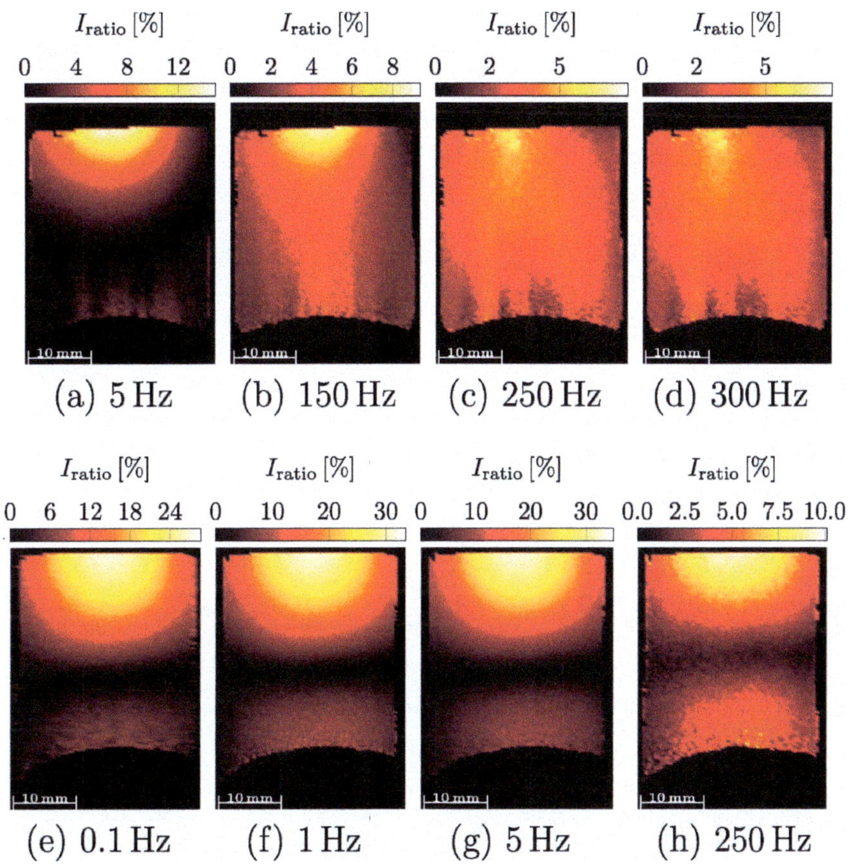

Figure 17.11 Relative measurements of the oscillating fields generated by 50 mA AC current in two Li-ion battery types at different driving AC frequencies. Slices are parallel to the battery at a distance of ~7.5 mm from the center with the battery tabs being located at the top of the image. (a)–(d) are maps from a 250 mAh Rochester Institute of Technology (RIT) cell, which has a stacked electrode design. The images in (e)–(h) are taken from a 600-mAh PowerStream (PS) jelly-rolled cell. Sixteen experiments with different parts of the waveform have been employed for reconstruction of each point of the PS images, while 32 were used for the RIT images. [79] Benders et al. (2020), figure 05[p.006]/with permission from Elsevier.

17.7 Employing Magnetometry to Investigate Commercial Batteries

Lessons learned with MRI have led to the idea that the magnetic fields around batteries could also be detected and mapped using advanced magnetometer sensors. Magnetometers have gained increasing popularity in recent years due to their increasingly high sensitivity (down to $fT/\sqrt{Hz}$ levels) and affordability. Hu et al. performed a first proof-of-principle demonstration of such an approach to reveal battery processes [58]. Employing the same principle as the io-MRI approach, the field caused by susceptibility in an external field is measured. For that purpose, the cell is placed in a solenoid and the sensitive sensor is placed outside. The solenoid coil is used to generate a polarizing field within which the battery components develop the induced fields on the basis of their magnetic susceptibilities. At the same time, the sensor can be placed in relatively close proximity to the cell, while also residing in an area of small magnetic field such that the sensor is not saturated. Figure 17.12 shows the setup, where in (a) the field alignment can be observed.

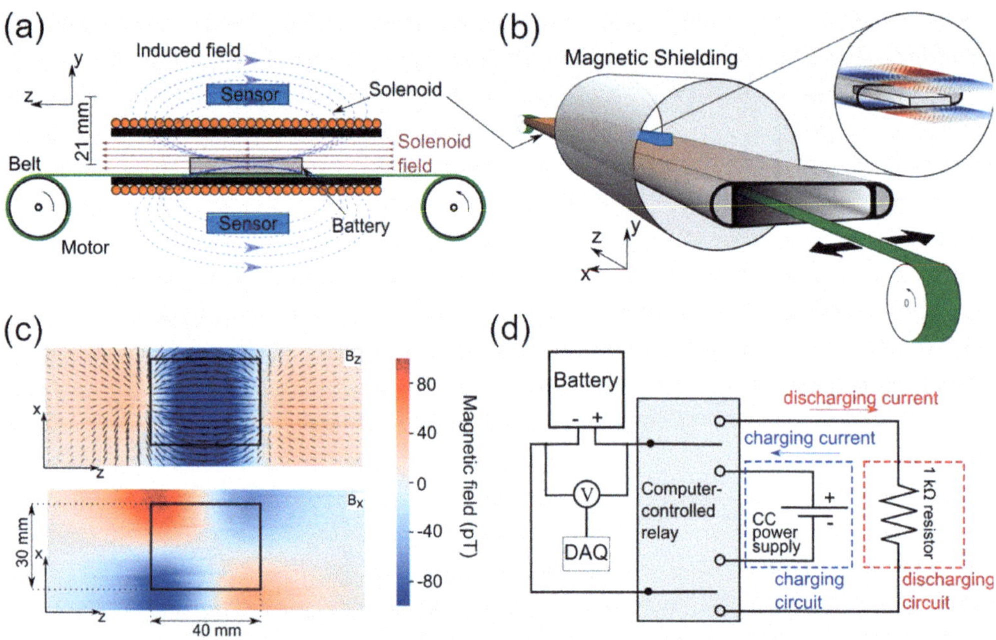

Figure 17.12 Experimental susceptometry setup. (a) Side view of the experimental arrangement. A battery is placed on a motor-driven "conveyor belt" that moves the cell through a custom-designed solenoid, which provides a constant internal magnetic field. Magnetic field sensors are placed in the ultralow-field region, above and below the solenoid. (b) Three-dimensional drawing of the experimental arrangement. The solenoid is shown extending beyond the Twinleaf MS-2 magnetic shielding (only the innermost shield is shown – a total of four concentric cylinders were used). The conveyor belt moves the battery back and forth along the *z*-axis, while translation stages on each end move the entire belt with 0.1-mm precision along the *x*-axis. (Inset) Illustration of the location of the resulting 2D field maps acquired by this setup. (c) Sample-induced magnetic field map for a healthy battery charged to 90% capacity, obtained from a single sensor 21 mm above the battery. Upper: color map of the *z*-component of the measured field, with an overlaid vector map of the total field in the *x*–*z*-sensor plane. Lower: the *x*-component color map of the measured field. The battery outline is shown as a black rectangle. (d) Circuit diagram for charging and discharging. The computer-controlled relay disconnects the battery cell from the charging/discharging circuits during measurements. However, the data acquisition (DAQ) unit remained connected throughout the measurement. V: Voltmeter. CC: constant current. Reprinted (adapted) from Hu et al. [58].

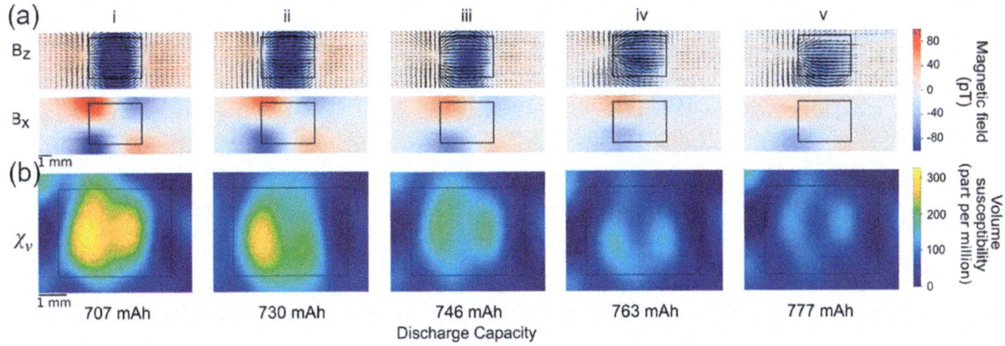

Figure 17.13 Magnetic field and corresponding susceptibility maps above a cell. The corresponding discharge capacity is given at the bottom. (a) Upper: field map in the z-direction, with an x–z field vector map overlaid. (a) Lower: field in the x-direction. The black rectangular outline represents the battery position. Battery leads (not indicated) are to the left of the black rectangle. (b) Two-dimensional magnetic susceptibility maps (indicated in ppm) obtained from a regularized inversion of the measured magnetic field. [58] Hu et al. (2020), figure 03 [p.05] / with permission from National Academy of Sciences.

A linear stage and a translation belt (b) allow the acquisition of spatially resolved field maps (c). The field pattern looks typical for a rectangular component with uniform susceptibility.

The field changes with discharge capacity as expected, and the pattern can be employed in a regularized inversion to obtain volume susceptibility maps (Figure 17.13). This technique is very similar to the io-MRI technique, but currently also has some drawbacks. While it is potentially less expensive, it does not currently perform very well in the presence of strong background magnetic fields. This work is rapidly developing, and it is anticipated that MRI and magnetometry will develop into complementary tools, and that both measurement modalities can support each other.

17.8 Measuring NMR Signals From the Inside of a Fully Enclosed Commercial Cell

While the inside-out approach has many potential applications in noninvasive battery diagnostics, obtaining information directly from the inside of commercial cells remains a challenge. There have been reports of some early in-cell NMR and MRI work, where a toroid cavity was used in which a metal rod is used simultaneously as the working electrode of a compression coin cell and as the central conductor of the toroid cavity [44,80]. This approach cannot be used immediately for pouch cells, however, and would also not be applicable to standard coin cells.

Benders et al. have recently shown that one may use the aluminum pouch of a cell as a resonator in order to bypass the skin depth and radiofrequency penetration issue [50]. To this end, two copper pads are brought into contact with the aluminum casing on opposite sides of a jelly-rolled pouch cell and then integrated into a resonant circuit (Figure 17.14). In so doing, the casing and the cell are incorporated into the tuned circuit and tuned and matched such that opposite sides can be driven out of phase, and thus allow for radiofrequency irradiation to enter into the inside compartment. This approach was demonstrated for ^{7}Li tuning with a lithium-ion battery. To accommodate for the atypical resonator shape, a custom probe was built in this case.

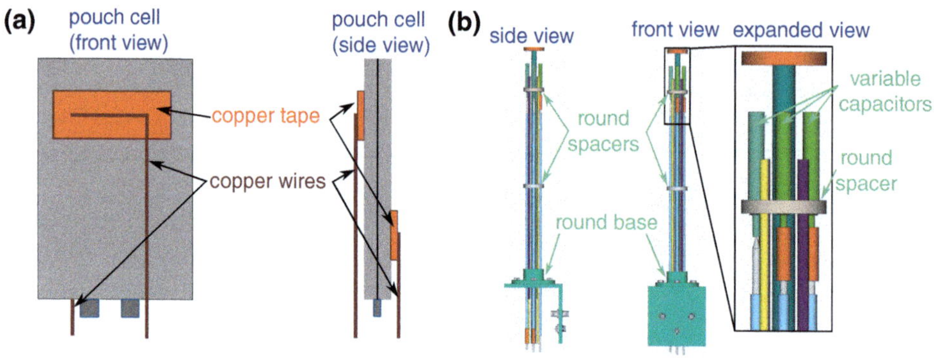

Figure 17.14 Schematic of a pouch cell battery (gray) (a) with pads attached for feeding the radiofrequency into the cell via capacitive coupling. (b) Schematic of the home-built probe for this setup. Adapted from Benders et al. [50].

In this work, ^{7}Li signals from three key environments can be observed: electrolyte lithium, intercalated lithium, and metallic lithium (Figure 17.15). Monitoring electrolyte lithium in such a way could be used to test the state of the electrolyte. The observation of intercalated lithium would indicate the level of anode participation in electron transfer processes, and tracking the amount of metallic lithium (in this graphite-based cell) allows determining the level of anodic microstructure growth and plating due to overcharging, for example. The approach taken in this work can potentially be extended to capacitive coupling and can provide essential insights into processes in commercial lithium cells.

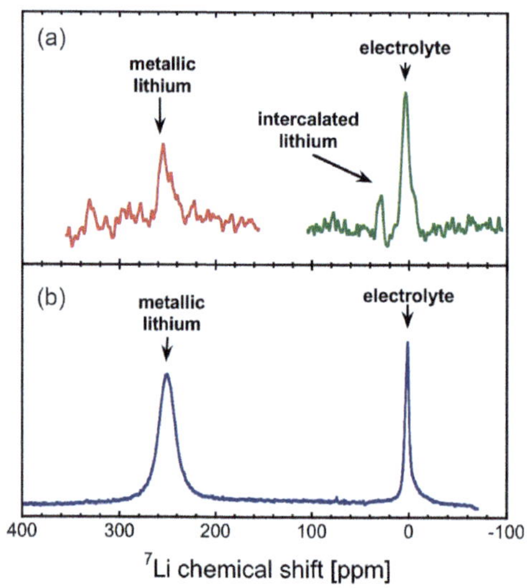

Figure 17.15 ^{7}Li nuclear magnetic resonance spectra obtained for the battery-as-coil setup using a Hahn echo preparation (a) and for a reference battery cell using a traditional solenoid coil and single pulse acquisition (b). Due to limited excitation bandwidth, the spectra in (a) were obtained from two separate measurements, one where the spectrometer frequency was at 255 ppm (left) and one where the spectrometer frequency was at 5 ppm (right).
Adapted from Benders et al. [50].

17.9 Conclusions and Outlook

In situ NMR spectroscopy and MRI imaging have come a long way in producing new insights into electrochemical processes inside cells. One aspect that has been largely missing in these efforts has been the ability to assess full cells in commercial-type or scaled up versions such as pouch cells. Many battery materials change their properties significantly once the first scale-up is performed from coin cell to pouch cell, and hence it is of great interest to study cells in their deployable designs. For a long time, it was not possible to obtain much information from the cells' inside, because of the inability to access the interior volume of a cell with radiofrequency irradiation. Recent work on io-MRI approaches, whereby the induced magnetic fields are mapped in the surroundings of cells, has provided new observables that can be used for battery characterization, diagnostics, and classification, as well as for monitoring specific electrochemical processes, without needing immediate access to the interior.

With these methods, it has been shown that the state-of-charge distribution can be mapped, and certain defects identified. Furthermore, the current distribution during charging and discharging can be mapped, which provides an additional opportunity for cell and process characterization. Extensions to this approach include techniques that allow imaging in the presence of strong background gradients (SPRITE), and solid detection media. In addition, an ultrafast variant of this approach has been demonstrated to show powerful cell classification based on a simple spectral method that does not require magnetic field gradients.

Recent work also includes efforts to map alternating magnetic field distributions in order to provide information similar to EIS, but in a localized fashion. Finally, progress has also been made in accessing the cell's interior with radiofrequency irradiation for the direct observation of electrolyte lithium, plated lithium, and intercalated lithium from within a standard pouch cell. All these recent demonstrations have provided new observables for cell examination, and are expected to aid in all stages of battery research, from cell development, to quality control, high-stress testing, and potentially in recycling, as key cell parameters can now be determined which are not available by any other means.

Acknowledgments

Alexej Jerschow acknowledges funding from the US National Science Foundation under award CBET 1804723.

References

1 Goodenough, J.B. and Kim, Y. (2010). Challenges for rechargeable Li batteries. *Chemistry of Materials: A Publication of the American Chemical Society* 22 (3): 587–603.

2 Goodenough, J.B. (2018). How we made the Li-ion rechargeable battery. *Nature Electronics* Mar 1 (3): 204–204.

3 Yoshio, M., Brodd, R.J., and Kozawa, A. (2010). *Lithium-Ion Batteries: Science and Technologies*. New York: Springer Science & Business Media. 460.

4 Verma, P., Maire, P., and Novák, P. (2010). A review of the features and analyses of the solid electrolyte interphase in Li-ion batteries. *Electrochimica Acta* 55 (22): 6332–6341.

5 Jorcin, J.-B., Orazem, M.E., Pébère, N. et al. (2006). CPE analysis by local electrochemical impedance spectroscopy. *Electrochimica Acta* 51 (8): 1473–1479.

6 Deng, Z., Zhang, Z., Lai, Y. et al. (2013). Electrochemical impedance spectroscopy study of a lithium/sulfur battery: Modeling and analysis of capacity fading. *Journal of the Electrochemical Society* 160 (4): A553.

7 Yufit, V., Shearing, P., Hamilton, R.W. et al. (2011). Investigation of lithium-ion polymer battery cell failure using X-ray computed tomography. *Electrochemistry Communications* 13 (6): 608–610.

8 Wu, Y., Saxena, S., Xing, Y. et al. (2018). Analysis of manufacturing-induced defects and structural deformations in lithium-ion batteries using computed tomography. *Energies* 11 (4): 925.

9 Gallaway, J.W., Erdonmez, C.K., Zhong, Z. et al. (2014). Real-time materials evolution visualized within intact cycling alkaline batteries. *Journal of Materials Chemistry A* 2 (8): 2757–2764.

10 Eastwood, D.S., Bayley, P.M., Jung Chang, H. et al. (2015). Three-dimensional characterization of electrodeposited lithium microstructures using synchrotron X-ray phase contrast imaging. *Chemical Communications* 51 (2): 266–268.

11 Hsieh, A.G., Bhadra, S., Hertzberg, B.J. et al. (2015). Electrochemical-acoustic time of flight: In operando correlation of physical dynamics with battery charge and health. *Energy & Environmental Science* 8 (5): 1569–1577.

12 Eckhard, K. and Schuhmann, W. (2008). Alternating current techniques in scanning electrochemical microscopy (AC-SECM). *Analyst* 133 (11): 1486–1497.

13 Bandarenka, A.S., Eckhard, K., Maljusch, A. et al. (2013). Localized electrochemical impedance spectroscopy: Visualization of spatial distributions of the key parameters describing solid/liquid interfaces. *Analytical Chemistry* 85 (4): 2443–2448.

14 Davenport, A.J., Forsyth, M., and Britton, M.M. (2010). Visualisation of chemical processes during corrosion of zinc using magnetic resonance imaging. *Electrochemistry Communications* 12 (1): 44–47.

15 Bray, J.M., Davenport, A.J., Ryder, K.S. et al. (2016). Quantitative, in situ visualization of metal-ion dissolution and transport using 1H magnetic resonance imaging. *Angewandte Chemie International Edition* 55 (32): 9394–9397.

16 Britton, M.M., Bayley, P.M., Howlett, P.C. et al. (2013). In situ, real-time visualization of electrochemistry using magnetic resonance imaging. *Journal of Physical Chemistry Letters* Sep 5 4 (17): 3019–3023.

17 Monzon, L.M.A. and Coey, J.M.D. (2014). Magnetic fields in electrochemistry: The Lorentz force. *A Mini-review. Electrochemistry Communications* 42: 38–41.

18 Monzon, L.M.A. and Coey, J.M.D. (2014). Magnetic fields in electrochemistry: The Kelvin force. *A Mini-review. Electrochemistry Communications* 42: 42–45.

19 Benders, S., Gomes, B.F., Carmo, M. et al. (2020). In-situ MRI velocimetry of the magnetohydrodynamic effect in electrochemical cells. *Journal of Magnetic Resonance* 1 (312): 106692.

20 Gomes, B.F.B., Ferreira Da Silva, P., Silva Lobo, C.M. et al. (2017). Strong magnetoelectrolysis effect during electrochemical reaction monitored in situ by high-resolution NMR spectroscopy. *Analytica Chimica Acta* 29 (983): 91–95.

21 Truong, T.-K., Avram, A., and Song, A.W. (2008). Lorentz effect imaging of ionic currents in solution. *Journal of Magnetic Resonance* 191 (1): 93–99.

22 Pourtaheri, N., Truong, T.-K., and Henriquez, C.S. (2013). Electromagnetohydrodynamic modeling of Lorentz effect imaging. *Journal of Magnetic Resonance* 1 (236): 57–65.

23 Eroğlu, H.H., Sadighi, M., and Eyüboğlu, B.M. (2019). Magnetohydrodynamic flow imaging of ionic solutions using electrical current injection and MR phase measurements. *Journal of Magnetic Resonance* 1 (303): 128–137.

24 Balasubramanian, M., Mulkern, R.V., Wells, W.M. et al. (2015). Magnetic resonance imaging of ionic currents in solution: The effect of magnetohydrodynamic flow. *Magnetic Resonance in Medicine* 74 (4): 1145–1155.

25 Codd, S.L., Howe, D.T., Seymour, J.D. et al. (2007). Magnetic resonance microscopy of heterogeneity in polymer electrolyte membranes. *Applied Magnetic Resonance* 32 (1–2): 13–24.

26 Baker R, T., Naji, L., Lochhead, K. et al. (2003). In situ magnetic resonance imaging of electrically-induced water diffusion in a Nafion ionic polymer film. *Chemical Communications* 0 (8): 962–963.

27 Feindel, K.W., LaRocque, L.P.-A., Starke, D. et al. (2004). In situ observations of water production and distribution in an operating H_2/O_2 PEM fuel cell assembly using 1H NMR microscopy. *Journal of the American Chemical Society* 126 (37): 11436–11437.

28 Feindel, K.W., Bergens, S.H., and Wasylishen, R.E. (2006). The use of 1H NMR Microscopy to study proton-exchange membrane fuel cells. *ChemPhysChem* 7 (1): 67–75.

29 Zhang, Z., Martin, J., Wu, J. et al. (2008). Magnetic resonance imaging of water content across the Nafion membrane in an operational PEM fuel cell. *Journal of Magnetic Resonance* 193 (2): 259–266.

30 Tsushima, S. and Hirai, S. (2009). Magnetic resonance imaging of water in operating polymer electrolyte membrane fuel cells. *Fuel Cells* 9 (5): 506–517.

31 Cattaneo, A.S., Villa, D.C., Angioni, S. et al. (2015). Operando electrochemical NMR microscopy of polymer fuel cells. *Energy & Environmental Science* 8 (8): 2383–2388.

32 Hope, M.A., Rinkel, B.L.D., Gunnarsdóttir, A.B. et al. (2020). Selective NMR observation of the SEI–metal interface by dynamic nuclear polarisation from lithium metal. *Nature Communications* 11 (1): 2224.

33 Chandrashekar, S., Trease, N.M., Chang, H.J. et al. (2012). ^{7}Li MRI of Li batteries reveals location of microstructural lithium. *Nature Materials* 11 (4): 311–315.

34 Ilott, A.J., Chandrashekar, S., Klöckner, A. et al. (2014). Visualizing skin effects in conductors with MRI: ^{7}Li MRI experiments and calculations. *Journal of Magnetic Resonance* 245: 143–149.

35 Chang, H.J., Ilott, A.J., Trease, N.M. et al. (2015). Correlating microstructural lithium metal growth with electrolyte salt depletion in lithium batteries using 7Li MRI. *Journal of the American Chemical Society* 137 (48): 15209–15216.

36 Klett, M., Giesecke, M., Nyman, A. et al. (2012). Quantifying mass transport during polarization in a Li ion battery electrolyte by in situ 7Li NMR imaging. *Journal of the American Chemical Society* 134 (36): 14654–14657.

37 Krachkovskiy, S.A., Pauric, A.D., Halalay, I.C. et al. (2013). Slice-selective NMR diffusion measurements: A robust and reliable tool for in situ characterization of ion-transport properties in lithium-ion battery electrolytes. *Journal of Physical Chemistry Letters* 4 (22): 3940–3944.

38 Sethurajan, A.K., Krachkovskiy, S.A., Halalay, I.C. et al. (2015). Accurate characterization of ion transport properties in binary symmetric electrolytes using in situ NMR imaging and inverse modeling. *Journal of Physical Chemistry B* 119 (37): 12238–12248.

39 Bazak, J.D., Krachkovskiy, S.A., and Goward, G.R. (2017). Multi-temperature in situ magnetic resonance imaging of polarization and salt precipitation in lithium-ion battery electrolytes. *Journal of Physical Chemistry C* 121 (38): 20704–20713.

40 Klamor, S., Zick, K., Oerther, T. et al. (2015). 7 Li in situ 1D NMR imaging of a lithium ion battery. *Physical Chemistry Chemical Physics* 17 (6): 4458–4465.

41 Ilott, A.J., Mohammadi, M., Chang, H.J. et al. (2016). Real-time 3D imaging of microstructure growth in battery cells using indirect MRI. *PNAS* 113 (39): 10779–10784.

42 Bray, J.M., Doswell, C.L., Pavlovskaya, G.E. et al. (2020). Operando visualisation of battery chemistry in a sodium-ion battery by 23 Na magnetic resonance imaging. *Nature Communications* 11 (1): 2083.

43 Serša, I. and Mikac, U. (2018). A study of MR signal reception from a model for a battery cell. *Journal of Magnetic Resonance* 1 (294): 7–15.

44 Gerald, R.E., Johnson, C.S., Rathke, J.W. et al. (2000). 7Li NMR study of intercalated lithium in curved carbon lattices. *Journal of Power Sources* 89 (2): 237–243.

45 Jackson, J.D. (1998). *Classical Electrodynamics*, 3e, 832. New York: Wiley.

46 Hoult, D.I. (2000). The principle of reciprocity in signal strength calculations – A mathematical guide. *Concepts in Magnetic Resonance* 12 (4): 173–187.

47 Hoult, D.I. (2011). The principle of reciprocity. *Journal of Magnetic Resonance* 213 (2): 344–346.

48 Ilott, A.J. and Jerschow, A. (2018). Aspects of NMR reciprocity and applications in highly conductive media. *Concepts in Magnetic Resonance Part A* 47A (2): e21466.

49 Ilott, A.J. and Jerschow, A. (2017). Super-resolution surface microscopy of conductors using magnetic resonance. *Scientific Reports* 7 (1): 5425.

50 Benders, S., Mohammadi, M., Klug, C.A. et al. (2020). Nuclear magnetic resonance spectroscopy of rechargeable pouch cell batteries: Beating the skin depth by excitation and detection via the casing. *Scientific Reports* 10 (1): 13781.

51 Ilott, A.J., Mohammadi, M., Schauerman, C.M. et al. (2018). Rechargeable lithium-ion cell state of charge and defect detection by in-situ inside-out magnetic resonance imaging. *Nature Communications* 9 (1): 1776.

52 Haase, A., Frahm, J., Matthaei, D. et al. (1986). FLASH imaging. Rapid NMR imaging using low flip-angle pulses. *Journal of Magnetic Resonance* 67 (2): 258–266.

53 Robinson, S., Schödl, H., and Trattnig, S. (2014). A method for unwrapping highly wrapped multi-echo phase images at very high field: UMPIRE. *Magnetic Resonance in Medicine* 72 (1): 80–92.

54 Chernova, N.A., Ma, M., Xiao, J. et al. (2007). Layered Li$_x$Ni$_y$Mn$_y$Co$_{1-2y}O_2$ cathodes for lithium ion batteries: Understanding local structure via magnetic properties. *Chemistry of Materials: A Publication of the American Chemical Society* 19 (19): 4682–4693.

55 Hertz, J.T., Huang, Q., McQueen, T. et al. (2008). Magnetism and structure of Li$_x$CoO$_2$ and comparison to Na$_x$CoO$_2$. *Physical Review B* 77 (7): 075119.

56 DiSalvo, F.J., Safran, S.A., Haddon, R.C. et al. (1979). Large anisotropy and stage dependence of the magnetic susceptibility of alkali-graphite intercalation compounds. *Physical Review B* 20 (12): 4883–4888.

57 Haacke, E.M., Liu, S., Buch, S. et al. (2015). Quantitative susceptibility mapping: Current status and future directions. *Magnetic Resonance Imaging* 33 (1): 1–25.

58 Hu, Y., Iwata, G.Z., Mohammadi, M. et al. (2020). Sensitive magnetometry reveals inhomogeneities in charge storage and weak transient internal currents in Li-ion cells. *PNAS* 117 (20): 10667–10672.

59 Pigliapochi, R., Benders, S., Silletta, E.V. et al. Ultrafast inside-out NMR assessment of rechargeable cells. *Batteries & Supercaps* 4: 322–326.

60 Romanenko, K. and Jerschow, A. (2019). Distortion-free inside-out imaging for rapid diagnostics of rechargeable Li-ion cells. *PNAS* 116 (38): 18783–18789.

61 Romanenko, K., Kuchel, P.W., and Jerschow, A. (2020). Accurate visualization of operating commercial batteries using specialized magnetic resonance imaging with magnetic field sensing. *Chemistry of Materials: A Publication of the American Chemical Society* 32 (5): 2107–2113.

62 Mohammadi, M., Silletta, E.V., Ilott, A.J. et al. (2019). Diagnosing current distributions in batteries with magnetic resonance imaging. *Journal of Magnetic Resonance* 1 (309): 106601.

63 Galeotti, M., Cinà, L., Giammanco, C. et al. (2015). Performance analysis and SOH (state of health) evaluation of lithium polymer batteries through electrochemical impedance spectroscopy. *Energy* 1 (89): 678–686.

64 Barsoukov, E. and Macdonald, J.R. (2018). *Impedance Spectroscopy: Theory, Experiment, and Applications*, 498. Hoboken, NJ: John Wiley & Sons.

65 Estrela Da Silva, J., Marques De Sá, J.P., and Jossinet, J. (2000). Classification of breast tissue by electrical impedance spectroscopy. *Medical & Biological Engineering & Computing* 38 (1): 26–30.

66 Glatthaar, M., Riede, M., Keegan, N. et al. (2007). Efficiency limiting factors of organic bulk heterojunction solar cells identified by electrical impedance spectroscopy. *Solar Energy Materials and Solar Cells* 91 (5): 390–393.

67 Halter, R.J., Schned, A., Heaney, J. et al. (2008). Electrical impedance spectroscopy of benign and malignant prostatic tissues. *Journal of Urology* 179 (4): 1580–1586.

68 Singh, P., Vinjamuri, R., Wang, X. et al. (2006). Fuzzy logic modeling of EIS measurements on lithium-ion batteries. *Electrochimica Acta* 51 (8): 1673–1679.

69 Eddahech, A., Briat, O., Woirgard, E. et al. (2012). Remaining useful life prediction of lithium batteries in calendar ageing for automotive applications. *Microelectronics Reliability* 52 (9): 2438–2442.

70 Pastor-Fernández, C., Dhammika Widanage, W., Marco, J., et al. (2016).Identification and quantification of ageing mechanisms in lithium-ion batteries using the EIS technique. 2016 IEEE Transportation Electrification Conference and Expo (ITEC), Dearborn, Michigan (27–29 June), 1–6.

71 Westerhoff, U., Kurbach, K., Lienesch, F. et al. (2016). Analysis of lithium-ion battery models based on electrochemical impedance spectroscopy. *Energy Technology* 4 (12): 1620–1630.

72 Eroğlu, H.H., Sadighi, M., and Eyüboğlu, B.M. (2018). Induced current magnetic resonance electrical conductivity imaging with oscillating gradients. *IEEE Transactions on Medical Imaging* 37 (7): 1606–1617.

73 Hamamura, M.J., Muftuler, L.T., Birgul, O. et al. (2006). Measurement of ion diffusion using magnetic resonance electrical impedance tomography. *Physics in Medicine and Biology* 51 (11): 2753–2762.

74 Mikac, U., Demšar, F., Beravs, K. et al. (2001). Magnetic resonance imaging of alternating electric currents. *Magnetic Resonance Imaging* 19 (6): 845–856.

75 Buračas, G.T., Liu, T.T., Buxton, R.B. et al. (2008). Imaging periodic currents using alternating balanced steady-state free precession. *Magnetic Resonance in Medicine* 59 (1): 140–148.

76 Halpern-Manners, N.W., Bajaj, V.S., Teisseyre, T.Z. et al. (2010). Magnetic resonance imaging of oscillating electrical currents. *PNAS* 107 (19): 8519–8524.

77 Yang, H., Cook, G.G., and Paley, M.N.J. (2003). Mapping of periodic waveforms using the ghost reconstructed alternating current estimation (GRACE) magnetic resonance imaging technique. *Magnetic Resonance in Medicine* 50 (3): 633–637.

78 Kim, K.H., Heo, H.-I., and Park, S.-H. (2018). Detection of fast oscillating magnetic fields using dynamic multiple TR imaging and Fourier analysis. *PLoS ONE* 13(1). https://www.ncbi.nlm.nih.gov/pmc/articles/PMC5761850

79 Benders, S., Mohammadi, M., Ganter, M.J. et al. (2020). Mapping oscillating magnetic fields around rechargeable batteries. *Journal of Magnetic Resonance* 319: 106811.

80 Gerald, R.E., Sanchez, J., Johnson, C.S. et al. (2001). In situ nuclear magnetic resonance investigations of lithium ions in carbon electrode materials using a novel detector. *Journal of Physics: Condensed Matter* 13 (36): 8269–8285.

18

Magnetic Resonance Imaging of Sodium-Ion Batteries

Claire L. Doswell[1], *Galina E. Pavlovskaya*[2,3], *Thomas Meersmann*[2,3], *and Melanie M. Britton*[1]

[1] *School of Chemistry, University of Birmingham, Edgbaston, Birmingham, UK*
[2] *Sir Peter Mansfield Imaging Centre, School of Medicine, University of Nottingham, Nottingham, UK*
[3] *NIHR Nottingham Biomedical Centre, Nottingham, UK*

18.1 Introduction

The ever-increasing use of portable electronic devices and electric vehicles is fueling a significant increase in research into energy storage. This interest is also driving the development of novel battery chemistries, which go beyond the current market leader – lithium-ion batteries (LIBs) [1]. As such, sodium-ion batteries (NIBs) are a promising new energy storage solution, owing to their lower cost and greater sustainability [2]. Despite chemical similarities between lithium and sodium, many materials employed as electrodes and electrolytes in LIBs do not perform well in NIBs [3], driving the search for new materials [4–6]. To identify suitable materials, and optimize their performance, new analytical techniques are required, which can characterize the performance of battery components, in situ and under working conditions (operando). This is particularly the case for NIBs, which are known to form a variety of metastable states [7, 8] that cannot be observed using ex situ experiments. Furthermore, as species evolve over time, it is important to be able to observe their evolution in real time, during charge cycling.

A variety of imaging techniques have been developed for operando characterization of batteries, such as X-ray tomography and electron microscopy. More recently, magnetic resonance imaging (MRI) has been used, which is noninvasive, able to provide spatially resolved physical and chemical information, and is complementary to other imaging techniques [9,10]. MRI has the advantage of being able to detect MR active nuclei (e.g. ^{1}H, ^{19}F, ^{6,7}Li, and ^{23}Na) within electrodes and electrolyte, enabling a more holistic view of behavior in a battery [9,11]. In LIBs and NIBs, electroactive species can be detected directly, across both electrolytic and metallic environments [12–14]. However, because radiofrequency radiation is not able to penetrate metals beyond a skin depth (~10^{-5} m) [15], it is essential that operando MRI studies employ cells that are not encased in metal [16]. A variety of plastic, nuclear magnetic resonance (NMR) compatible battery designs have been reported in the literature [16–21]. In addition to changes in the construction of the cell, the switching of radiofrequency (RF) and magnetic field gradient pulses, which can induce eddy currents, can cause variations in RF ($\boldsymbol{B}_1$) and magnetic field ($\boldsymbol{B}_0$) fields, leading to image artifacts [16,22,23]. Hence, it is necessary to align metallic components, such as current collectors and electrodes, with respect to both $\boldsymbol{B}_1$ and $\boldsymbol{B}_0$ fields [16].

Magnetic Resonance Microscopy: Instrumentation and Applications in Engineering, Life Science and Energy Research, First Edition. Edited by Sabina Haber-Pohlmeier, Bernhard Blümich, and Luisa Ciobanu.

In LIBs, ^{6,7}Li NMR spectroscopy has been used to investigate the growth of lithium dendrite structures. This has been achieved by monitoring dendrites using ^{6}Li and ^{7}Li NMR [14] and visualizing them directly using ^{7}Li MRI [24,25]. In particular, it has been shown that lithium dendrites have a different chemical shift compared with the metal electrode. This enables the use of chemical shift imaging to map the location of dendrite formation and other microstructural changes [25]. Three-dimensional (3D) ^{7}Li MRI is able to reveal the distribution of dendrites (Figure 18.1). However, spatial resolution is limited by the relatively low sensitivity and typically short relaxation times of ^{7}Li [25]. To overcome this, higher-resolution images of dendritic structures can be acquired, indirectly, using the ^{1}H NMR signal from the electrolyte [26]. Using this method, Ilott et al. were able to generate "negative" images of dendrite growth (Figure 18.2), by assuming an absence of signal intensity in the electrolyte image indicated the presence of Li microstructures [26].

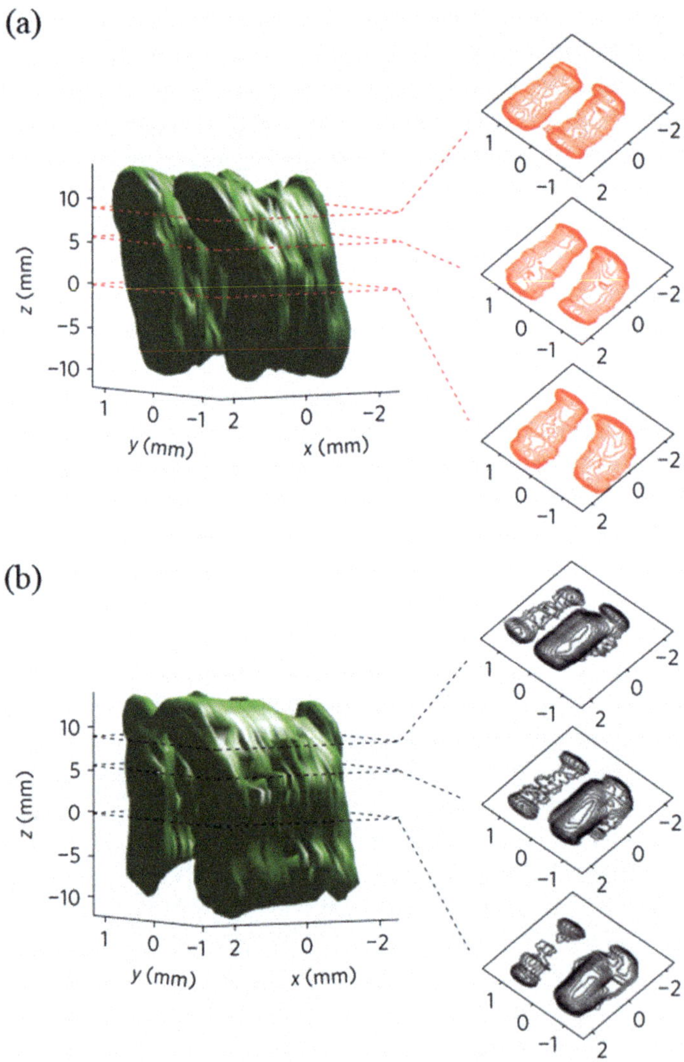

Figure 18.1 Three-dimensional ^{7}Li magnetic resonance image of a bag cell before (a) and after charging (b). Amended with permission from reference [25]. Copyright 2012 Nature Publishing Group.

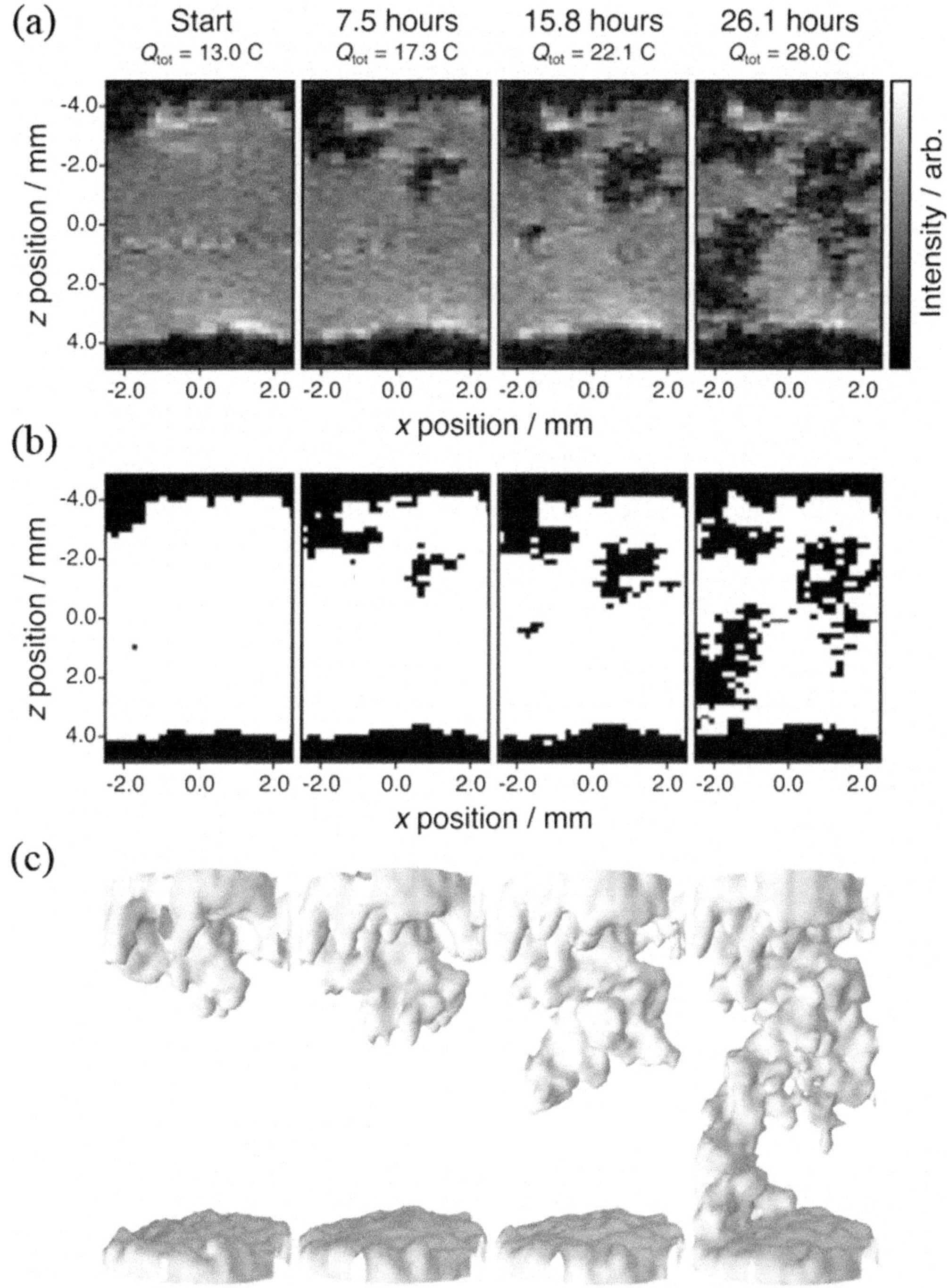

Figure 18.2 3D ^{1}H Fast Low Angle SHot (FLASH) imaging of lithium dendrites. Two-dimensional slices (a) at four stages of charging. Segmented images (b) from (a) with a threshold applied to visualize dendrite location. Three-dimensional images (c) of a dendrite growth with Gaussian filter applied. [26] Ilott et al. (2016), figure 01[p.02]/with permission from National Academy of Sciences.

Mass transport and cell polarization, in LIBs, have been monitored in situ using a combination of indirect detection, by ^{1}H and ^{19}F MRI, of the electrolyte and direct detection, by ^{7}Li MRI, of the charge carrier [21,27–30]. Other battery chemistries have been investigated using indirect detection of the electrolyte. This is particularly useful where the electroactive species is either not MR active or its MR properties make its use impractical. For example, ^{1}H MRI has been used to investigate the distribution of zinc species during the operation of a zinc–air cell [31].

In the case of NIBs, NMR investigations have largely relied on in situ and ex situ ^{23}Na NMR spectroscopy [32–34]. Ex situ ^{23}Na NMR spectroscopy has been used to investigate cluster sizes of quasimetallic species formed in different hard carbon materials [34]. Ex situ ^{23}Na NMR spectroscopy, combined with in situ X-ray diffraction mapping and electron paramagnetic resonance, has been used to study the sodiation mechanism of hard carbon in a sodium metal cell [35]. Recently, in situ ^{23}Na NMR spectroscopy has been used to investigate sodium metal deposition and the formation of high-surface-area mossy and dendritic sodium microstructures during cell cycling [32], identify different sodium crystallographic sites [36,37], and sodium states in hard carbon materials [33]. Operando ^{23}Na NMR spectroscopy has been used, with ex situ solid-state NMR and X-ray analysis techniques, to provide insight into the insertion of sodium into hard carbon [8]. In this study, Stratford et al. determined that the sodium first moves into defect sites, pore walls, and interlayer regions within the hard carbon structure giving rise to a signal with a negative chemical shift with respect to the sodium in the electrolyte in the ^{23}Na NMR spectrum. This is then followed by intercalation between planes in the carbon structure, correlating with the appearance of a ^{23}Na environment with positive chemical shift as compared with the sodium in the electrolyte. Finally, sodium forms clusters within the pores, which causes an increase in the chemical shift toward the metallic sodium due to an increase in the Knight shift [8].

While ^{23}Na NMR spectroscopy has proved useful for identifying species during charge cycling, it is important to be able to spatially locate species, to better understand battery chemistry. However, the use of ^{23}Na MRI, to directly visualize the distribution of electroactive species, comes with some technical challenges. ^{23}Na has a lower receptivity compared with ^{1}H [38], resulting in a lower signal-to-noise ratio (SNR) and greater signal averaging. Moreover, ^{23}Na typically has significantly shorter relaxation times than ^{1}H and even ^{7}Li. Despite these challenges, ^{23}Na MR techniques are now starting to be used to study the chemistry of sodium batteries. Battery chemistry can be monitored directly, using the ^{23}Na MR signal of the electroactive species [12,39], or indirectly, using the ^{1}H or ^{19}F MR signal of the electrolyte [21,26,40]. Sodium dendrites have been visualized directly using signal from metallic ^{23}Na in an electrochemical cell [12,39,41]. A study by Xiang et al. [39] demonstrated extensive growth of sodium microstructures and change in chemical shift of the sodium in solvents with and without an additive, as seen in Figure 18.3. Dendrite growth was also monitored using two-dimensional (2D) T_2 relaxation time-weighted ^{23}Na MRI in an all-solid-state cell, in combination with X-ray tomography and scanning electron microscopy [41] and in a sodium metal cell [12]. Two-dimensional ^{23}Na MRI has been combined with indirect ^{1}H MRI of the electrolyte to provide a "negative" image of the sodium microstructures [12]. From this study, the sodium metal dendrites can be seen to develop over time and have an irregular structure and distribution across the carbon electrode. This chapter explores the use of 2D ^{23}Na MRI and ^{23}Na NMR spectroscopy, in combination with 3D ^{1}H MRI in investigating dendrite growth in a sodium metal cell and the evolution of sodium species in a sodium-ion full cell.

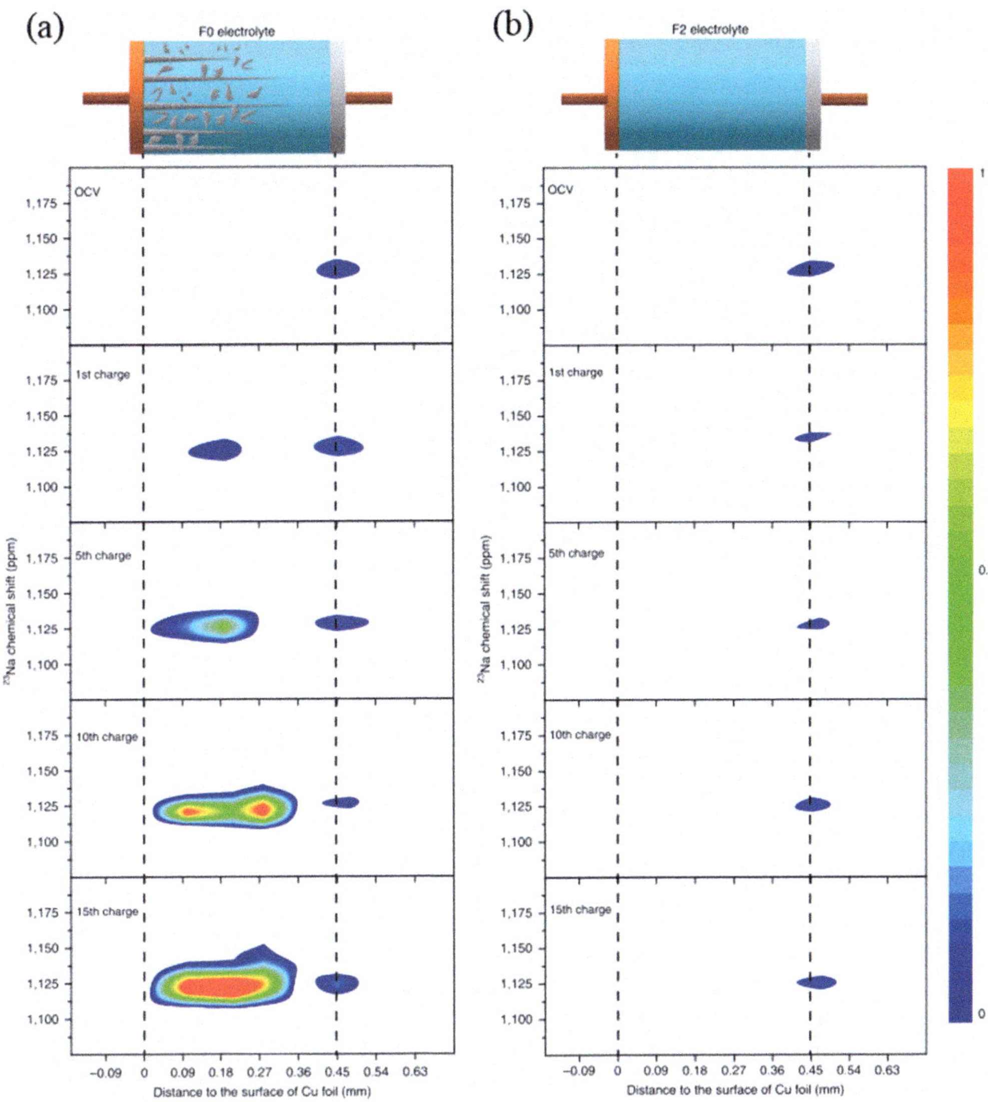

Figure 18.3 One-dimensional ^{23}Na magnetic resonance images of metallic sodium (1130 ppm) in cells with Cu and Na electrodes with electrolyte without (a) and with (b) an additive at difference charge cycles. The Cu electrode is at 0 mm. Amended with permission from reference [39]. Copyright 2020 Nature Publishing Group.

18.2 Experimental

The operando cell employed in this work was based on a Swagelok® union (i.d. 3/8″ = 9.5 mm) shown in Figure 18.4, which is made from polytetrafluoroethene (PTFE). The commercial Swagelok components are made from PTFE and the center of the union has been milled to maintain the diameter throughout the length of the Swagelok union. The electrode contact is made using copper wire threaded through purpose-built polyether ether ketone (PEEK) rods to ensure firm and even electrode contact and cell compression. In the metal cell experiments, these rods

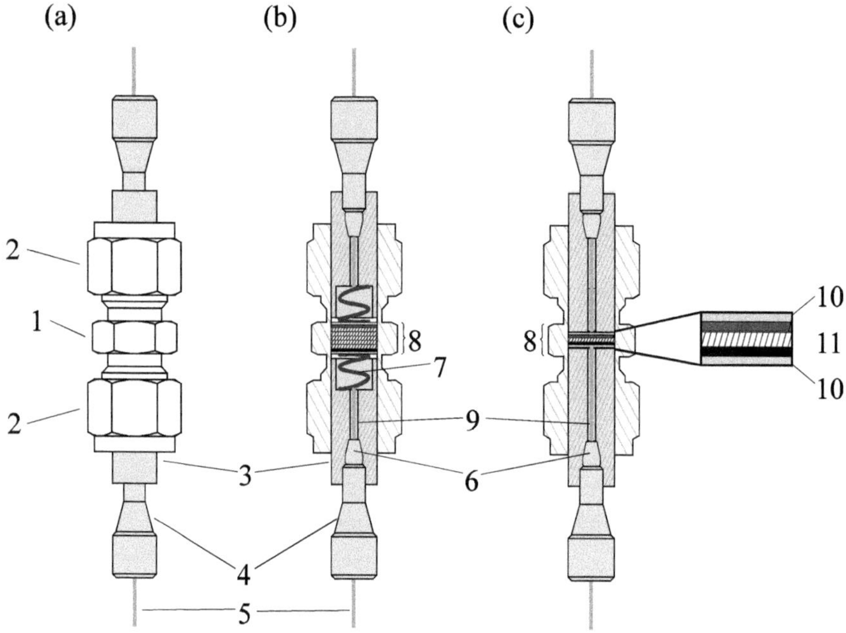

Figure 18.4 Exterior (a) and cutaway of the sodium metal (b) and full cell (c) Swagelok cells with an expansion of the electrode and separator stack. Components are: (1) 3/8′′ polytetrafluoroethene (PTFE) Swagelok union, (2) PTFE Swagelok nut and ferrule assembly, (3) custom polyether ether ketone (PEEK) rod, (4) 1/16′′ PEEK high performance liquid chromatography (HPLC) nut, (5) insulated copper wire, (6) 1/16′′ PEEK HPLC ferrule, (7) brass springs, (8) electrode and separator stack, (9) copper wire with PFTE sheath, (10) aluminum current collectors, (11) expansion of the electrode and separator stack.

contain a recess in which a brass spring (Small-Order Springs & Pressings, Ltd., UK) is positioned to provide cell compression. The copper wire is secured in position by 1/16′′ PEEK high-performance liquid chromatography (HPLC) nuts and ferrules (i.d. 1/16′′ = 1.6 mm, VICI®), which compress a PTFE sheath fitted to the wire to ensure airtightness.

Details on materials synthesis and preparation can be found in reference [12]. Sodium metal cells were constructed in an argon-filled glove box (MBraun, $H_2O < 0.5$ ppm, $O_2 < 0.5$ ppm) using a sodium metal counter electrode (CE) and hard carbon working electrode (WE), separated by either five or two Whatman GF/B glass microfiber separators with 1 M $NaPF_6$ in 1:1 ethylene carbonate (EC): dimethyl carbonate (DMC) with aluminum current collectors on either side (Figure 18.4b). In order to ensure compression across the electrode stack, brass springs were used. However, these were not used in the full cell configuration, which comprised a $NaNi_{1/3}Fe_{1/3}Mn_{1/12}Sn_{1/12}O_2$ (NaNFMSO) cathode and a hard carbon anode, separated by one Whatman GF/B glass microfiber separator with 1 M $NaPF_6$ in 1:1 EC: diethyl carbonate (DEC) with aluminum current collectors on either side (Figure 18.4c), and assembled in a dry room (dew point of −50°C).

Electrochemical experiments on both sodium metal cell and full cells were performed using a two-electrode configuration on an Ivium Octostat 5000. In order to minimize noise in the MR data while cycling cells within the MRI magnet, shielding and low-pass filtering were used as described previously [40]. For the sodium metal cell, an initial charge–discharge (formation) cycle was performed at 30 mA/g outside the magnet prior to MR imaging. The second charge cycle, performed at 20 mA/g, was executed with the cell inside the magnet. Galvanostatic plating at 150 mA/g for 137 min was then used to plate sodium metal onto the hard carbon WE. This was continued at up to 600 mA/g for 210 min. For the sodium-ion full cell, a formation cycle was performed at 30 mA/g inside the magnet during the acquisition of operando ^{23}Na NMR spectra every 33 min. A charge cycle was also performed outside the magnet at 86 mA/g before the collection of in situ ^{23}Na NMR spectra.

All MR data were acquired on a Bruker 9.4-T spectrometer equipped with either 25 mm ^{23}Na or ^{1}H single-resonance RF coils or a ^{1}H/^{23}Na dual-resonance RF coil. The cell was positioned in a WB40 microimaging probe using a universal specimen pot with holes drilled through the lid and bottom, to enable wires to connect to the potentiostat.

^{23}Na NMR spectra were acquired separately for electrolytic and metallic ^{23}Na environments, using pulse-acquire or spin echo sequences, at 105.8561019 MHz and 105.9757449 MHz, respectively. This large frequency difference, between metallic and electrolytic ^{23}Na, is the result of the Knight shift [32], and allows selective acquisition of either environment. Spectra were acquired with 1024 points, 2048 averages, a spectral width of 50 kHz, and a repetition time T_R=0.1 s and processed with zero-filling to 4096 points. ^{23}Na NMR spectra, covering the full spectral range (0–1200 ppm), were acquired using a spin echo sequence, with a spectral width of 250 kHz, a transmitter frequency of 105.9130825 MHz (positioned between the electrolyte and metallic signal), collecting 512 points, 32 768 averages, and a repetition time T_R = 0.05 s. Short excitation and refocusing pulse durations were used, 2 μs and 4 μs, respectively, to enable the excitation of a sufficiently broad frequency range. Spectra were Fourier transformed with zero-filling to 2048 points and line broadening of 1800 Hz, and baseline correction. Spectral fitting was performed with DMFIT software [42] using a combination of Gaussian and Lorentzian functions, using the minimum number of peaks to adequately fit spectra.

^{23}Na images of the electrolyte, in a pristine metal cell, were acquired using a gradient echo sequence with read encoding along the vertical (z) axis of the cell and phase encoding across the horizontal axis of the cell. Images were acquired with 1024 averages, a spectral width of 200 kHz, comprising 256 read and 16 phase points with a field of view (FOV) of 59 mm × 19.5 mm. Data were Fourier transformed using a sine bell squared filter and zero-filled to 512 × 64 points, which resulted in voxels of 115 μm × 305 μm. ^{23}Na images of sodium metal, in a pristine metal cell, were acquired using a spin echo sequence, with the same orientation as the electrolytic images. The images were acquired with 1024 averages, a spectral width of 200 kHz, and 256 read and 16 phase points with a FOV of 59 mm × 16.9 mm. Data were Fourier transformed using a sine bell squared filter and zero-filled to 512 × 64 points, which resulted in voxels of 115 μm× 264 μm. ^{23}Na images of the full cell were acquired using a spin echo pulse sequence with the same orientation as the sodium metal cells. Images were acquired with 2048 averages, a spectral width of 50 kHz, comprising 64 read and 16 phase points with a FOV of 26.9 mm × 94.2 mm. Data were Fourier transformed using a sine bell squared filter and zero-filled to 256 × 64 points, which resulted in voxels of 105 μm × 1472 μm.

Three-dimensional ^{1}H MR images of the electrolyte in the sodium metal cell were acquired using a multi-slice multi- (spin) echo sequence with horizontal frequency encoding, 15 mm isotropic FOV, with a 117 μm × 117 μm × 469 μm voxel size and a 1.0 s repetition time. Zero-filling in the y-direction was used during processing to produce 3D images with isotropic voxels.

Negative images, to enable observation of sodium microstructures, were produced from the ^{1}H 3D MRI data, by only displaying voxels below a threshold intensity, which was set to 7.4% of the maximum image intensity. This enabled regions to be identified where the electrolyte was below level of detection and either absent or depleted (compared to the pristine cell) following the formation of metallic sodium microstructures within the separator.

18.3 Results

18.3.1 Distribution and Evolution of Sodium Species

Figure 18.5 shows 3D ^{1}H MR images of the electrolyte in a pristine sodium metal cell. These images show that, while the electrolyte fills the separator, there is also a significant amount of electrolyte that is distributed throughout the Swagelok cell, predominantly between the PEEK rod and the PTFE Swagelok components. These different electrolyte locations can be observed in the ^{23}Na NMR spectra, for the electrolyte, in Figure 18.6a where the spectrum of electrolytic sodium shows two peaks, one corresponding to the sodium within the separator (blue) and the other from sodium in the electrolyte that is between the PEEK rod and the Swagelok casing (gray). Figure 18.6b shows a spectrum of the sodium metal, with one environment as expected.

In Figure 18.7a, 2D ^{23}Na MR images for the electrolytic (blue) and metallic (red) sodium are shown. The signal intensity for the electrolytic sodium can be seen to increase in the center of the cell, which is due to the increases in signal arising from a large slice thickness and the cylindrical shape of the cell. The signal from the sodium metal appears as two narrow strips of signal, which is because of the skin depth penetration of RF [15].

^{23}Na NMR spectra were acquired for the sodium metal cell, after a formation cycle, and shown in Figure 18.7b and c. No change is observed in the spectrum for metallic sodium; however, in the spectrum for the electrolyte, there are two new environments: one downfield (−7 ppm) believed to arise from sodium in the vicinity of the hard carbon electrode (red) and the other (1 ppm) believed to arise from sodium with reduced mobility (green), for instance the sodium in the solid electrolyte interphase (SEI).

In Figure 18.8a, ^{23}Na 2D MR images of the sodium metal (red) and sodium in the electrolyte (blue), for a sodium metal cell following a formation cycle, are presented. These images indicate the distribution of sodium in the electrolyte is not homogeneous in this cell. It is not clear what

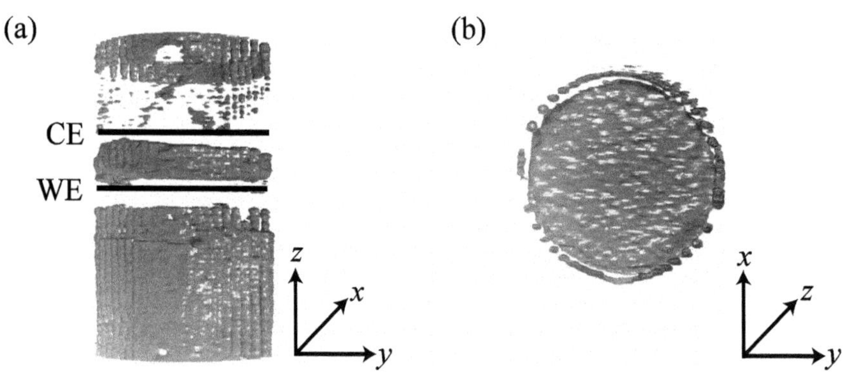

Figure 18.5 3D ^{1}H magnetic resonance images of the 1 M $NaPF_6$ in ethylene carbonate (EC): dimethyl carbonate (DMC) (EC/DMC) electrolyte, within a Swagelok sodium metal cell containing five separators, from the side (a) and above (b) the cell.

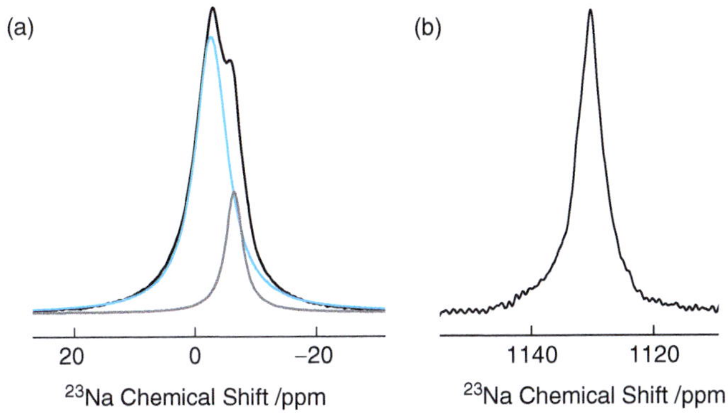

Figure 18.6 ^{23}Na nuclear magnetic resonance spectra of sodium in the 1 M $NaPF_6$ in ethylene carbonate (EC): dimethyl carbonate (DMC) (EC/DMC) electrolyte (a) and metallic sodium (b) for a pristine sodium metal cell. Two peaks are observed for the sodium electrolyte (a), arising from electrolyte inside the stack of five separators (0 ppm, blue) and outside the electrode stack (−5 ppm, gray).

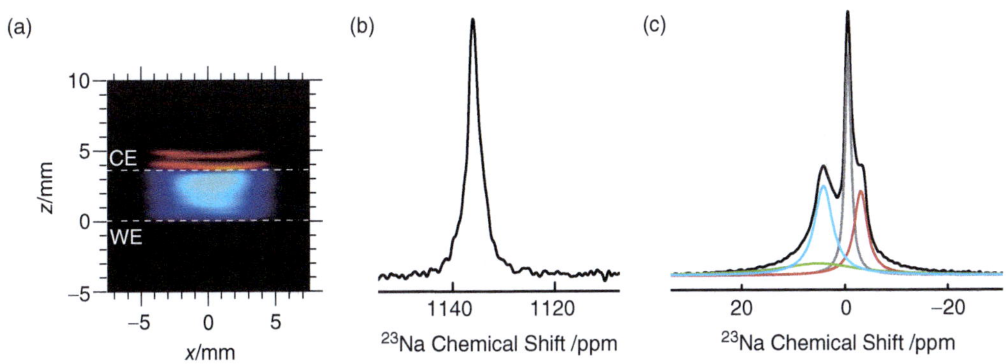

Figure 18.7 2D ^{23}Na magnetic resonance images (a) of sodium in the 1 M $NaPF_6$ in ethylene carbonate (EC): dimethyl carbonate (DMC) (EC/DMC) electrolyte (blue) and sodium metal (red) in a pristine sodium metal cell containing five separators. ^{23}Na nuclear magnetic resonance (NMR) spectra of the sodium metal (b) and sodium in the 1 M $NaPF_6$ in EC/DMC electrolyte (c) showing the electrolyte inside the separator (0 ppm, blue), outside of the separator stack (−5 ppm, gray), sodium in the vicinity of the hard carbon working electrode (−7 ppm, red) and sodium with reduced mobility (1 ppm, green) of a sodium metal cell after a formation cycle has been performed.

the origins of this inhomogeneity are, but they could be a consequence of cell assembly, uneven compression across the electrodes leading to inhomogeneity in the cell chemistry, or heterogeneous degradation of the electrolyte. However, observation of such factors is essential to assist with the interpretation of electrochemical behavior for a cell.

To investigate sodiation mechanisms further, full cells were investigated, which did not contain metallic sodium CEs, enabling observation of the formation of metallic sodium during charge cycling. Figure 18.8b and c shows ^{23}Na NMR spectra of the electrolyte, within a full cell, before and after a formation cycle. From these spectra, four environments can be observed following peak fitting: signal from sodium in the electrolyte inside the separator stack (blue), sodium in the electrolyte outside the separator stack (gray), sodium in the vicinity of the hard carbon (red), and sodium with reduced mobility (green). The peak from sodium with reduced

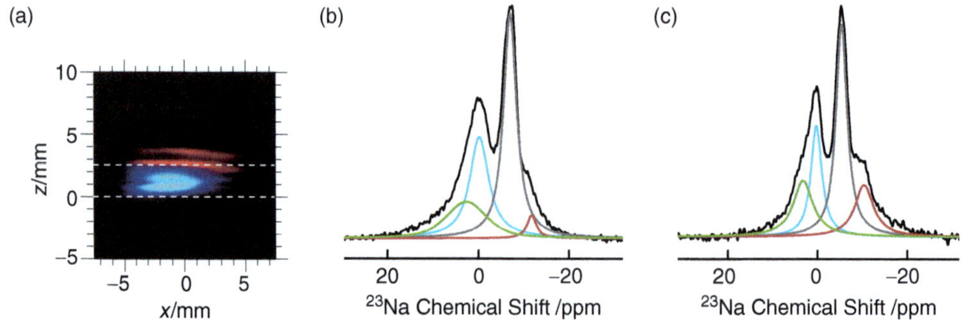

Figure 18.8 2D ^{23}Na magnetic resonance images (a) of the sodium in the 1 M $NaPF_6$ in ethylene carbonate (EC): dimethyl carbonate (DMC) (EC/DMC) electrolyte (blue) and sodium metal (red) in a sodium metal cell with two separators after a formation cycle has been performed. ^{23}Na nuclear magnetic resonance spectra of sodium in the 1 M $NaPF_6$ in EC/DEC electrolyte in a full cell, containing a single separator before (b) and after (c) the formation cycle showing sodium in the electrolyte outside the electrode stack (−5 ppm, gray) and inside the separator (0 ppm, blue), sodium in the vicinity of the hard carbon electrode (−13 ppm, red) and sodium with reduced mobility (7 ppm, green).

mobility required fitting to a combined Gaussian and Lorentzian model, suggesting this peak is a sum of a variety of similar environments. These peaks are believed to arise from sodium within the SEI or degradation of the electrolyte upon contact with the electrode. Of note, is the difference in range of sodium species observed in full and metal cells prior to the formation cycle. In the metal cell, only one sodium environment is observed within the electrode stack (Figure 18.6b). However, in the full cell (Figure 18.8b), at least three distinct environments are observed in the electrolytic sodium. It is possible that the differences are simply because of the changes in electrode materials used. However, it is also possible that an increase in compression in the full cell, at the time of assembly, may have promoted reaction between the sodium electrolyte and electrodes, giving rise to these additional sodium environments. Following the formation cycle in the full cell (Figure 18.8c), a similar range of sodium environments are observed. However, the relative intensities are changed. In particular, the peak at −13 ppm, associated with sodium in the vicinity of the hard carbon, increases in intensity, which could be due to irreversible incorporation of sodium into the hard carbon. Additionally, the width and Gaussian weighting of the peak at 7 ppm decreases, which may reflect a change in the range of sodium environments.

18.3.2 Metallic Sodium Formation

The formation of sodium dendrites is an area of interest in NIBs and was investigated in the sodium metal cell, during galvanostatic plating. Figure 18.9 shows ^{23}Na NMR spectra of the electrolyte following the formation cycle (Figure 18.9a) and after galvanostatic plating (Figure 18.9b). Three peaks can be observed in both spectra that are from sodium in the electrolyte outside the electrode–separator stack (gray), sodium electrolyte in the separator (blue), and reacted sodium in the vicinity of the hard carbon electrode (red). After the growth of dendrites, the peak from sodium with reduced mobility (green) disappears below the level of detection, suggesting that it is replaced by the formation of dendrites. ^{23}Na NMR spectra of metallic sodium (Figure 18.10) show a range of sodium environments corresponding to the sodium metal electrode (red) and new metallic sodium environments. These additional environments are believed to be associated

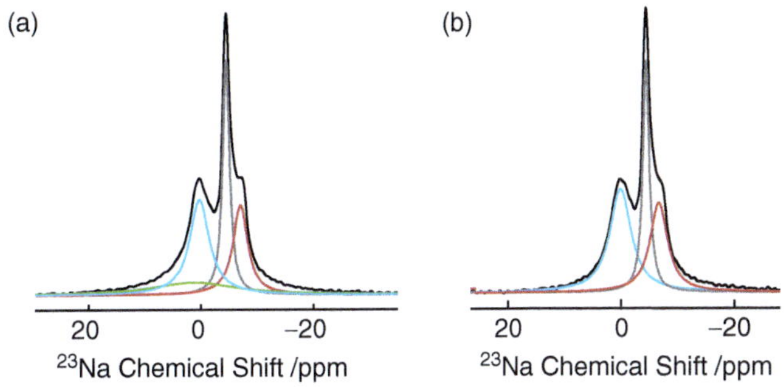

Figure 18.9 ^{23}Na nuclear magnetic resonance spectra of the sodium in the 1 M $NaPF_6$ in ethylene carbonate (EC): dimethyl carbonate (DMC) (EC/DMC) electrolyte after the formation cycle (a) and after the formation of dendrites (b) in a sodium metal cell. Peaks are observed for electrolyte inside the separator (0 ppm, blue), electrolyte outside the separator stack (−5 ppm, gray), and reacted sodium in the vicinity of the hard carbon working electrode (−7 ppm, red). However, the peak for sodium with reduced mobility (1 ppm, green) is only observed in the cell prior to dendrite formation.

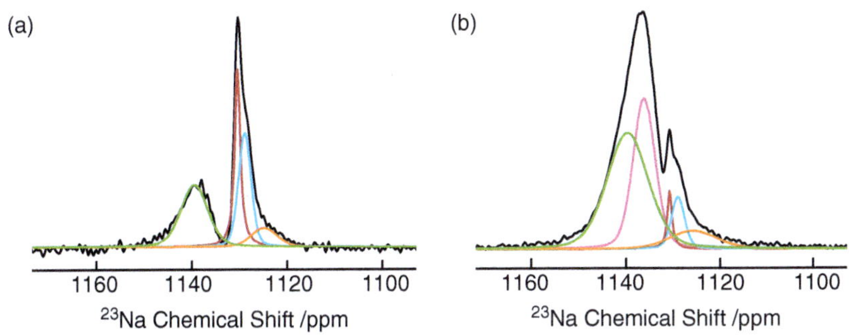

Figure 18.10 ^{23}Na nuclear magnetic resonance spectra of the sodium metal before (a) and after (b) the growth of dendrites in a sodium metal cell showing the signal from the sodium metal electrode (1131 ppm, red), sodium microstructures at 90° to the $\boldsymbol{B}_0$ field (1136 ppm, 1141 ppm, green and pink) and sodium plated on the hard carbon electrode or pitted on the sodium metal electrode at 0° to the $\boldsymbol{B}_0$ field (1129 ppm, 1126 ppm, blue and orange).

with sodium microstructures, or dendrites. The chemical shifts for these new metallic sodium microstructures are sensitive to the orientation with respect to $\boldsymbol{B}_0$. The peak at 1141 ppm is believed to arise from sodium microstructures at 0° to the $\boldsymbol{B}_0$ field (light blue and green), which are growing perpendicular to the sodium CE. However, there are additional peaks (1129–1125 ppm) at a lower chemical shift to the sodium CE (1131 ppm, red), which could arise from microstructures growing at different orientations to the WE, or from pitting on the CE, at an angle close to 90° to $\boldsymbol{B}_0$ (dark blue and orange). As galvanostatic plating continues, the peak at 1141 ppm increases in intensity, and is greater than the sodium CE. This increase in intensity arises from the higher surface area expected on the dendritic sodium, compared with the sodium CE. Furthermore, the number of peaks above 1131 ppm increases, suggesting a greater range of orientations for these metallic surfaces as the dendrites grow. However, a similar evolution is not observed for the sodium environments with lower chemical shift to the sodium CE, which do not increase significantly in intensity, possibly further supporting the supposition that these peaks arise from pitting on the CE, as sodium is removed from the electrode.

The formation of sodium dendrites can be observed in the ^{23}Na MR images of sodium metal (red) and electrolyte (blue) in Figure 18.11a. The additional sodium environment, believed to arise from sodium dendrites on the hard carbon, can indeed be observed on the surface of the WE. The signal intensity for this metallic dendritic sodium appears to be greater than that of the CE, due to the greater surface area for these microstructures compared with the CE, which agrees with what is observed in the ^{23}Na NMR spectra (Figure 18.10). Unfortunately, these images are unable to reveal the structure of these dendrites and it is not possible to visualize in 3D, because of the low SNR and high line widths for the ^{23}Na signal. However, by using ^{1}H MRI of the EC/DMC electrolyte it is possible to visualize the shape and location of the dendrites (Figure 18.11b).

Using the ^{1}H MR images of the electrolyte, it was possible to generate negative images, which are able to indicate the regions where the electrolyte is below the limit of detection because it is not present or has been displaced by the growth of sodium dendrites. Figure 18.12 shows images of the surface of the hard carbon electrode of the pristine cell and after dendrite growth generated using the ^{1}H 3D electrolyte images and correlate to the large signal intensity seen on the WE in Figure 18.11a. The dendrites are not uniformly distributed across the electrode, with larger structures observed near the left on the left side. Furthermore, they appear to have rounded structures, which are supported by the variety of peaks observed in the metallic ^{23}Na NMR spectrum after dendrite growth (Figure 18.10b).

While sodium metal cells are most commonly employed for operando studies of battery chemistry and dominate operando NMR and MRI studies of LIBs and NIBs, the presence of the sodium CE can give rise to increased internal resistance in the cell, as well as mask the development of new metallic sodium environments during charge cycling. By moving to a full cell configuration, it is possible to prevent both of these issues. Figure 18.13 shows operando ^{23}Na NMR spectra, and corresponding electrochemistry, during the formation cycle of a full cell. Prior to the formation cycle, the pristine cell displays only a single electrolytic environment, as expected. As the formation cycle proceeds, a peak develops below 0 ppm, which then disappears after 3 h, before reappearing as the cell continues to charge. This peak indicates the presence of sodium which has inserted within the hard carbon or SEI formation. Formation of quasimetallic nanoparticles is observed for the first time in a sodium full cell as a signal that develops out of the electrolytic signal during the potential plateau, after 2.5 h of charging. This environment shifts further

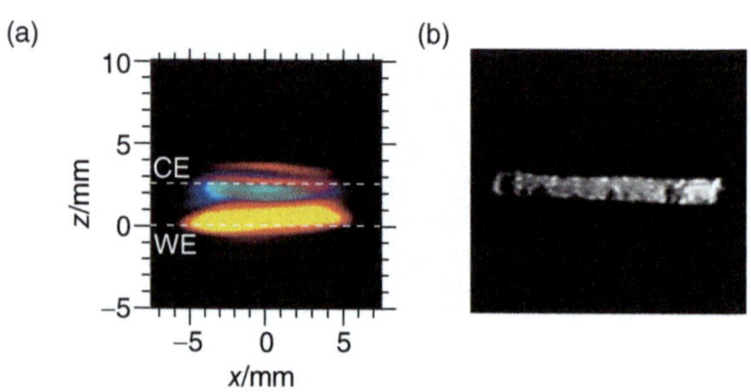

Figure 18.11 2D ^{23}Na magnetic resonance (MR) images (a) of the sodium in the 1 M $NaPF_6$ in ethylene carbonate (EC): dimethyl carbonate (DMC) (EC/DMC) electrolyte (blue) and sodium metal (red) in a sodium metal cell after dendrites have formed. (b) Selected slice from a 3D ^{1}H MR image of showing the location of the 1 M $NaPF_6$ in EC/DMC electrolyte in the sodium metal cell after the formation of dendrites.

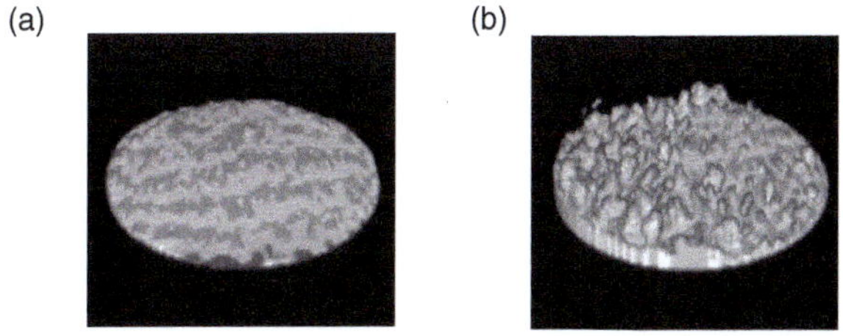

Figure 18.12 Images of the surface of the hard carbon working electrode of a sodium metal cell before (a) and after (b) the growth of dendrites. Generated using the regions where the signal intensity in a 3D ^{1}H image of the 1 M $NaPF_6$ in ethylene carbonate (EC): dimethyl carbonate (DMC) (EC/DMC) electrolyte falls below the limit of detection.

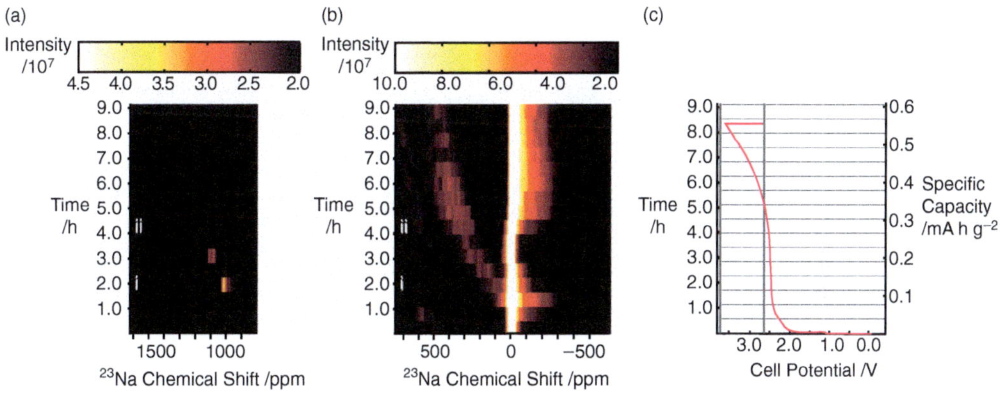

Figure 18.13 ^{23}Na nuclear magnetic resonance spectra acquired in operando during a formation cycle of a sodium-ion full cell showing the formation of a metallic species and the evolution of quasimetallic species. Labeled spectra i and ii are shown in Figure 18.9a and b, respectively.

upfield and increases in peak width as the cell continues to charge. This suggests that the nanoparticles are becoming larger and more metallic as more sodium is incorporated within the hard carbon electrode. Figure 18.14 shows selected ^{23}Na NMR spectra that were acquired during the formation cycle of a sodium full cell. Figure 18.14a shows multiple environments in the region associated with the electrolyte signal; these are sodium in the vicinity of the hard carbon electrode (red, green), such as in the SEI or intercalated sodium, and initial formation of quasimetallic nanoparticles (orange). Figure 18.14b shows the development of the quasimetallic species, which has moved further upfield.

After 2 h of charging, at the potential plateau around 2.5 V, a metallic signal (pink) is observed in the spectra at over 1000 ppm (Figure 18.14a). The development of metallic sodium has not been previously reported in the literature, but has been hypothesized [43]. However, it is possibly not surprising that this new metallic sodium has not been observed previously, as the majority of operando studies employ sodium metal CEs, which would swamp the small signal arising from this new metallic environment. As the full cell progresses in the formation cycle, the intensity of

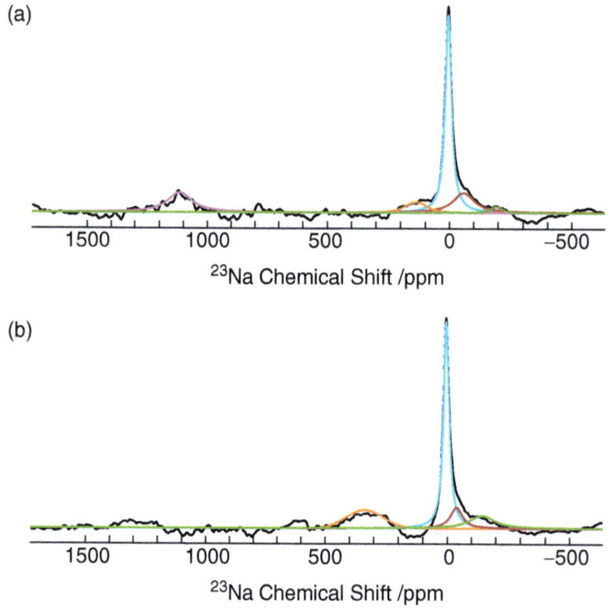

Figure 18.14 ^{23}Na nuclear magnetic resonance spectra covering the full spectral range acquired in operando during the formation cycle of a sodium-ion full cell showing the formation of a metallic signal (pink) at low electrochemical potential with the electrolyte (blue) and species in the vicinity of the hard carbon electrode or solid electrolyte interphase (red and green) (a) and subsequent quasimetallic species (orange) (b).

the metallic peak initially decreases in intensity before shifting further upfield and disappearing at the end of the potential plateau.

18.4 Conclusions

This chapter demonstrates what information can be acquired from a sodium battery using off-the-shelf MRI equipment. It is possible to detect multiple sodium species using ^{23}Na NMR spectroscopy before, during, and after cycling batteries. Differences in the sodium species present before charging in sodium metal and full cells could be due to differences in the hard carbon used or cell compression. Electrolyte is visualized in 3D using ^{1}H MRI and is distributed both inside and outside of the electrode stack. The electrolyte outside of the electrode stack is not electroactive and comprises a large amount of the sodium signal observed in the ^{23}Na NMR spectra. Future cell designs should aim to minimize this excess electrolyte in addition to providing even compression across the battery. Formation of metallic dendrites has been observed using ^{23}Na MRI, and their distribution and shape in 3D confirmed using ^{1}H MRI. To reduce potential issues with cell internal resistance and possible masking of smaller metallic signals, a full cell configuration was used. Quasimetallic nanoparticles were formed during the formation cycle of a full cell and their development monitored in operando using ^{23}Na NMR spectroscopy. Additionally, a new metallic species has been detected that forms during the formation cycle. This species may not have been observed in the sodium metal electrodes as the signal from the metal may obscure new metallic environments formed during cycling. Further investigations into the nature of the metallic species observed during the formation cycle should be conducted, in addition to optimization of the SNR; for example, by increasing the sample filling factor in the coil and reducing the requirement for the sample holder.

References

1 Grey, C.P. and Tarascon, J.M. (2016). Sustainability and in situ monitoring in battery development. *Nature Materials* 16 (1): 45–56.

2 Hwang, J.Y., Myung, S.T. and Sun, Y.K. (2017). Sodium-ion batteries: Present and future. *Chemical Society Reviews* 46 (12): 3529–3614.

3 Pascal, G. and Mireille, F. (1988). Electrochemical intercalation of sodium in graphite. *Solid State Ionics* 28–30: 1172–1175.

4 Hou, H., Qiu, X., Wei, W. et al. (2017). Carbon anode materials for advanced sodium-ion batteries. *Advanced Energy Materials* 7 (24): 1602898.

5 Irisarri, E., Ponrouch, A. and Palacin, M.R. (2015). Review—Hard carbon negative electrode materials for sodium-ion batteries. *Journal of the Electrochemical Society* 162 (14): A2476–A2482.

6 Li, Y., Lu, Y., Zhao, C. et al. (2017). Recent advances of electrode materials for low-cost sodium-ion batteries towards practical application for grid energy storage. *Energy Storage Materials* 7: 130–151.

7 Marbella, L.E., Evans, M.L., Groh, M.F. et al. (2018). Sodiation and desodiation via helical phosphorus intermediates in high-capacity anodes for sodium-ion batteries. *Journal of the American Chemical Society* 140 (25): 7994–8004.

8 Stratford, J.M., Allan, P.K., Pecher, O. et al. (2016). Mechanistic insights into sodium storage in hard carbon anodes using local structure probes. *Chemical Communication (Cambridge, UK)*. 52 (84): 12430–12433.

9 Britton, M.M. (2017). MRI of chemical reactions and processes. *Progress in Nuclear Magnetic Resonance Spectroscopy* 101: 51–70.

10 Britton, M.M. (2010). Magnetic resonance imaging of chemistry. *Chemical Society Reviews* 39 (11): 4036–4043.

11 Pecher, O., Carretero-González, J., Griffith, K.J. et al. (2017). Materials' methods: NMR in battery research. *Chemical Materials* 29 (1): 213–242.

12 Bray, J.M., Doswell, C.L., Pavlovskaya, G.E. et al. (2020). Operando visualisation of battery chemistry in a sodium-ion battery by (23)Na magnetic resonance imaging. *Nature Communications* 11 (1): 2083.

13 Chang, H.J., Ilott, A.J., Trease, N.M. et al. (2015). Correlating microstructural lithium metal growth with electrolyte salt depletion in lithium batteries using (7)Li MRI. *Journal of the American Chemical Society* 137 (48): 15209–15216.

14 Chang, H.J., Trease, N.M., Ilott, A.J. et al. (2015). Investigating Li microstructure formation on Li anodes for lithium batteries by in situ6li/7Li NMR and SEM. *The Journal of Physical Chemistry C* 119 (29): 16443–16451.

15 Ilott, A.J., Chandrashekar, S., Klockner, A. et al. (2014). Visualizing skin effects in conductors with MRI: (7)Li MRI experiments and calculations. *Journal of Magnetic Resonance* 245: 143–149.

16 Britton, M.M. (2014). Magnetic resonance imaging of electrochemical cells containing bulk metal. *Chemical Physics Chemistry* 15 (9): 1731–1736.

17 Wiemers-Meyer, S., Winter, M. and Nowak, S. (2017). A battery cell for in situ NMR measurements of liquid electrolytes. *Physical Chemistry Chemical Physics* 19 (7): 4962–4966.

18 Ni, Z.-R., Cui, X.-H., Cao, S.-H. et al. (2017). A novel in situ electrochemical NMR cell with a palisade gold film electrode. *AIP Advances* 7 (8): 085205.

19 Benders, S., Gomes, B.F., Carmo, M. et al. (2020). In-situ MRI velocimetry of the magnetohydrodynamic effect in electrochemical cells. *Journal of the Magnetic Resonance* 312: 106692.

20 Bazak, J.D., Krachkovskiy, S.A. and Goward, G.R. (2017). Multi-temperature in situ magnetic resonance imaging of polarization and salt precipitation in lithium-ion battery electrolytes. *Journal of Physical Chemistry C* 121 (38): 20704–20713.

21 Krachkovskiy, S.A., Bazak, J.D., Werhun, P. et al. (2016). Visualization of steady-state ionic concentration profiles formed in electrolytes during Li-Ion battery operation and determination of mass-transport properties by in situ magnetic resonance imaging. *Journal of the American Chemical Society* 138 (25): 7992–7999.

22 Graf, H., Steidle, G., Martirosian, P. et al. (2006). Effects on MRI due to altered rf polarization near conductive implants or instruments. *Medical Physics* 33 (1): 124–127.

23 Vashaee, S., Goora, F., Britton, M.M. et al. (2015). Mapping B(1)-induced eddy current effects near metallic structures in MR images: A comparison of simulation and experiment. *Journal of Magnetic Resonance* 250: 17–24.

24 Marbella, L.E., Zekoll, S., Kasemchainan, J. et al. (2019). 7Li NMR chemical shift imaging to detect microstructural growth of lithium in all-solid-state batteries. *Chemistry of Materials* 31 (8): 2762–2769.

25 Chandrashekar, S., Trease, N.M., Chang, H.J. et al. (2012). 7Li MRI of Li batteries reveals location of microstructural lithium. *Nature Materials* 11 (4): 311–315.

26 Ilott, A.J., Mohammadi, M., Chang, H.J. et al. (2016). Real-time 3D imaging of microstructure growth in battery cells using indirect MRI. *Proceedings of the National Academy of Sciences of the United States of America* 113 (39): 10779–10784.

27 Klett, M., Giesecke, M., Nyman, A. et al.(2012). Quantifying mass transport during polarization in a Li ion battery electrolyte by in situ 7Li NMR imaging. *Journal of the American Chemical Society* 134 (36): 14654–14657.

28 Chandrashekar, S., Oparaji, O., Yang, G. et al. (2016). Communication – 7Li MRI Unveils concentration dependent diffusion in polymer electrolyte batteries. *Journal of the Electrochemical Society* 163 (14): A2988–A2990.

29 Chien, P.H., Feng, X., Tang, M. et al. (2018). Li distribution heterogeneity in solid electrolyte Li10GeP2S12 upon electrochemical cycling probed by (7)Li MRI. *Journal of Physical Chemistry Letters* 9 (8): 1990–1998.

30 Klamor, S., Zick, K., Oerther, T. et al. (2015). (7)Li in situ 1D NMR imaging of a lithium ion battery. *Physical Chemistry Chemical Physics* 17 (6): 4458–4465.

31 Britton, M.M/, Bayley, P.M., Howlett, P.C. et al. (2013). In situ, real-time visualization of electrochemistry using magnetic resonance imaging. *Journal of Physical Chemistry Letters* 4 (17): 3019–3023.

32 Bayley, P.M., Trease, N.M. and Grey, C.P. (2016). Insights into electrochemical sodium metal deposition as probed with in situ (23)Na NMR. *Journal of the American Chemical Society* 138 (6): 1955–1961.

33 Morita, R., Gotoh, K., Fukunishi, M. et al. (2016). Combination of solid state NMR and DFT calculation to elucidate the state of sodium in hard carbon electrodes. *Journal of Materials Chemistry A* 4 (34): 13183–13193.

34 Morita, R., Gotoh, K., Kubota, K. et al. (2019). Correlation of carbonization condition with metallic property of sodium clusters formed in hard carbon studied using 23Na nuclear magnetic resonance. *Carbon* 145: 712–715.

35 Qiu, S., Xiao, L.F., Sushko, M.L. et al. (2017). Manipulating adsorption-insertion mechanisms in nanostructured carbon materials for high-efficiency sodium ion storage. *Advance Energy Materials* 7 (17): 11.

36 Smiley, D.L., Carlier, D. and Goward, G.R. (2019). Combining density functional theory and (23) Na NMR to characterize Na2FePO4F as a potential sodium ion battery cathode. *Solid State Nuclear Magnetic Resonance* 103: 1–8.

37 Li, Q., Liu, Z., Zheng, F. et al. (2018). Identifying the structural evolution of the sodium ion battery Na2FePO4F cathode. *Angewandte Chemie International Edition* 57 (37): 11918–11923.

38 Madelin, G., Lee, J.S., Regatte, R.R. et al. (2014). Sodium MRI: Methods and applications. *Progress in Nuclear Magnetic Resonance Spectroscopy* 79: 14–47.

39 Xiang, Y., Zheng, G., Liang, Z. et al. (2020). Visualizing the growth process of sodium microstructures in sodium batteries by in-situ (23)Na MRI and NMR spectroscopy. *Nature Nanotechnology*: 15: 883–890.

40 Bray, J.M., Davenport, A.J., Ryder, K.S. et al. (2016). Quantitative, in situ visualization of metal-ion dissolution and transport using (1) H magnetic resonance imaging. *Angewandte Chemie International Edition England* 55 (32): 9394–9397.

41 Rees, G.J., Spencer Jolly, D., Ning, Z. et al. (2021). Imaging sodium dendrite growth in all-solid-state sodium batteries using (23) Na T2-weighted magnetic resonance imaging. *Angewandte Chemie International Edition England* 60 (4): 2110–2115.

42 Massiot, D., Fayon, F., Capron, M. et al. (2002). Modelling one- and two-dimensional solid-state NMR spectra. *Magnetic Resonance in Chemistry* 40 (1): 70–76.

43 Dugas, R., Ponrouch, A., Gachot, G. et al. (2016). Na reactivity toward carbonate-based electrolytes: The effect of FEC as additive. *Journal of the Electrochemical Society* 163 (10): A2333–A2339.

19

The Fun of Applications – a Perspective

Y.-Q. Song

Athinoula A. Martinos Center for Biomedical Imaging, Department of Radiology, Massachusetts General Hospital, Charlestown, MA, USA
John A. Paulson School of Engineering and Applied Sciences, Harvard University, Cambridge, MA, USA
Schlumberger-Doll Research, Cambridge, MA, USA

19.1 Introduction

In the past few decades, tremendous advances in magnetic resonance (MR) technologies have been focused on achieving high magnetic fields, for spectroscopy, medical imaging, and MR microscopy (MRM). An equally exciting new area of lower-field (1–2 T) spectroscopy grade instruments creates novel uses of nuclear magnetic resonance (NMR) for chemistry experiments, e.g. [1–3]. Much progress has also been made to develop ex situ sensors for mobile measurement [4,5]. However, one has to acknowledge that NMR remains a technique that requires experts, and applications primarily are still in the NMR labs as research tools. One reason for the lack of motivation for application development is the impression that applications are technically simple, and thus not scientifically rewarding. In this chapter, I will try to explore the topics of the MR research opportunities that application development provides as a contribution to MR science/technology.

One interesting case study is MR technology for well logging, a commercial service that is deployed globally and is highly integrated with other exploration technologies. The NMR technique satisfies the basic requirement of being noninvasive, similar to other key measurement methods, such as electrical, acoustic, and radiation methods. The main uniqueness of MR is that the signal directly originates from the fluids, instead of the solid materials. In contrast, other well-logging methods rely on models and calibrations to separate signals from fluid and solids, which results in significant uncertainties [6–8].

The key event in the development of NMR well logging is likely be the first successful downhole-worthy tool and the first downhole measurement. However, equally important was the first laboratory demonstration of the viability of a prototype bearing the relevant tool design, such as the appropriate tool size, and the ability to obtain sufficient accuracy in the relevant sample properties (e.g. porosity) [9–12]. After the early demonstration of the technology in the 1990s, continuing development has advanced the methodology in every technical area, including hardware, MR methods, data analysis, data interpretation, and integration with other measurements. Much of the MR research was carried out in universities and research labs around the world in addition to the developers of the downhole tools.

I will develop two lists below to highlight the subject areas (i) that the MR well logging has contributed back to MR science and technology, and (ii) that it still needs further development. Many of the developments have already expanded outside well logging, e.g. into medical magnetic

Magnetic Resonance Microscopy: Instrumentation and Applications in Engineering, Life Science and Energy Research, First Edition. Edited by Sabina Haber-Pohlmeier, Bernhard Blümich, and Luisa Ciobanu.

resonance imaging (MRI) and MRM. The lists will not include aspects that, although essential for the success of the downhole tools, such as specific electronics, high temperature and high pressure, and deployment, may be less relevant for most academic MR research. References are provided as an illustration instead of a comprehensive review of the subjects.

19.2 List 1: Areas of MR Research Motivated by Well-logging Applications

- *Spin dynamics in grossly inhomogeneous field.* In high-field NMR spectroscopy, magnetic fields are uniform and radiofrequency pulses may cover the entire sample. For well-logging tools, the magnetic field is generated from permanent magnets and the sweet spot of detection is outside the tool body. As a result, the magnetic field is grossly nonuniform given the large extent of the sample. Empirically, CPMG (Carr–Purcell–Meiboom–Gill) sequence was used successfully to detect signals [13]. However, the spin dynamics were not understood until Hürlimann developed the theory of effective rotation for CPMG and similar pulse sequences [14–16]. Detailed understanding of the spin dynamics for CPMG is essential to quantify the effects of relaxation and diffusion [17]. This behavior is very complex as the spin dynamics cannot be described by a single coherence pathway as in the majority of MR experiments. This development has been continued in recent years to the full simulation of spin dynamics [18].

  One interesting recent development is the broadband radiofrequency pulses [19–21] and the broadband hardware [22] to improve signal-to-noise ratio. Will these pulses be useful for MRI and MRM?
- *Permanent magnet design and optimization.* Various magnet designs have been published and many may find their roots in the design of the well-logging tools. For example, a 2D saddle point magnet was designed for CMR [11] and such saddle point design remains sound for strong magnetic field, low gradient, and good sensitivity. The recent barrel magnet can be viewed as a 3D saddle point [23] that enjoys a similar benefit. Such a magnet has shown a very interesting use in probing tomato juice within commercial containers [24]. Many interesting ideas have been published on the theoretical approach to magnet design, e.g. [25,26].
- *Data analysis and inversion.* While the Fourier transformation is a well-known analysis method for conventional high-field spectroscopy, it is not appropriate for the analysis of relaxation and diffusion data in order to obtain the spectrum of relaxation and diffusion. Because the relaxation and diffusion behaviors can often be expressed as an exponential decay function, a numerical Laplace inversion is often employed. For many years, this subject has been a focus of academic groups and research papers (a recent example is [27]). With the wide use of 2D experiments of relaxation and diffusion in recent years, more research has been targeted to the 2D inversion and has improved algorithms (a recent example is [28]). Many other related 2D techniques have been published for a variety of applications [29].

  In recent years, MR spectroscopy and MRI have experimented with compressed sensing techniques and the data matrix may not be compatible with conventional Fourier transformation. In this case, the analysis is mathematically similar to the inversion techniques used for relaxation and diffusion analyses.
- *MR physics of porous media.* MR behavior in porous media has become a significant field of research to study materials with pore sizes from nanometers, such as zeolites, to many microns, such as in rock and soil [7]. The realization of the impact of pore space on relaxation and diffusion has provided the motivation for much research since the 1970s [30–32]. For example, the

concept of surface relaxation was a critical theoretical advance that underpinned much of the later development and applications. The physics of restricted diffusion has had an impact on the study of rocks and other inorganic porous materials, and now it is increasingly applied to medical MRI to understand tissue microstructure, and MRM for other materials and applications [33,34]. It continues as a subject of new physics, e.g. [35–37], to probe and understand the connectivity between pores. A related topic is the development of numerical simulation of porous media behavior based on high-resolution tomography. MR measurements have contributed significantly to its progress, e.g. [38,39].

- *MR physics of complex fluids.* Crude oil is a mixture of a large range of molecules of different molecular sizes and moieties. Even though it is mostly composed of simple hydrocarbons, the minority amount of asphaltenes, acids, and surface-active molecules results in complex viscosity behavior, temperature dependence, rheology, and surface activities. Such complexity has been the motivation of much research [40], new experimental techniques, and new theories facing a significant challenge [41–43]. The recent work by Singer et al. on the molecular simulation demonstrated detailed molecular dynamics [44] in complex samples.
- *Two-dimensional NMR pulse sequences, theory.* CPMG has been used from the early years of MR well logging. However, its resolving power for different fluids is limited. The introduction of 2D/3D NMR of relaxation and diffusion in early 2000 [45] improved the fluid resolution by combining T_1, T_2, and diffusion encodings, and it has become a routine measurement protocol. The adoption of multidimensional sequences promoted many studies of its measurement protocols, inversion methods, and interpretation for experiments in the lab and downhole. Numerous papers have been published in both petroleum-focused and academic journals. Recently, ultrafast 2D measurement has been demonstrated [46] using hyperpolarization.

  Relaxation and diffusion are widely used in medical MRI as a way to achieve tissue differentiation. A combination of these parameters in a multidimensional fashion will surely enhance resolution (e.g. [47,48]). Such an experiment presents a unique challenge to clinical MRI and MRM applications due to the long acquisition time.

Continuous development over the past three decades has made MR well logging an indispensable technology for subsurface exploration. However, it also faces challenges from the exploration of geologies such as unconventional and ultrahigh temperature reservoirs, as well as the need to improve measurement accuracy and efficiency.

19.3 List 2: Further Research Areas Motivated by MR Well-logging Applications

- *Interpretation of NMR data.* NMR data are often used to obtain estimates of permeability, bound fluid fraction, capillary curves, and other petrophysical quantities. However, the methods and equations used for such interpretation are mostly empirical and often require calibration. Many widely used models (such as the Katz–Thompson equation [49]) are overly idealistic and do not take into consideration the heterogeneity of natural materials. For example, permeability is to a large extent controlled by the pore throats; however, NMR relaxation does not directly identify pore throat, let alone its distribution. Thus, it is possible that the T_2 data alone are insufficient to characterize permeability. For this topic, spatially resolved relaxation and diffusion techniques (e.g. MRM) could be very helpful in quantifying heterogeneity.

- *Emulsions.* Emulsification occurs often in many stages of crude oil transport and processing. Crude oils contain surfactant molecules so that they can be easily emulsified. Drilling mud is a complex mixture of many components, such as water, oil, salts, solid particles, as well as many additives for reducing friction and tuning viscosity. Solid particles (e.g. barite) are added to increase mud density; however, the mud has to exhibit sufficient strength to hold the particles without sedimentation. Many diffusion methods [50] have been developed to obtain oil/water drop sizes for clean emulsions. However, application of these methods to samples of mud or crude oil will face significant challenges due to the complexity of the mixture and the unavoidable magnetic susceptibility of the barite particles.
- *Optimization of pulse sequences.* During MR well logging, the tool typically passes the formation at a constant speed and the measurement is made during the constant motion. As a result, the total time devoted to a particular depth of the formation is short and inversely proportional to the logging speed. This is a particular challenge for performing 2D experiments while multiple scans are needed to complete a measurement suite. As a result, optimization of the 2D sequences for high efficiency will be extremely useful. The well-logging pulse sequence may be quite complex with many segments, each with a different wait time, number of acquired echoes, and number of averages. Development of mathematics to optimize such experiments is likely quite a challenge [51]. The same situation is encountered in reaction monitoring when acquiring NMR spectra of the reaction fluid flowing in a bypass through the magnet. A similar challenge exists for medical MRI and MRM applications of 2D experiments.
- *Tool movement.* The motion of the logging tools provides another challenge to MR because almost all other MR experiments are conducted with no movement between the sample and NMR magnet/coils. The relative movement of the sample and the instrument presents a significant time-varying magnetic field during the pulse sequence and can exhibit dramatically different spin dynamics [18].
- *Ringing, short echo time.* Subsurface organic materials range from light oil to highly viscous bitumen, and further to solid kerogen. As a result, T_2 ranges from seconds to 10 μs. Since the T_2 is often distributed continuously from short to long, it is very useful to measure short T_2 signals even for estimating porosity. Substantial improvement of instrument deadtime to allow measurements with echo time (TE) much less than 200 μs is highly desirable.
- *Miniaturization.* Well-logging tools are all large and heavy, primarily due to the large electronics components. In fact, this is also true of most NMR spectrometers. With the diversity of permanent magnet-based NMR systems, small NMR devices in the form of a sensor are possible [52]. Many papers have discussed miniaturized spectrometers, and more are being investigated (e.g. [53]).

19.4 Conclusions

To conclude, I would like to emphasize that a real-world application almost always comes with unique technical and scientific challenges. These challenges can become an exciting opportunity to explore MR (and other disciplines) in many different directions. What is required of us is an open mind and willingness to collaborate. The result will be an opportunity to learn from others and competing techniques, the fun and excitement of pushing MR into unfamiliar areas, and a great potential to make an impact beyond our laboratories. Sometimes, this can be literally outside. While the examples given mostly relate to porous media studies the methodical argument is readily expanded to apply to the voxels of an image.

Acknowledgment

The author thanks Alex Song for reading and editing the manuscript, and Dr. Isik Kizilyalli, program director of the Advanced Research Projects Agency-Energy (ARPA-E), for the support of this research under contract DE-AR0001063.

References

1 Blümich, B. and Singh, K. (2017). Desktop NMR and its applications from materials science to organic chemistry. *Angewandte Chemie International Edition* 57: 6996–7010.
2 Danieli, E., Perlo, J., Duchateau, A.L.L. et al. (2014). On-line monitoring of chemical reactions by using bench-top nuclear magnetic resonance spectroscopy. *Chemphyschem* 15: 3060–3066.
3 Sans, V. and Cronin, L. (2016). Towards dial-a-molecule by integrating continuous flow, analytics and self-optimisation. *Chemical Society Reviews* 45: 2032–2043.
4 Blümich, B., Perlo, J., and Casanova, F. (2008). Mobile single-sided NMR. *Progress in Nuclear Magnetic Resonance Spectroscopy* 52: 197–269.
5 Hürlimann, M.D. and Heaton, N.J. (2015). NMR well-logging. In: *Mobile NMR and MRI: Developments and Applications* (eds. M. Johns, E.O. Fridjonsson, S. Vogt, and A. Haber), Cambridge, UK: Royal Society of Chemistry.
6 Kenyon, W.E. (1992). Nuclear magnetic resonance as a petrophysical measurement. *Nuclear Geophysics* 6: 153–171.
7 Coates, G.R., Xiao, L., and Prammer, M.G. (1999). *NMR Logging Principles and Applications*. Houston, TX: Halliburton Energy Services.
8 Ellis, D.V. and Singer, J.M. (2007). *Well Logging for Earth Scientists*. Dordrecht, the Netherlands: Springer Netherlands.
9 Jackson, J.A., Burnett, L.J., and Harmon, J.F. (1980). Remote (inside-out) NMR. III. Detection of nuclear magnetic resonance in a remotely produced region of homogeneous magnetic field. *Journal of Magnetic Resonance* 41: 411–421.
10 Jackson, J.A. (2001). Los Alamos NMR well logging project. *Concepts in Magnetic Resonance* 13: 368–378.
11 Kleinberg, R.L., Sezginer, A., Griffin, D.D. et al. (1992). Novel NMR apparatus for investigating an external sample. *Journal of Magnetic Resonance* 97: 466–485.
12 Kleinberg, R.L. and Jackson, J.A. (2001). An introduction to the history of NMR well logging. *Concepts in Magnetic Resonance* 13: 340–342.
13 Goelman, G. and Prammer, M.G. (1995). The CPMG pulse sequence in strong magnetic field gradients with applications to oil-well logging. *Journal of Magnetic Resonance A* 113: 11–18.
14 Hürlimann, M.D. (2001). Carr-Purcell sequences with composite pulses. *Journal of Magnetic Resonance* 152: 109–123.
15 Hürlimann, M.D. (2001). Optimization of timing in the Carr-Purcell-Meiboom-Gill sequence. *Magnetic Resonance Imaging* 19: 375–378.

16 Hürlimann, M.D. and Griffin, D.D. (2000). Spin dynamics of Carr-Purcell-Meiboom-Gill-like sequences in grossly inhomogeneous B0 and B1 fields and application to NMR well logging. *Journal of Magnetic Resonance* 143: 120–135.

17 Hürlimann, M.D. (2001). Diffusion and relaxation effects in general stray field NMR experiments. *Journal of Magnetic Resonance* 148: 367–378.

18 Duan, C., Ryan, C., Utsuzawa, S. et al. (2017). Effect of off-resonance on T1 saturation recovery measurement in inhomogeneous fields. *Journal of Magnetic Resonance* 281: 31–43.

19 Koroleva, V.D.M., Mandal, S., Song, Y.-Q. et al. (2013). Broadband CPMG sequence with short composite refocusing pulses. *Journal of Magnetic Resonance* 230: 64–75.

20 Mandal, S., Borneman, T.W., Koroleva, V.D.M. et al. (2014). Direct optimization of signal-to-noise ratio of CPMG-like sequences in inhomogeneous fields. *Journal of Magnetic Resonance* 247: 54–66.

21 Casabianca, L.B., Mohr, D., Mandal, S. et al. (2014). Chirped CPMG for well-logging NMR applications. *Journal of Magnetic Resonance* 242: 197–202.

22 Mandal, S., Utsuzawa, S., Cory, D.G. et al. (2014). An ultra-broadband low-frequency magnetic resonance system. *Journal of Magnetic Resonance* 242: 113–125.

23 Utsuzawa, S. and Fukushima, E. (2017). Unilateral NMR with a barrel magnet. *Journal of Magnetic Resonance* 282: 104–113.

24 Pinter, M., Harter, T., McCarthy, M. et al. (2014). Towards using NMR to screen for spoiled tomatoes stored in 1,000 L, aseptically sealed, metal-lined totes. *Sensors* 14: 4167–4176.

25 Marble, A.E., Mastikhin, I.V., Colpitts, B.G. et al. (2005). An analytical methodology for magnetic field control in unilateral NMR. *Journal of Magnetic Resonance* 174: 78–87.

26 Judeinstein, P., Ferdeghini, F., Oliveira-Silva, R. et al. (2017). Low-field single-sided NMR for one-shot 1D-mapping: Application to membranes. *Journal of Magnetic Resonance* 277: 25–29.

27 Fricke, S.N., Seymour, J.D., Battistel, M.D. et al. (2020). Data processing in NMR relaxometry using the matrix pencil. *Journal of Magnetic Resonance* 313: 106704.

28 Guo, J., Xie, R., Xiao, L. et al. (2019). Nuclear magnetic resonance T1-T2 inversion with double objective functions. *Journal of Magnetic Resonance* 308: 106562.

29 Galvosas, P. and Callaghan, P.T. (2010). Multi-dimensional inverse Laplace spectroscopy in the NMR of porous media. *Comptes Rendus Physique* 11: 172–180.

30 Senturia, S.D. and Robinson, J.D. (1970). Nuclear spin lattice relaxation of liquid confined in porous solids. *Society of Petroleum Engineers Journal* 10: 237–244.

31 Cohen, M.H. and Mendelson, K.S. (1982). Nuclear magnetic relaxation and the internal geometry of sedimentary rocks. *Journal of Applied Physics* 53: 1127–1135.

32 De Gennes, P.G. (1982). Surface and interphase physics – Excitation transfer in random media. *Comptes Rendus De l'Académie Des Sciences III* 295: 1061–1064.

33 Li, M., Xiao, D., Romero, Z.L. et al. (2016). Mapping three-dimensional oil distribution with pi-EPI MRI measurements at low magnetic field. *Journal of Magnetic Resonance* 269: 13–23.

34 Groß, D., Zick, K., and Guthausen, G. (2017). Recent MRI and diffusion studies of food structures. *Annual Reports on NMR Spectroscopy* 90: 145–197.

35 Song, Y.-Q., Carneiro, G., Schwartz, L.M. et al. (2014). Experimental identification of diffusive coupling using 2D NMR. *Physical Review Letters* 113: 235503.

36 Thrane, L.W., Seymour, J.D., and Codd, S.L. (2019). Probing diffusion dynamics during hydrate formation by high field NMR relaxometry and diffusometry. *Journal of Magnetic Resonance* 303: 7–16.

37 Gao, Y. and Blümich, B. (2020). Analysis of three-site T2-T2 exchange NMR. *Journal of Magnetic Resonance* 315: 106740.

38 Arns, C.H., AlGhamdi, T., and Arns, J.-Y. (2011). Numerical analysis of nuclear magnetic resonance relaxation–diffusion responses of sedimentary rock. *New Journal of Physics* 13 (13): 015004.

39 Shikhov, I., d'Eurydice, M.N., Arns, J.Y. et al. (2017). An experimental and numerical study of relative permeability estimates using spatially resolved NMR. *Transport in Porous Media* 118: 225–250.

40 Lo, S.-W., Hirasaki, G.J., Kobayashi, R. et al. (1998). Relaxation time and diffusion measurements of methane and n-decane mixtures. *The Log Analyst* 39: 43–47.

41 Freed, D., Burcaw, L., and Song, Y.-Q. (2005). Scaling laws for diffusion coefficients in mixtures of alkanes. *Physical Review Letters* 94: 067602.

42 Korb, J.P. (2018). Multiscale nuclear magnetic relaxation dispersion of complex liquids in bulk and confinement. *Progress in Nuclear Magnetic Resonance Spectroscopy* 104: 12–55.

43 Zielinski, L., Ramamoorthy, R., Minh, C.C. et al. (2010). Restricted diffusion effects in saturation estimates from 2D diffusion-relaxation NMR maps. *Society of Petroleum Engineers Journal* 134841: 1–8.

44 Singer, P.M., Parambathu, A.V., Wang, X. et al. (2020). Elucidating the 1H NMR relaxation mechanism in polydisperse polymers and bitumen using measurements, MD simulations, and models. *Journal of Physical Chemistry* 124: 4222–4233.

45 Song, Y.-Q., Venkataramanan, L., Hürlimann, M.D. et al. (2002). T1-T2 correlation spectra obtained using a fast two-dimensional Laplace inversion. *Journal of Magnetic Resonance* 154: 261–268.

46 Ahola, S., Zhivonitko, V.V., Mankinen, O. et al. (2015). Ultrafast multidimensional Laplace NMR for a rapid and sensitive chemical analysis. *Nature Communications* 6: 1–7.

47 Benjamini, D. and Basser, P.J. (2016). Use of marginal distributions constrained optimization (MADCO) for accelerated 2D MRI relaxometry and diffusometry. *Journal of Magnetic Resonance* 271: 40–45.

48 de Almeida Martins, J.P. and Topgaard, D. (2018). Multidimensional correlation of nuclear relaxation rates and diffusion tensors for model-free investigations of heterogeneous anisotropic porous materials. *Scientific Reports* 8: 1–12.

49 Katz, A.J. and Thompson, A.H. (1986). Quantitative prediction of permeability in porous rock. *Physical Review B* 34: 8179–8181.

50 Johns, M.L. (2009). NMR studies of emulsions. *Current Opinion in Colloid & Interface Science* 14: 178–183.

51 Song, Y.-Q., Tang, Y., Hürlimann, M.D. et al. (2018). Real-time optimization of nuclear magnetic resonance experiments. *Journal of Magnetic Resonance* 289: 72–78.

52 Song, Y.-Q., Utsuzawa, S., and Tang, Y. (2019). Low fields but high impact: Ex-situ NMR and MRI. *Journal of Magnetic Resonance* 306: 109–111.

53 Michal, C.A. (2018). A low-cost multi-channel software-defined radio-based NMR spectrometer and ultra-affordable digital pulse programmer. *Concepts in Magnetic Resonance* 48B: e21401–e21414.

Index

Note: Page numbers followed by *f* lead to figures and *t* to tables.

Magnetic Resonance Microscopy: Instrumentation and Applications in Engineering, Life Science and Energy Research, First Edition. Edited by Sabina Haber-Pohlmeier, Bernhard Blümich, and Luisa Ciobanu.

d

j

k

l

m